Modeling and Control of Static Converters for Hybrid Storage Systems

Arezki Fekik
Akli Mohand Oulhadj University, Bouira, Algeria

Nacereddine Benamrouche
University of Tizi Ouzou, Algeria

A volume in the Advances in Environmental Engineering and Green Technologies (AEEGT) Book Series

IGI Global
PUBLISHER of TIMELY KNOWLEDGE

Published in the United States of America by
IGI Global
Engineering Science Reference (an imprint of IGI Global)
701 E. Chocolate Avenue
Hershey PA, USA 17033
Tel: 717-533-8845
Fax: 717-533-8661
E-mail: cust@igi-global.com
Web site: http://www.igi-global.com

Library of Congress Cataloging-in-Publication Data

Names: Fekik, Arezki, 1990- editor. | Benamrouche, Nacereddine, 1961- editor.
Title: Modeling and control of static converters for hybrid storage systems / Arezki Fekik and Nacereddine Benamrouche, editors.
Description: Hershey, PA : Engineering Science Reference, [2021] | Includes bibliographical references and index. | Summary: "This book bridges the gap between modeling and control of Static converters for hybrid storage systems offering various interdisciplinary applications starting from mathematical concepts, modeling, analysis, and until the realization and the experimental work"-- Provided by publisher.
Identifiers: LCCN 2021004525 (print) | LCCN 2021004526 (ebook) | ISBN 9781799874478 (h/c) | ISBN 9781799874485 (s/c) | ISBN 9781799874492 (eISBN)
Subjects: LCSH: Electric current converters. | Hybrid power systems--Equipment and supplies. | Electric power systems--Control--Simulation methods. | Energy storage.
Classification: LCC TK7872.C8 M63 2021 (print) | LCC TK7872.C8 (ebook) | DDC 621.31/26--dc23
LC record available at https://lccn.loc.gov/2021004525
LC ebook record available at https://lccn.loc.gov/2021004526

This book is published in the IGI Global book series Advances in Environmental Engineering and Green Technologies (AEEGT) (ISSN: 2326-9162; eISSN: 2326-9170)

British Cataloguing in Publication Data
A Cataloguing in Publication record for this book is available from the British Library.

For electronic access to this publication, please contact: eresources@igi-global.com.

Advances in Environmental Engineering and Green Technologies (AEEGT) Book Series

Sang-Bing Tsai
Zhongshan Institute, University of Electronic Science and Technology of China, China & Wuyi University, China
Ming-Lang Tseng
Lunghwa University of Science and Technology, Taiwan
Yuchi Wang
University of Electronic Science and Technology of China Zhongshan Institute, China

ISSN:2326-9162
EISSN:2326-9170

Mission

Growing awareness and an increased focus on environmental issues such as climate change, energy use, and loss of non-renewable resources have brought about a greater need for research that provides potential solutions to these problems. Research in environmental science and engineering continues to play a vital role in uncovering new opportunities for a "green" future.

The **Advances in Environmental Engineering and Green Technologies (AEEGT)** book series is a mouthpiece for research in all aspects of environmental science, earth science, and green initiatives. This series supports the ongoing research in this field through publishing books that discuss topics within environmental engineering or that deal with the interdisciplinary field of green technologies.

Coverage

- Radioactive Waste Treatment
- Green Transportation
- Water Supply and Treatment
- Cleantech
- Air Quality
- Sustainable Communities
- Contaminated Site Remediation
- Electric Vehicles
- Biofilters and Biofiltration
- Pollution Management

IGI Global is currently accepting manuscripts for publication within this series. To submit a proposal for a volume in this series, please contact our Acquisition Editors at Acquisitions@igi-global.com or visit: http://www.igi-global.com/publish/.

The Advances in Environmental Engineering and Green Technologies (AEEGT) Book Series (ISSN 2326-9162) is published by IGI Global, 701 E. Chocolate Avenue, Hershey, PA 17033-1240, USA, www.igi-global.com. This series is composed of titles available for purchase individually; each title is edited to be contextually exclusive from any other title within the series. For pricing and ordering information please visit http://www.igi-global.com/book-series/advances-environmental-engineering-green-technologies/73679. Postmaster: Send all address changes to above address.

Titles in this Series

For a list of additional titles in this series, please visit: www.igi-global.com/book-series

Handbook of Research on Microbial Remediation and Microbial Biotechnology for Sustainable Soil
Junaid Ahmad Malik (Government Degree College, Bijbehara, ndia)
Engineering Science Reference • © 2021 • 806pp • H/C (ISBN: 9781799870623) • US $295.00

Handbook of Research on Strategic Management for Current Energy Investments
Serhat Yüksel (Istanbul Medipol University, Turkey) and Hasan Dinçer (Istanbul Medipol University, Trkey)
Engineering Science Reference • © 2021 • 422pp • H/C (ISBN: 9781799883357) • US $295.00

Forest Fire Danger Prediction Using Deterministic-Probabilistic Approach
Nikolay Viktorovich Baranovskiy (National Research Tomsk Polytechnic University, Russia)
Engineering Science Reference • © 2021 • 297pp • H/C (ISBN: 9781799872504) • US $195.00

Role of IoT in Green Energy Systems
Vasaki Ponnusamy (Universiti Tunku Abdul Rahman, Malaysia) Noor Zaman (Taylor's University, Malaysia) Low Tang Jung (Universiti Teknologi PETRONAS, Malaysia) and Anang Hudaya Muhamad Amin (Higher Colleges of Technology, UAE)
Engineering Science Reference • © 2021 • 405pp • H/C (ISBN: 9781799867098) • US $195.00

Combating and Controlling Nagana and Tick-Borne Diseases in Livestock
Caleb Oburu Orenge (Department of Veterinary Anatomy and Physiology, Faculty of Veterinary Medicine and Surgery, Egerton University, Kenya)
Engineering Science Reference • © 2021 • 440pp • H/C (ISBN: 9781799864332) • US $185.00

Artificial Intelligence and IoT-Based Technologies for Sustainable Farming and Smart Agriculture
Pradeep Tomar (Gautam Buddha University, India) and Gurjit Kaur (Delhi Technological University, India)
Engineering Science Reference • © 2021 • 400pp • H/C (ISBN: 9781799817222) • US $215.00

Handbook of Research on Waste Diversion and Minimization Technologies for the Industrial Sector
Ashok K. Rathoure (M/s Akone Services, Lucknow, India)
Engineering Science Reference • © 2021 • 455pp • H/C (ISBN: 9781799849216) • US $295.00

Smart Farming Techniques and Knowledge-Based Agriculture Emerging Research and Opportunities
Petraq Papajorgji (European University of Tirana, Albania) and Francois Pinet (Irstea Centre de Clermont-Ferrand, France)
Engineering Science Reference • © 2021 • 230pp • H/C (ISBN: 9781799845409) • US $195.00

IGI Global
PUBLISHER of TIMELY KNOWLEDGE

701 East Chocolate Avenue, Hershey, PA 17033, USA
Tel: 717-533-8845 x100 • Fax: 717-533-8661
E-Mail: cust@igi-global.com • www.igi-global.com

Table of Contents

Preface xiv

Chapter 1
Robust Decentralized Voltage Tracker of Islanded Multi-DG AC Microgrids Using Invariant Ellipsoids 1
Ehab Hassan Eid Bayoumi, Egyptian Chinese University, Egypt
Hisham Soliman, faculty of Engineering, Cairo University, Egypt
Farag El-Sheikhi, Istanbul Esenyurt University, Istanbul, Turkey

Chapter 2
Design, Control, Management, and Performance Analysis of PV-Battery Supercapacitor DC-System Using Buck Converter 36
Yasmine Amara, Electronics Department, Blida 1 University, Algeria
Zoulikha Tebri, Electrotechnique Department, LATAGE Laboratory, University Mouloud Mammeri of Tizi-Ouzou, Algeria
Zina Larabi, Electrotechnique Department, University Mouloud Mammeri of Tizi-Ouzou, Algeria

Chapter 3
True Power Loss Reduction by Enhanced Tree Squirrel Search, Enhanced Salp Swarm, and Swim Bladder Operation-Based Shark Optimization Algorithms 65
Lenin Kanagasabai, Prasad V. Potluri Siddhartha Institute of Technology, India

Chapter 4
Extension of Aspiration Level Model for Optimal Planning of Fast Charging Stations 91
Manikanta Surya Narayana Suri, Amrita School of Engineering, Bengaluru, India
Deepa Kaliyaperumal, Amrita School of Engineering, Bengaluru, India

Chapter 5
Power Quality Improvement for Grid-Connected Photovoltaic Panels Using Direct Power Control. 107
Arezki Fekik, Akli Mohand Oulhadj University, Bouira, Algeria
Mohamed Lamine Hamida, Electrical Engineering Advanced Technology Laboratory (LATAGE), Mouloud Mammeri University, Algeria
Hamza Houassine, Akli Mohand Oulhadj University, Bouira, Algeria
Ahmad Taher Azar, Prince Sultan University, Saudi Arabia
Nashwa Ahmad Kamal, Faculty of Engineering, Cairo University, Egypt
Hakim Denoun, Electrical Engineering Advanced Technology Laboratory (LATAGE), Mouloud Mammeri University, Algeria
Sundarapandian Vaidyanathan, Vel Tech University, India
Aceng Sambas, Universitas Muhammadiyah Tasikmalaya, Indonesia

Chapter 6
Comparative Analysis of Various Types of Control Techniques for Wind Energy Conversion System 143
Bibhu Prasad Ganthia, IGIT, Sarang, India
Subrat Kumar Barik, KIIT University (Deemed), India
Byamakesh Nayak, KIIT University (Deemed), India

Chapter 7
Enhanced Symbiotic Organisms Search and Hydrological Cycle Algorithms for Real Power Loss Diminution and Voltage Stability Enhancement 175
Lenin Kanagasabai, Prasad V. Potluri Siddhartha Institute of Technology, India

Chapter 8
Operation and Startup of Three-Phase Grid-Connected PWM Inverter for an Experimental Test Bench With DSPACE Real-Time Implementation of PQ Control: Active and Reactive Powers Control Strategy of the Grid Side Converter With Experimental Integration 207
Kamal Elyaalaoui, EMI, Mohammed V University in Rabat, Morocco
Moussa Labbadi, EMI, Mohammed V University in Rabat, Morocco
Khalid Chigane, EMI, Mohammed V University in Rabat, Morocco
Mohammed Ouassaid, EMI, Mohammed V University in Rabat, Morocco
Mohamed Cherkaoui, EMI, Mohammed V University in Rabat, Morocco

Chapter 9
Multilevel Inverter for Hybrid Fuel Cell/PV Energy Conversion System 233
Arezki Fekik, Akli Mohand Oulhadj University, Bouira, Algeria
Mohamed Lamine Hamida, Electrical Engineering Advanced Technology Laboratory (LATAGE), Mouloud Mammeri University, Algeria
Hakim Denoun, Electrical Engineering Advanced Technology Laboratory (LATAGE), Mouloud Mammeri University, Algeria
Ahmad Taher Azar, Prince Sultan University, Riyadh, Saudi Arabia
Nashwa Ahmad Kamal, Faculty of Engineering, Cairo University, Egypt
Sundarapandian Vaidyanathan, Vel Tech University, India
Amar Bousbaine, University of Derby, UK
Nacereddine Benamrouche, Electrical Engineering Technology Laboratory (LATAGE), Mouloud Mammeri University, Algeria

Chapter 10
Power Analysis Using Various Types of Wind Turbines 271
Bibhu Prasad Ganthia, Indira Gandhi Institute of Technology, Sarang, India
Monalisa Mohanty, ITER, SIksha 'O' Anusandhan (Deemed), India
Jai Kumar Maherchandani, College of Technology and Engineering, Maharana Pratap University of Agriculture and Technology, India

Chapter 11
Flying Capacitor Inverter Integration in a Renewable Energy System 287
Mohamed Lamine Hamida, LATAGE Laboratory, Mouloud Mammeri University, Tizi Ouzou, Algeria
Arezki Fekik, Akli Mohand Oulhadj University, Bouira, Algeria
Hakim Denoun, LATAGE Laboratory, Mouloud Mammeri University, Tizi Ouzou, Algeria
Aghiles Ardjal, L2CSP Laboratory, Mouloud Mammeri University, Tizi Ouzou, Algeria
Aicha Aissa Bokhtache, GEER Laboratory, Hassiba Benbouali University, Chlef, Algeria

Compilation of References 307

About the Contributors 348

Index 354

Detailed Table of Contents

Preface xiv

Chapter 1
Robust Decentralized Voltage Tracker of Islanded Multi-DG AC Microgrids Using Invariant Ellipsoids 1
Ehab Hassan Eid Bayoumi, Egyptian Chinese University, Egypt
Hisham Soliman, faculty of Engineering, Cairo University, Egypt
Farag El-Sheikhi, Istanbul Esenyurt University, Istanbul, Turkey

This chapter develops a robust decentralized voltage tracker for islanded MGs. The proposed controller is robust against the plug and play operation of the MG, loads, and line parameter uncertainties. The problem is solved in the framework of linear matrix inequality (LMI). The proposed robust control represents the load changes and the parameter variations of lines connecting the DGs as a norm-bounded uncertainty. The proposed controller utilizes local measurements from DGs (i.e., it is totally decentralized). Control decentralization is accomplished by decomposing the global system into subsystems. The effect of the rest of the system on a specific subsystem is considered as a disturbance to minimize (disturbance rejection control). The controller is designed by the invariant-sets (approximated by the invariant ellipsoids). Different time-domain simulations are carried out as connecting and disconnected one or more DGs, connecting and disconnecting local loads DGs and transmission line parameters variation.

Chapter 2
Design, Control, Management, and Performance Analysis of PV-Battery Supercapacitor DC-System Using Buck Converter 36
Yasmine Amara, Electronics Department, Blida 1 University, Algeria
Zoulikha Tebri, Electrotechnique Department, LATAGE Laboratory, University Mouloud Mammeri of Tizi-Ouzou, Algeria
Zina Larabi, Electrotechnique Department, University Mouloud Mammeri of Tizi-Ouzou, Algeria

The intermittent nature of photovoltaic energy necessitates the incorporation of storage devices to ensure the continuality of loads feeding. In addition, it is important to model, control, and verify the operating of the designed system before implementation. Furthermore, the integration of power electronic interfaces plays a significant role in protecting the system and benefiting from solar energy. To this end, a buck converter is chosen to charge the battery and supply the supercapacitor. The control strategy is based on the maximum power point tracking techniques when the management algorithm recommends MPPT function mode. Otherwise, a feedback constant voltage PI controller is designed. Indeed, perturb and observe and incremental conductance are implemented and compared to analyze the system efficiency within the management strategy to charge the battery, switch between the controllers, and feed a supercapacitor in case of full battery charge. The obtained results using MATLAB/SIMULINK platform confirm the behaviour of the proposed strategy.

Chapter 3
True Power Loss Reduction by Enhanced Tree Squirrel Search, Enhanced Salp Swarm, and Swim Bladder Operation-Based Shark Optimization Algorithms 65
Lenin Kanagasabai, Prasad V. Potluri Siddhartha Institute of Technology, India

In this chapter, enhanced tree squirrel search optimization algorithm (ETSS), enhanced salp swarm algorithm (ESS), and swim bladder operation-based shark algorithm (SBS) have been applied to solve the power loss reduction problem. Enhanced tree squirrel search optimization algorithm (ETSS) utilizes the jumping exploration method and progressive exploration technique—both possess winter search strategy—in order to preserve the population diversity and to perk up the convergence speed. A new-fangled winter exploration strategy is implemented in the jumping exploration technique. In enhanced salp swarm algorithm (ESS) an inertia weight $\omega \in [0, 1]$ is applied, which picks up the pace of convergence during the period of exploration. Then swim bladder operation-based shark algorithm (SBS) is proposed to solve the problem. Based on contracting and expanding actions of the swim bladder in shark, SBS algorithm has been modelled.

Chapter 4
Extension of Aspiration Level Model for Optimal Planning of Fast Charging Stations 91
Manikanta Surya Narayana Suri, Amrita School of Engineering, Bengaluru, India
Deepa Kaliyaperumal, Amrita School of Engineering, Bengaluru, India

Electric vehicles will play a dominant role in future transportation due to their friendliness towards the present day environment. The battery which drives the vehicle can be refilled using battery charging and battery swapping techniques. Fast charging stations provide faster service to the customers. Though battery swapping method outperforms battery charging in many ways, the heavy infrastructure requirement of the former requires time in integrating with the real world. Queuing models are used to depict the real-time behavior of service stations. The aspiration level model provides the optimal value of charging piles for the given system capacity in a fast-charging station. The parameters in the aspiration level model can be formulated to an optimization problem. In the present work, the optimal planning for an fast charging station in Beijing is carried out using genetic algorithm. The simulation work is carried out in MATLAB/Simulink.

Chapter 5
Power Quality Improvement for Grid-Connected Photovoltaic Panels Using Direct Power Control. 107
Arezki Fekik, Akli Mohand Oulhadj University, Bouira, Algeria
Mohamed Lamine Hamida, Electrical Engineering Advanced Technology Laboratory (LATAGE), Mouloud Mammeri University, Algeria
Hamza Houassine, Akli Mohand Oulhadj University, Bouira, Algeria
Ahmad Taher Azar, Prince Sultan University, Saudi Arabia
Nashwa Ahmad Kamal, Faculty of Engineering, Cairo University, Egypt
Hakim Denoun, Electrical Engineering Advanced Technology Laboratory (LATAGE), Mouloud Mammeri University, Algeria
Sundarapandian Vaidyanathan, Vel Tech University, India
Aceng Sambas, Universitas Muhammadiyah Tasikmalaya, Indonesia

This chapter displays a control strategy for a photovoltaic system (PV) linked to the network with two phases of a PWM converter, where the first phase is a DC-DC converter linked among the photovoltaic source and the DC-AC converter. The second phase is a DC-AC converter linked to the grid. The maximum power point (MPP) is tracked by DC-DC converter, which increases the DC bus voltage. The P&O (perturbation and observation) technique is utilized as a direct current (DC-DC) converter controller to make the PV arrays work at greatest value of power under changing weather conditions. The DC-AC converter transfers the maximum power extracted from the PV cell into the grid. To improve the energy quality produced by the photovoltaic field other than the performance of the pulse width modulation (PWM) inverter, direct power control (DPC) is used to achieve these improvements. The simulation results showed a good performance of the suggested controller. Decoupled power control is achieved successfully, and a good power quality with low harmonic distortion rate (THD) is obtained.

Chapter 6
Comparative Analysis of Various Types of Control Techniques for Wind Energy Conversion System.......... 143
Bibhu Prasad Ganthia, IGIT, Sarang, India
Subrat Kumar Barik, KIIT University (Deemed), India
Byamakesh Nayak, KIIT University (Deemed), India

The interest towards renewable energy has been enhanced due to zero pollutant emission. Considering the present scenario, wind as a renewable source of energy is highly recommended. As it is freely available and free from pollution, wind can effectively play a role for energy generation. This can produce quality power during grid integrations as the load demands. Due to rapid variations in wind speed, wind energy systems need highly synchronized and powerful controller techniques for power regulations to overcome transients, voltage sags, and swells. A suitable and responsive controller is essential for power generation from wind energy. The controllers for wind energy system are categorized into five controller designs according to their locations to control the demand of the turbine system during grid integrations. In this chapter, various controller designs and implementations are highlighted with reference to previous works and existing studies.

Chapter 7
Enhanced Symbiotic Organisms Search and Hydrological Cycle Algorithms for Real Power Loss Diminution and Voltage Stability Enhancement.. 175
Lenin Kanagasabai, Prasad V. Potluri Siddhartha Institute of Technology, India

In this chapter, enhanced symbiotic organisms search (ESOS) algorithm and hydrological cycle (HC) algorithm are projected to solve factual power loss lessening problem. Symbiotic search algorithm is based on the actions between two different organisms in the ecosystem: mutualism, commensalism, and parasitism. Exploration procedure has been initiated arbitrarily, and each organism indicates a solution with fitness value. Quasi-oppositional-based learning and chaotic local search have been applied to augment the performance of the algorithm. In this work, hydrological cycle (HC) algorithm has been utilized to solve the optimal reactive power problem. It imitates the circulation of water form land to sky and vice versa. Only definite number of water droplets is chosen for evaporation, and it is done through roulette-wheel selection method. In the condensation stage, water drops move closer, combine, and also collusion occurs as the temperature decreases.

Chapter 8
Operation and Startup of Three-Phase Grid-Connected PWM Inverter for an Experimental Test Bench With DSPACE Real-Time Implementation of PQ Control: Active and Reactive Powers Control Strategy of the Grid Side Converter With Experimental Integration 207
Kamal Elyaalaoui, EMI, Mohammed V University in Rabat, Morocco
Moussa Labbadi, EMI, Mohammed V University in Rabat, Morocco
Khalid Chigane, EMI, Mohammed V University in Rabat, Morocco
Mohammed Ouassaid, EMI, Mohammed V University in Rabat, Morocco
Mohamed Cherkaoui, EMI, Mohammed V University in Rabat, Morocco

The main objective of this chapter is the experimental validation of active and reactive power control at the connection point for a three-phase grid connected inverter. It gives an overview on the adopted vector control strategy, regulation of the angle of orientation of the blades (pitch control), synchronization grid side converter to the power network using phase closed loop (PLL). Once the experimental test bench is described, the authors devote a first part to the design of the block circuit diagram of the experimental platform and the control strategy implemented in the DSPace DS1104, and they suggest some steps to associate the inverter to the electrical network. Subsequently, they discuss the experimental results validating the proposed power control. The purpose of this experimental results is the DSPACE real-time implementation of PQ control using three-phase inverter and development of a startup algorithm of the experimental test bench.

Chapter 9
Multilevel Inverter for Hybrid Fuel Cell/PV Energy Conversion System 233
Arezki Fekik, Akli Mohand Oulhadj University, Bouira, Algeria
Mohamed Lamine Hamida, Electrical Engineering Advanced Technology Laboratory (LATAGE), Mouloud Mammeri University, Algeria
Hakim Denoun, Electrical Engineering Advanced Technology Laboratory (LATAGE), Mouloud Mammeri University, Algeria
Ahmad Taher Azar, Prince Sultan University, Riyadh, Saudi Arabia
Nashwa Ahmad Kamal, Faculty of Engineering, Cairo University, Egypt
Sundarapandian Vaidyanathan, Vel Tech University, India
Amar Bousbaine, University of Derby, UK
Nacereddine Benamrouche, Electrical Engineering Technology Laboratory (LATAGE), Mouloud Mammeri University, Algeria

Power converters assume a significant part in fuel cell power generation systems and solar power conversion systems which are an alternative to fossil fuel production systems. There is therefore a demand for high quality power conditioning used in PEMFC systems and photovoltaic panels. This chapter proposes a hybrid electric power (FC/PV) production strategy with the use of converter topology as the power interface and also introduces a three-level inverter topology for different operating levels. The converter increases the input voltage to the rated voltage and turns into a DC bus; the multi-level inverter converts the voltage to AC and supplies AC loads. This chapter develops a hybrid electric power generation strategy, which can produce output with positive and zero sequences. Integrating the three-stage inverter into the hybrid renewable energy (FC/PV) production system allows for near sinusoidal current with low THD. The topology of hybrid energy production using the multi-level converter is tested on Matlab.

Chapter 10
Power Analysis Using Various Types of Wind Turbines 271
Bibhu Prasad Ganthia, Indira Gandhi Institute of Technology, Sarang, India
Monalisa Mohanty, ITER, SIksha 'O' Anusandhan (Deemed), India
Jai Kumar Maherchandani, College of Technology and Engineering, Maharana Pratap University of Agriculture and Technology, India

This chapter highlights on the design, operation, and comparative analysis of different types of wind turbine systems with respect to steady state and transient phenomenal activities under rapid wind speed variations. Here, Type I, which is fixed speed induction generator based, and Type II, which is DFIG based, variable speed operated systems are initially compared. In the next part, Type III wind turbine system is presented, which uses DFIG; later, it is compared with the Type IV WT system, which uses permanent magnet synchronous generator. This chapter provides a comparative overview on existing wind power systems including an analytic discussion of key principles and innovations for wind turbines. In this energy conversion system, various designs of wind turbines, pitch angle controlled based variable speed wind turbines governed by help of electronic power converters, were preferred. This scope of dynamic simulation-based study is implemented using MATLAB Simulink to convey the feasibility of the proposed wind turbine models.

Chapter 11
Flying Capacitor Inverter Integration in a Renewable Energy System 287
Mohamed Lamine Hamida, LATAGE Laboratory, Mouloud Mammeri University, Tizi Ouzou, Algeria
Arezki Fekik, Akli Mohand Oulhadj University, Bouira, Algeria
Hakim Denoun, LATAGE Laboratory, Mouloud Mammeri University, Tizi Ouzou, Algeria
Aghiles Ardjal, L2CSP Laboratory, Mouloud Mammeri University, Tizi Ouzou, Algeria
Aicha Aissa Bokhtache, GEER Laboratory, Hassiba Benbouali University, Chlef, Algeria

This chapter presents a three-cell flying capacitor converter photovoltaic (PV) system. This system consists of a DC-DC boost power converter connected in series with a multicell inverter. The perturb and observe MPPT technique has been used to extract the maximum power from the solar panel and generate the duty signal to control the switch of the DC-DC converter. The three-cell flying capacitor inverter ensures the conversion of the output voltage of the boost chopper to the alternative voltage. This topology is made up of hybrid association of commutation cells, which makes it possible to share the voltage constraint on several switches. A closed loop control based on PWM has been proposed to control the capacitor voltages of the inverter. The output current is controlled using a PI regulator. The aim of the proposed three cell inverter is to produce an approximate sinusoidal output current with a very low THD. The simulation results assess the effectiveness of the control.

Compilation of References 307

About the Contributors 348

Index 354

Preface

The energy transition initiated in recent years has enabled the growing integration of renewable production into the energy mix. This renewable production is often decentralized and its primary energy random. Microgrids make it possible to maximize the efficiency of energy transmission from source to consumer by bringing the latter together geographically and therefore by reducing losses linked to transport. But the lack of inertia and the micro-grid support system makes it weak and energy storage is necessary to ensure its proper functioning. It has also been shown that current storage technologies do not make it possible to provide both a large capacity of energy and of power at the same time, whether for technological or economic reasons. Hybrid storage is therefore a solution that combines the advantages of several technologies and also reduces their disadvantages. Several topologies allow for hybrid storage.

This book places particular emphasis on the Modeling and control of Static converters for hybrid storage systems. This book will have chapters with a current focus on research and innovative solutions to so many problems in this area. This book will cover most of the modeling, control theorems and optimization techniques that will solve many scientific problems for researchers in the field of power converter control for renewable energy hybrid storage and can be considered as a good reference for young researchers.

Many multidisciplinary applications will be addressed in this book with their new fundamental principles, modeling, analysis, design, realization and experimental results.

This book will include new circuits and systems, which will help many researchers to solve practically many nonlinear problems. Simply, this book will make it possible to bridge the gap between various interdisciplinary applications starting from mathematical concepts, modeling, analysis, and until the realization and the experimental work.

ORGANIZATION OF THE BOOK

This well-structured book consists of 11 full chapters.

Book Features

- The book presents the dissemination of advanced research in the field of renewable energies and the control of static converters.
- The chapters of the book are clearly illustrated with numerical examples and simulations.
- The chapters of the book deal with details of engineering applications and future areas of research.

- The chapters of the book give a good study of the literature with a long list of references.
- The chapters deal with recent research problems in the fields of renewable energies and control of static converters, control engineering, control of static converters, energy storage systems and applications.

SUMMARY OF THE BOOK CHAPTERS

Robust Decentralized Voltage Tracker of Islanded Multi-DG AC Microgrids Using Invariant Ellipsoids

Chapter 1 develops a robust decentralized voltage tracker for islanded MGs. The proposed controller is robust against the plug and play operation of the MG, loads and lines parameters uncertainties. The problem is solved in the frame work of linear matrix inequality (LMI). The proposed robust control represents the load changes, and the parameters variations of lines connecting the DGs as a norm-bounded uncertainty. The proposed controller utilizes local measurements from DGs, i.e., it is totally decentralized. Control decentralization is accomplished by decomposing the global system into subsystems. The effect of the rest of the system on a specific subsystem is considered as a disturbance to minimize (disturbance rejection control). The controller is designed by the invariant-sets (approximated by the invariant ellipsoids). Different time-domain simulations are carried out as: connecting and disconnected one or more DGs, connecting and disconnecting local loads DGs and Transmission line parameters variation.

Design, Control, Management, and Performance Analysis of PV-Battery Supercapacitor DC-System Using Buck Converter

In Chapter 2, the intermittent nature of photovoltaic energy necessitates the incorporation of storage devices to ensure the continuality of loads feeding. In addition, it is important to model, control, and verify the operating of the designed system before implementation. Furthermore, the integration of power electronic interfaces plays a significant role in protecting the system and benefiting from solar energy. To this end, a buck converter is chosen to charge the battery and supply the Supercapacitor. The control strategy is based on the Maximum Power Point Tracking techniques when the management algorithm recommends MPPT function mode. Otherwise, a feedback constant voltage PI controller is designed. Indeed, Perturb and observe and Incremental Conductance are implemented and compared to analyze the system efficiency within the management strategy to charge the battery, switch between the controllers, and feed a Supercapacitor in case of full battery charge. The obtained results using MATLAB/SIMULINK platform confirm the behavior of the proposed strategy.

True Power Loss Reduction by Enhanced Tree Squirrel Search, Enhanced Salp Swarm, and Swim Bladder Operation-Based Shark Optimization Algorithms: True Power Loss Reduction

In Chapter 3, Enhanced Tree Squirrel Search Optimization Algorithm (ETSS), Enhanced Salp Swarm Algorithm (ESS) and Swim bladder operation-based shark algorithm (SBS) has been applied to solve the power loss reduction problem. Enhanced Tree Squirrel Search Optimization Algorithm (ETSS) utilizes the jumping exploration method and progressive exploration technique, both possess winter search strategy, in- order to preserve the population diversity and to perk up the convergence speed. A new-fangled winter exploration strategy implemented in the jumping exploration technique. In Enhanced Salp Swarm Algorithm (ESS) an inertia weight $\omega \in [0, 1]$ is applied, which pick up the pace of convergence during the period of exploration. Then Swim bladder operation-based shark algorithm (SBS) is proposed to solve the problem. Based on contracting and expanding actions of the swim bladder in shark, SBS algorithm has been modelled.

Extension of Aspiration Level Model for Optimal Planning of Fast Charging Stations

In Chapter 4, electric vehicles will play a dominant role in future transportation due to their friendliness towards the present day's environment. The battery which drives the vehicle can be refilled using battery charging and battery swapping techniques. Fast Charging Stations provides faster service to the customers. Though battery swapping method outperforms battery charging in many ways, the heavy infrastructure requirement of the former requires time in integrating with the real world. Queuing models are used to depict the real-time behavior of service stations. The aspiration level model provides the optimal value of charging piles for the given system capacity in a Fast-Charging Station. The parameters in the aspiration level model can be formulated to an optimization problem. In the present work, the optimal planning for a Fast Charging Station in Beijing is carried out using Genetic algorithm. The simulation work is carried out in MATLAB/Simulink.

Power Quality Improvement for Grid-Connected Photovoltaic Panels Using Direct Power Control

Chapter 5 displays a control strategy for a photovoltaic system (PV), linked to the network with two phases of a PWM converter, where the first phase is a DC-DC converter linked among the photovoltaic source and the DC-AC converter. The second phase is a DC-AC converter linked to the grid. The Maximum Power Point (MPP) is tracked by DC-DC converter, which increases the DC bus voltage. The P&O (perturbation and Observation) technique is utilized as a direct current (DC-DC) converter controller to make the PV arrays work at greatest value of power under changing weather conditions. The DC-AC converter transfers the maximum power extracted from the PV cell into the grid. To improve the energy quality produced by the photovoltaic field other than the performance of the Pulse Width Modulation (PWM) inverter, Direct Power Control (DPC) is used to achieve these improvements. The simulation results showed a good performance of the suggested controller. Decoupled power control is achieved successfully and a good power quality with low harmonic distortion rate (THD) is obtained.

Comparative Analysis of Various Types of Control Techniques for Wind Energy Conversion System

In Chapter 6, interests towards renewable energy enhance its demand due to zero pollutant emission. Considering present scenario wind as renewable source of energy is highly recommended. As it is freely available and free from pollution, this wind can be effectively play highly potential role for energy generations. This can produce quality power during grid integrations as the load demands. Due to rapid variations in wind speed wind energy system needs highly synchronized and powerful controller techniques for power regulations to overcome transients, voltage sags and swells. A suitable and fast responsive controller is essential for power generation from wind energy. The controllers for wind energy system categorized into five controller designs according to its locations to control the demand of the turbine system during grid integrations. In this chapter various controller designs and its implementations are highlighted with reference to previous works and existing researches.

Enhanced Symbiotic Organisms Search and Hydrological Cycle Algorithms for Real Power Loss Diminution and Voltage Stability Enhancement

In Chapter 7, Enhanced Symbiotic Organisms Search (ESOS) Algorithm and Hydrological Cycle (HC) algorithm is projected to solve factual power loss lessening problem. Symbiotic search algorithm is based on the actions between two different organisms in the ecosystem- mutualism, commensalism and parasitism. Exploration procedure has been initiated arbitrarily and each organism indicates a solution with fitness value. Quasi-oppositional based learning and chaotic local search has been applied to augment the performance of the algorithm. Then In this work Hydrological Cycle (HC) algorithm has been utilized to solve the optimal reactive power problem. It imitates the circulation of water form land to sky and vice versa. Only definite number of water droplets is chosen for evaporation and it done through Roulette-wheel selection method. In the condensation stage water drops are move closer, combine and also collusion will occur as the temperature decreases.

Operation and Startup of Three-Phase Grid-Connected PWM Inverter for an Experimental Test Bench With DSPACE Real-Time Implementation of PQ Control: Active and Reactive Powers Control Strategy of the Grid Side Converter With Experimental Integration

The main objective of Chapter 8 is the experimental validation of active and reactive power control at the connection point for a three-phase grid connected inverter. It gives an overview on the adopted vector control strategy, regulation of the angle of orientation of the blades (Pitch control), synchronization grid side converter to the power network using phase closed loop (PLL). Once the experimental test bench is described, we will devote in a first part to the design of the Block circuit diagram of the experimental platform and the control strategy implemented in the DSPace DS1104, and we suggest some steps to associate the inverter to the electrical network. Subsequently, we discuss the experimental results validating the proposed power control. The purpose of this experimental results is the DSPACE Real-Time Implementation of PQ control using three-phase inverter and development of a startup algorithm of the experimental test bench.

Multilevel Inverter for Hybrid Fuel Cell/PV Energy Conversion System

In Chapter 9, power converters assume a significant part in fuel cell power generation systems and solar power conversion systems which are an alternative to fossil fuel production systems. There is therefore a demand for high quality power conditioning used in PEMFC systems and photovoltaic panels. This Chapter proposes a hybrid electric power (FC / PV) production strategy with the use of converter topology as the power interface and also introduces a three-level inverter topology for different operating levels. The converter increases the input voltage to the rated voltage and turns into a DC bus, the multi-level inverter converts the voltage to AC and supplies AC loads. This chapter develops a hybrid electric power generation strategy, which can produce output with positive and zero sequences. Integrating the three-stage inverter into the Hybrid Renewable Energy (FC / PV) production system allows for near sinusoidal current with low THD. The topology of hybrid energy production using the multi-level converter is tested on Matlab.

Power Analysis Using Various Types of Wind Turbines

Chapter 10 highlights on the design, operation and comparative analysis of different types of Wind Turbine system with respect to steady state and transient phenomenal activities under rapid wind speed variations. Here Type I which is fixed speed induction generator based and Type II which is DFIG based variable speed operated system are initially compared. In the next part Type III wind turbine system presented which uses DFIG; later it is compared with the Type IV WT system which uses Permanent Magnet Synchronous Generator. This chapter provides a comparative overview on existing wind power systems including an analytic discussion of key principles and innovations for wind turbines. In this energy conversion system various designs of wind turbines, pitch angle controlled based variable speed wind turbines governed by help of electronic power converters were preferred. This scope of dynamic simulation based study is implemented using MATLAB Simulink to convey the feasibility of the proposed Wind Turbine models.

Application of Flying Capacitor Inverter for Renewable Energy System

Chapter 11 presents a three-cell flying capacitor converter photovoltaic (PV) system. This system consists of a DC-DC boost power converter connected in series with a multicell inverter. The boost converter consists of increasing the output voltage of the solar panel. The Perturb & Observe MPPT technique is used to extract the maximum power from the solar panel and generate the duty signal to control the switch of the DC-DC converter. The three-cell flying capacitor inverter ensures the conversion of the output voltage of the boost chopper to the alternative voltage, this topology is made up of hybrid association of commutation cells, which makes it possible to share the voltage constraint on several switches. With this inverter the output voltage and current can be improved with regarding the THD. A closed loop control based on PWM is proposed to control the capacitor voltages of the multi-level inverter. The output current is controlled using a PI regulator. The whole PV system generation scheme has been implemented using MATLAB/Simulink software. The simulation results verified the effectiveness of the control. In this way, the proposed solar power-fed three cell inverter produces an approximate sinusoidal output current with a very low THD.

AUDIENCE

The book is primarily intended for academic and industrial researchers, who work in the fields of research - renewable energies and control of electronic power converters. The book can also be used at higher or advanced level as a textbook or major reference for courses such as modeling of renewable energy sources, numerical simulation, electronic power conversion systems, control systems, electrical energy storage systems and many others.

Arezki Fekik
Akli Mohand Oulhadj University, Bouira, Algeria

Nacereddine Benamrouche
University of Tizi Ouzou, Algeria

ACKNOWLEDGMENT

As the editors, we hope that the chapters in this book will stimulate further research in renewable energy and theirs control, and utilize them in real-world applications. We hope that this book, covering so many different aspects, will be of value for all readers. We would like to thank also the reviewers for their diligence in reviewing the chapters. Special thanks go to IGI Global, especially the book editorial team.

Chapter 1
Robust Decentralized Voltage Tracker of Islanded Multi-DG AC Microgrids Using Invariant Ellipsoids

Ehab Hassan Eid Bayoumi
Egyptian Chinese University, Egypt

Hisham Soliman
faculty of Engineering, Cairo University, Egypt

Farag El-Sheikhi
Istanbul Esenyurt University, Istanbul, Turkey

ABSTRACT

This chapter develops a robust decentralized voltage tracker for islanded MGs. The proposed controller is robust against the plug and play operation of the MG, loads, and line parameter uncertainties. The problem is solved in the framework of linear matrix inequality (LMI). The proposed robust control represents the load changes and the parameter variations of lines connecting the DGs as a norm-bounded uncertainty. The proposed controller utilizes local measurements from DGs (i.e., it is totally decentralized). Control decentralization is accomplished by decomposing the global system into subsystems. The effect of the rest of the system on a specific subsystem is considered as a disturbance to minimize (disturbance rejection control). The controller is designed by the invariant-sets (approximated by the invariant ellipsoids). Different time-domain simulations are carried out as connecting and disconnected one or more DGs, connecting and disconnecting local loads DGs and transmission line parameters variation.

DOI: 10.4018/978-1-7998-7447-8.ch001

INTRODUCTION

A microgrid (MG) is a small-scale power grid composed of various interconnected local loads, distributed generation (DGs), storage devices, master controller, power electronics converters, protective devices, along with communication, and automation systems. The growth of MGs permit extensive penetration of renewable energy sources (RESs) into the power systems. Consequently, MGs can diminish transmission, distribution system losses, global warming gas emissions. It also enhance reliable and secure service of power to customers (Bayoumi E.H.E., et al. 2020; Olivares D. E., et al. 2014).

MGs have two modes of operations: grid-connected and islanded modes. In the grid-connected mode, MGs transfer power to the utility grid at point of common coupling (PCC). In the islanded mode, the MG works autonomously without connection to the utility grid. The characteristics of the microgrid are: bidirectional power transfer, and power electronics devices installed with DGs. The control of MG in each operation mode along with the switching between the modes are the challenge that require to be resolved in order to use MGs effectively and successfully (Sadabadi M. S., et al. 2017).

The islanded mode of operation is able to be scheduled or unscheduled and it arises when MGs are separated from the utility grid (Yazdanian M., et al. 2014). The unscheduled islanded mode of operation is started when the utility grid faces a continual abnormal condition. When the MG is disconnected, real and reactive power mismatch between the load and generation causes deviation in voltage and frequency. The main function of the control system in islanded mode of operation is to reestablish the balance to preserve the MG stability (Olivares D. E., et al. 2014). To guarantee stability during the islanded mode of operation, new control techniques has to be implemented to ensure both voltage and frequency stability through suitable real and reactive power sharing arrangements. One of the well-known control hicrarchy used for the MGs is the Energy Management System (EMS). This control technique specifies the DGs power and voltage set points (Bidram A., et al. 2017; Xuand Y., 2020).

Sustaining the stability of MGs operation in both grid-connected and islanded modes along with the switching between the two modes introduces excessive challenge as a result of the increasing penetration of renewable energy-based DGs (Shafiee G., et al. 2014). This challenge is due to low- inertia inverter-interfaced DGs compared to large rotational inertia of the conventional rotating synchronous generators (Schiffer J., et al. 2014). The DGs with low-inertia inverters do not improve grid stability. One of the main important challeges in MGs is plug -and-play (PnP) operation of DGs due to the inherently discontinuous nature of renewable energy sources.

Plug-and-play (PnP) characteristic must be impeded in any ideal MG system. It should permit normal integration of any new loads or sources without the necessity to redesign the MG control system. There are some efforts in achieving this target. Generally, most of these efforts can be classified into two main categories: stability, and Communications. MG system stability must be guaranteed when it is operating Plug-and-play (a new DG unit is added/removed due to any abnormal condition or the intermittency of renewable energy resources) (Xuand Y., et al. 2020). Under this performance, the frequency and voltage stability of MGs have to be preserved even when communication links are lost without the need to retune the MG control system. To attain this performance, a decentralized control strategy is required. The droop control is normally applied to MGs control. This strategy is based on the power balance as of synchronous generators (Olivares D. E., et al. 2014; Sadabadi M. S., et al. 2017). The frequency or the rotor speed of a synchronous generator decreases linearly with its output power (Kundur P., et al. 1994). Although the primary droop control strategy delivers a proportional power sharing between DGs, its implementation is limited because it has only one tunable parameter. Extra challenges related to this control technique involve the coupling between frequency and voltage deviations taking into consideration the real and reactive power dynamics. Furthermore, the droop control technique has a poor behaviour in the case of parameters' changes of lines connecting the DGs (Etemadi A. H., et al. 2014).

Non-droop-based methods have been established to control both the voltage and frequency of the islanded MGs (Etemadi A. H., et al. 2014). Various voltage controller design approaches have been applied: robust servomechanism (Kundur P., et al. 1994), full-order H∞ control (Moradi R., et al. 2010), two-degree-of-freedom robust control (Babazadeh M. et al. 2013), voltage control employing the loop-shaping technique (Bahrani B., et al. 2013), decentralized state feedbacks (Riverso S., et al. 2015a), and robust fixed-order decentralized H∞ method (Sadabadi M. S., et al. 2015). These control approaches are utilized for the voltage of a single-DG and a multi-DG MGs (Etemadi A. H., et al. 2014). The control method of (Etemadi A. H., et al. 2014) applies the open-loop scheme to regulate the frequency of each DG unit via an internal oscillator with $\omega_0 = 2\pi f_0$, where f_0 is the nominal system frequency. A global positioning system (GPS) is employed to synchronize all DGs through a common time reference (Etemadi A. H., et al. 2014). A power management system (PMS) is utlized to share the load by solving an optimal power flow (OPF) problem. The PMS then transmits reference points to DGs (Etemadi A. H., et al. 2014). The load sharing's accuracy depends on how regular the OPF problem is solved. To preserve the DGs plug-and-play operation, the non-droop-based local controllers has to be re-tuned to ensure system stability when the MG topology change. According to the decentralized control approach established in (Riverso S., et al. 2015b), local controllers of all DGs must be re-tuned once a DG unit is plugged in/out.

In spite of the above-mentioned research work on isladed MG control strategies, the present paper focuses on plug-and-play performance of inverter-based DGs. The main contributions of this paper are summarized as follows:

1. An effective modelling of the MGs' plug and-play operation (PnP), load changes, and lines (connecting the DGs) parameters uncertainty.
2. The control decentralization is achieved by decomposing the global system into subsystems. The interaction effect of the rest of the system on a specific subsystem is treated as a disturbance to be rejected by the proposed control.
3. Based on the linear matrix inequalities, a novel theorem is introduced to guarantee the solvability of the optimal decentralized trackers (in which the output follows the desired inputs) achieving external disturbance rejection.
4. The suggested trackers are also robust against plug and play MGs' operation, load changes, lines parameters variations.
5. A simple controller design is presented, based on the invariant sets.

The paper is organized as follows. The MG control problem is introduced in section 1. Section 2 describes the MG study-system. The mathematical model of the MGs is described in section 3.

This manuscript is organized as follows. In Section 2, The dynamic models of the islanded distributed generated microgrids are introduced. In Section 3, The mathematical setup of the robust decentralized control problem is stated. In Section 4, a servomechanism control design based on the invariant ellipsoids is developed. In the proposed decentralized design decomposes the global sysytem into subsystems, for each the interaction on every susbystem is considered as disturbances. In Section 5, simulation results is used to test the validity of the proposed robust decentralized tracker scheme.

Notations: The notations are standard. R^m is the set of vectors of $m \times 1$ dimension, $R^{r \times q}$ is the set of real matrices of dimension $r \times q$, and $(.)'$ denotes the transpose of a vector or a matrix. For a matrix P, $P > 0$ (<0) means that P is a symmetric positive (negative) definite matrix. Additionally, (*) in a matrix means the symmetric part. Similarly, $(M+N+*)$ means $(M+N+M'+N')$. Finally, 0 and I denote the zero matrices and the identity matrix, respectively.

THE DYNAMIC MODEL OF ISLANDED MICROGRID

A block diagram of a MG network with five-Distributed Energy Resources (DER) is given in Fig. 1(a). The islanded MGs control scheme and the power management system (PMS) that contain many DGs is depicted in Fig. 1(a), (Etemadi A. H., et al. 2012). The voltage set points at each DG unit terminals are determined and communicated by PMS to the DGs' local controllers through a low bandwidth communication medium. The local controllers afford a robust and decentralized control of the DGs terminal voltage. An open-loop control is used to control the DGs frequency employing an domestic crystal oscillator for each DG. The DGs are synchronized with a global positioning system (GPS), as shown in Fig. 1(b).

Figure 1. (a) The proposed utility microgrid, (b) The PMS and the non-droop control

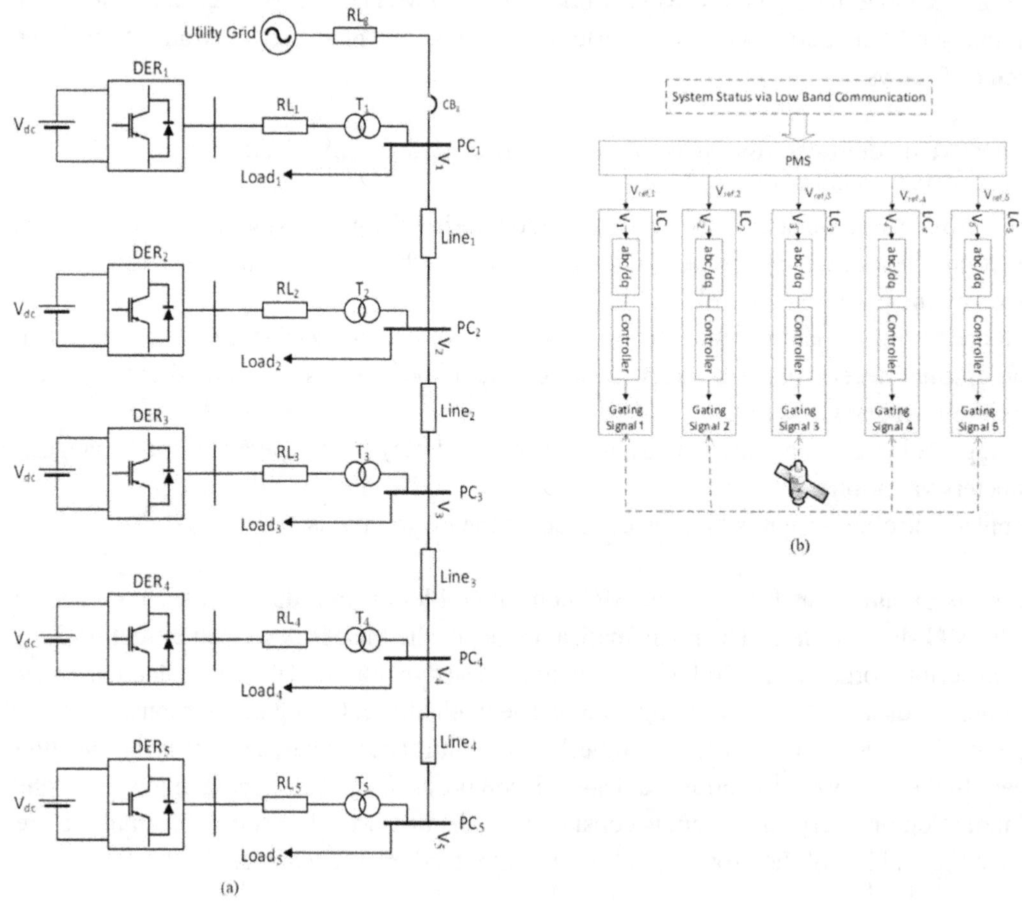

All the five 0.6kV rated DGs in the MG have power ratings of 1.6, 1.2, 0.8, 1 and 0.75 MVA. The MG has four 13.8kV distribution line segments and five local loads. Each DG unit is interfaced by a voltage sourced converter (VSC) having 1.5-kV DC voltage source. The converter is connected to the grid via a series RL filter and a 0.6-kV/13.8-kV step-up transformer. The main utility grid is modeled by an AC voltage source behind series inductive impedance. The status of circuit breaker CBg will decide whether the MG is in grid-connected or the islanded mode of operation. Table 1 summarizes the MG parameters.

Table 1. Parameters of the five-DGs microgrid

Base Values				
S_{base} = 1.6 MVA $V_{base,\ low}$ = 0.6 kV $V_{base,\ high}$ = 13.8 kV				
Transformers				
0.6/13.8 kV Δ/ Y_g Xt = 8%				
Load Parameters				
Load 1	**Load 2**	**Load 3**	**Load 4**	**Load 5**
R_1 350 Ω 2.94 pu	R_2 375 Ω 3.15 pu	R_3 380 Ω 3.19 pu	R_4 370 Ω 3.11 pu	R_5 400 Ω 3.36 pu
X_{L1} 41.8 Ω 0.35 pu	X_{L2} 37.7 Ω 0.32 pu	X_{L3} 36.5 Ω 0.31 pu	X_{L4} 37.2 Ω 0.31 pu	X_{L5} 45.2 Ω 0.38 pu
X_{C1} 44.2 Ω 0.37 pu	X_{C2} 40.8 Ω 0.34 pu	X_{C3} 39.2 Ω 0.33 pu	X_{C4} 40.1 Ω 0.34 pu	X_{C5} 48.2 Ω 0.4 pu
R_{l1} 2Ω 0.02 pu	R_{l2} 2Ω 0.02 pu	R_{l3} 2Ω 0.02 pu	R_{l4} 2Ω 0.02 pu	R_{l5} 2Ω 0.02 pu
Line Parameters				
R = 0.34 Ω/ km X = 0.31 Ω/ km				
Line 1	**Line 2**	**Line 3**	**Line 4**	
5 km	10 km	12 km	10 km	
Filter Parameters (based on DER_i Ratings)				
X_f = 15% Quality Factor = 50				

The linearized MG model shown in Fig. 1(a) is utilized to acquire a decentralized control based in a rotating dq-frame. Fig 2 shows the line diagram used for deriving the state space equations.

Figure 2. The line diagram of the proposed microgrid

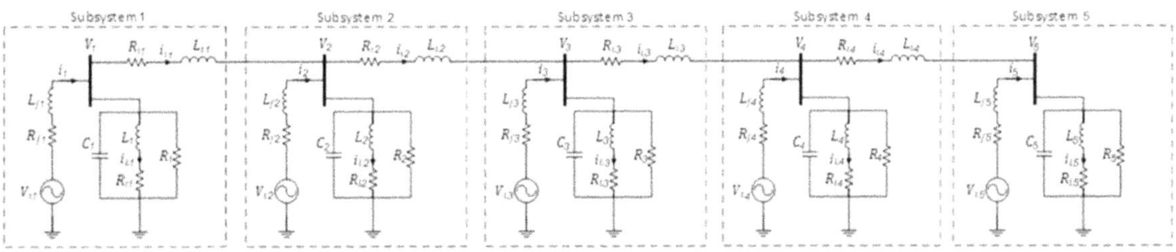

Neglecting the harmonics, the system's fundamental frequency part is employed in controller design. The DGs are modeled by a three-phase controlled voltage source in series with inductive RL impedance of the filter. The equivalent RLC parallel circuits represent the loads. Distribution lines are modeled as inductive RL impedances.

The study MG is composed of N subsystems, where N=5. The subsystem 1 is modeled in the *abc* frame as follows:

$$\begin{aligned} i_{1,abc} &= i_{t1,abc} + C_1\dot{v}_{1,abc} + i_{L1,abc} + \frac{v_{1,abc}}{R_1} \\ v_{t1,abc} &= L_{f1}\frac{di_{1,abc}}{dt} + R_{f1}i_{t1,abc} + v_{1,abc} \\ v_{1,abc} &= L_1\frac{di_{L1,abc}}{dt} + R_{l1}i_{L1,abc} \\ v_{1,abc} &= L_{t1}\frac{di_{t1,abc}}{dt} + R_{t1}i_{t1,abc} + v_{2,abc} \end{aligned} \qquad (1)$$

Where x_{abc} is a 3x1 vector. Considering a three-wire system, subsystem 1 (1) can be transformed to dq-frame, Fig. 3, (Etemadi A. H., et al. 2012) as follows

$$f_{dq} = \frac{2}{3}\begin{bmatrix} cos\theta & \cos\left(\theta - \frac{2}{3}\pi\right) & \cos\left(\theta - \frac{4}{3}\pi\right) \\ -sin\theta & -\sin\left(\theta - \frac{2}{3}\pi\right) & -\sin\left(\theta - \frac{4}{3}\pi\right) \\ \frac{1}{\sqrt{2}} & \frac{1}{\sqrt{2}} & \frac{1}{\sqrt{2}} \end{bmatrix} \qquad (2)$$

The phase angle $\theta(t) = \int_0^t w(\tau)d\tau + \theta_0$. w is the angular frequency of the DGs internal oscillator. The mathematical model of subsystem 1 in the dq- frame, based on (1) and (2), is as follows:

$$\begin{aligned} \dot{V}_{1,dq} &= \frac{1}{C_1}I_{1,dq} - \frac{1}{C_1}I_{t1,dq} - \frac{1}{C_1}I_{L1,dq} - \frac{V_{1,dq}}{C_1R_1} - jwV_{1,dq} \\ \dot{I}_{1,dq} &= \frac{1}{L_{f1}}V_{t1,dq} - \frac{R_{f1}}{L_{f1}}I_{1,dq} - \frac{1}{L_{f1}}V_{1,dq} - jwI_{1,dq} \\ \dot{I}_{t1,dq} &= \frac{1}{L_1}V_{1,dq} - \frac{R_{l1}}{L_1}I_{L1,dq} - jwI_{L1,dq} \\ \dot{I}_{t1,dq} &= \frac{1}{L_{t1}}V_{1,dq} - \frac{R_{t1}}{L_{t1}}I_{t1,dq} - \frac{1}{L_{t1}}V_{2,dq} - jwI_{t1,dq} \end{aligned} \qquad (3)$$

Figure 3. Synchronously rotating dq- frame

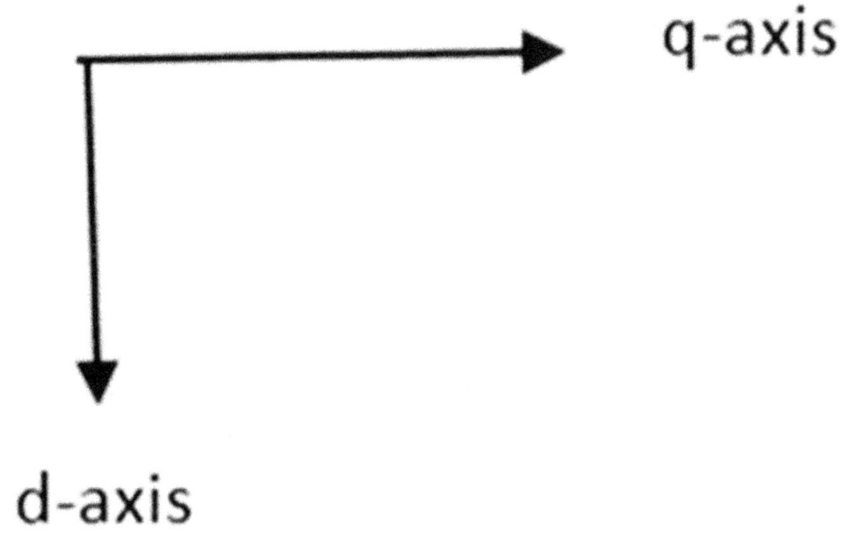

Similar to Subsystem 1, Subsystems 2 .. 5 dq-frame based models can be constructed. These models are used to form the overall system's state-space model.

$$\dot{x} = Ax + Bu, y = Cx \tag{4}$$

where

$x' = [V_{1,d}, V_{1,q}, I_{1,d}, I_{1,q}, I_{L1,d}, I_{L1,q}, I_{t1,d}, I_{t1,q},$

$V_{2,d}, V_{2,q}, I_{2,d}, I_{2,q}, I_{L2,d}, I_{L2,q}, I_{t2,d}, I_{t2,q},$

....

$V_{5,d}, V_{5,q}, I_{5,d}, I_{5,q}, I_{L5,d}, I_{L5,q}]$

$u' = \left[V_{t1,d}, V_{t1,q}, V_{t2,d}, V_{t2,q}, \ldots V_{t5,d}, V_{t5,q}\right],$

$$y' = \left[V_{1,d}, V_{1,q}, V_{2,d}, V_{2,q}, \ldots V_{5,d}, V_{5,q}\right]$$

$A \in R^{38\times38}$, $B \in R^{38\times10}$ and $C \in R^{10\times38}$ are the state matrices as given in the appendix.

DESIGN OF DECENTRALIZED TRACKING CONTROL

In control system design, the output has to follow the input. If the input is constant, the control problem is termed a regulator problem. If the input is time varying, it is called a servomechanism (tracking) problem. In this section, a decentralized servomechanism controller for the system (4) is designed. The designed controllers are decentralized in the sense that it uses only the local information of its subsystem. Centralized vs. decentralized control can be summarized as follows: The benefit of using decentralized control via local subsystem information is avoiding using a hub computer (controller) whose failure will cause losing the global system stability. It also avoids a costly communication network and its associated delay to transmit information of the whole system to the centralized controller. The interconnected composite system (4) can be decomposed into three subsystems. Later, local controllers can be used for each subsystem (Etemadi A. H., et al. 2014). Splitting the matrix $A = \{A_{ij}, i,j=1,2,3\}$ in (4) into a two parts: diagonal, A_d, and off diagonal D, one gets

$$\dot{x} = A_d x + Bu + Dx, y = Cx \tag{5}$$

where

$$A_d = \begin{bmatrix} A_{11} & 0 & \ldots & 0 \\ 0 & A_{22} & \ldots & 0 \\ \ldots & \ldots & \ddots & \ldots \\ 0 & 0 & \ldots & A_{55} \end{bmatrix}, B = \begin{bmatrix} B_1 & 0 & \ldots & 0 \\ 0 & B_2 & \ldots & 0 \\ \ldots & \ldots & \ddots & \ldots \\ 0 & 0 & \ldots & B_5 \end{bmatrix}, D = \begin{bmatrix} 0 & A_{12} & \ldots & A_{15} \\ A_{21} & 0 & \ldots & A_{25} \\ \ldots & \ldots & \ddots & \ldots \\ A_{51} & A_{52} & \ldots & 0 \end{bmatrix}, C = \begin{bmatrix} C_1 & 0 & \ldots & 0 \\ 0 & C_2 & \ldots & 0 \\ \ldots & \ldots & \ddots & \ldots \\ 0 & 0 & \ldots & C_5 \end{bmatrix}$$

In (5), the impact of interconnecting the rest of the system on a specific subsystem is regarded as an external disturbance, Dx, which is to be rejected by the controller proposed. The vector x is assumed to be an external bounded disturbance w (Nazin S. A., et al. 2007).

A decentralized tracker with three control agents for each DG unit is considered with the following structure:

$$u = Kx + K_I\xi, \dot{\xi} = r - Cx \tag{6}$$

Where

$$K = \begin{bmatrix} K_1 & 0 & \dots & 0 \\ 0 & K_2 & \dots & 0 \\ \dots & \dots & \ddots & \dots \\ 0 & 0 & \dots & K_5 \end{bmatrix}, K_I = \begin{bmatrix} K_{I1} & 0 & \dots & 0 \\ 0 & K_{I2} & \dots & 0 \\ \dots & \dots & \ddots & \dots \\ 0 & 0 & \dots & K_{I5} \end{bmatrix}$$

The augmented closed-loop system is represented by the following equation:

$$\begin{bmatrix} \dot{x} \\ \dot{\xi} \end{bmatrix} = \begin{bmatrix} A_d + BK & BK_I \\ -C & 0 \end{bmatrix} \begin{bmatrix} x \\ \xi \end{bmatrix} + \begin{bmatrix} D \\ 0 \end{bmatrix} w + \begin{bmatrix} 0 \\ I \end{bmatrix} r \quad (7)$$

MG voltage control problem is to design a decentralized controller which the output voltage tracks the reference voltage. The tracker must be robust against the load variations and to suppress the interconnection effect of the rest of the network on the specific subsystem's output voltage. This is termed disturbance rejection control.

The definition of the invariant ellipsoid is first presented as it is used in the sequel.

Invariant Ellipsoid

Consider a continuous-time stationary dynamic linear system.

$$\dot{x} = Ax + Dw, \; z = C_z x \quad (8)$$

Where the $x(t) \in R^n$ is the system state vector, $z(t) \in R^l$ is the system output to optimize. It is assumed that the output for feedback y is equal to the output to optimize, z. Hence $C=C_z$.
The external disturbances $w(t) \in R^m$ are bounded i.e. subject to the following constraint:

$$w(t) \leq 1, \forall t \geq 0 \quad (9)$$

The Euclidean norm of the vector(.) is denoted by with (.). The control target is to minimize the impact of disturbance Dw on output z. As matrix D can be scaled to satisfy (9), the generality is not lost.

No other constraints are assumed to be random or harmonic and L∞ is considered a bounded external disturbance. The system (8) is assumed to be stable, the pair (A, D) is controllable, and C has a maximum-rank. The goal is to find the system's family of invariant ellipsoids.

Consider an ellipsoid E with origin as the center given by the following equation:

$$E = \{x : x' P^{-1} x \leq 1\}, \; P > 0 \quad (10)$$

where the matrix of the ellipsoid P is a symmetric positive definite matrix. This ellipsoid confine and approximates the invariant sets of the system (8) subject to the disturbance constraint (9).

When for any initial state $x(0)$ *inside* E, the trajectory $x(t)$ remains inside the ellipsoid for the future moment $t > 0$ for all permissible disturbances (9), E is called a state invariant. If the initial state $x(0)$

begins *outside* the ellipsoid and the trajectory $x(t)$ is attracted to it for $t > 0$, E is termed attractive. This happens if the Lyapunov function $V(x) = x'P^{-1}x,\ P > 0$ does not increase outside this ellipsoid. In this case,

$$\dot{V}(x) \le 0 \text{ for all } x(t) \text{ subject to } V(x) \ge 1$$

Ref. [17] shows that the ellipsoid E is invariant (and attractive) for system (8) if and only if

$$AP + PA' + \alpha P + \frac{1}{\alpha} DD' \le 0.$$

The effect of external disturbance on the dynamic system trajectories is represented by invariant ellipsoid. The ellipsoid volume must be minimal to minimize this effect. To achieve this target, an objective function is formulated in terms of the ellipsoid matrix f(*P*). Due to its linearity, the trace function is selected as the objective function.

$$f(P) = tr(P) \tag{11}$$

This objective function resembles to the sum of the squares of the semi-axes of the state-invariant ellipsoid of the original system. The *tr(.)* denotes the sum of the diagonal elements of (.). As the trace function provides the semi-definite program (SDP), it is used in this paper. MATLAB-based software can be used to solve this optimization problem because it is convex.

The ellipsoid of (10) will be replaced by $E_z = z'P^{-1}z,\ z = Cx$ as the performance z must be optimized against the disturbances. Minimizing the volume of E_z provides less impact of the disturbance on MG voltage.

The suggested controller design can be represented by a constrained minimization problem. This problem can be solved by a state-feedback as described below.

Invariant Ellipsoid Design of State-Feedback + Integral Tracker

As a starting point, the controller designed for good disturbance rejection to offer adequate transients to disturbances such as $w(t)$. Later, the design will be extended to include reference-tracking.

Regulator Invariant Set Design

The controllable linear system (8) is modified to (12):

$$\dot{x} = Ax + Bu + Dw,$$

$$z = Cx + B_2u, \tag{12}$$

where $x \in R^n$ is the system phase state, $z \in R^l$ is the system output, $u \in R^p$ is the control signal, and $w \in R^m$ is the external disturbance that satisfies (9). A linear proportional state feedback controller K is desired.

$$u = Kx, \tag{13}$$

The controller should stabilize the closed-loop system and also optimally rejects the impact of the external disturbances $w(t)$. It is worth noting that the inclusion of the positive component B_2u in (12) prevents large values of control. Considering this controller, the closed-loop form is given as follows:

$$\dot{x} =(A + B.K)x+ Dw,\ z =(C + B_2K)x. \tag{14}$$

The optimal control is obtained via minimal invariant ellipsoid as determined by Theorem 1. The problem is reduced to a semidefinite programming and one-dimensional convex minimization.

Theorem 1[17]. Let the external disturbances be L∞-bounded and the pair (A, B) controllable for the system (11). Then, the problem of designing a state feedback controller by state (12) that optimally rejects the external disturbances (in the sense of the trace that is output-invariant to the ellipsoid) is equivalent to

minimize $tr\left[CPC' + B_2ZB_2'\right]$ subject to the constraints

$$\left(AP + BY + *\right) + \alpha P + \frac{1}{\alpha}DD' \le 0, \alpha > 0 \tag{15}$$

$$\begin{bmatrix} Z & Y \\ * & P \end{bmatrix} \ge 0,\ P > 0 \tag{16}$$

where $Y = KP$, and minimization is carried out with respect to the variables α, $P = P' \in R^{n\times n}$, $Y \in R^{p\times n}$ and $Z = Z' \in R^{p\times p}$.

The above design strategy does not consider a reference input for the regulator problem. In addition, it does not provide command following shown by a good transient response. Therefore, when developing a control system attention must be given to both good disturbance dismissal and good command following.

Invariant Ellipsoid Design of Servomechanism Control

Since the MG model (5) has no integrator, a steady state error for a step input is expected. An integrator is connected in the feedforward path between the error comparator and the plant to eliminate the steady state error, as shown in Figure 4.

From Fig. 2, we obtain

$$\dot{x} = A_d x + Bu + Dw, y = Cx, u = Kx + k_i\xi, \dot{\xi} = r - Cx \tag{17}$$

Figure 4. Servo system

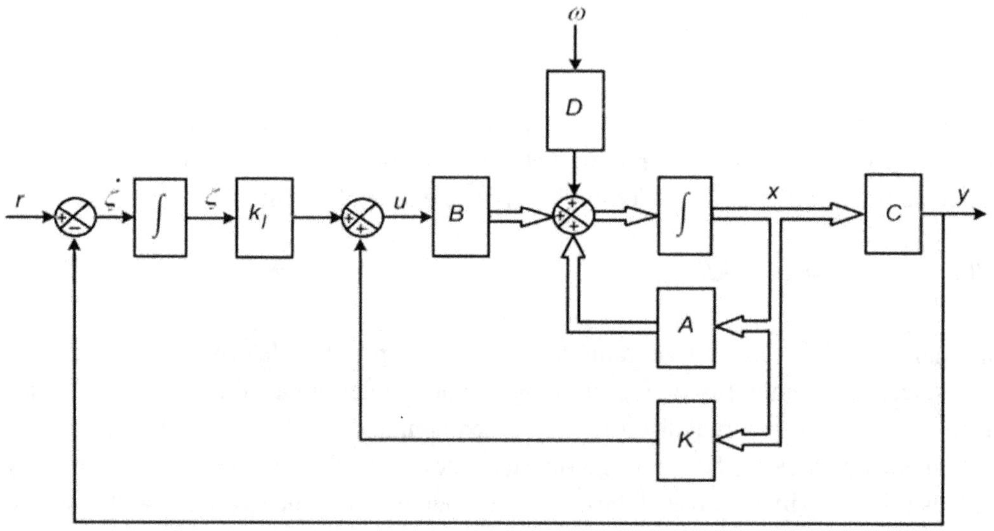

The augmented system is therefore as follows:

$$\begin{bmatrix}\dot{x}\\ \dot{\xi}\end{bmatrix}=\begin{bmatrix}A & 0\\ -C & 0\end{bmatrix}\begin{bmatrix}x\\ \xi\end{bmatrix}+\begin{bmatrix}B\\ 0\end{bmatrix}u+\begin{bmatrix}D\\ 0\end{bmatrix}w+\begin{bmatrix}0\\ I\end{bmatrix}r, z=\begin{bmatrix}C & 0\end{bmatrix}\begin{bmatrix}x\\ \xi\end{bmatrix}+B_2u \tag{18}$$

or

$$\begin{bmatrix}\dot{x}\\ \dot{\xi}\end{bmatrix}=\widehat{A}\begin{bmatrix}x\\ \xi\end{bmatrix}+\widehat{B}u+\widehat{D}w+\begin{bmatrix}0\\ I\end{bmatrix}r, z=\widehat{C}\begin{bmatrix}x\\ \xi\end{bmatrix}+B_2u \tag{19}$$

Where the augmented matrices are

$$\widehat{A}=\begin{bmatrix}A & 0\\ -C & 0\end{bmatrix},\widehat{B}=\begin{bmatrix}B\\ 0\end{bmatrix},\widehat{D}=\begin{bmatrix}D\\ 0\end{bmatrix},\widehat{C}=\begin{bmatrix}C & 0\end{bmatrix}$$

Theorem 1 for regulator design can be generalized to the tracker ellipsoidal design. The proposed controller has to be decentralized and roust against load variation (uncertainty) of the MG. For this, (19) is modified to

$$\begin{bmatrix} \dot{x} \\ \dot{\xi} \end{bmatrix} = \left(\widehat{A} + \Delta\widehat{A}\right)\begin{bmatrix} x \\ \xi \end{bmatrix} + \widehat{B}u + \left(\widehat{D} + \Delta\widehat{D}\right)w + \begin{bmatrix} 0 \\ I \end{bmatrix} r, z = \widehat{C}\begin{bmatrix} x \\ \xi \end{bmatrix} + B_2 u \tag{20}$$

The system time varying uncertainties $\Delta\widehat{A}(t), \Delta\widehat{D}(t)$ have the norm-bounded form

$$\Delta\widehat{A} = M\Delta_A N, \Delta\widehat{D} = F\Delta_D H, \Delta_A(t) \leq 1, \Delta_D(t) \leq 1 \tag{21}$$

The following theorem is developed.

Theorem 2: Let the external disturbances be L∞-bounded and the pair ($\widehat{A}, \widehat{B}$), controllable for the system (20). Then, the issue of developing decentralized robust state feedback with integral controller $u = \begin{bmatrix} K & k_i \end{bmatrix}\begin{bmatrix} x \\ \xi \end{bmatrix}$ which rejects optimally the external disturbances (in the sense of the trace that is output-invariant to the ellipsoid) is equivalent to minimize $tr\left[\widehat{C}\widehat{P}\widehat{C}' + B_2 Z B_2'\right]$ subject to the following constraints

$$\begin{bmatrix} \left(\widehat{A}\widehat{P} + \widehat{B}\widehat{Y} + *\right) + \alpha\widehat{P} + \int MM' + \rho FF' & \widehat{D} & \widehat{P}N' & 0 \\ * & -\alpha I & 0 & H' \\ * & * & -\int I & 0 \\ * & * & * & -\rho I \end{bmatrix} \leq 0, \alpha, \int, \rho > 0$$

$$\begin{bmatrix} Z & \widehat{Y} \\ * & \widehat{P} \end{bmatrix} \geq 0, \widehat{P} > 0 \tag{22}$$

where $\widehat{Y} = \widehat{K}\widehat{P}, \widehat{K} = \begin{bmatrix} K & k_i \end{bmatrix}$, and the minimization is carried out with respect to the variables $\alpha, \varepsilon, \rho, P = P' \in R^{(n+1).(n+1)}$, $\widehat{Y} \in R^{p.(n+l)}$ and $Z = Z' \in R^{p \times p}$.

Proof

For the tracking problem, the matrices in (15) are replaced by the augmented system matrices. Using Fact 2, we get

$$\begin{bmatrix} \left(\widehat{A}\widehat{P} + \widehat{B}\widehat{Y} + *\right) + \alpha\widehat{P} & \widehat{D} \\ * & -\alpha I \end{bmatrix} \leq 0, \alpha > 0 \tag{23}$$

To cope with load variation (uncertainty) in the MG,

$$\widehat{A} \text{ is replaced with} \left(\widehat{A}+\Delta\widehat{A}\right), \text{ and } \widehat{D} \text{ with} \left(\widehat{D}+\Delta\widehat{D}\right)$$

in (23). The resulting system is

$$\begin{bmatrix} \left(\left\{\widehat{A}+\Delta\widehat{A}\right\}\widehat{P}+\widehat{B}\widehat{Y}+*\right)+\alpha\widehat{P} & \widehat{D}+\Delta\widehat{D} \\ * & -\alpha I \end{bmatrix} \le 0, \alpha > 0$$

Substituting for the norm-bounded uncertainty (21), one gets

$$\begin{bmatrix} \left(\left\{\widehat{A}+M\Delta N'\right\}\widehat{P}+\widehat{B}\widehat{Y}+*\right)+\alpha\widehat{P} & \widehat{D}+F\Delta_D H \\ * & -\alpha I \end{bmatrix} \le 0, \alpha > 0$$

Separating the uncertainty terms, we obtain

$$\begin{bmatrix} \left(\widehat{A}\widehat{P}+\widehat{B}\widehat{Y}+*\right)+\alpha\widehat{P} & \widehat{D} \\ * & -\alpha I \end{bmatrix} + \left(\begin{bmatrix} M \\ 0 \end{bmatrix}\Delta_A\begin{bmatrix} N\widehat{P} & 0 \end{bmatrix}+*\right) + \left(\begin{bmatrix} F \\ 0 \end{bmatrix}\Delta_D\begin{bmatrix} 0 & H \end{bmatrix}+*\right) \le 0 \tag{24}$$

Using Fact 1 to eliminate $\Delta_A(t)$, $\Delta_D(t)$, (24) is satisfied if the following matrix equation is satisfied

$$\begin{bmatrix} \left(\widehat{A}\widehat{P}+\widehat{B}\widehat{Y}+*\right)+\alpha\widehat{P} & \widehat{D} \\ * & -\alpha I \end{bmatrix} + \int \begin{bmatrix} M \\ 0 \end{bmatrix}\begin{bmatrix} M \\ 0 \end{bmatrix}' + \int^{-1} \begin{bmatrix} \widehat{P}N' \\ 0 \end{bmatrix}\begin{bmatrix} \widehat{P}N' \\ 0 \end{bmatrix}'$$

$$\rho\begin{bmatrix} F \\ 0 \end{bmatrix}\begin{bmatrix} F \\ 0 \end{bmatrix}' + \rho^{-1}\begin{bmatrix} 0 \\ H' \end{bmatrix}\begin{bmatrix} 0 \\ H' \end{bmatrix}' \le 0,\ \alpha, \epsilon, \rho > 0$$

Using Fact 2, theorem is proved.

For +/- 10% changes in load resistance, lines parameters, the norm-bounded uncertainty

$$\left(\Delta A_i = M.\Delta_A(t).N, \left\|\Delta_A(t)\right\| < 1, \Delta D = F.\Delta_D(t).H, \left\|\Delta_D(t)\right\| < 1\right)$$

for the subsystems DERS are calculated by the singular value decomposition. The proposed decentralized robust tracker is given by Table 2:

Table 2. The proposed decentralized robust tracker

subsys	K	K_I
1	[-0.080152, 0.12222, -17.266,1.2503e-05, -0.76065, 4.2797,7.7335, -2.5312; -0.12222. -0.080152, -1.2505e-05,-17.266, -4.2797, 0.76065, 2.5312,7.7335]	[265.1000, -153.6725; 153.6725, 265.1000]
2	[-0.15399, 0.28054, -27.455, 0.00012262, 5.5384, 6.651, 11.815, -4.1284; -0.28054, -0.15399, -0.00012269, -27.455, -6.651, 5.5384, 4.1284, 11.815]	[253.0500, -124.0850; 124.0850, 253.0500]
3	[-0.14444, 0.37611, -33.416, 0.00025113, 11.449, 5.7181, 13.001, -5.8362; -0.37611, -0.14444, -0.00025115, -33.416, -5.7181, 11.449, 5.8362, 13.001]	[264.3750, -149.9075; 149.9075, 264.3750]
4	[-0.09195, 0.3694, -30.952, 0.00020276, 13.39, 3.2103, 11.174, -6.3282; -0.3694, -0.09195, -0.0002025, -30.952, -3.2103, 13.39, 6.3282, 11.174]	[347.1750, -245.1100; 245.1100, 347.1750]
5	[-0.72926 0.49461 -90.683 6.0814e-05 54.48 9.0184; -0.49461 -0.72926 -6.0813e-05 -90.683 -9.0184 54.48]	[232.85, -89.373; 89.373, 232.85]

SIMULATION RESULTS

The simulation is performed to check the validity of the proposed decentralized robust tracker scheme. A case with five distributed generation sources, as shown in Fig. 1.

The designed control tracker operation is established and assessed by using MATLAB/SimPower System Toolbox. The consider generalized islanded PWM-inverter MG consisting of five-DGs as given in Fig.2. Each DG supplies a local load and all DGs parameters (transmission lines and local loads) are specified in Table 1. In Figs. 5-27, the dynamic performance of the proposed designed trackers is evaluated via several Plug and Play capabilities test cases, including connecting and disconnecting one or more DGs to the proposed microgrid, connecting and disconnecting one or more DGs local load (e.g: during electric vehicles charging or after fully charged), and transmission lines parameter variation.

Case #1: Disconnect DG/DGs from the Microgrid

Disconnect One-DG from the Microgrid

In this scenario, one-DG is disconnected from the 5-DGs Microgrid given in Fig. 2. To clarify which state vector of subsystem is zero is detailed in Table 3. Is summarizing all the possibilities of disconnected any of the five-DGs. From Table 3, one states are selected to be simulated (state 3). In state 3, the DG3 is disconnected to the microgrid given in Fig. 2. The dynamic performance for the proposed MG system during this scenario is given in Figs. 5, and 6. At t= 2 s the DG3 is disconnected from the proposed MG system, and it is reconnected back at t = 3 s. The dynamic response for the V_{d-q} components of the five-DGs are shown in Fig. 5 (a, b) and (c, d) respectively. The instantaneous voltage of PC3 is illustrated in Fig 5 (e, f). The active and the reactive power of the five-DGs are given in Fig. 6 (a, b) and (c, d) respectively.

Figure 5. The dynamic performance of the proposed MG system under plug and play operation (DG3 is disconnected and connected back). (a, b) Vd-component of the five-DGs, (c, d) Vq-component of the five-DGs, and (e, f) instantaneous voltages of the PCC3.

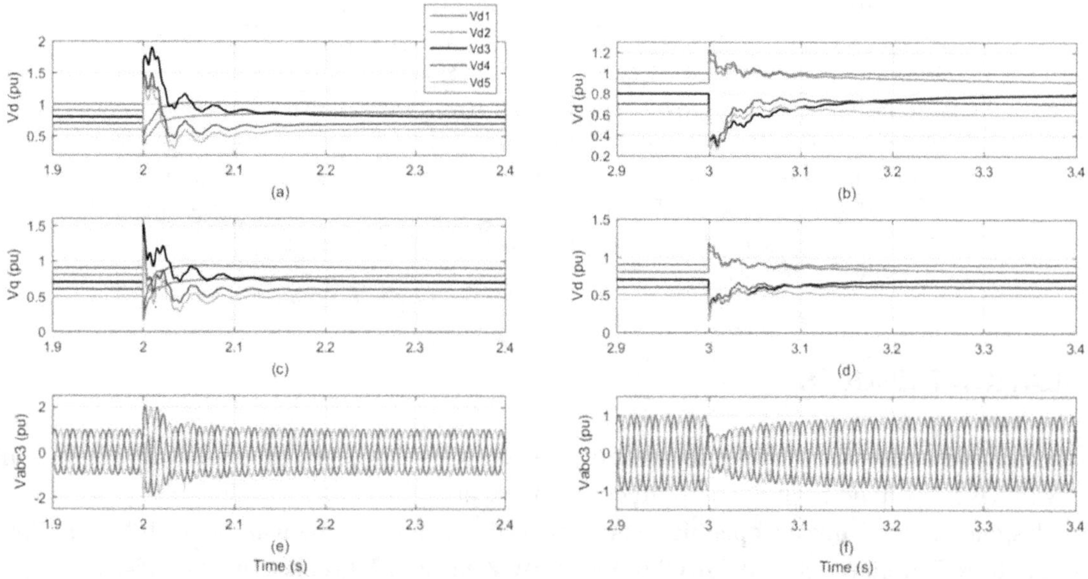

Figure 6. The dynamic responses of the proposed MG system under plug and play operation (DG3 is disconnected and connected back). (a, b) Output active power of the five-DGs, and (c, d) Output reactive power of the five-DGs.

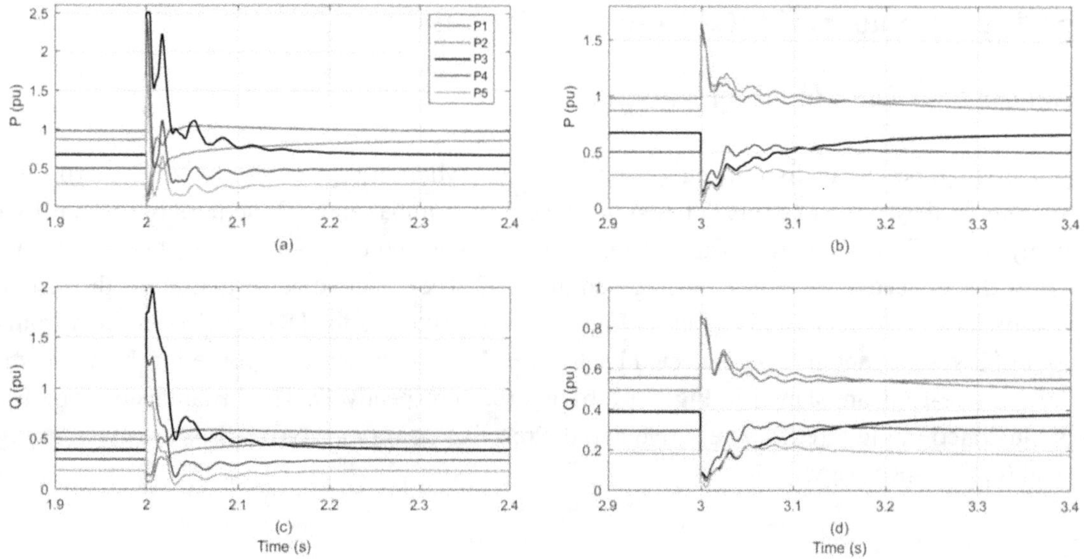

Table 3. One DG is disconnected from the microgrid

	Disconnected DG	Zero Matrix
1	DG_1	A_{21}
2	DG_2	A_{12}, A_{32}
3	DG_3	A_{23}, A_{43}
4	DG_4	A_{34}, A_{54}
5	DG_5	A_{45}

Disconnect Two-DGs from the Microgrid

In this scenario, two-DGs are disconnected from the 5-DGs Microgrid given in Fig. 2. To justify which state vector of subsystem is zero, Table 4 is showing all the probabilities of disconnected two-DGs from the five-DGs MG system. From Table 4, Two states are selected to be simulated (states 3, and 6). In state 3, the DG1 and DG4 are disconnected from the microgrid. While, in state 6, the DG2 and DG4 are disconnected from the microgrid. The dynamic performance for the proposed MG system during this scenario is given in Figs. 7, 8 and 9 for the first state (state 3). In the second state (state 6), the performance is presented in Fig 10, 11 and 12. At t = 2 s the DG1 and DG4 are disconnected from the proposed MG system, and it is reconnected back at t = 3 s. In the first state (state 3), the dynamic response for $V_{d\text{-}q}$ components of the five-DGs are demonstrated in Fig. 7 (a, b) and (c, d) respectively. The instantaneous voltage of PC1 and PC4 are given in Fig 8 (a, b), and (c, d) correspondingly. The active and the reactive power of the five-DGs are presented in Fig. 9 (a, b) and (c, d) respectively. In the second state (state 6), the dynamic response for $V_{d\text{-}q}$ components of the five-DGs are established in Fig. 10 (a, b) and (c, d) respectively. The instantaneous voltage of PC2 and PC4 are shown in Fig 11 (a, b), and (c, d) correspondingly. The active and the reactive power of the five-DGs are presented in Fig. 12 (a, b) and (c, d) respectively.

Table 4. Two DGs are disconnected from the microgrid

	Disconnected DGs	Zero Matrices
1	DG1 and DG2	A_{12}, A_{21} and A_{32}
2	DG1 and DG3	A_{21} and A_{23}, A_{43}
3	DG1 and DG4	A_{21} and A_{34}, A_{54}
4	DG1 and DG5	A_{21} and A_{45}
5	DG2 and DG3	A_{12}, A_{23}, A_{32} and A_{43}
6	DG2 and DG4	A_{12}, A_{32} and A_{34}, A_{54}
7	DG2 and DG5	A_{12}, A_{32} and A_{45}
8	DG3 and DG4	A_{23}, A_{34}, A_{43} and A_{54}
9	DG3 and DG5	A_{23}, A_{43} and A_{45}
10	DG4 and DG5	A_{34}, A_{45} and A_{54}

Figure 7. The dynamic performance of the proposed MG system under plug and play operation (DG1 and DG4 are disconnected and connected back). (a, b) Vd-component of the five-DGs, and (c, d) Vq-component of the five-DGs.

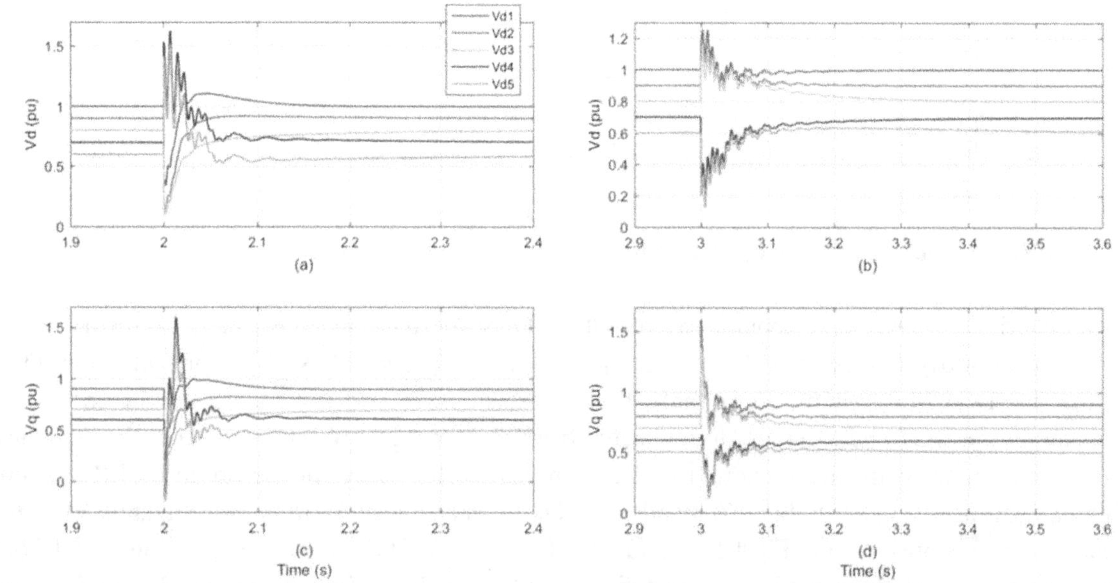

Figure 8. The dynamic performance of the proposed MG system under plug and play operation (DG1 and DG4 are disconnected and connected back). (a, b) instantaneous voltages of the PCC1, and (c, d) instantaneous voltages of the PCC4.

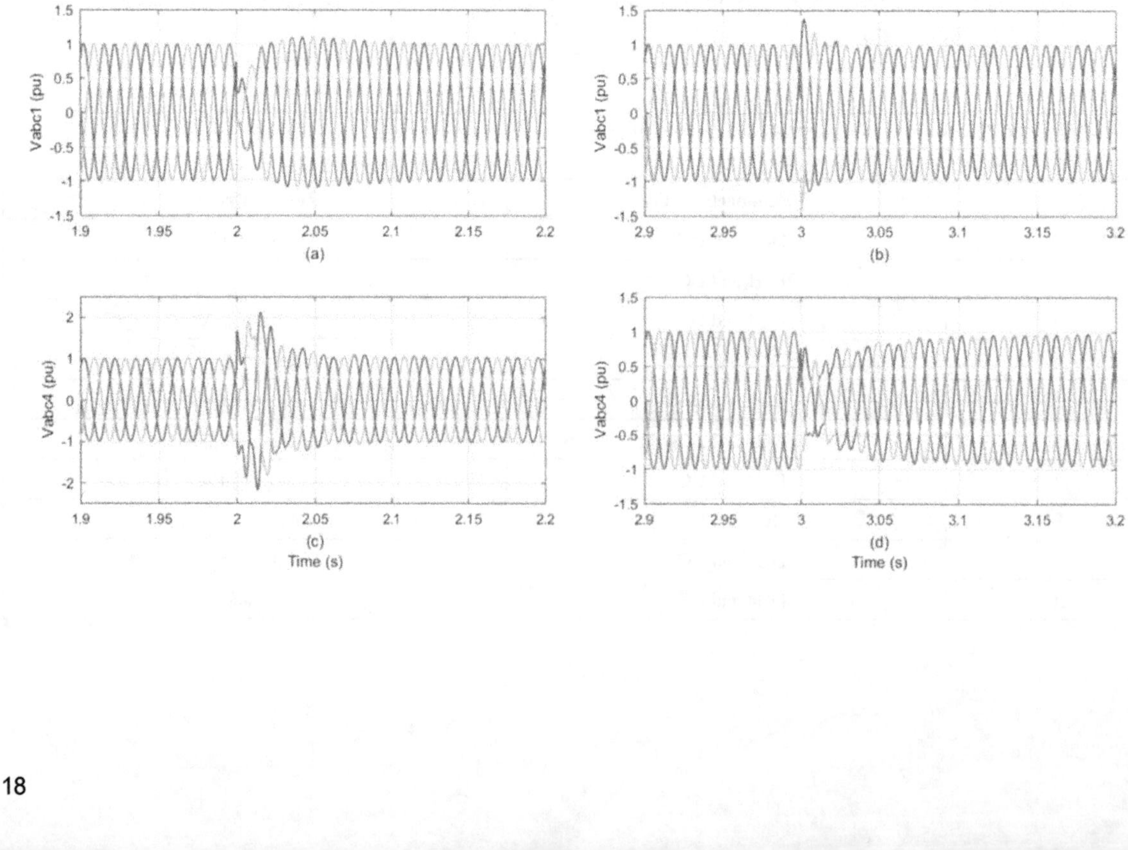

Figure 9. The dynamic performance of the proposed MG system under plug and play operation (DG1 and DG4 are disconnected and connected back). (a, b) output active-power of the five-DGs, and (c, d) Output reactive-power of the five-DGs.

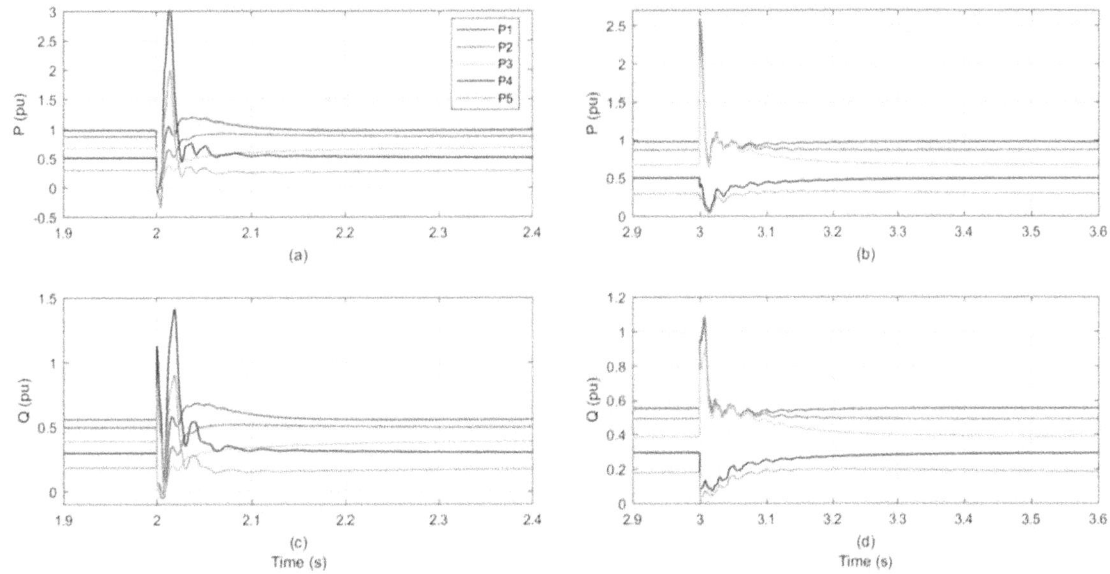

Figure 10. The dynamic performance of the proposed MG system under plug and play operation (DG2 and DG3 are disconnected and connected back). (a, b) Vd-component of the five-DGs, and (c, d) Vq-component of the five-DGs.

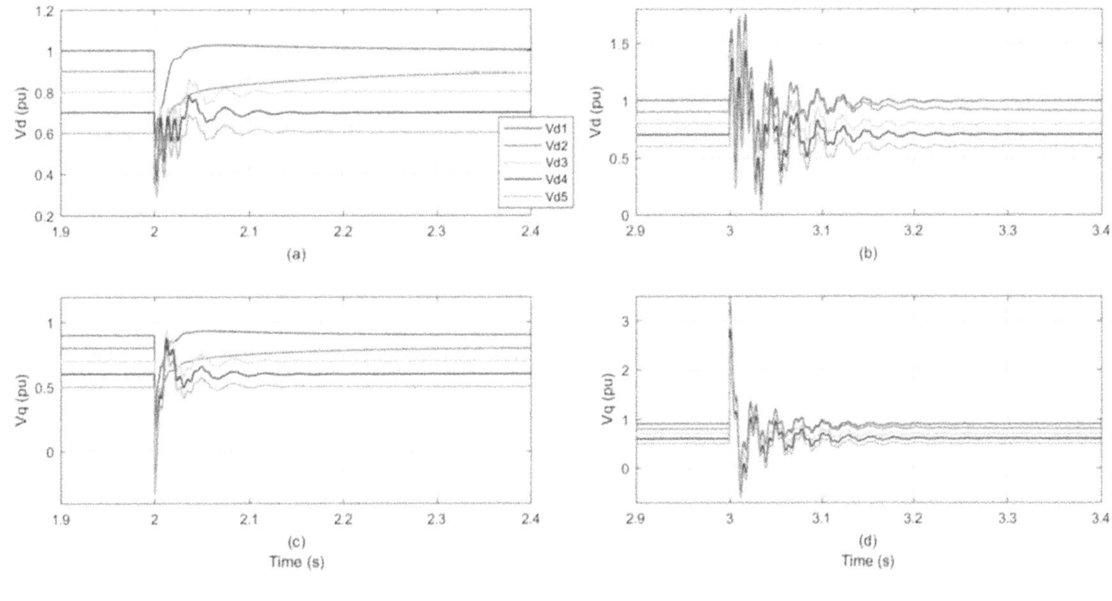

Figure 11. The dynamic performance of the proposed MG system under plug and play operation (DG2 and DG3 are disconnected and connected back). (a, b) instantaneous voltages of the PCC1, and (c, d) instantaneous voltages of the PCC4.

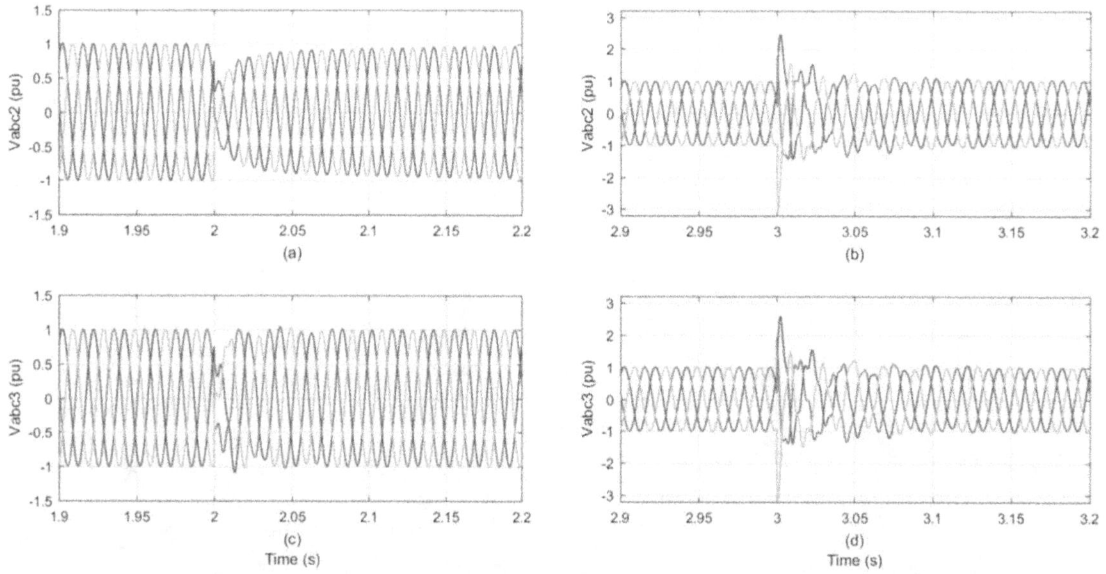

Figure 12. The dynamic performance of the proposed MG system under plug and play operation (DG2 and DG3 are disconnected and connected back). (a, b) output active-power of the five-DGs, and (c, d) Output reactive-power of the five-DGs.

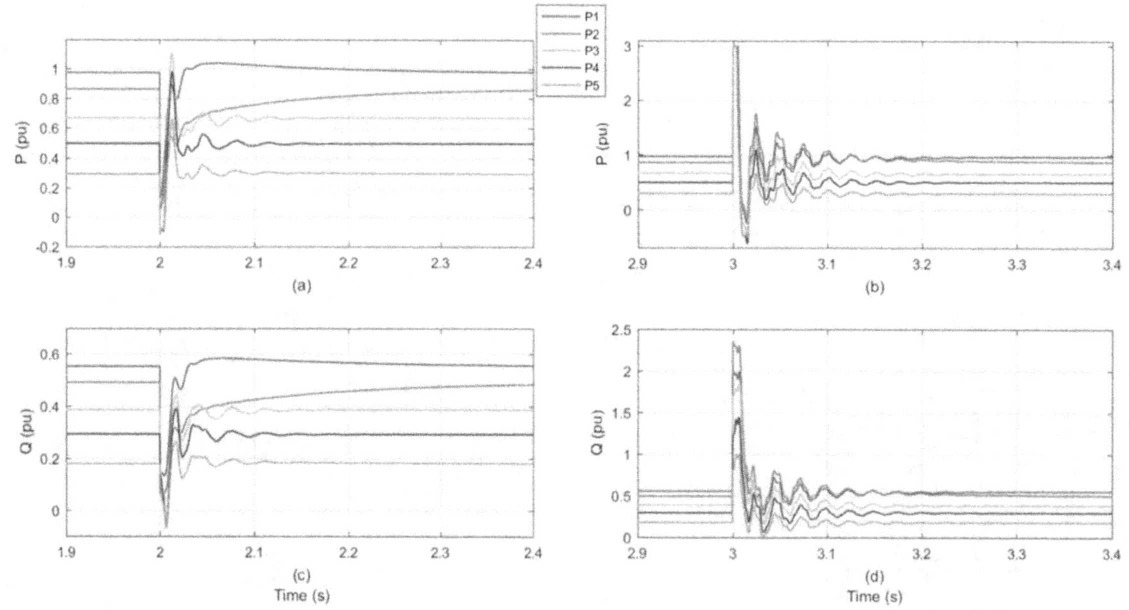

Disconnected Three-DGs from the Microgrid

In this scenario, three-DGs are disconnected from the 5-DGs Microgrid as given in Fig. 2. To make it clear which state vector of subsystem is zero, Table 5 is showing all the options of disconnected three-DGs from the five-DGs MG system. From Table 5, state 5 is selected to be simulated. In this state DG1, DG3 and DG5 are disconnected from the five-MG system at t = 2 s. The three-disconnected DGs are start reconnected back randomly as given in Table 6. The dynamic performance for the proposed MG system during this scenario is given in Figs. 13, 14 and 15. At t = 2 s the DG1, DG3 and DG5 are disconnected from the proposed MG system, and it is reconnected back randomly at t = 3 s, 4 s and 5 s as illustrated in Table 6. The dynamic response for the V_d component of the five-DGs are demonstrated in Fig. 13 (a, b, c, d) and Vq-component in Fig. 13 (e, f, g, h). The instantaneous voltage of PC1, PC3 and PC5 are given in Fig 14 (a, b, c, d), (e, f, g, h) and (I, j, k, l) respectively. The active and the reactive power of the five-DGs are presented in Fig. 15 (a, b, c, d) and (e, f, g, h) respectively.

Table 5. Three DGs are disconnected from the microgrid

	Disconnected DGs	**Zero Matrices**
1	DG_1, DG_2 and DG_3	A_{12}, A_{21}, A_{23}, A_{32} and A_{43}
2	DG_1, DG_2 and DG_4	A_{12}, A_{21}, A_{32} and A_{34}
3	DG_1, DG_2 and DG_5	A_{12}, A_{21}, A_{32} and A_{45}
4	DG_1, DG_3 and DG_4	A_{21}, A_{23}, A_{34} and A_{54}
5	DG_1, DG_3 and DG_5	A_{21}, A_{23} $A_{43}$and A_{45}
6	DG_1, DG_4 and DG_5	A_{21}, A_{34}, A_{45} and A_{54}
7	DG_2, DG_3 and DG_4	A_{12}, A_{23}, A_{32}, A_{34}, A_{43} and A_{54}
8	DG_2, DG_3 and DG_5	A_{12}, A_{23}, A_{32} $A_{43}$and A_{45}
9	DG_2, DG_4 and DG_5	A_{12} A_{32}, A_{34}, A_{45} and A_{54}
10	DG_3, DG_4 and DG_5	A_{23}, A_{34}, A_{43}, A_{45} and A_{54}

Table 6. Timing of figures 13, 14 1nd 15

	t = [0, 2[s	**t = 2 s**	**t = 3 s**	**t = 4 s**	**t = 5 s**
DG1	Connected	Disconnected	Disconnected	Reconnected	Connected
DG2	Connected	Connected	Connected	Connected	Connected
DG3	Connected	Disconnected	Disconnected	Disconnected	Reconnected
DG4	Connected	Connected	Connected	Connected	Connected
DG5	Connected	Disconnected	Reconnected	Connected	Connected

Figure 13. The dynamic performance of the proposed MG system under plug and play operation (DG1, DG3 and DG5 are disconnected and connected back). (a, b, c, d) Vd-component of the five-DGs, and (e, f, g, h) Vq-component of the five-DGs.

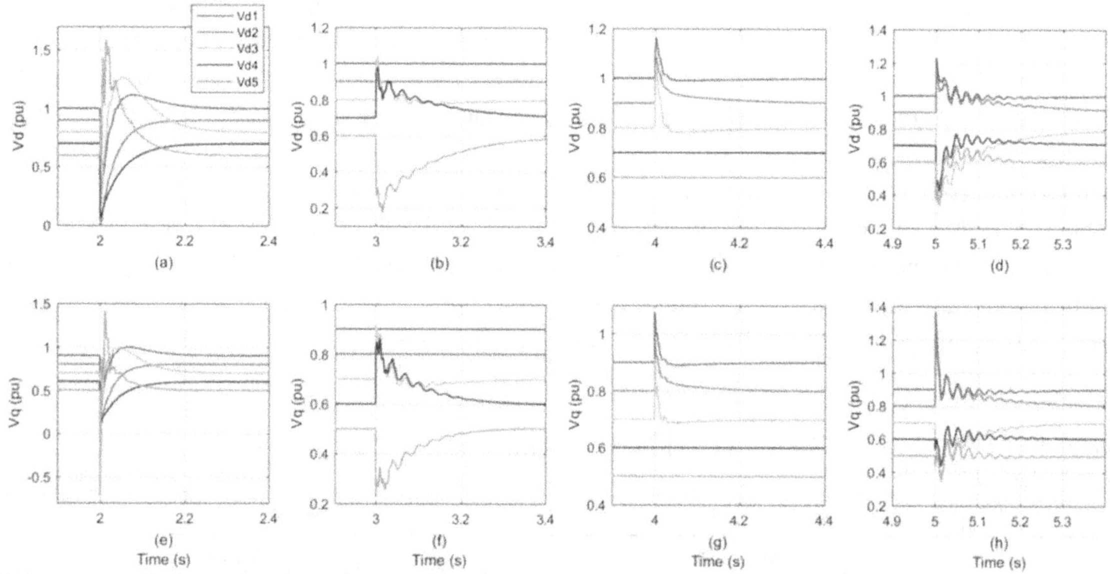

Figure 14. The dynamic performance of the proposed MG system under plug and play operation (DG1, DG3 and DG5 are disconnected and connected back). (a, b, c, d) instantaneous voltages of the PCC1, (e, f, g, h) instantaneous voltages of the PCC3, and (I, j, k, l) instantaneous voltages of the PCC5.

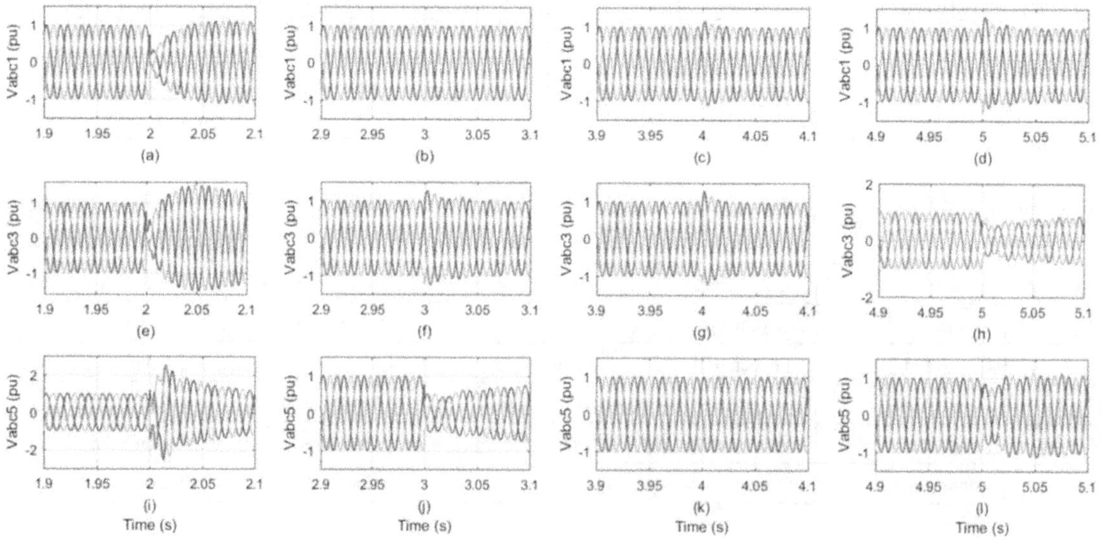

Figure 15. The dynamic performance of the proposed MG system under plug and play operation (DG1, DG3 and DG5 are disconnected and connected back). (a, b, c, d) output active-power of the five-DGs, and (e, f, g, h) Output reactive-power of the five-DGs.

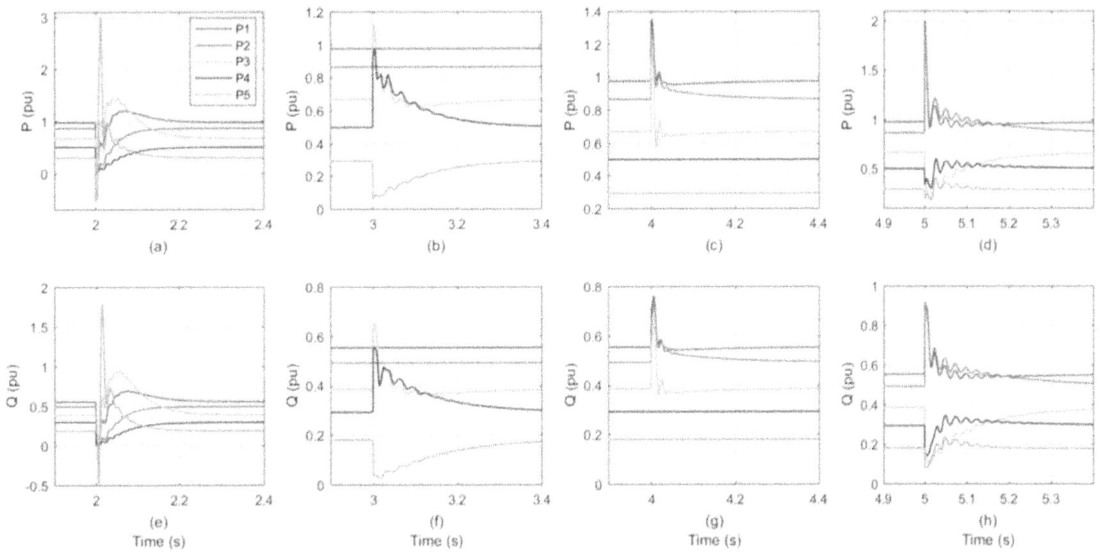

Disconnect Four-DGs from the Microgrid

In this scenario, four-DGs are disconnected from the 5-DGs Microgrid given in Fig. 2. To clarify which state vector of subsystem is zero, Table 7 displays all the selections of disconnected four-DGs from the five-DGs MG system. From Table 7, state 4 is selected to be simulated. In this state DG1, DG3 DG4 and DG5 are disconnected from the five-MG system at t = 2 s. The three-disconnected DGs are start reconnected back randomly as given in Table 8. The dynamic performance for the proposed MG system during this scenario is given in Figs. 16, 17 and 18. At t = 2 s the DG1, DG3, DG4 and DG5 are disconnected from the proposed MG system, and it is reconnected back randomly at t = 3 s, 4 s and 5 s as illustrated in Table 8. The dynamic response for the V_d component of the five-DGs are demonstrated in Fig. 16 (a, b, c, d) and Vq-component in Fig. 16 (e, f, g, h). The instantaneous voltage of PC1, PC3 and PC5 are given in Fig 17 (a, b, c, d), (e, f, g, h), (I, j, k, l) and (m, n, o, p) respectively. The active and the reactive power of the five-DGs are presented in Fig. 18 (a, b, c, d) and (e, f, g, h) respectively.

Table 7. Four DGs are disconnected from the microgrid

	Disconnected DGs	Zero Matrices
1	DG_1, DG_2, DG_3 and DG_4	A_{12}, A_{21}, A_{23}, A_{32}, A_{34}, A_{43} and A_{54}
2	DG_1, DG_2, DG_3 and DG_5	A_{12}, A_{21}, A_{23}, A_{32}, A_{43} and A_{45}
3	DG_1, DG_2, DG_4 and DG_5	A_{12}, A_{21}, A_{32}, A_{34}, A_{45} and A_{54}
4	DG_1, DG_3, DG_4 and DG_5	A_{21}, A_{23}, A_{34}, A_{43}, A_{45} and A_{54}
5	DG_2, DG_3, DG_4 and DG_5	A_{12}, A_{23}, A_{32}, A_{43}, A_{34}, A_{45} and A_{54}

Figure 16. The dynamic performance of the proposed MG system under plug and play operation (DG1, DG3, DG4 and DG5 are disconnected and connected back). (a, b, c, d) Vd-component of the five-DGs, and (e, f, g, h) Vq-component of the five-DGs.

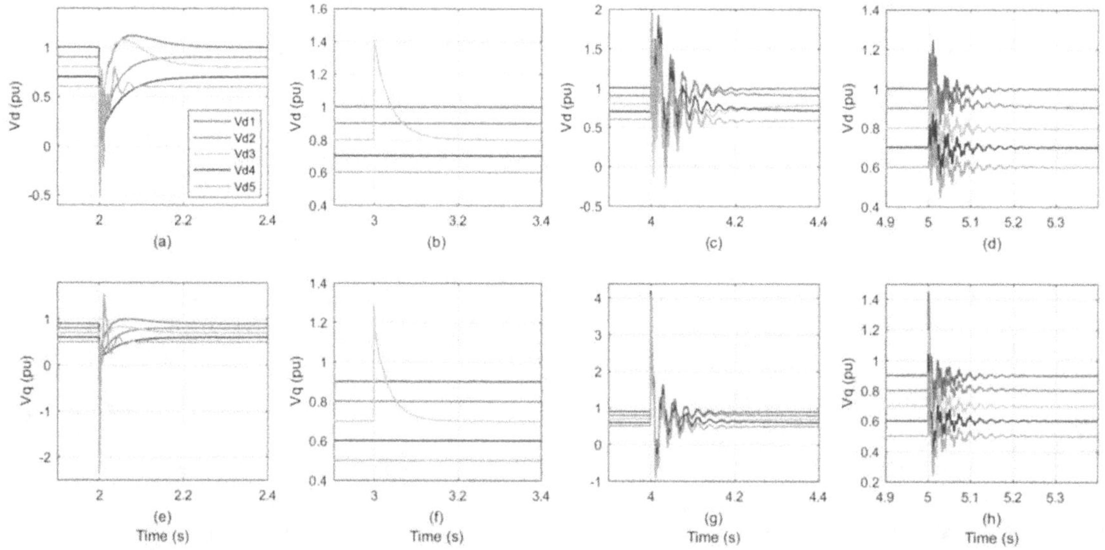

Figure 17. The dynamic performance of the proposed MG system under plug and play operation (DG1, DG3, DG4 and DG5 are disconnected and connected back). (a, b, c, d) instantaneous voltages of the PCC1, (e, f, g, h) instantaneous voltages of the PCC3, (I, j, k, l) instantaneous voltages of the PCC4, and (m, n, o, p) instantaneous voltages of the PCC5.

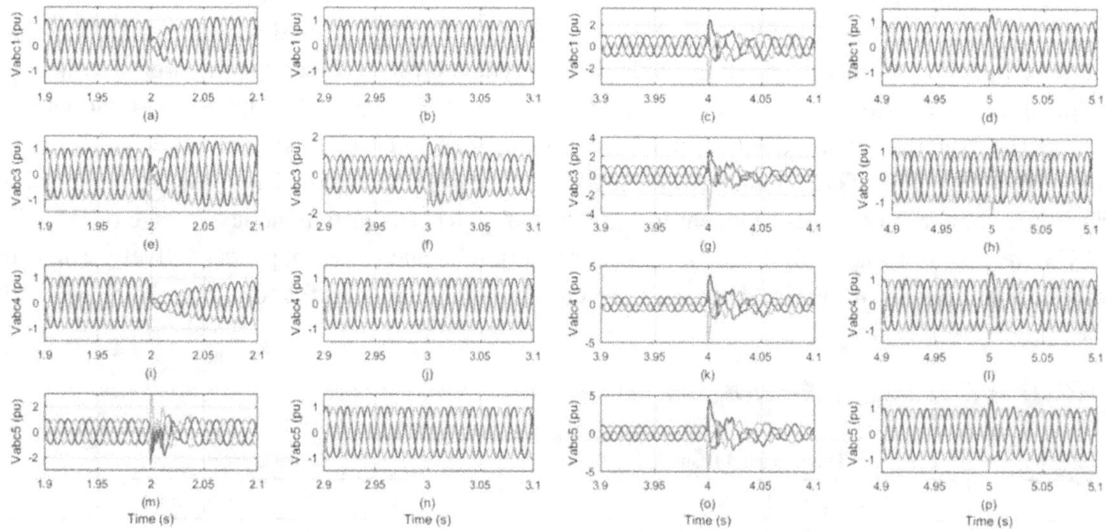

Figure 18. The dynamic performance of the proposed MG system under plug and play operation (DG1, DG3, DG4 and DG5 are disconnected and connected back). (a, b, c, d) output active-power of the five-DGs, and (e, f, g, h) Output reactive-power of the five-DGs.

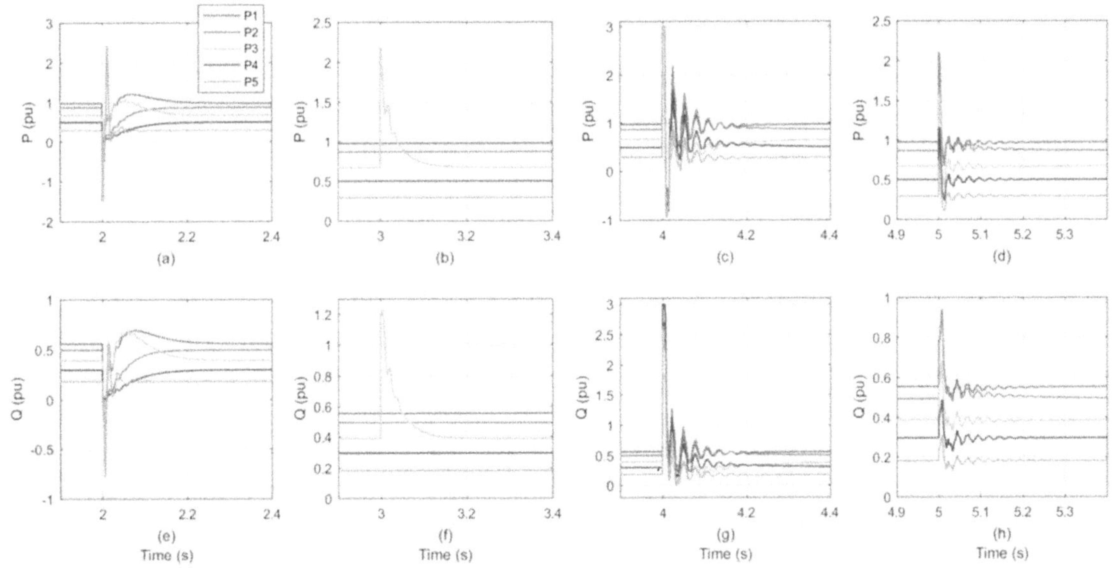

Table 8. Timing of figures 16, 17 1nd 18

	t = [0, 2[s	t = 2 s	t = 3 s	t = 4 s	t = 5 s
DG1	Connected	Disconnected	Disconnected	Disconnected	Reconnected
DG2	Connected	Connected	Connected	Connected	Connected
DG3	Connected	Disconnected	Disconnected	Reconnected	Connected
DG4	Connected	Disconnected	Reconnected	Connected	Connected
DG5	Connected	Disconnected	Disconnected	Reconnected	Connected

Dynamic responses of the five-DGs as a function of Plug and Play operation are shown in Figs. 5-18. All the possible scenarios are illustrated. One/two case study has been selected in each scenario. The results demonstrate the robust performance of the proposed control technique to Plug and Play functionality of all the DGs in the MG system. The results authenticate that the voltage controllers engaged to the desired values in a short time, guarantee stability, fast response, and provide a required performance according to the IEEE standards (IEEE Recommended Practice for Monitoring Electric Power Quality, et al. 2009).

Case # 2: DGs Local Loads are Connected or Disconnected

In this case the DG local load are connected and disconnected e.g. during electric vehicles batteries charging and after fully charged. To disconnect the local load of any DG, this can be done by considering the load resistance is equal or approximately to infinity (≅∞). There are four-scenarios in this case:

1. One DG load are connected or disconnected.
2. Two-DGs loads are connected or disconnected.
3. Three-DGs loads are connected or disconnected.
4. Four-DGs loads are connected or disconnected.

The third scenario is selected to be simulated (three-DGS load are connected or disconnected). This scenario has been implemented to select the connected and the disconnected DG randomly as given in Table 9. The dynamic performance for the proposed MG system during this scenario is dedicated in Figs. 19, 20 and 21. At t = 2 s the DG4 local load is disconnected, and at t = 3 s, DG1 and DG3 local load are disconnected from the proposed MG system. At t = 4 s the DG1 and DG4 are reconnected and at t = 5 s DG3 is reconnected back to the MG system. The dynamic response for the V_d component of the five-DGs are demonstrated in Fig. 19 (a, b, c, d) and Vq-component in Fig. 19 (e, f, g, h). The instantaneous voltage of PC1, PC3 and PC4 are presented in Fig 20 (a, b, c, d), (e, f, g, h), and (I, j, k, l) respectively. The active and the reactive power of the five-DGs are shown in Fig. 21 (a, b, c, d) and (e, f, g, h) respectively.

In this scenario the voltage controllers are tested against loads uncertainties. The results indicate that the designed controllers are robust against load uncertainties with a very short transient period (less 1.5 s). Furthermore, the dynamic responses of the MG system reject the second uncertainty case in a swiftly manner. Also, the designed controllers can extremely manage with the second plug and play case mentioned in this paper.

Figure 19. The dynamic performance of the proposed MG system under DGs local loads are connected or disconnected. (a, b, c, d) Vd-component of the five-DGs, and (e, f, g, h) Vq-component of the five-DGs.

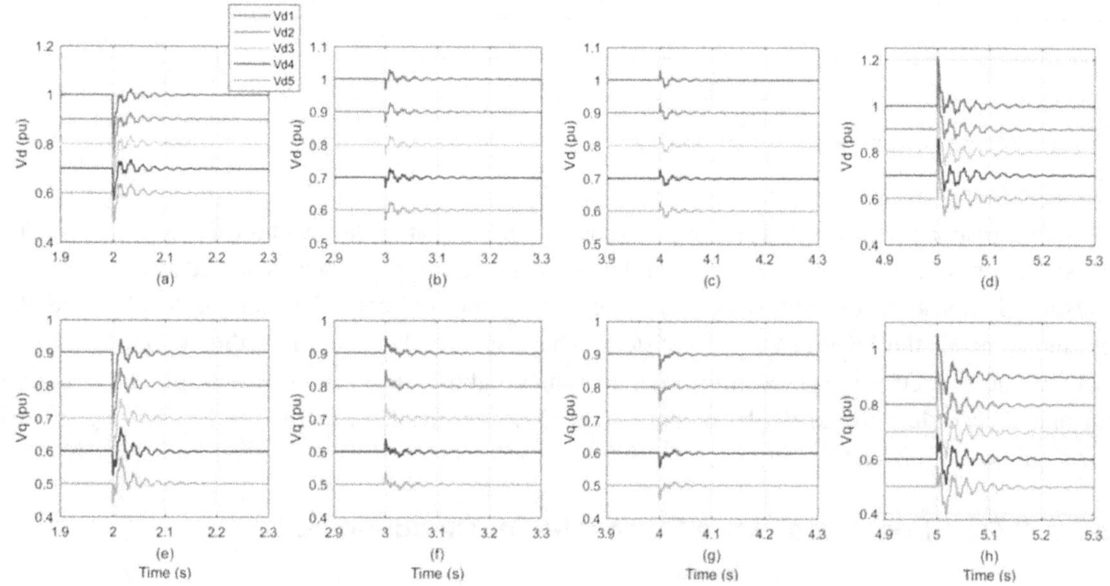

Figure 20. The dynamic performance of the proposed MG system under DGs local loads are connected or disconnected. (a, b, c, d) instantaneous voltages of the PCC1, (e, f, g, h) instantaneous voltages of the PCC3, (I, j, k, l) and instantaneous voltages of the PCC4.

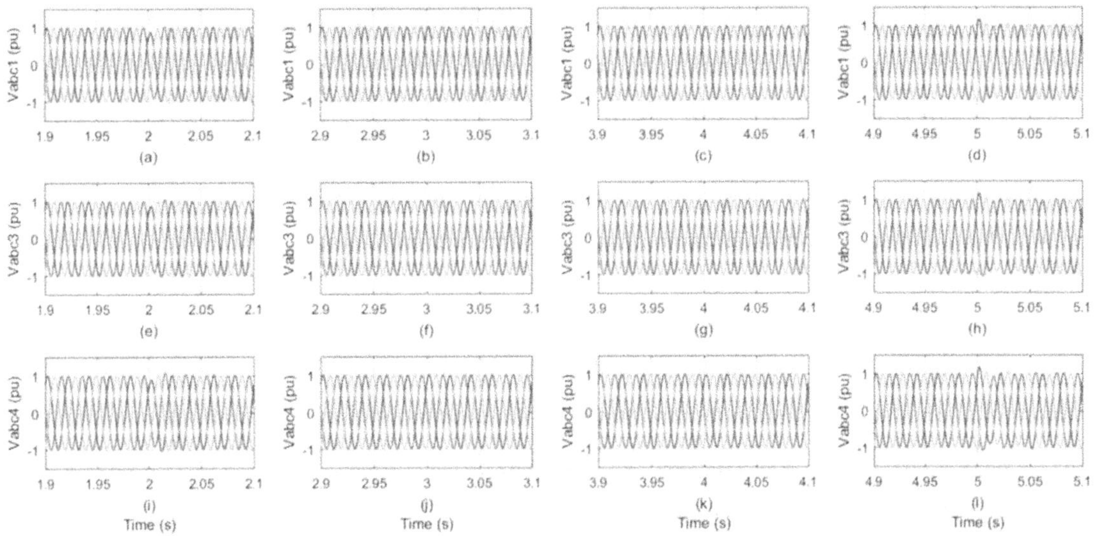

Figure 21. The dynamic performance of the proposed MG system under DGs local loads are connected or disconnected. (a, b, c, d) output active-power of the five-DGs, and (e, f, g, h) Output reactive-power of the five-DGs.

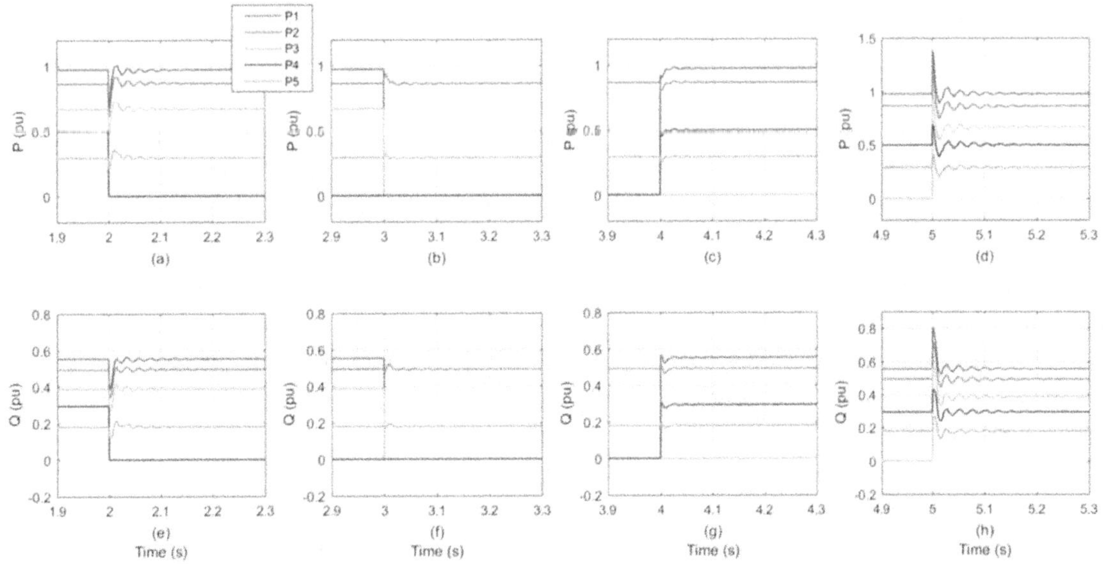

Table 9. Timing of figures 19, 20 and 21

	t = [0, 2[s	t = 2 s	t = 3 s	t = 4 s	t = 5 s
DG1	Load Connected	Load Connected	Load Disconnected	Load Reconnected	Load Connected
DG2	Load Connected	Load Connected	Load Connected	Load Connected	Load Connected
DG3	Load Connected	Load Connected	Load Disconnected	Load Disconnected	Load Reconnected
DG4	Load Connected	Load Disconnected	Load Disconnected	Load Reconnected	Load Connected
DG5	Load Connected	Load Connected	Load Connected	Load Connected	Load Connected

Case #3: Changing the Transmission Line (TL) Parameters

The transmission line connected the DG to the microgrid has two parameters R_l and X_l. In this case we are going to test the proposed system against the transmission line parameter variations. Therefore, there are two-scenarios for this case:

1. Change of the R_l of one of the transmission lines parameter.
2. Change of X_l of one of the transmission lines parameter.

In the first scenario the TL 1 is selected (between PCC1 and PCC2) to test the proposed voltage controllers against R-change in the TL. The test is conducted as given in Table 10. The dynamic response for the V_d component of the five-DGs are displayed in Fig. 22 (a, b, c) and Vq-component in Fig. 22 (d, c, e). The instantaneous voltage of PC1 and PC2 are given in Fig 23 (a, b, c), and (d, e, f) respectively. The active and the reactive power of the five-DGs are shown in Fig. 24 (a, b, c) and (d, e, f) respectively.

Results demonstrate that the robustness and the fast responses of the designed controllers are to cope with uncertainty. Moreover, the MG capability to reject and absorb this uncertainty is done rapidly and swiftly.

In the second scenario the TL 3 is selected (between PCC3 and PCC4) to test the proposed voltage controllers against L-change in the TL. The test is conducted as given in Table 11. The dynamic response for the V_d component of the five-DGs are displayed in Fig. 25 (a, b, c) and Vq-component in Fig. 25 (d, c, e). The instantaneous voltage of PC3 and PC4 are given in Fig 26 (a, b, c), and (d, e, f) respectively. The active and the reactive power of the five-DGs are shown in Fig. 27 (a, b, c) and (d, e, f) respectively.

The impact of changing the L-parameter in MG TL is way more critical in the dynamics of MG system compared to changing the R-parameter in MG TL [need Ref]. This fact has been appeared clearly in the MG dynamic responses shown in Fig 25-27. Even though, the proposed controllers have cope with this uncertainty efficiently, rapidly and guarantees a remarkable response.

The results of the three-cases reveal that the designed voltage controllers are robust against all the disturbances caused by the plug and play in all scenario. Moreover, the dynamic responses of the MG system decline the uncertainties in a swift and too short transient manner. Finally, we can deliberate that the designed voltage controllers can guarantees a remarkable response that can manage with all the plug and play scenarios mentioned in this paper and corresponds with the IEEE recommendation and standard (IEEE Recommended Practice for Monitoring Electric Power Quality, et al. 2009).

Table 10. Timing of figures 22, 23 and 24

	t = [0, 2[s	t = 2 s	t = 3 s	t = 4 s
TL1	R_l = 100% nominal value	R_l = 50% nominal value	R_l = 150% nominal value	R_l = 100% nominal value

Figure 22. The dynamic performance of the proposed MG system due to change in R-parameter of the transmission line. (a, b, c) Vd-component of the five-DGs, and (d, e, f) Vq-component of the five-DGs.

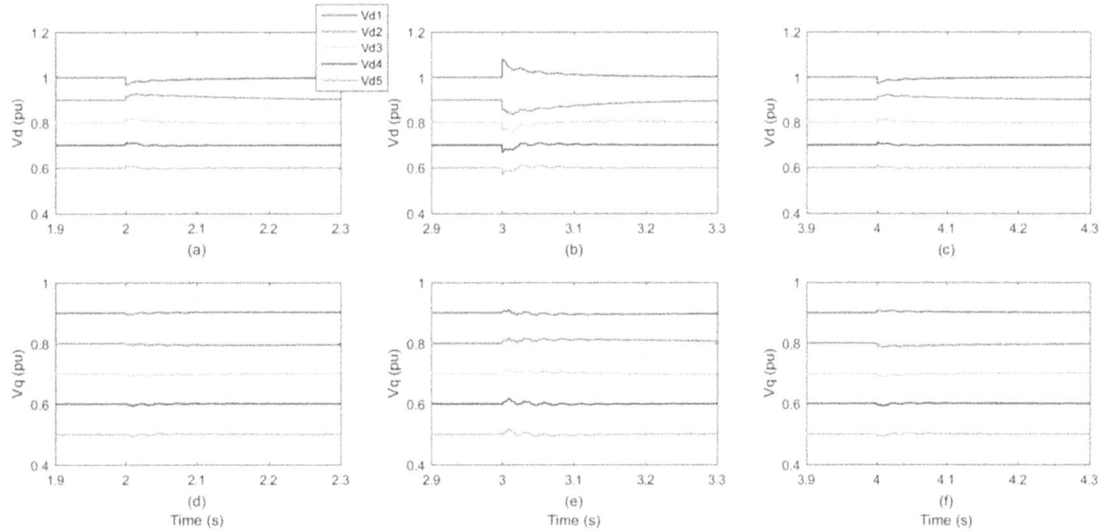

Figure 23. The dynamic performance of the proposed MG system due to change in R-parameter of the transmission line. (a, b, c) instantaneous voltages of the PCC1, and (d, e, f) instantaneous voltages of the PCC2.

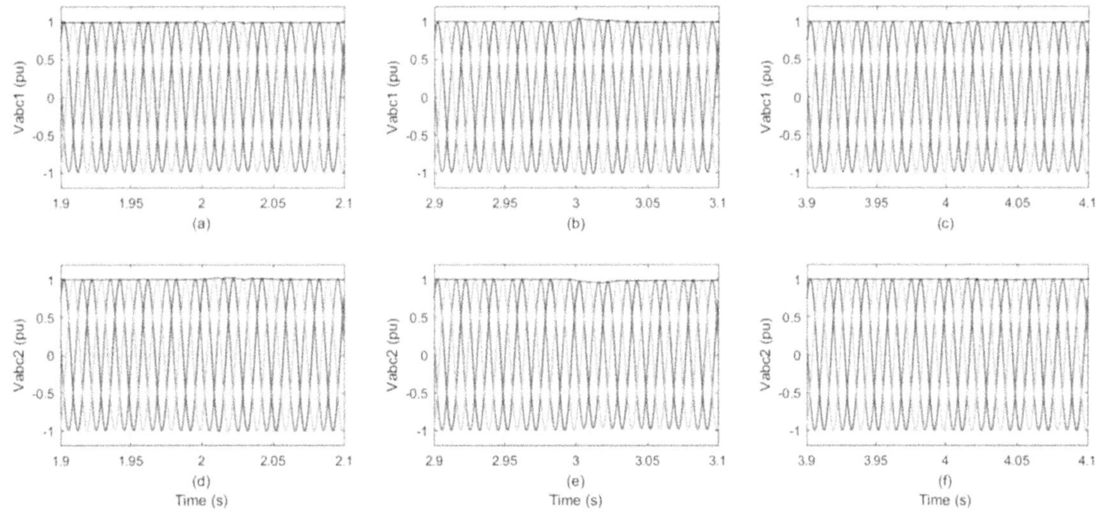

Figure 24. The dynamic performance of the proposed MG system due to change in R-parameter of the transmission line. (a, b, c) output active-power of the five-DGs, and (d, e, f) Output reactive-power of the five-DGs.

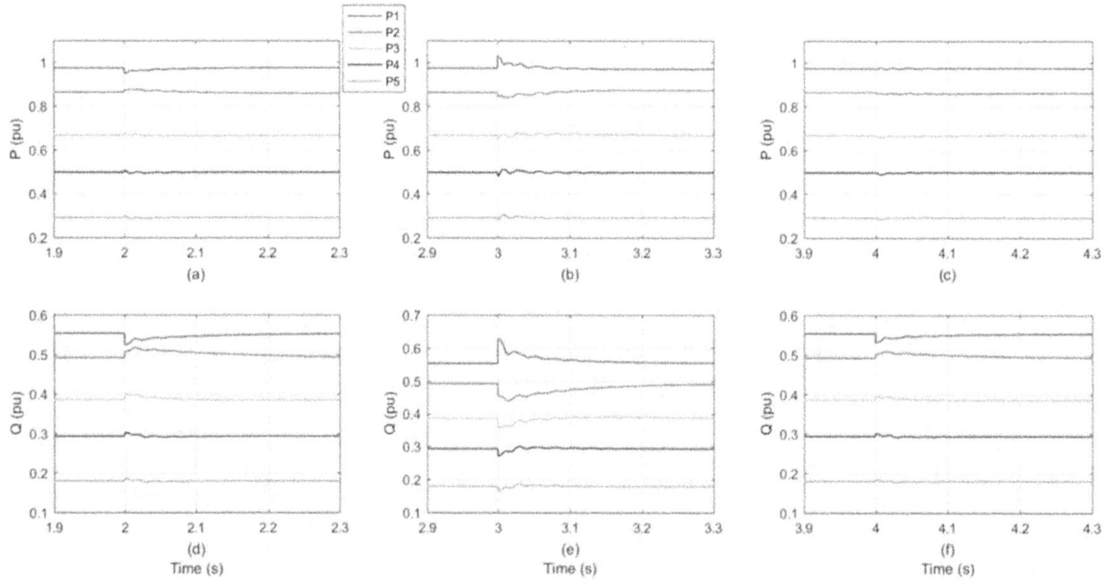

Table 11. Timing of figures 25, 26 and 27

	t = [0, 2[s	t = 2 s	t = 3 s	t = 4 s
TL3	L_1 = 100% nominal value	L_1 = 50% nominal value	L_1 = 150% nominal value	L_1 = 100% nominal value

Figure 25. The dynamic performance of the proposed MG system due to change in L-parameter of the transmission line. (a, b, c) Vd-component of the five-DGs, and (d, e, f) Vq-component of the five-DGs.

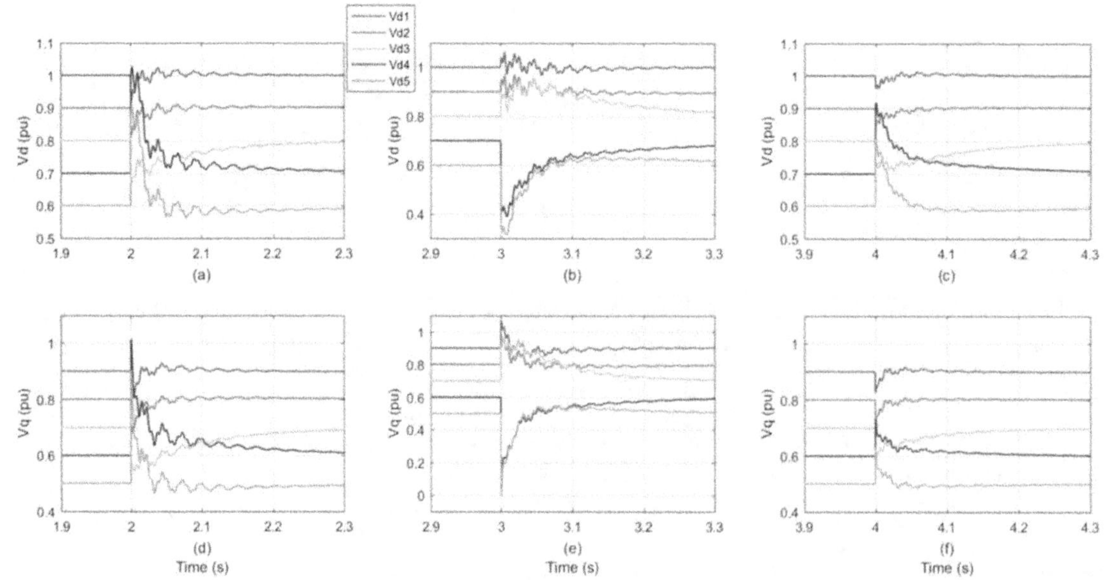

Figure 26. The dynamic performance of the proposed MG system due to change in L-parameter of the transmission line. (a, b, c) instantaneous voltages of the PCC3, and (d, e, f) instantaneous voltages of the PCC4.

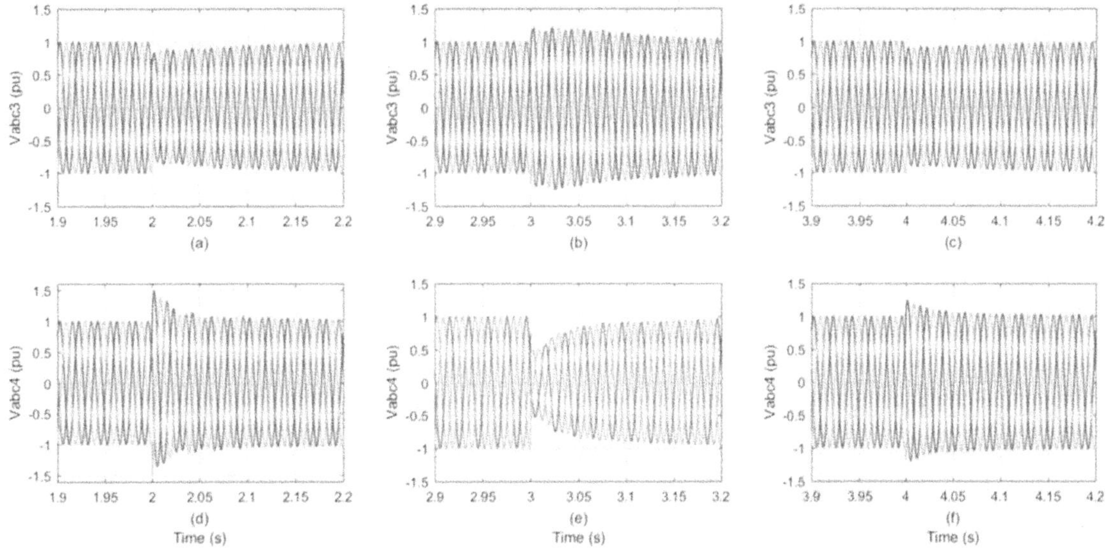

Figure 27. The dynamic performance of the proposed MG system due to change in L-parameter of the transmission line. (a, b, c) output active-power of the five-DGs, and (d, e, f) Output reactive-power of the five-DGs.

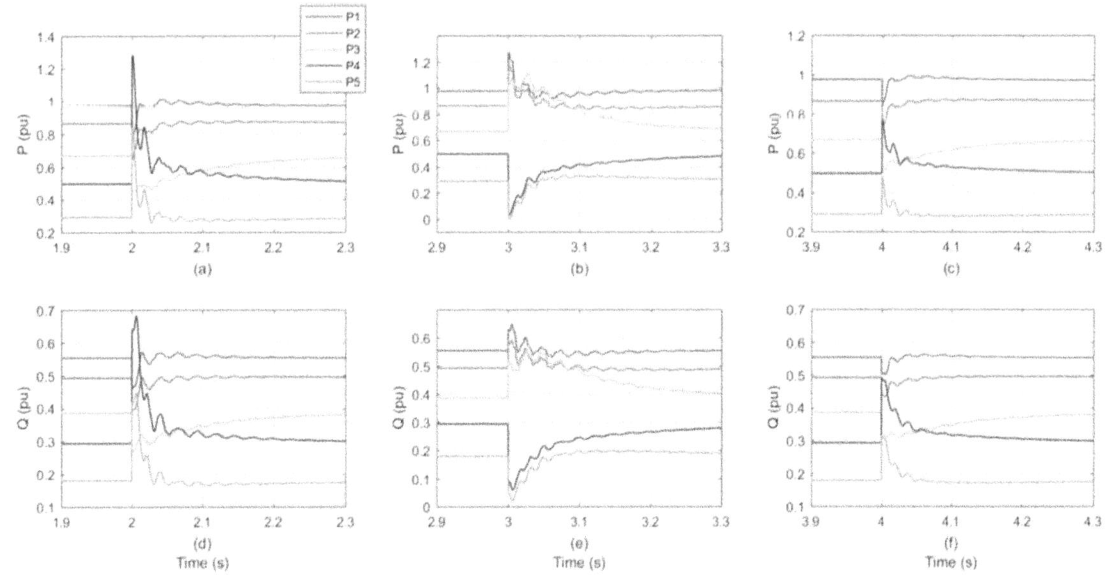

CONCLUSION

Nowadays a Microgrid (MG) penetration is currently a growing problem across the world. This is leading to several challenges. The plug and play operations of MGs is consider to be one of these challenges. In this paper, we utilize a new technique for modeling and control of islanded MGs. Every DG's local control is established on a decentralized robust approach that uses the MG's linear state-space model. The design is straightforward and based on the ellipsoidal approximation to invariant sets. The local controller for each DG considers the interaction effect of the rest of the system as an external disturbance to be rejected. Regardless of uncertainties in the transmission line parameters and plug-and-play operation, the recommended tracker affords robust stability and fast transient response. Furthermore, the proposed control technique does not create any steady state error and negligible interaction between the DGs. To assess the proposed optimal tracker, a group of case studies are carried out using MATLAB/SimPowerSystems Toolbox. These results show that:

- A fast reference tracking for the proposed control strategy.
- The load voltage magnitude is robustly preserved despite the transmission line parameter uncertainties and plug-and-play operations.

$$A = \begin{matrix} A_{11} & A_{12} & \cdots & A_{15} \\ A_{21} & A_{22} & \cdots & A_{25} \\ \cdots & \cdots & \ddots & \cdots \\ A_{51} & A_{52} & \cdots & A_{55} \end{matrix}$$

$$A_{11} = \begin{matrix} \frac{-1}{R_1C_1} & w & \frac{1}{C_1} & 0 & \frac{-1}{C_1} & 0 & \frac{-1}{C_1} & 0 \\ -w & \frac{-1}{R_1C_1} & 0 & \frac{1}{C_1} & 0 & \frac{-1}{C_1} & 0 & \frac{-1}{C_1} \\ \frac{-1}{L_{f1}} & 0 & \frac{-R_{f1}}{2} & w & 0 & 0 & 0 & 0 \\ 0 & \frac{-1}{L_{f1}} & -w & \frac{-R_{f1}}{L_{f1}} & 0 & 0 & 0 & 0 \\ \frac{1}{L_{t1}} & 0 & & & \frac{-R_{l1}}{L_1} & w & 0 & 0 \\ 0 & \frac{1}{L_1} & 0 & 0 & -w & \frac{-R_{l1}}{L_1} & 0 & 0 \\ \frac{1}{L_{t1}} & 0 & 0 & 0 & 0 & 0 & \frac{-R_{t1}}{L_{t1}} & w \\ 0 & \frac{1}{L_{t1}} & 0 & 0 & 0 & 0 & -w & \frac{-R_{t1}}{L_{t1}} \end{matrix}$$

$$A_{12} = \frac{-1}{L_{t1}} \times \begin{bmatrix} 0_{6\times2} & 0_{6\times6} \\ 1_{2\times2} & 0_{2\times6} \end{bmatrix}, A_{13}, \ldots, A_{14} = 0_{8\times8}, A_{15} = 0_{8\times6}$$

$$A_{22} = \begin{bmatrix} \frac{-1}{R_2C_2} & w & \frac{1}{C_2} & 0 & \frac{-1}{C_2} & 0 & \frac{-1}{C_2} & 0 \\ -w & \frac{-1}{R_2C_2} & 0 & \frac{1}{C_2} & 0 & \frac{-1}{C_2} & 0 & \frac{-1}{C_2} \\ \frac{-1}{L_{f2}} & 0 & \frac{-R_{f2}}{L_{f2}} & w & 0 & 0 & 0 & 0 \\ 0 & \frac{-1}{L_{f2}} & -w & \frac{-R_{f2}}{L_{f2}} & 0 & 0 & 0 & 0 \\ \frac{1}{L_{t2}} & 0 & 0 & 0 & \frac{-R_{l2}}{L_2} & w & 0 & 0 \\ 0 & \frac{1}{L_2} & 0 & 0 & -w & \frac{-R_{l2}}{L_2} & 0 & 0 \\ \frac{1}{L_{t2}} & 0 & 0 & 0 & 0 & 0 & \frac{-R_{t2}}{L_{t2}} & w \\ 0 & \frac{1}{L_{t2}} & 0 & 0 & 0 & 0 & -w & \frac{-R_{t2}}{L_{t2}} \end{bmatrix}$$

$$A_{21} = \frac{1}{C_2} \times \begin{bmatrix} 0_{2\times6} & I_{2\times2} \\ 0_{6\times6} & 0_{6\times2} \end{bmatrix}, A_{23} = \frac{-1}{L_{t2}} \begin{bmatrix} 0_{6\times2} & 0_{6\times6} \\ I_{2\times2} & 0_{2\times6} \end{bmatrix}, A_{24}, A_{25} = 0_{8\times6},$$

$$A_{55} = \begin{bmatrix} \frac{-1}{R_5C_5} & \omega & \frac{1}{C_5} & 0 & \frac{-1}{C_5} & 0 \\ -\omega & \frac{-1}{R_5C_5} & 0 & \frac{1}{C_5} & 0 & \frac{-1}{C_5} \\ \frac{-1}{L_{f5}} & 0 & \frac{-R_{f5}}{L_{f5}} & \omega & 0 & 0 \\ 0 & \frac{-1}{L_{f5}} & -\omega & \frac{-R_{f5}}{L_{f5}} & 0 & 0 \\ \frac{1}{L_5} & 0 & 0 & 0 & \frac{-R_{l5}}{L_5} & \omega \\ 0 & \frac{1}{L_5} & 0 & 0 & -\omega & \frac{-R_{l5}}{L_5} \end{bmatrix}$$

$$A_{51}, \ldots A_{53} = 0_{6\times8}, A_{54} = \frac{1}{C_3} \times \begin{bmatrix} 0_{2\times6} & I_{2\times2} \\ 0_{4\times6} & 0_{4\times2} \end{bmatrix}$$

Nonzero entries of B_{38x10} are $b_{3,1} = b_{4,2} = \dfrac{1}{L_{f1}},\ldots,b_{35,9} = \dfrac{1}{L_{f5}},\ and\ b_{36,10} = \dfrac{1}{L_{f5}}$

Non-zero entries of $C_{10\times38}$, i.e., $C_{1,1}$, $C_{2,2}$, ..., $C_{9,33}$, and $C_{10,34}$ are unity.

REFERENCES

Babazadeh, M., & Karimi, H. (2013). A robust two-degree-of-freedom control strategy for an islanded microgrid. *IEEE Transactions on Power Delivery*, *28*(3), 1339–1347.

Bahrani, B., Saeedifard, M., Karimi, A., & Rufer, A. (2013). A multivariable design methodology for voltage control of a single-DG-unit microgrid. *IEEE Transactions on Industrial Informatics*, *9*(2), 589–599.

Bayoumi, E. H. E., Soliman, H., & Soliman, M. (2020). Deadbeat PQ Control for Islanded Smart Grids. *EJECE. European Journal of Electrical Engineering and Computer Science*, *4*(5), 1–9. doi:10.24018/ejece.2020.4.5.242

Bidram, A., Nasirian, V., Davoudi, A., & Lewis, F. L. (2017). *Cooperative Synchronization in Distributed Microgrid Control*. Springer.

Etemadi, A. H., Davison, E. J., & Iravani, R. (2012). A Decentralized Robust Control Strategy for Multi-DER Microgrids—Part I: Fundamental Concept. *IEEE Transactions on Power Delivery*, *27*(4), 1843–1853.

Etemadi, A. H., Davison, E. J., & Iravani, R. (2014). A generalized decentralized robust control of islanded microgrids. *IEEE Transactions on Power Systems*, *29*(6), 3102–3113.

IEEE Recommended Practice for Monitoring Electric Power Quality. (2009). *IEEE Standard 1159*.

Kundur, P. (1994). *Power System Stability and Control*. McGraw-Hill.

Moradi, R., Karimi, H., & Karimi-Ghartemani, M. (2010) Robust decentralized control for islanded operation of two radially connected DG systems. *Proc. IEEE Int. Symp. Ind. Electron. (ISIE)*, 2272–2277.

Nazin, S. A., Polyak, B. T., & Topunov, M. V. (2007). Rejection of Bounded Exogenous Disturbances by the Method of Invariant Ellipsoids. *Automation and Remote Control*, *68*(3), 467–486.

Olivares, D. E., Mehrizi-Sani, A., Etemadi, A. H., Canizares, C. A., Iravani, R., Kazerani, M., Hajimiragha, A. H., Gomis-Bellmunt, O., Saeedifard, M., Palma-Behnke, R., Jimenez-Estevez, G. A., & Hatziargyriou, N. D. (2014). Trends in microgrid control. *IEEE Transactions on Smart Grid*, *5*(4), 1905–1919. doi:10.1109/TSG.2013.2295514

Riverso, S., Sarzo, F., & Ferrari-Trecate, G. (2015a). Plug-and-play voltage and frequency control of islanded microgrids with meshed topology. *IEEE Transactions on Smart Grid*, *6*(3), 1176–1184.

Riverso, S., Sarzo, F., & Ferrari-Trecate, G. (2015b). Plug-and-play voltage and frequency control of islanded microgrids with meshed topology. *IEEE Transactions on Smart Grid*, *6*(3), 1176–1184.

Sadabadi, M. S., Karimi, A., & Karimi, H. (2015). Fixed-order decentralized/distributed control of islanded inverter-interfaced microgrids. *Control Engineering Practice*, *45*, 174–193.

Sadabadi, M. S., Shafiee, Q., & Karimi, A. (2017). Plug-and-Play Voltage Stabilization in Inverter-Interfaced Microgrids via a Robust Control Strategy. *IEEE Transactions on Control Systems Technology*, *25*(3), 781–791. doi:10.1109/TCST.2016.2583378

Schiffer, J., Ortega, R., Astolfi, A., Raisch, J., & Sezi, T. (2014). Conditions for stability of droop-controlled inverter-based microgrids. *Automatica*, *50*(10), 2457–2469.

Shafiee, G., Guerrero, J. M., & Vasquez, J. C. (2014). Distributed secondary control for islanded microgrids—A novel approach. *IEEE Transactions on Power Electronics*, *29*(2), 1018–1031.

Xuand, Y., & Yi, Z. (2020b). *Optimal distributed secondary control for an islanded microgrid, in the book:Distributed Control Methods and Cyber Security Issues in Microgrids*. Academic Press.

Yazdanian, M., & Mehrizi-Sani, A. (2014). Distributed control techniques in microgrids. *IEEE Transactions on Smart Grid*, *5*(6), 2901–2909. doi:10.1109/TSG.2014.2337838

Chapter 2
Design, Control, Management, and Performance Analysis of PV-Battery Supercapacitor DC-System Using Buck Converter

Yasmine Amara
Electronics Department, Blida 1 University, Algeria

Zoulikha Tebri
Electrotechnique Department, LATAGE Laboratory, University Mouloud Mammeri of Tizi-Ouzou, Algeria

Zina Larabi
Electrotechnique Department, University Mouloud Mammeri of Tizi-Ouzou, Algeria

ABSTRACT

The intermittent nature of photovoltaic energy necessitates the incorporation of storage devices to ensure the continuality of loads feeding. In addition, it is important to model, control, and verify the operating of the designed system before implementation. Furthermore, the integration of power electronic interfaces plays a significant role in protecting the system and benefiting from solar energy. To this end, a buck converter is chosen to charge the battery and supply the supercapacitor. The control strategy is based on the maximum power point tracking techniques when the management algorithm recommends MPPT function mode. Otherwise, a feedback constant voltage PI controller is designed. Indeed, perturb and observe and incremental conductance are implemented and compared to analyze the system efficiency within the management strategy to charge the battery, switch between the controllers, and feed a supercapacitor in case of full battery charge. The obtained results using MATLAB/SIMULINK platform confirm the behaviour of the proposed strategy.

DOI: 10.4018/978-1-7998-7447-8.ch002

INTRODUCTION

Solar activity is the main source of renewable energy sources. The earth receives an energy capacity of the order of 1.8×10^{14} MW per year, which is several thousand times higher than the actual electricity consumption. So, solar energy can meet our current and future energy needs (Sheina,2019).

The global power growth of renewable energies made its investigation competitive with conventional fossil fuel sources. According to the statistics of 2017, the added capacity for solar energy is 98GW and is about 52GW for wind energy, however gas and coal grown only with 38GW and 35GW, respectively. According to the same statistics, the investigation in solar photovoltaic energy has increased significantly and taken the top of development and installation compared to the other renewable energies with 32%, followed by wind energy by 10%. China takes the head of global PV energy installation with 23%, followed by the USA with 14%, Japan with 13%, Germany with 13%, Italy with 6%, however, the rest of the world takes only 30% of the total PV installations (Ahmad,2020).

It is worth mentioning that thanks to the French scientist Edmund Becquerel, the conversion of solar electromagnetic radiation into electricity (i.e. known as the photovoltaic effect) has been successfully conducted in 1839 (Lofthouse,2015).

In the recent few years, large installations of PV systems have been realised due to the availability of solar energy all around the earth and the development of new solar cell technologies. In addition, the decrease in the PV technologies and PV equipment prices, the increase in the power efficiency encouraged investigators to explore more and more solar energy to produce electricity. Moreover, Photovoltaic Generators (PVGs) do not contain any mechanical parts that emit noise when generating electricity. This makes the PV system the most suitable for buildings and residential cities (Jha, 2017).

To get high performance with Solar Photovoltaic energy, it is important to well design the installation components and controllers. During the last decades, researchers proposed several solutions to enhance the extracted, transferred, and supplied energy and guarantee its continuality, decrease costs, and ensure the reliability of the system. To this end, diverse materials technologies are tested and used to constitute the PV cell which differs on the efficiency of conversion, costs, and robustness, such as, mono and poly Crystalline Silicon (C-Si), Thin Film, Cadmium Telluride (CdTe), Copper Indium Gallium Selenide (CIGS) technologies (Konagai,2011; Guha,2000). In fact, a Small power value is obtained from the solar cells. So, a combination of several cells in series or parallel is important to generate a required power value. This connection forms a solar module or namely a PV panel (Ngyen,2008). The series combination of several panels constitutes the PV string, while, the parallel connection of strings constitute the PV array. In fact, the number of series-connected modules depends on the required voltage value, while, the number of parallel-connected strings is related to the required output current. It is well known that the obtained current-voltage (Ipv, Vpv) and Power-voltage (Ppv, Vpv) curves at the PVG terminals had a non-linear form, moreover, the direct connection of the module to load is not beneficial for the installation since no protection is presented and the power generated by the PVGs has an intermittent and non-linear nature (Mohmoud,2006). In practice, power electronic devices (converters and inverters) are incorporated between the PVGs and loads to deal with such troubles. The choice of Power devices to use depends on the load requirement. For instance, if the installation contains AC-loads the single-phase, three-phase, or multilevel inverters are required to transfer the DC generated current to an AC form (Kasper,2013). Else, if the installed loads are of DC nature, a DC-DC converter is sufficient to feed them with the desired DC power (Behjati,2014). The DC-DC converters may be divided into three large categories which defer on the manner that the magnetic elements are maneuvering, arranged, and controlled. These DC-DC power converters consist of an inductor, a capacitor, and a switching device (IGBT, MOSFET,...), that transfer a DC voltage to a different form of DC voltage by stepping it down or up depending on the designed system, and which are: (1) DC-DC buck (step down) converters: which give a low voltage value at their terminals. (2) DC-DC boost (step-up) Converters: permits high voltage at the loads' side. Whereas, the DC-DC buck-boost (step-up/ step down) converters provide high or low voltage value depending on the load requirement (Li,2008; Coelho,2009).

Indeed, the control of the switching device ON-OFF is very important to ensure high power yield and transfer. To this end, regulators known as Maximum Power Point Tracking (MPPT) techniques are designed and incorporated within the photovoltaic system to get the desired signal of the switching gate (Amara,2018). Several MPPT techniques are developed and reviewed (Hassen,2019; El-Khozondar,2016) . They differ on the number of the used sensors, the control process, the complexity of implementation, cost, efficacy, and response time. The MPPT techniques may be divided into three categories namely:

1. Conventional techniques: such as Perturb and Observe (P&O) (Ali,2014), Incremental Conductance (Inc-Cond) (Dhaouadi,2019), Constant Current (CC-MPPT) (Chain,2009) Constant Voltage (CV-MPPT) (Masoum,2002).
2. (2) Intelligent techniques: as Fuzzy Logic Control (FLC) based MPPT (Ezzouitine,2019), Neural Network (NN) (Faria,2019), and meta-heuristic techniques i.e. Particle Swarm Optimization (PSO) (Renaudineau,2014), Slap Swarm Optimization (SSO) (Mirjalili,2017), Genetic Algorithms (GA) (Kumar,2015), Artificial Bee Colony (ABC)(Salmi,2016).
3. (3) Hybrid techniques: obtained combining the conventional and intelligent techniques for more efficiency and adapt the system to different phenomena that may degrade the operating of the system such as, the variations in weather conditions (Irradiance and temperature), partial shading events and the non-linear energy demand i.e. Fuzzy-Inc-Cond (Yasmine (1),2020), HC-PSO (Yasmine(2),2020).

The use of energy storage devices is very beneficial in photovoltaic installation especially in rural areas and during nights or shaded periods. The batteries differ in the manner of storing energy such as Lead-acid technology (Kyriakopoulos,2016), Nickel-Cadmium (Ni-Cd) technology (Luo,2015), and Lithium-ion (Li-ion) technology (Tsang,2011) that may be considered as the most promising challenger. Its lifetime and performance are among the best electrochemical accumulators. The cost is still significant compared to Lead technology but a major drop in the next few years is expected so that Li-ion becomes the most attractive technology for PV applications (Etxeberria,2010; Lee,2011).

Flywheel, Supercapacitor (SC), compressed air energy, and some other storage devices are also included to maintain the continuality of load feeding during low production and store the surplus power during peak production (Guasch,2003). In fact, the hybridization of the batteries with a Supercapacitor will enhance the sustainability of charge and lifetime of the devices. Moreover, it presents an economic benefit for both consumers and electrical grid users. The control and the management of the energy flowed to the battery and SC is very important to maintain their operating and guarantee their security and long lifetime. This may be possible through the preservation of their State of Charge (SoC) within acceptable levels by avoiding the over-charge and deep discharge states (Pathak,2019).

In this work, the design of a photovoltaic DC system is investigated. A DC-DC buck converter is used as an interface between the panel and the storage device. The control of the switching states is ensured using a P&O compared to Inc-Cond technique and PI controller throughout the management process. The objectives of the present work may be listed as follows:

- Model a standalone PV system including buck converter, battery storage and a super-capacitor ;
- Control the DC-DC buck converter switching On/OFF states using the most widely adopted MPPT techniques which are P&O and Inc-Cond, or through a feedback PI controller using the voltage delivered to energy storage devices;
- Manage the power flow to the energy storage system through the control of the SoC level and switch between MPPT control mode and PI control mode;
- Ensure a high efficiency of the system under standard conditions, variable irradiance levels and during the full battery charge.

Accordingly, the rest of this chapter may be recapitulated as follows: First, a background of the related works with the present work is done. Then, the designed system is well described including the design of the PV panel, battery, Supercapacitor and DC-DC buck converter. Besides, the used P&O, Inc-Cond and PI controllers are well explained, and then the management algorithm is detailed. Furthermore, Simulations are fulfilled through MATLAB/ SIMULINK and the important results are drawn and discussed. Finally, future research direction and conclusion are done.

BACKGROUND

These last few years, the hybridization of the photovoltaic generators with other sources and the integration of hybrid energy storage devices are largely studied and implemented. However, the voltage stability and power efficiency are largely affected by the sudden changes which may take place during the operating process causing faults and disturbing the system. To this end, several kinds of research are developed focusing on either enhancing the production or the accumulation of energy through the management and improvement of the controllers.

In (Bagalini,2019), the authors detailed the features of integrating the battery storage systems within PV generation in grid-connected residential areas. It is noted that the techno-economic benefit of this system is higher, as it reduces the electricity tariffs through the self-consumption in residential areas and the import of electricity from the electrical grid, and ensure the continuality of loads feeding.

To cover load energy demand through the analysis of annual wind speed and solar radiation, the authors of (Moghaddam,2019), proposed an optimized meta-heuristic technique based on Improved Crow Search Algorithm (ICSA) to design a PV/wind/ battery system, then, analyzed the benefits of this hybrid system. High generation and reliability are obtained compared to the conventional Craw Search Algorithm (CSA) and PSO. Besides, in (Tak,2018), a hybrid system composed of PV, wind, and battery is simulated using a Neuro-Fuzzy controller and tested under variable load. The obtained results demonstrate the benefit of the hybrid generation in overcoming the intermittent renewable sources and decreasing the utility grid trouble. Several other pieces of research focus on hybrid generation combined with storage either in the standalone system (Kumar,2018), or grid-connected mode (Khadkikar,2009).

Because of the noise produced by the wind turbines during the production, their integration in residential areas will be a source of several problems. So, the hybrid storage system will be a solution to deal with problems meet during operating of the PV generation system and cover consumers with an uninterruptable power supply.

In (Yasmine (2),2020), a PV/Battery standalone system is proposed to tackle the problems of partial shading. In fact, the management strategy is accompanied by the improvement of the PSO controller and the preservation of the SoC within accepted limits. The results proved the high efficiency of the method. Indeed, this system may be improved more by integrating others kinds of sources such as wind and fuel cells, or others energy storage devices i.e. Supercapacitor, fly-well, etc.

In (Glavin,2008), the Supercapacitor bank is combined with the Lead-acid battery bank in high power standalone PV generation system. The hybrid storage is simulated and managed using the load requirement and SoC to optimize the power, reduce the battery size and enhance its lifetime, and avoid the load shutdown during low generation. Tests under variable loads profiles shown high reliability and power quality than using battery storage alone.

In (Wang,2014), a hybrid battery Supercapacitor energy storage system is integrated into a PV plant. The management strategy is applied under normal, deficit, and excess power-sharing modes and respecting the SoC constraints. The output power is maintained near to its maximum ensuring an increase in the efficiency and performance of the hybrid system.

In (Bhattachajee,2012), comparisons between three charge control strategies for battery in photovoltaic system which are constant current (CC), constant voltage (CV) and Constant Current/Constant voltage (CC-CV) are done. Where, the design of the controller is related to the application and depend on the management applied. Other works focus on integrating intelligent and smart controllers within these control strategies in the hybrid generation (Naresh,2019; Al-Barazanchi,2015). Or hybrid storage energy system (Kollimalla,2014; Zhu,2013). Whereas, others integrate different power electronic interfaces and improve them to smooth the system and minimize faults (Chahwan,2007).

Based on the previews works and to benefit from the advantages associated with the proposed system of (Yasmine(2).,2020) in managing the power flow to the load side, the proposed system used a buck converter to control the battery charge and a Supercapacitor as an additional storage during peak production and full battery charge. However, in (Yasmine(2),2020), a separate controllers and power interfaces are used for the energy extraction and storage.

DESIGN PROCESS

Figure 1, illustrates the simulated PV-Battery Supercapacitor standalone system which is composed of a PVG of the characteristics listed in Table 1. A buck converter is inserted between the PV panel and the battery to step down the voltage to the battery required level and MPPT control accompanied by the management strategy to supervise the whole system.

The design of PV Panel, battery storage, Supercapacitor and the buck converter is as follows:

Photovoltaic Panel Modelling

A photo-generator named a photovoltaic cell is an electrical device capable of transferring the sunlight energy through the photovoltaic effect to direct electricity. Indeed, it is manufactured using a semiconductor materiel with two junctions P-N. When light falls on its surface, charges and gaps are created at the P and N junctions due to the high moving energy of the electrons. Consequently, a direct current is created. Unfortunately, a solar cell produced a low power value. So, to get high power, cells are organized in series (to enhance the output voltage) and parallel (to enhance the output current value) to get the desired power quantity in a device named a Photovoltaic module or panel (Begalini,2019).

Figure 1. PV-battery supercapacitor standalone system

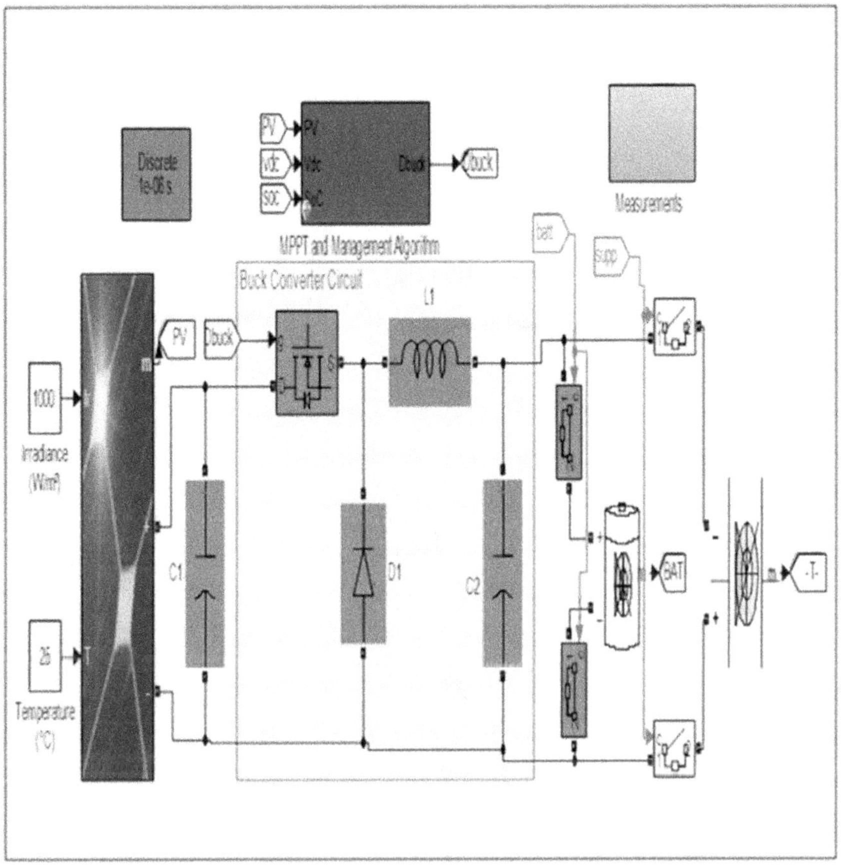

Table 1. Electrical specification of the photovoltaic generator

Parameters	Symbols	values
Maximum Power	P_{mpp}	250.29 W
Open Circuit Voltage	V_{oc}	36.6 V
Short circuit current	I_{sc}	8.75 A
Voltage at maximum power point	V_{mpp}	30.9 V
Current at maximum power point	I_{mpp}	8.1 A

It is important to note, that a single diode solar cell configuration is composed of the combination of a current source (I_{ph}), a diode (I_d) and resistor in shunt connection (R_{sh}), that are in turn attached to a series resistor (R_s) as depicted in Figure.2. Clearly, a low value is affected to the series resistor, whereas, the value of the parallel resistor is infinity(Khadkikar,2009).

Nonlinear quantities are given at the output, because of the dependence of the photovoltaic cell efficiency on the weather conditions of temperature and irradiance.

Figure 2. Single diode solar cell

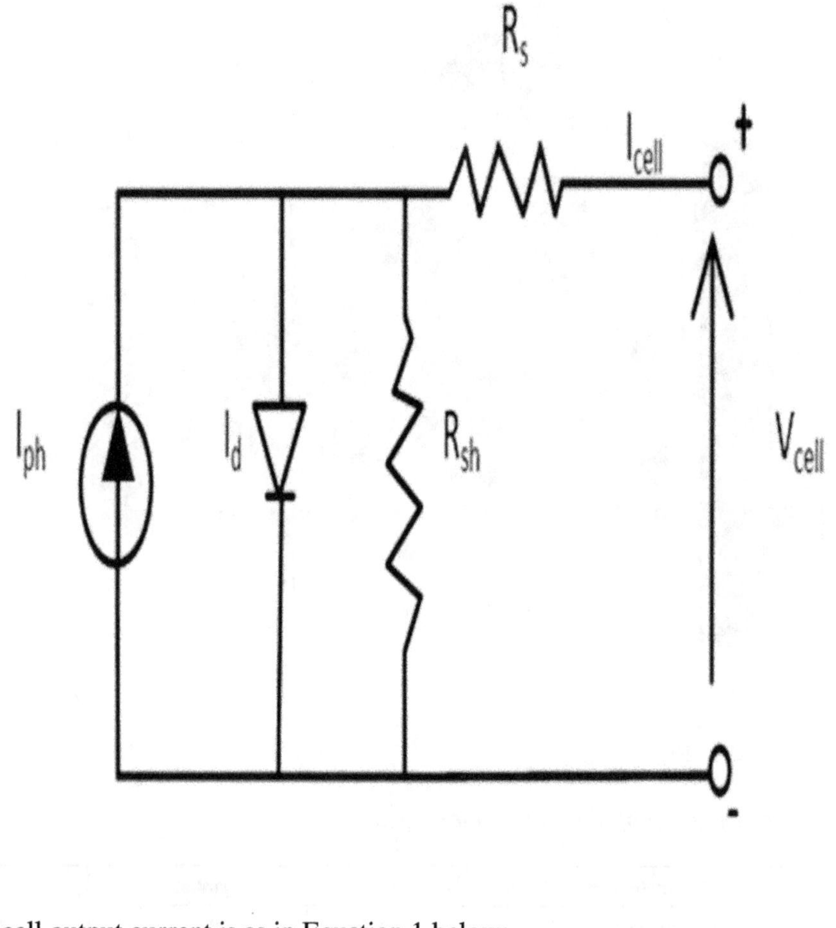

The Solar cell output current is as in Equation.1 below:

$$I_{cell} = I_{ph} - I_0\left(e^{\left(\frac{q(V_{cell}+R_s I_{cell})}{AKT}\right)} - 1\right) - \frac{V_{cell} + R_s I_{cell}}{R_{sh}} = I_{ph} - I_d - I_{sh} \quad (1)$$

Where V_{cell}, I_{cell} are the output cell voltage and current, respectively. q is the electron charge, K represents the Boltzmann's constant. T is the operating temperature (kelvin), I_0 is the diode reserve saturation current, A is ideality factor.

An enhancement on the output current (I_{pv}) is obtained when several solar cells are connected in parallel . Whereas the panel output voltage (V_{pv}) is rising when the solar cells are connected in series . In practice, bypass diodes are connected in parallel with cells to avoid the destruction of the panel when a cell is damaged, and blocking diodes are inserted in series with PV cells to avoid the circular current in the opposite direction(Li,2008; Coalho,2009) . The Panel is devoted to generate sufficient voltage required by the load, besides; the parallel connection of cells is not secured. So, the majority of commercial panels are made of the series connected cells (Kumar,2018).

The direct connection of the solar Panel to a resistive load permits the draw of the PV panel (Current-voltage) and (Power-voltage) characteristics as in Figure.3. When the output is short circuited the Open Circuit Voltage is zero (V_{oc}=0), in this case, the Short Circuit Current (I_{sc}) is at its maximum value (I_{sc}=I_{ph}) .But, when the V_{oc} is at its maximum value, the current is equal to zero (I_{sc}=0)(Tak, 2018).

Figure 3. Solar cell current-voltage curve

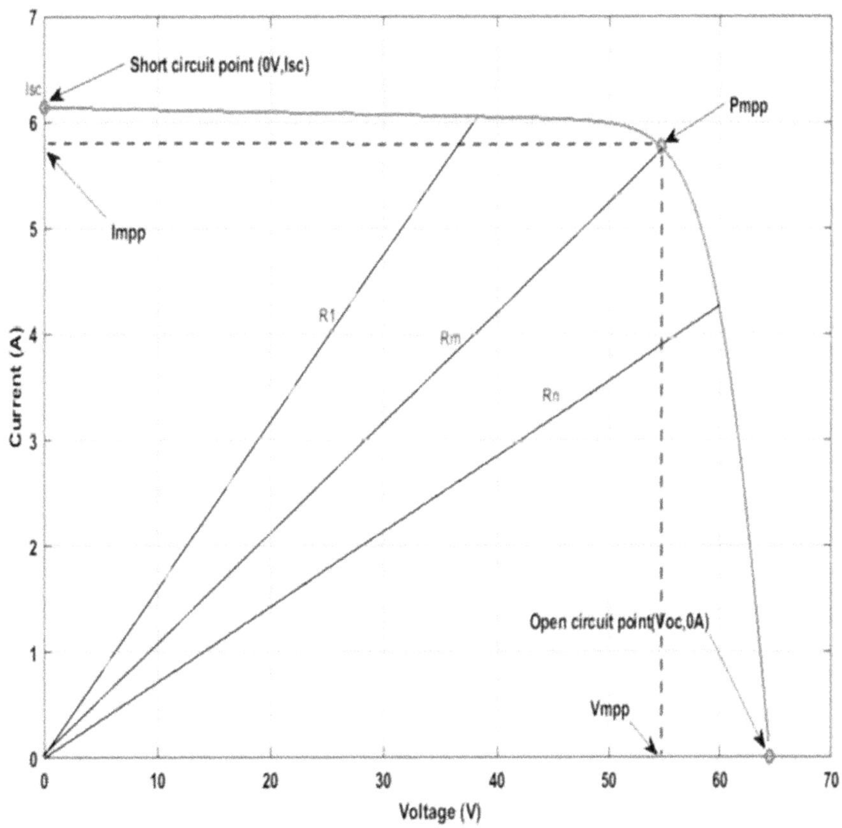

The Fill Factor and efficiency of the device are defined as in Equation.2.and Equation.3.It is very important to ameliorate and improve these parameters to maximize the extraction of energy.

$$Fillfactor = \frac{V_{MPP} I_{MPP}}{V_{oc} I_{sc}} \tag{2}$$

$$\mu = \frac{P_{MPP}}{insolation \times area\ \exp lored} \tag{3}$$

The design of the number of the panels to integrate in the installation (N_P) depend on the Load requirement of the energy (E_L) (W.h) and the energy produced by a module (E_{PV}) (W.h) (Khadkikar,2009)

$$N_P = \frac{E_L}{E_{PV}} \tag{4}$$

Variable Irradiance Effect

As depicted in Figure 4, The application of the irradiance values of 0.2KW/m², 0.4 KW/m², 0.6 KW/m², 0.8 KW/m² and 1 KW/m², respectively, to the a panel connected to a variable resistive load, shows that the generated Power will increase with the increase of solar irradiance.

Figure 4. Panel current-voltage and power-voltage curves under variable irradiance levels

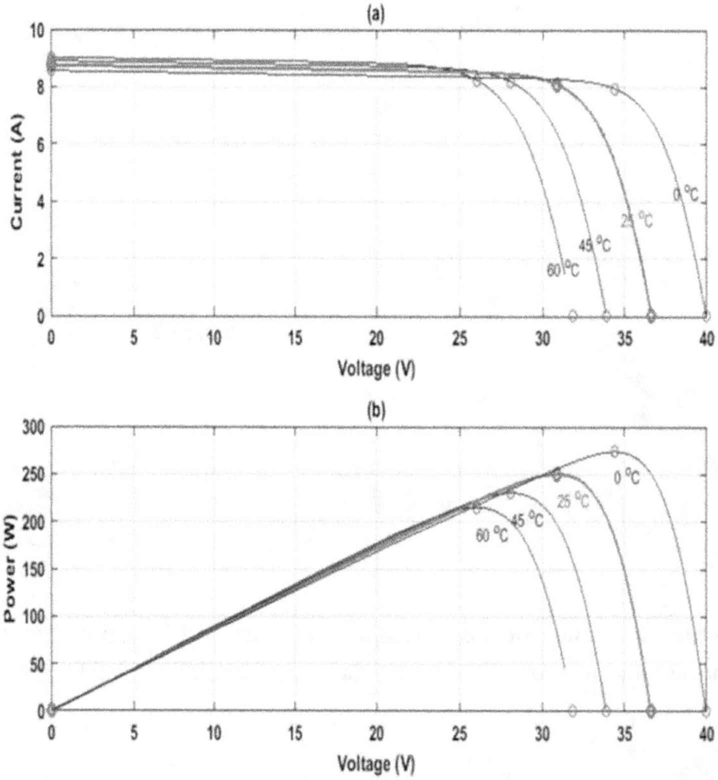

Variable Temperature Effect

The temperature has a negligible influence on the value of the short-circuit current. The output power of the panel decreases by about 0.5% with each degree of temperature rise above 25 ° C. Indeed, it is important to note that when the illumination is lower than 0.2KW / m^2, the generated voltage varies in turn, as in Figure 5.

Figure 5. Panel current-voltage and power-voltage curves under variable temperature

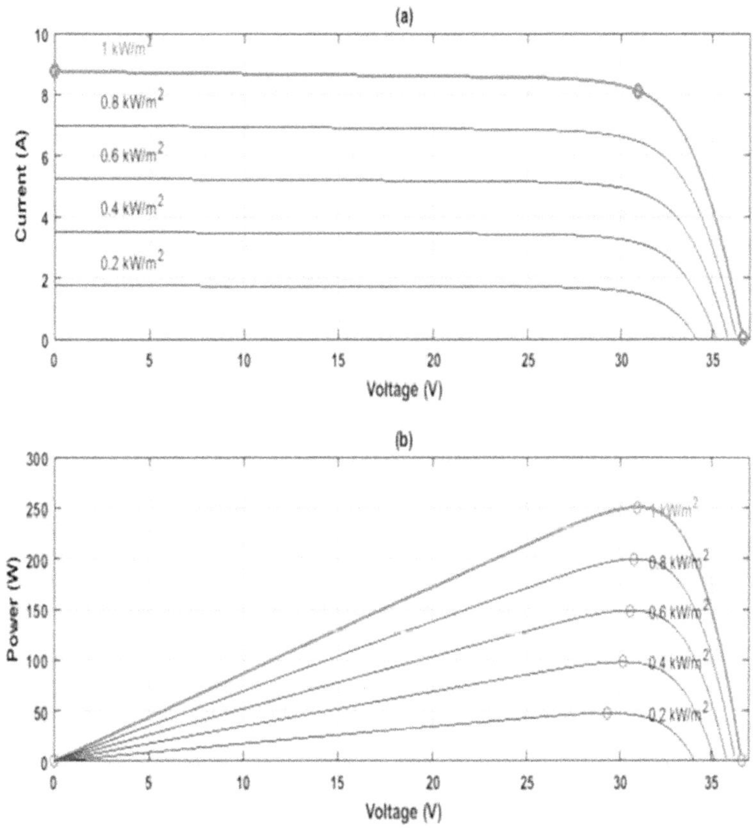

Battery Storage Design

During varying irradiance, partial or total shading conditions, the energy produced by the PVGs is insufficient. So, it is important to accompany the installation with energy storage devices to store the energy during peak production and compensate the installation requirement. The characteristics of the energy storage devices and their performances differ from technology to other depending on the manufacturers. Nowadays, a variety of storage devices are fabricated, each one has its defined design for specific applications. They differ on size, cost, efficiency, performance, lifetime and the way how to store energy (i.e. chemical, electrical...). In photovoltaic applications, the most widely energy storage adopted are Lead acid, Nickel Cadmium (NiCd), Li-Ion, Fly-well and Supercapacitor (Kyriakopoulos,2016; Tsang,2011; Etxeberria,2010).

The choice of storage technology will depend basically on the storage system price, electricity tariff, rate return, maintenance cost, limited charge rate and the battery life within the operating conditions corresponding to the selected application. Indeed, in the photovoltaic application Li-Ion batteries are the most widely applied and adapted due to their advantages compared to the others technologies. Accordingly, it is used to fulfill the objective of this work. The model of this latter is available in MATLAB/ SIMULINK library, with the discharge characteristics of Figure.6.a bellow

Figure 6. a Battery discharge characteristics, b. Supercapacitor equivalent circuit

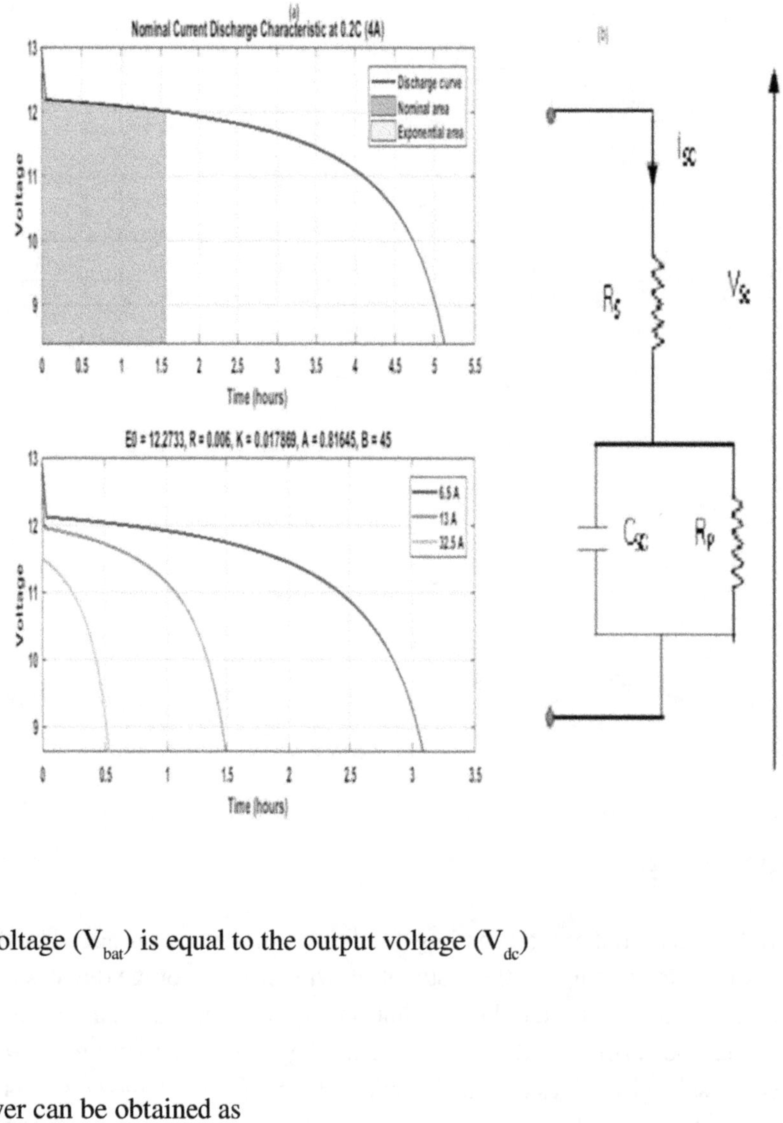

The battery voltage (V_{bat}) is equal to the output voltage (V_{dc})

$$V_{bat} = V_{dc} \tag{5}$$

Whereas the power can be obtained as

$$P_{bat} = -V_{bat} \times I_{bat} \tag{6}$$

In fact, the negative power value indicates that the battery is in the charging mode; otherwise it is in discharging mode.

During discharging mode the battery voltage is obtained with (Chahwan,2007):

$$V_{bat} = E_0 - \frac{KQ}{Q - \int i_{bat} dt} i_{bat}^* - \frac{KQ}{Q - \int i_{bat} dt} \int i_{bat} dt + A.\exp(-B \int i_{bat} dt) \tag{7}$$

During charging mode the voltage is expressed as (Chahwan,2007):

$$V_{bat} = E_0 - \frac{KQ}{\int i_{bat} dt + 0.1Q} i_{bat}^* - \frac{KQ}{Q - \int i_{bat} dt} \int i_{bat} dt + A.\exp(-B \int i_{bat} dt) \tag{8}$$

where: E_0 is the battery constant voltage, K is the polarization constant, Q corresponds to the battery capacity, i_{bat} and i_{bat}^* denote the battery current and filtered current, respectively. A and B are exponential voltage and capacity, respectively.

The battery SoC is determined as depending on the initial SoC value (SoC_0),the battery capacity Q and the output battery current i_{bat} as

$$SoC = SoC_0 - \frac{1}{Q} \int i_{bat} dt \tag{9}$$

The storage capacity required in any PV installation must be superior then the requirement of the installation for reasons of security and low system efficiency.

It is essential to design the number of required batteries (N_b) in a system before integrating them. This design is related to the total power required by the load (P_L in Watt), the number of operating hours (t in h) (i.e. during nights periods for 4h), the efficiency (μ, %) battery voltage (V_{bat} Volte), the battery depth of discharge (DoD(%)), and the used battery rating (C_b(Ah)) as expressed in Equation.10 bellow

$$N_{bat} = \frac{P_L \times t}{\mu \times V_{bat} \times DoD \times C_{bat}} \tag{10}$$

Supercapacitor Design

The Supercapacitor (SC) is adopted to surmount the problems integrated between the potential of power and energy storage with high quality and quantity compared to conventional capacitor and battery. SC is used in many domains all such as digital and telecommunication devices, electrical vehicles, and transportation. In photovoltaic applications, the SC is incorporated either alone or combined with battery storage under variable working conditions to improve the power quality by ensuring the continuality of the load feeding. The SC can be modelled using the equivalent circuit via electrical components divided into two parts namely resistive parts (represents the energy loss) and capacitive part (represents the energy storage) (Figure 6.b), or electrochemical models which are based on the electrochemical reactions of the electrode to generate electrical energy (Glavin,2008; Kollimalla,2014) .

The SC output voltage (V_{sc}) can be expressed using the stern equation (V_T) which is a function of series and parallel SCs, the SC current (i_{sc}), and the SC resistor (R_{sc}) as (Zhou,2011)

$$V_{sc} = V_T - i_{sc} \times R_{sc} \tag{11}$$

The electric charge of the SC Q_T is in function of i_{sc} during the operating mode of SC, and current of discharge $i_{sc\text{-}dis}$ when i_{sc}=0 as in Equation 12 bellow

$$Q_T = \begin{cases} \int i_{sc} dt \\ \int i_{sc-disch} dt \rightarrow i_{sc} = 0 \end{cases} \tag{12}$$

The SC state of charge (SoC_{sc}) is obtained using the initial SC amount (Q_{init}), i_{sc} and Q_T as

$$SoC_{SC} = \frac{Q_{init} - \int i_{sc} dt}{Q_T} \tag{13}$$

DC-DC Buck Converter Design

The aim of the DC-DC converters is not to limited on gating the required voltage level at its output but also, choose a better control approach to maximize the power extracted from the PVGs and minimize the power loss during its transfer to the loads through the ON/OFF gate states control (Kasper,2014; Behjati,2014; Li,2008). Besides, the MPPT controller is designed to accomplish this goal.

The main objectives of incorporating the DC-DC converters in a photovoltaic system are (Kasper,2014; Behjati,2014; Li,2008):

- Eliminate the high frequency harmonics in the PV current ;
- Protect the panel by separating it from the load;
- Transfer the required power to load with the desired voltage level (high or low);
- Aid on extracting maximum power from the panel.

Basically, the switching cycle must be modified to control the steamed power to the battery. Several charge controllers are revealed to control the switching dc-dc converter (Behjati,2014). A buck category converter is devoted to step down the input voltage to an inferior quantity is labelled in Figure.7.

The role of the diode in the circuit is to guarantee the unidirectional power flow, whereas, the MOSFET switch is controlled through ON/OFF states. The inductor and capacitor store, filter and transfer the current and voltage to loads.

Figure 7. Buck converter circuit

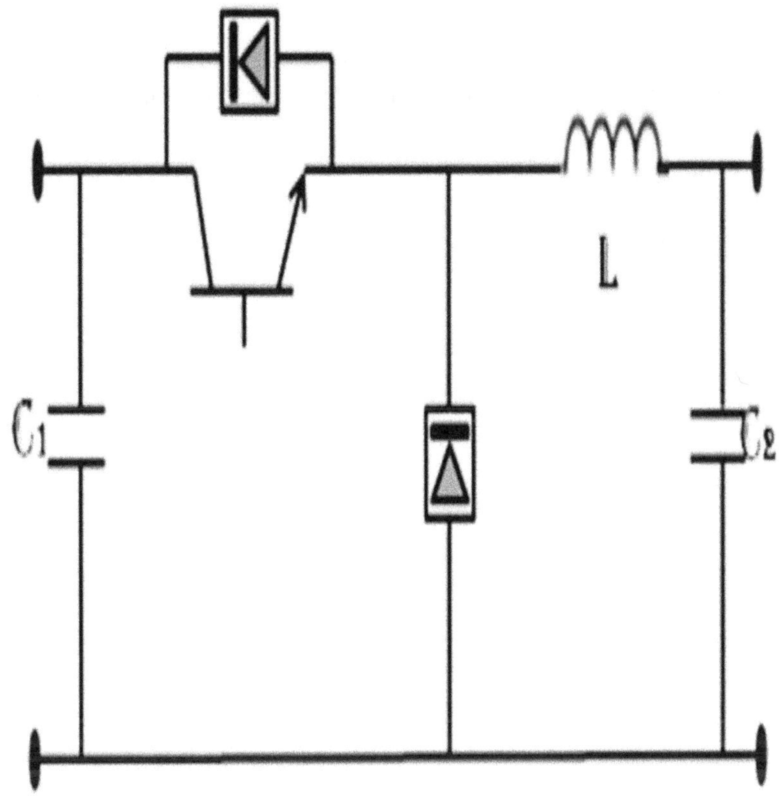

During, the ON switching state $T_{ON} = DT$,the transfer function can be obtained as

$$L\frac{dI_L}{dt} = V_{pv} - V_{dc} \tag{14}$$

The diode is in the reverse biased state, and the current (Ipv) passes through the inductor. Where

$$V_L = L\frac{dI_L}{dt}$$

In this case

$$I_L = \frac{V_{PV} - V_{dc}}{L} \times DT + I_{Li} \tag{15}$$

where I_{Li} the inductor initial current value. The ripple current ΔI_L can be obtained as

$$\Delta I_L = I_L - I_{Li} = \frac{V_{PV} - V_{dc}}{L} \times DT \tag{16}$$

During the OFF state ($T_{OFF} = (1-D)T$) the inductor stored the current (I_L) flows through the diode. The voltage is given as

$$V_L = -V_{dc} \tag{17}$$

So

$$I_{L0} = -\frac{-V_{dc}}{L}(1-D) \times t + I_L \tag{18}$$

The current ripple in this case is

$$\Delta I_L = I_L - I_{L0} = \frac{V_{dc}}{L}(1-D) \times T \tag{19}$$

Considering the Equation 16 and Equation 18, we can deduce that:

$$V_{dc} = D \times V_{PV} \tag{20}$$

The buck converter is designed through the use of the previous equations and formula indicated in Table I bellow (Behjati,2014; Kasper2014; Li,2008):

Table 2. Buck converter parameters design

Parameter	Symbol	formula
PV output Capacitor	C_{PV}	$\frac{(1-D)I_{pv}}{f\Delta P_{pv}}$
Buck inductor	L_{buck}	$\frac{D(1-D)V_{pv}}{f\Delta I_L}$
Buck capacitor	C_{buck}	$\frac{(1-D)V_{dc}}{8L\Delta I_L f^2}$
Duty cycle	D	$\frac{V_{dc}}{V_{pv}}$

CONTROL DESIGN

Perturbe and Obsreve (P&O)

The simplest method to permit maximum extraction of the power from the photovoltaic generator is the Perturb and Observe (P&O) which is adapted in several applications due to its simple structure and ease of implementation (Ali,2014; Kandermir,2017).

The function of the MPPT can be fulfilled following the steps below:

First, the PV current and voltage ($I_{pv}(k)$ and $V_{pv}(_k)$) at each instant are sensed, then, the PV power is calculated as:

$$P_{pv}(k) = I_{pv}(k) \times V_{pv}(k) \tag{21}$$

Analysing the nonlinear curve V_{pv} (k) -P_{pv} (k) at the PVG output, the variations of the PVG voltage ($\Delta V_{pv}(k)$) and power ($\Delta P_{pv}(k)$), and at two consecutive instants (k-1) and (k) can be calculated as

$$\Delta V_{pv}(k) = V_{pv}(k) - V_{pv}(k-1) \tag{22}$$

$$\Delta P_{pv}(k) = P_{pv}(k) - P_{pv}(k-1) \tag{23}$$

Three cases may be considered to reach the MPP as follows

1. $\Delta P_{pv}(k) > 0 \rightarrow V_{PV}(k) = V_{pv}(k-1) + \Delta V_{pv}(k)$ (at the curve left side)
2. Else if $\Delta P_{pv}(k) \approx 0 \rightarrow V_{PV}(k) = V_{pv}(k-1)$ (at the lap of the (V_{pv} (k) -P_{pv} (k)) curve), the MPP in this case is reached.
3. Else if ($\Delta P_{pv}(k) < 0 \rightarrow V_{PV}(k) = V_{pv}(k-1) - \Delta V_{pv}(k)$)(at the right side of the (V_{pv} (k) -P_{pv} (k)) curve)

The obtained voltage value through this process is compared to the output panel voltage before passing it through a PI control in order to get the current gate signal.

The process is repeated until reaching a power value near the MPP. The described P&O MPPT algorithm control principal is given as in the Figure 8 below:

Incremental Conductance (Inc-cOND)

The algorithm is developed to handle the low efficiency and high oscillations and other limits of the P&O techniques. The incremental conductance (Inc-Cond) is based on incrementing or decrementing instantaneous conductance (I_{pv}(k) /V_{pv}(k)) depending on the slop of power (dP/dV). The MPP is reached when dP/dV is zero, increased on the left of the (P_{pv},V_{pv}) curve and decreased on the Right side of this curve (Dhaouadi,2019) as in Figure 9.

Figure 8. P&O algorithm

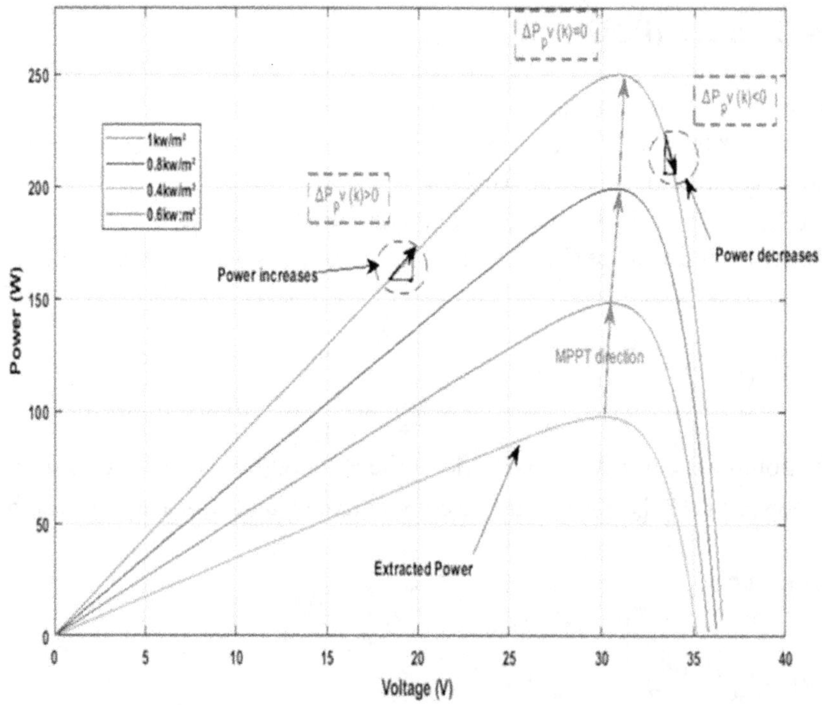

Figure 9. Inc-cond algorithm

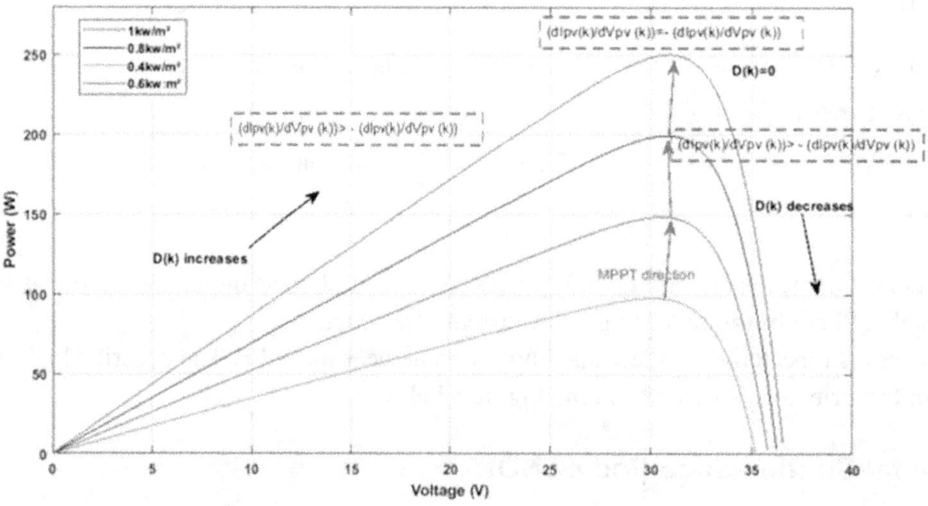

Pi Control

To get optimum performance a proportional integral controller (PI) is used in a feedback constant voltage loop, the PI output signal is then compared to a carrier triangular signal with the appropriate switching frequency value to get a pulse signal to control the MOSFET ON-OFF states.

Figure 10. Feedback PI control

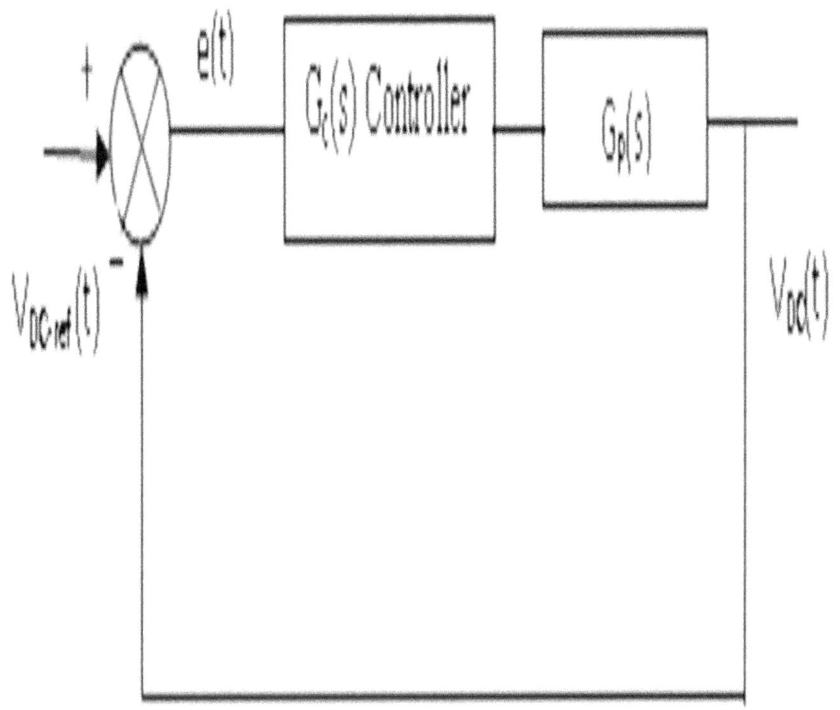

As in Figure.10. the output Voltage of the buck converter Vdc(t) is compared with the reference voltage Vdc-ref(t) then the error between these two values (e(t)) is the input of the PI control. The transfer function of the PI control is obtained as

$$G_C(s) = K_p \times \frac{K_i}{s} = \frac{K_p s + K_i}{s} \tag{24}$$

The buck converter close loop transfer function is as

$$G(s)=\frac{\frac{V_{pv}(s)}{LC}G_C(s)}{s^2+\frac{1}{RC}s+\frac{1}{LC}+\frac{V_{pv}(s)}{LC}G_c(s)}=\frac{\frac{V_{pv}(s)}{LC}K_ps+\frac{V_{pv}(s)}{LC}K_i}{s^3+\frac{1}{RC}s^2+(\frac{1}{LC}+\frac{V_{pv}(s)}{LC}K_p)s+\frac{V_{pv}(s)}{LC}K_i} \tag{25}$$

MANAGEMENT PROCESS

To manage the standalone PV-battery Supercapacitor system, a management control strategy is shown in Figure.11. and implemented to guarantee higher consistency of this system beneath different conditions. The management strategy proposed is fulfilled based on two conventional MPPTs (P&O and Inc-Cond), PI controller, and depending on the battery state of charge (SoC). In this study, the P_{pv} is considered sufficient to supply the required energy. So, the management algorithm will consider only the battery SoC to switch between the controllers.

Figure 11. Flowchart of the proposed management strategy

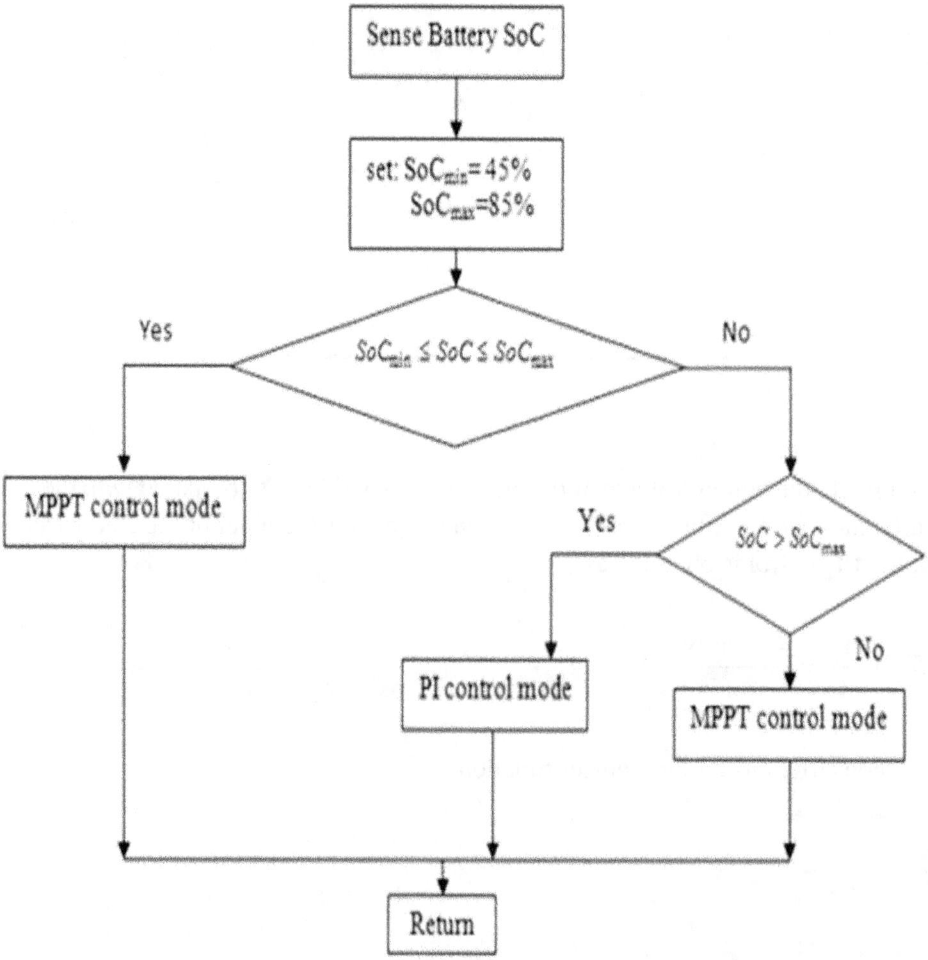

It is very important to control the discharge and charge of the battery and keep them at their limits. Certainly, the limit of the charging state is designed as SoC_{max}=85% to stay away from the overcharging of the battery. Whereas, the permitted low SoC_{min} value is 45% in order to avoid the deep discharge state. Otherwise, the battery may be burned, feel down, and can also break down the system (Dzung, 2014). Moreover, a Supercapacitor is added as second storage to store energy when the battery is full-charged.

When the SoC is between SoC_{min} and SoC_{max} the management algorithm orders the system to work on the MPPT control mode. While, the system works on PI control mode when the SoC > SoC_{max}, in this case, the battery current is equal to zero, while, the SC is in charging mode, and the SC current is proportional to the PV current.

6. SIMULATIONS AND DISCUSSIONS

The system shown in Figure 1 is simulated and analysed in MATLAB/SIMULINK 2017.b platform to show its reliability and test its performance. First, the system is tested under standard conditions of temperature and irradiance (25°C, 1kw/m^2). After that, it is tested under variable irradiance level, then, under over charging battery state.

The Table 3 below shows the designed parameters during the simulation process

Table 3. The designed simulation parameters

System	Parameter	Symbol	Value
Buck converter	PV output Capacitor	C_{PV}	100 µf
	Buck inductor	L_{buck}	2 mH
	Buck capacitor	C_{buck}	100µf
	Switching frequency	f_s	5Khz
Battery	Nominal voltage	V_{bat}	12V
	Rated capacity	C_{bat}	20 Ah
	Initial state of charge	SoC_{in}	50%
PI Controller	Proportional gain	K_p	0.1
	Integrator gain	K_i	0.1

Simulation Under Standard Test Conditions (STD)

Under Standard Test Conditions (STD), slight perturbation is shown in the P_{pv} curves when starting the simulation of the system, this perturbation takes 0.02s and 0.04s to stabilize at the MPP for the Inc-Cond and P&O techniques, respectively. As given in the zoom-in Figure 12 below

In this case, the battery storage is in charging mode. Analysing the obtained characteristics drawn in Figure.13, the V_{bat}, I_{bat}, and also P_{bat} are at their nominal value respecting the battery datasheet of Table. 3. The negative sign of the current value indicates that the Battery storage is in the charging mode of operating. Where, the SoC is raising because the power generated by the photovoltaic generator is sufficient to feed the battery. The V_{bat}, SoC curves are identical using either P&O or Inc-Cond. while the I_{bat} using Inc-Cond achieves its maximum faster than during the use of P&O. as depicted in Figure.13.

Figure 12. Power harvested under standard conditions using P&O and Inc-Cond

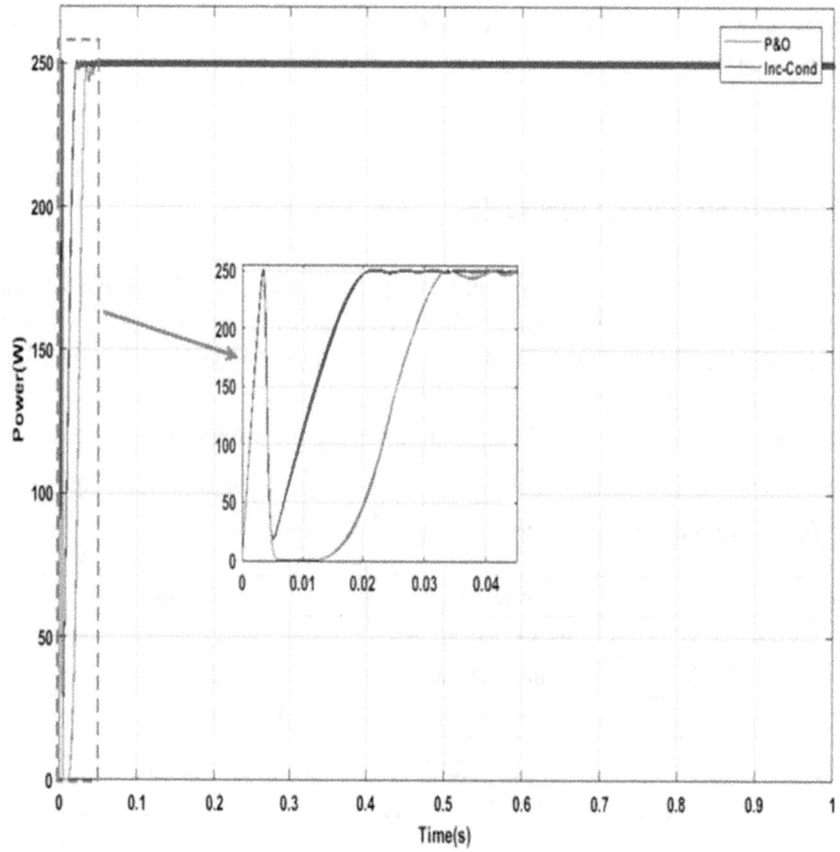

The efficiency of the extracted power from the panel is equal to 99.73% and 99.93% for P&O and Inc-Cond, respectively indicating an efficient MPP extraction.

Moreover, the efficiency of the transferred power to the battery is equal to 94.15% and 94.04% through these techniques. Consequently, high extracted and transferred power efficiency for both algorithms under STD where the Inc-Cond outperforms P&O (proving what is done in the literature). In this case, switches between the SC and the system are open respecting the management strategy. So, no current is delivered to the SC.

Simulation Under Variable Irradiance Level

Both efficiencies of the produced and transferred power are reduced compared to that obtained under STDs. Applying the irradiance curve of Figure.14, the resulted output power is depicted in Figure.15. where the Inc-Cond outperforms the P&O in response time and the harvested power as clarified in the zoom-in of Figure.15 at different time periods. Besides, the efficiency analysis of the extracted power using P&O and Inc-Cond is equal to 81.13% and 89.78%, respectively. whereas, the efficiency of the transferred power using these techniques are 94.57% and 94.7%, respectively.

The analysis of the battery characteristics of Figure.16, shows identical voltage and SoC curves, whereas a small variation is indicated in the current curves leads to a variation in the battery power.

Figure 13. Battery current, voltage, power and SoC under STD conditions

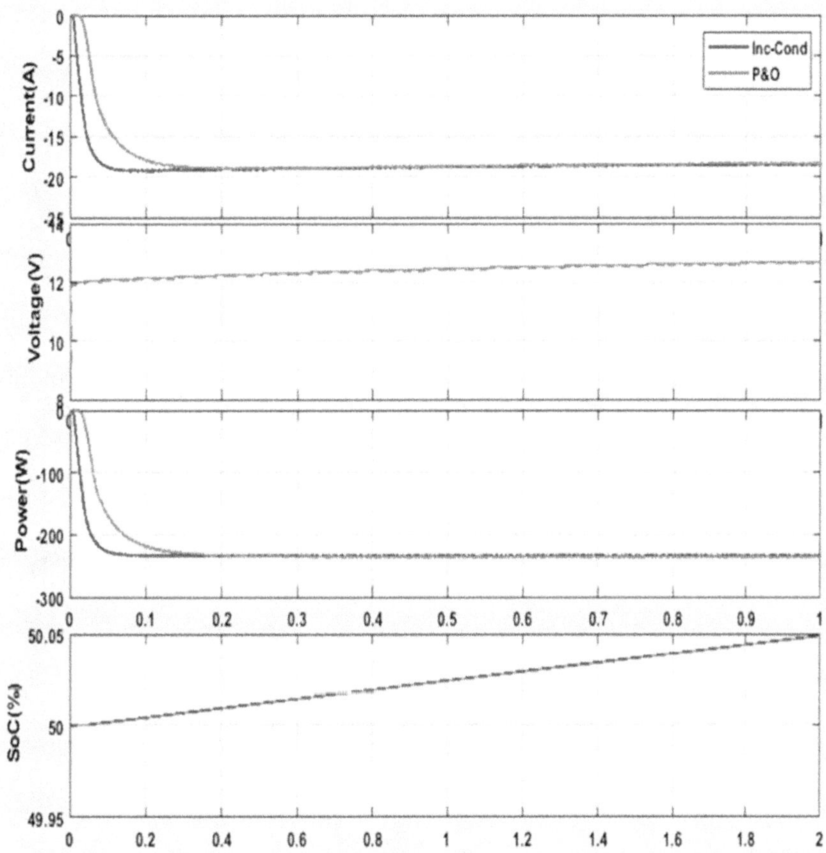

Simulation Under Over Battey Charge

In this case, the management algorithm will order the controller to switch OFF the battery since SoC > 85% and the power is sufficient to feed the SC. Clearly, the controller will switch from MPPT control mode to PI controller, at the same time, the switches separating the SC from the installation will close to supply it. According to Figure.17.a, the curves of the current follow the change of the irradiance curve shown in Figure.17.b, the currents follow the change in the irradiance level. Moreover, the SC current is proportional to the PV output current, while the battery current is equal to zero.

The efficiency of the extracted power is 82.69%, which is low compared to that obtained during the MPPT mode. whereas, the efficiency of the transferred power is equal to 96.21%.

Figure 14. The applied irradiance curve

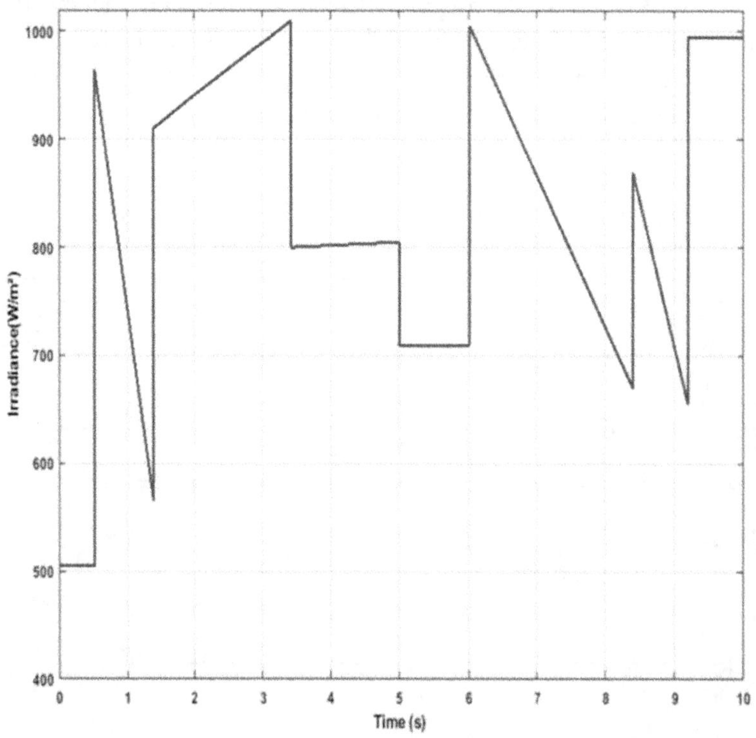

Figure 15. Harvested power under variable weather conditions using P&O and Inc-Cond

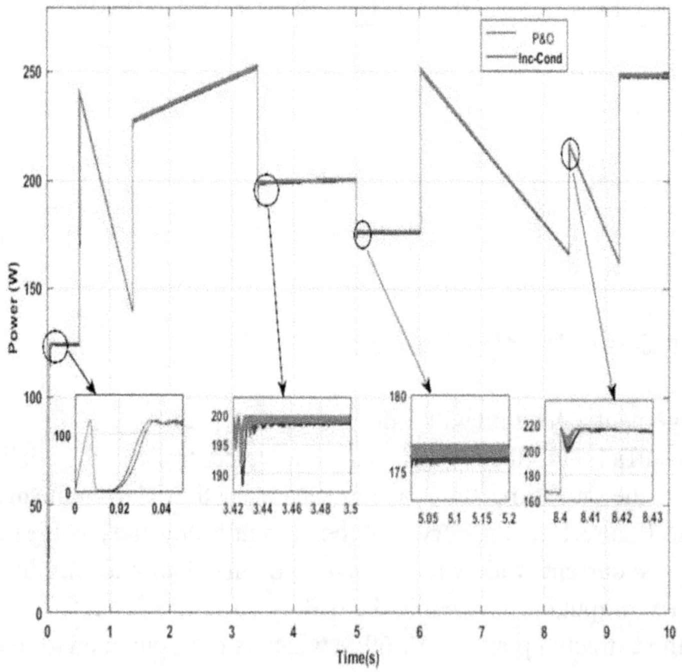

Figure 16. Battery current, voltage, power and SoC under variable conditions

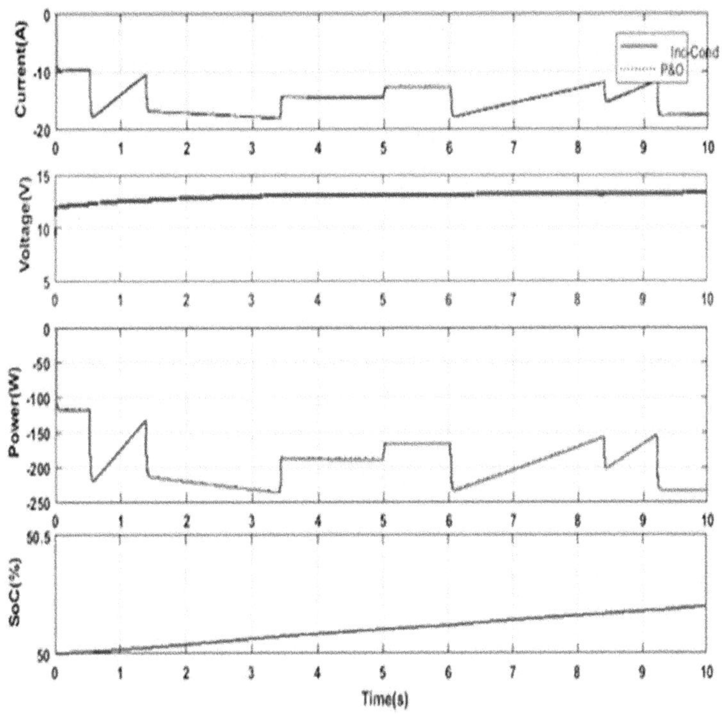

Figure 17. a. Applied irradiance curve, b. currents under over battery charge

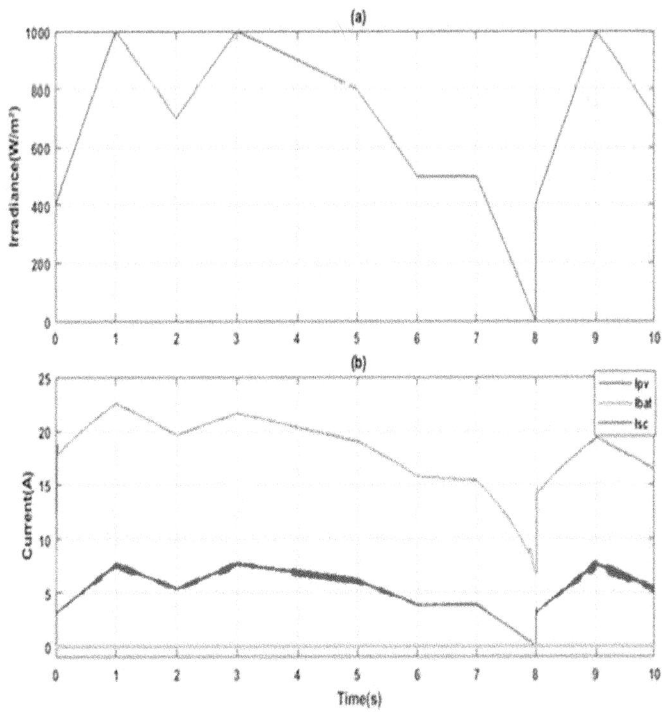

CONCLUSION

The design of the power components of the photovoltaic installation is important to extract and transfer the maximum power from the PV panel connected to the battery storage and SC through a buck converter. The power management strategy is given to control and protect the power flow and preserve the battery SoC within its limits under all applied conditions.

The P&O algorithm within the PI controller and the management algorithms provided maximum extracted power from the PV Panel. A detailed design of the installation for a standalone PV Battery Supercapacitor system. Also, the integration of the SC to store the surplus energy when the battery is fully charged is investigated. Several simulations have been affected to demonstrate the performance of the system under STD and irradiance change. Based on the obtained results, it can be concluded that the proposed system offers a reliable and alternative solution of renewable energy to charge the battery, and avoid its overcharge state by incorporating SC to store the surplus power and managing the whole system, surmounting any intermitting challenge and permitting an extensive increase of such systems mainly in islanded places where the grid is not available.

FUTURE RESEARCH DIRECTION

for the drawbacks meet during their operating modes. Moreover, the proposed system will be investigated integrating a hybrid storage system including other storage devices with buck converter and high power generation, then, test its performances and quality under variable weather conditions and load variations. Besides, The other dc-dc converter topologies i.e. boost and buck-boost converter with resistive, capacitive, inductive loads or grouping of these loads, will be also explored under the achievement of the designed control scheme, in order to, reduce faults and minimize the power dissipated during the transmission process, while undertaking the robustness and the stability and offering optimal, low cost and flexible system.

The delay in the control response time and problems meet during the operating of the converters can be mitigated by designing suitable and intelligent controllers such as fuzzy logic controller, neural network, and meta-heuristic techniques (i.e. particle swarm optimization and genetic algorithm) with decreased structure complexity and high efficacy to reduce harmonics in the switching devices. The proposed scheme can be also incorporated with an inverter to feed AC loads and AC electrical grid. Hybrid generation is also suitable to be explored within this scheme to ensure the balance power to load under all applied surroundings.

REFERENCES

Ahmad, T., & Zhang, D. (2020). A critical review of comparative global historical energy consumption and future demand: The story told so far. *Energy Reports*, *6*, 1973–1991. doi:10.1016/j.egyr.2020.07.020

Al-Barazanchi, S. A. M., & Vural, A. M. (2015, December). Modeling and intelligent control of a stand-alone PV-Wind-Diesel-Battery hybrid system. In *2015 International Conference on Control, Instrumentation, Communication and Computational Technologies (ICCICCT)* (pp. 423-430). IEEE.

Ali, A., Hasan, A. N., & Marwala, T. (2014, October). Perturb and observe based on fuzzy logic controller maximum power point tracking (MPPT). In *2014 International Conference on Renewable Energy Research and Application (ICRERA)* (pp. 406-411). IEEE.

Amara, Y., Boukenoui, R., Bradai, R., & Salhi, H. (2018, December). Design and Control of Two-Stage Standalone Photovoltaic Generation System. In *2018 International Conference on Communications and Electrical Engineering (ICCEE)* (pp. 1-5). IEEE.

Bagalini, V., Zhao, B. Y., Wang, R. Z., & Desideri, U. (2019). Solar PV-Battery-Electric Grid-Based Energy System for Residential Applications: System Configuration and Viability. *Research*, *2019*, 3838603.

Behjati, H., Davoudi, A., & Lewis, F. (2014). Modular DC–DC converters on graphs: Cooperative control. *IEEE Transactions on Power Electronics*, *29*(12), 6725–6741. doi:10.1109/TPEL.2014.2303192

Bhattacharjee, A. (2012). Design and comparative study of three photovoltaic battery charge control algorithms in MATLAB/SIMULINK environment. *International Journal of Advanced Computer Research, 2*(3), 129.

Chahwan, J., Abbey, C., Chamberland, M., & Joos, G. (2007, August). Battery storage system modelling, design and operation for wind energy integration in power systems. In *CIGRE Canada Conference on Power Systems* (pp. 1-8). Academic Press.

Chen, Y. Y., & Man, Y. K. (2009, June). Constant current-based maximum-power-point tracking for photovoltaic power systems. In *2009 Chinese Control and Decision Conference* (pp. 3422-3425). IEEE.

Coelho, R. F., Concer, F., & Martins, D. C. (2009, October). A study of the basic DC-DC converters applied in maximum power point tracking. In *2009 Brazilian power electronics conference* (pp. 673-678). IEEE.

Dhaouadi, G. U. I. Z. A., Djamel, O. U. N. N. A. S., Youcef, S. O. U. F. I., & Salah, C. H. E. N. I. K. H. E. (2019, September). Implementation of Incremental Conductance Based MPPT Algorithm for Photovoltaic System. In *2019 4th International Conference on Power Electronics and their Applications (ICPEA)* (pp. 1-5). IEEE.

El-Khozondar, H. J., El-Khozondar, R. J., Matter, K., & Suntio, T. (2016). A review study of photovoltaic array maximum power tracking algorithms. *Renewables: Wind, Water, and Solar*, *3*(1), 3. doi:10.118640807-016-0022-8

Etxeberria, A., Vechiu, I., Camblong, H., & Vinassa, J. M. (2010, October). Hybrid energy storage systems for renewable energy sources integration in microgrids: A review. In *2010 Conference Proceedings IPEC* (pp. 532-537). IEEE.

Ezzouitine, K., Boulezhar, A., & El Afou, Y. (2019). A cascaded H-bridge multilevel inverter with photovoltaic MPPT control. *Periodicals of Engineering and Natural Sciences*, *6*(2), 415–425.

Faria, J., Pombo, J., Calado, M. D. R., & Mariano, S. (2019). Power management control strategy based on artificial neural networks for standalone PV applications with a hybrid energy storage system. *Energies*, *12*(5), 902.

Glavin, M. E., Chan, P. K., Armstrong, S., & Hurley, W. G. (2008, September). A stand-alone photovoltaic supercapacitor battery hybrid energy storage system. In *2008 13th International power electronics and motion control conference* (pp. 1688-1695). IEEE.

Guasch, D., & Silvestre, S. (2003). Dynamic battery model for photovoltaic applications. *Progress in Photovoltaics: Research and Applications*, *11*(3), 193–206.

Guha, S., Yang, J., & Banerjee, A. (2000). Amorphous silicon alloy photovoltaic research—Present and future. *Progress in Photovoltaics: Research and Applications*, *8*(1), 141–150. doi:10.1002/(SICI)1099-159X(200001/02)8:1<141::AID-PIP305>3.0.CO;2-I

Hassan, M. S., Mughal, S. N., Jarial, R. K., & Sood, Y. R. (2019). A Comparative Analysis of Different Maximum Power Point Tracking Algorithms of Solar Photovoltaic System. In *Applications of Computing, Automation and Wireless Systems in Electrical Engineering* (pp. 217–229). Springer. doi:10.1007/978-981-13-6772-4_20

Jha, S. K., Bilalovic, J., Jha, A., Patel, N., & Zhang, H. (2017). Renewable energy: Present research and future scope of Artificial Intelligence. *Renewable & Sustainable Energy Reviews*, *77*, 297–317. doi:10.1016/j.rser.2017.04.018

Kasper, M., Bortis, D., & Kolar, J. W. (2013). Classification and comparative evaluation of PV panel-integrated DC–DC converter concepts. *IEEE Transactions on Power Electronics*, *29*(5), 2511–2526. doi:10.1109/TPEL.2013.2273399

Khadkikar, V., Varma, R. K., & Seethapathy, R. (2009, October). Grid voltage regulation utilizing storage batteries in PV solar—Wind plant based distributed generation system. In 2009 IEEE Electrical Power & Energy Conference (EPEC) (pp. 1-6). IEEE.

Kollimalla, S. K., & Mishra, M. K. (2014, April). Coordinated control and energy management of hybrid energy storage system in PV system. In *2014 International Conference on Computation of Power, Energy, Information and Communication (ICCPEIC)* (pp. 363-368). IEEE.

Konagai, M. (2011). Present status and future prospects of silicon thin-film solar cells. *Japanese Journal of Applied Physics*, *50*(3R), 030001. doi:10.1143/JJAP.50.030001

Kumar, P. (2018, April). Power Management in Microgrid Integrated Wind-PV-Battery Generation System. In *2018 International Conference on Power Energy, Environment and Intelligent Control (PEEIC)* (pp. 746-750). IEEE.

Kumar, P., Jain, G., & Palwalia, D. K. (2015, August). Genetic algorithm based maximum power tracking in solar power generation. In *2015 International Conference on Power and Advanced Control Engineering (ICPACE)* (pp. 1-6). IEEE.

Kyriakopoulos, G. L., & Arabatzis, G. (2016). Electrical energy storage systems in electricity generation: Energy policies, innovative technologies, and regulatory regimes. *Renewable & Sustainable Energy Reviews*, *56*, 1044–1067.

Lee, J., Jeong, S., Han, Y. H., & Park, B. J. (2010). Concept of cold energy storage for superconducting flywheel energy storage system. *IEEE Transactions on Applied Superconductivity*, *21*(3), 2221–2224.

Li, Q., & Wolfs, P. (2008). A review of the single phase photovoltaic module integrated converter topologies with three different DC link configurations. *IEEE Transactions on Power Electronics*, *23*(3), 1320–1333. doi:10.1109/TPEL.2008.920883

Lofthouse, J., Simmons, R. T., & Yonk, R. M. (2015). *Reliability of renewable energy: Solar.* Institute of Political Economy, Utah State University.

Luo, X., Wang, J., Dooner, M., & Clarke, J. (2015). Overview of current development in electrical energy storage technologies and the application potential in power system operation. *Applied Energy*, *137*, 511–536.

Mahmoud, M. M. (2006). Transient analysis of a PV power generator charging a capacitor for measurement of the I–V characteristics. *Renewable Energy*, *31*(13), 2198–2206. doi:10.1016/j.renene.2005.09.019

Masoum, M. A., Dehbonei, H., & Fuchs, E. F. (2002). Theoretical and experimental analyses of photovoltaic systems with voltage and current-based maximum power-point tracking. *IEEE Transactions on Energy Conversion*, *17*(4), 514–522.

Mirjalili, S., Gandomi, A. H., Mirjalili, S. Z., Saremi, S., Faris, H., & Mirjalili, S. M. (2017). Salp Swarm Algorithm: A bio-inspired optimizer for engineering design problems. *Advances in Engineering Software*, *114*, 163–191.

Moghaddam, S., Bigdeli, M., Moradlou, M., & Siano, P. (2019). Designing of stand-alone hybrid PV/wind/battery system using improved crow search algorithm considering reliability index. *International Journal of Energy and Environmental Engineering*, *10*(4), 429–449.

Naresh, M., & Tripathi, R. K. (2019, March). *Intelligent control strategy for power management in hybrid renewable energy system. In 2019 Innovations in Power and Advanced Computing Technologies (i-PACT)* (Vol. 1). IEEE.

Nguyen, D., & Lehman, B. (2008). An adaptive solar photovoltaic array using model-based reconfiguration algorithm. *IEEE Transactions on Industrial Electronics*, *55*(7), 2644–2654. doi:10.1109/TIE.2008.924169

Pathak, P. K., & Yadav, A. K. (2019). Design of battery charging circuit through intelligent MPPT using SPV system. *Solar Energy*, *178*, 79–89.

Renaudineau, H., Donatantonio, F., Fontchastagner, J., Petrone, G., Spagnuolo, G., Martin, J. P., & Pierfederici, S. (2014). A PSO-based global MPPT technique for distributed PV power generation. *IEEE Transactions on Industrial Electronics*, *62*(2), 1047–1058.

Salmi, H., Badri, A., & Zegrari, M. (2016). Maximum Power Point Tracking (MPPT) Using Artificial Bee Colony Based Algorithm for Photovoltaic System. *Int J Intell Inf Syst*, *5*, 1–4.

Sheina, S. G., Pirozhnikova, A. P., & Priss, E. A. (2019, December). The concept of sustainable development of renewable energy in the modern world. *IOP Conference Series. Materials Science and Engineering*, *698*(5), 055010. doi:10.1088/1757-899X/698/5/055010

Tak, S. Z., & Garg, V. (2018). Simulation of Grid Supported PV/Wind/Battery Hybrid System with Variable Load. *2018 Fifth International Conference on Parallel, Distributed and Grid Computing (PDGC)*, 627-632.

Tsang, K. M., & Chan, W. L. (2011). Current sensorless quick charger for lithium-ion batteries. *Energy Conversion and Management*, *52*(3), 1593–1595.

Wang, G., Ciobotaru, M., & Agelidis, V. G. (2014). Power smoothing of large solar PV plant using hybrid energy storage. *IEEE Transactions on Sustainable Energy*, *5*(3), 834–842.

Yasmine, A., Rafik, B., Rachid, B., & Adel, M. (2020a). Grid connected photovoltaic system efficiency and quality improvement using fuzzy-incond MPPT. *International Journal of Power Electronics and Drive Systems*, *11*(3).

Yasmine, A., Rafik, B., Rachid, B., & Adel, M. (2020b, May). Management of Hybrid PV/Battery Standalone System under Partial Shading Conditions. In *1st International Conference on Communications, Control Systems and Signal Processing (CCSSP)* (pp. 429-434). IEEE.

Zhou, H., Bhattacharya, T., Tran, D., Siew, T. S. T., & Khambadkone, A. M. (2010). Composite energy storage system involving battery and ultracapacitor with dynamic energy management in microgrid applications. *IEEE Transactions on Power Electronics*, *26*(3), 923–930.

Zhu, Y., Zhuo, F., & Shi, H. (2013, June). *Power management strategy research for a photovoltaic-hybrid energy storage system. In 2013 IEEE ECCE Asia Downunder*. IEEE.

Chapter 3
True Power Loss Reduction by Enhanced Tree Squirrel Search, Enhanced Salp Swarm, and Swim Bladder Operation-Based Shark Optimization Algorithms

Lenin Kanagasabai
Prasad V. Potluri Siddhartha Institute of Technology, India

ABSTRACT

In this chapter, enhanced tree squirrel search optimization algorithm (ETSS), enhanced salp swarm algorithm (ESS), and swim bladder operation-based shark algorithm (SBS) have been applied to solve the power loss reduction problem. Enhanced tree squirrel search optimization algorithm (ETSS) utilizes the jumping exploration method and progressive exploration technique—both possess winter search strategy—in order to preserve the population diversity and to perk up the convergence speed. A new-fangled winter exploration strategy is implemented in the jumping exploration technique. In enhanced salp swarm algorithm (ESS) an inertia weight ω∈ [0, 1] is applied, which picks up the pace of convergence during the period of exploration. Then swim bladder operation-based shark algorithm (SBS) is proposed to solve the problem. Based on contracting and expanding actions of the swim bladder in shark, SBS algorithm has been modelled.

INTRODUCTION

Reactive power problem plays a key role in secure and economic operations of power system. Many computational techniques are utilized to solve the engineering problems. (Azar & Vaidyanathan, 2015a,b,c, 2016; Azar & Zhu, 2015; Azar & Serrano, 2015a,b,c,d, 2016a,b, 2017; Boulkroune et al, 2016a,b; Ghoudelbourk et al., 2016; Meghni et al, 2017a,b,c; Azar et al., 2017a,b,c,d; Azar 2010a,b, 2012; Mekki et al., 2015; Vaidyanathan & Azar, 2015a,b,c,d, 2016a,b,c,d,e,f,g, 2017a,b,c; Zhu & Azar, 2015;

DOI: 10.4018/978-1-7998-7447-8.ch003

Grassi et al., 2017; Ouannas et al., 2016a,b, 2017a,b,c,d,e,f,g,h,I,j; Singh et al., 2017; Vaidyanathan et al, 2015a,b,c; Wang et al., 2017; Soliman et al., 2017; Tolba et al., 2017). At first in this chapter, Enhanced Tree Squirrel Search Optimization Algorithm (ETSS) has been applied to solve optimal reactive power problem. Tree squirrel dynamic foraging tactic and gliding process are designed scientifically to solve the optimal reactive power problem. Exploration process commence when tree squirrels start the foraging. In the period of autumn season squirrels explore for food by gliding from one tree to another and it will modify their location and determine different areas of the forest. Projected Enhanced Tree Squirrel Search Optimization Algorithm utilizes the jumping exploration method and progressive exploration technique, both possess winter search strategy ; in primary evolutionary stage condition is $S_c^t \geq S_{min}$; summer exploration stratagem for the later evolutionary phase condition is $S_c^t < S_{min}$. Newfangled winter exploration strategy is implemented in the jumping exploration technique. With reference to the summer modernizing technique, only safe individuals which progress towards Ta are arbitrarily displaced; remaining will continue at their modernized positions devoid of any alteration. Then this paper proposes Enhanced Salp Swarm Algorithm (ESS) to solve optimal reactive power problem. Salp swarm algorithm replicate mechanism of foraging of salps (swarming) in oceans. Modelling of algorithm is based on salp chain and front of chain leader salp will be there and then remaining salps are followers. In the projected Enhanced Salp Swarm Algorithm (ESS) an inertia weight $\omega \in [0,1]$ is applied to perk up the pace of convergence during exploration. Both the exploration and exploitation are balanced by which the optimal solution is attained. Then this paper proposes Swim bladder operation based shark algorithm (SBS) to solve optimal reactive power problem. Function of the swim bladder in shark is perfectly imitated to model the algorithm. When shark rise towards the surface swim bladder will be in expanded mode, consequently it greatly diminish the density of the shark. Due to the Buoyancy force acting on the shark it pushes it in upward direction and also through that the shark can alter its altitude and then touches the surface of the water. Correspondingly, by contracting swim bladder at any time the shark will move down towards the base of the water, then, accordingly it will escalate the density of the shark which permits the gravity to obtain it closer to the base of the water. Fittest shark will possess the volume "0" and the flabbiest shark will possess the volume "1". Scalar value "x" will be multiplied by the velocity of the smallest amount fit shark by "x" and fix the velocity of the fittest shark value as "0". Proposed Enhanced Tree Squirrel Search Optimization Algorithm (ETSS), Enhanced Salp Swarm Algorithm (ESS) and Swim bladder operation based shark algorithm (SBS) has been tested in standard IEEE 14, 30 and 300 bus test systems and outcomes show the projected ETSS, ESS and SBS algorithms condensed the factual power loss proficiently.

This chapter is organized as follows: In section 2 problem formuation is introduced. In section 3, Enhanced Tree Squirrel Search Optimization Algorithm (ETSS) is described. In section 4, Enhanced Salp Swarm Algorithm (ESS) is presented. In section 5, Swim bladder operation based shark algorithm (SBS) has beendescribed. Results and simulation are shown in section 6. Finally in section 7, concluding remarks with future directions are given.

Problem Formulation

Mathematical formulation of Factual power loss diminution is defined as,

$$F = P_L = \sum_{k \in Nbr} g_k \left(V_i^2 + V_j^2 - 2V_i V_j \cos\theta_{ij} \right) \tag{1}$$

$$\mathrm{F} = \mathrm{P_L} + \omega_v \times \text{Voltage Deviation} \tag{2}$$

Then

$$\text{Voltage Deviation} = \sum_{i=1}^{Npq} |V_i - 1| \tag{3}$$

Equality constraints is defined as,

$$\mathrm{P_G} = \mathrm{P_D} + \mathrm{P_L} \tag{4}$$

In-Equality constraints is defined as,

$$P_{gslack}^{\min} \le P_{gslack} \le P_{gslack}^{\max} \tag{5}$$

$$Q_{gi}^{\min} \le Q_{gi} < Q_{gi}^{\max},\ i \in N_g \tag{6}$$

$$V_i^{\min} \le V_i \le V_i^{\max},\ i \in N \tag{7}$$

$$T_i^{\min} \le T_i \le T_i^{\max},\ i \in N_T \tag{8}$$

$$Q_c^{\min} \le Q_c \le Q_C^{\max},\ i \in N_C \tag{9}$$

ENHANCED TREE SQUIRREL SEARCH OPTIMIZATION ALGORITHM

Tree squirrel dynamic foraging tactic and gliding process are designed scientifically to solve the optimal reactive power problem. Exploration process commence when tree squirrels start the foraging. In the period of autumn season squirrels explore for food by gliding from one tree to another and it will modify their location and determine different areas of the forest.

In forest there will be "n" number of Tree squirrels (TS) and position of i_{th} Tree squirrel can be defined by a vector. Tree squirrels location is indicated by,

$$TS = \begin{bmatrix} TS_{1,1} & \cdots & TS_{1,d} \\ \vdots & \ddots & \vdots \\ TS_{n,1} & \cdots & TS_{n,d} \end{bmatrix} \tag{10}$$

In the forest preliminary position of each Tree squirrel is defined by,

$$TS_i = TS_L + U(0,1) \times (TS_U - TS_L) \tag{11}$$

For every Tree squirrel location fitness value is computed by,

$$f = \begin{bmatrix} \left| f_1\left(\left|TS_{1,1}, TS_{1,2}, .., TS_{1,d}\right|\right)\right| \\ \left| f_2\left(\left|TS_{2,1}, TS_{2,2}, .., TS_{2,d}\right|\right)\right| \\ \vdots \\ \left| f_n\left(\left|TS_{n,1}, TS_{n,2}, .., TS_{n,d}\right|\right)\right| \end{bmatrix} \tag{12}$$

Towards hickory nut tree, Tree squirrels will move from the acorn nut trees ($TS_{\alpha)}$ and new-fangled location of squirrels can be acquired by,

$$TS_{\alpha t}^{t+1} = \begin{cases} TS_{\alpha t}^{t+1} = TS_{\alpha t}^{t} + d_g \times G_c \times \left(Ts_{ht}^{t} - TS_{\alpha t}^{t}\right) R_1 \geq P_{dp} \\ \quad Random\ location \qquad otherwise \end{cases} \tag{13}$$

Towards acorn nut trees Tree squirrels will move from normal trees (TS_{nt}) may move to accomplish their everyday energy needs. Then the new-fangled location of squirrels is given by,

$$TS_{nt}^{t+1} = \begin{cases} TS_{nt}^{t+1} = TS_{nt}^{t} + d_g \times G_c \times \left(Ts_{\alpha t}^{t} - TS_{nt}^{t}\right) R_2 \geq P_{dp} \\ \quad Random\ location \qquad otherwise \end{cases} \tag{14}$$

In order to accumulate hickory nuts at the time of food shortage some squirrels which are in the normal trees which already consumed acorn nuts possibly progress towards hickory nut tree and new –fangled location is given by,

$$TS_{nt}^{t+1} = \begin{cases} TS_{nt}^{t+1} = TS_{nt}^{t} + d_g \times G_c \times \left(Ts_{ht}^{t} - TS_{nt}^{t}\right) R_3 \geq P_{dp} \\ \quad Random\ location \qquad otherwise \end{cases} \tag{15}$$

Tree squirrel gliding at steady and the frictional drag is defined as,

$$D = 1/2 \rho V^2 S C_D \tag{16}$$

At steady state glide angle is defined as,

$$\Phi = \arctan\left(\frac{D}{L}\right) \tag{17}$$

Gliding distance (dg) approximately computed by,

$$d_g = \left(\frac{h_g}{\tan\Phi}\right) \tag{18}$$

Foraging action of Tree squirrels is significantly affected by the Seasonal variations and it's defined by,

$$S_c^t = \sqrt{\sum_{k=1}^{d}\left(TS_{at,k}^t - TS_{ht,k}\right)^2} \tag{19}$$

Test the seasonal monitoring condition; $S_c^t < S_{minimum}$ and the value of $S_{minimum}$ is computed by.

$$S_{minimum} = \frac{10E^{-6}}{(365)^t / \left(t_m / 2.5\right)} \tag{20}$$

At the end of winter season repositioning of Tree squirrels is modelled by,

$$Ts_{nt}^{new} - Ts_L + Levy(n)\times\left(TS_U - TS_L\right) \tag{21}$$

Levy flight is a rank of non-Gaussian random procedure whose capricious walks are haggard from Levy stable distribution. Allocation by $L(s) \sim |s|^{-1-\beta}$ where $0 < \beta < 2$ is an index. Scientifically defined as,

$$L(s,\gamma,\mu) = \begin{cases} \sqrt{\frac{\gamma}{2\pi}} \quad exp\left[-\frac{\gamma}{2(s-\mu)}\right]\frac{1}{(s-\mu)^{3/2}} & if\ 0<\mu<s<\infty \\ 0 \ if\ s\le 0 \end{cases} \tag{22}$$

In terms of Fourier transform Levy distribution defined as

$$F(k) = exp\left[-\alpha|k|^{\beta}\right],\ 0<\beta\le 2, \tag{23}$$

Fresh state is calculated as,

$$X^{t+1} = X^t + \alpha \oplus Levy(\beta) \tag{24}$$

$$X^{t+1} = X^t + random\left(size\left(D\right)\right) \oplus Levy\left(\beta\right) \tag{25}$$

Non-trivial scheme of engendering step size by,

$$X^{t+1} = X^t + random\left(size\left(D\right)\right) \oplus Levy\left(\beta\right) \sim 0.01 \frac{u}{|v|^{1/\beta}} \left(x_j^t - gb\right) \tag{26}$$

$$u \sim N\left(0, \sigma_u^2\right) \quad v \sim N\left(0, \sigma_v^2\right) \tag{27}$$

With

$$\sigma_u = \left\{ \frac{\Gamma\left(1+\beta\right) \sin\left(\pi\beta / 2\right)}{\Gamma\left[\left(1+\beta\right)/2\right] \beta 2^{(\beta-1)/2}} \right\}^{1/\beta}, \quad \sigma_v = 1 \tag{28}$$

Tree Squirrel Search Optimization Algorithm

1. Commence
2. Describe the input parameters
3. Engender capricious locations for n number of Tree squirrels using $TS_i = TS_L + U(0,1) \times (TS_U - TS_L)$
4. For every Tree squirrel's location Calculate fitness value
5. In uphill order arrange the locations of Tree squirrels based on their fitness value
6. Affirm the Tree squirrels on common trees, acorn nuts trees and hickory nut tree
7. Arbitrarily choose a number of Tree squirrels which are on the common trees to move on the way to hickory nut tree and the left over will move in the direction of acorn nuts trees
8. While (end criterion is not satisfied)
9. For t = 1 to n1 (n1 = sum of Tree squirrels which are on the acorn trees and stirring towards hickory nut tree)
10. If $R_1 \geq P_{dp}$

$$TS_{at}^{t+1} = TS_{at}^t + d_g \times G_c \times \left(Ts_{ht}^t - TS_{at}^t\right)$$

11. Else

$Fs_{at}^{t+1} = a$ capricious position of search space

12. End
13. For t = 1 to n2 (n2 = sum of Tree squirrels in normal trees and moving towards acorn trees)
14. If $R_2 \geq P_{dp}$

$$TS_{nt}^{t+1} = TS_{nt}^{t} + d_g \times G_c \times \left(Ts_{at}^{t} - TS_{nt}^{t}\right)$$

15. Else
16. $Fs_{nt}^{t+1} = a$ capricious position of search space
17. End
18. For t = 1 to n3 (n3 = sum of Tree squirrels in normal trees and moving towards hickory nut tree)
19. If $R_3 \geq P_{dp}$

$$TS_{nt}^{t+1} = TS_{nt}^{t} + d_g \times G_c \times \left(Ts_{ht}^{t} - TS_{nt}^{t}\right)$$

20. Else
21. $Ts_{nt}^{t+1} = a$ capricious position of search space
22. End
23. Compute the seasonal constant (Sc)
24. If (Seasonal observing condition is satisfied)
25. Arbitrarily reposition Tree squirrels using $Ts_{nt}^{new} = Ts_L + Levy(n) \times \left(TS_U - TS_L\right)$
26. End
27. Modernize the minimum value of seasonal constant (S_{min}) using $S_{minimum} = \dfrac{10E^{-6}}{365^{t/(t_m/2.5)}}$
28. End
29. Final optimal solution is position of squirrel on hickory nut tree
30. End

Projected Enhanced Tree Squirrel Search Optimization Algorithm (ETSS) utilizes the jumping exploration method and progressive exploration technique, both possess winter search strategy ; in primary evolutionary stage condition is $S_c^t \geq S_{min}$; summer exploration stratagem for the later evolutionary phase condition is $S_c^t < S_{min}$.

In order to preserve the population diversity and to perk up the convergence speed, a new-fangled winter exploration strategy implemented in the jumping exploration technique.When $r \geq P_{dp}$, TS_i in secure mode, the position is modernized by,

$$TS_i^{t+1} = TS_i^t + d_g \times G_c \times \left(T_h^t - TS_i^t\right) \tag{29}$$

When $r<P_{dp}$, TS_i in dying out mode, and the new-fangled one is computed by,

$$TS_i^{t+1} = L + random(1,D) \times \left(U - L\right) \tag{30}$$

Those individuals are under threat will escape & search for new-fangled position by,

$$R = \frac{U - L}{2} \tag{31}$$

$$TS_j^{new} = TS_j^t + d_g \times G_c \times \left(T_{ai}^t - TS_j^t\right) - d_g \times G_c \times \left(TS_i^t - TS_j^t\right)\ i = 1,2,..,N_{fs} \tag{32}$$

$$TS_j^{t+1} = TS_j^{new} \times \left(0.5 + random\right) \tag{33}$$

With reference to the summer modernizing technique, only safe individuals which progress towards Ta are arbitrarily displaced; remaining will continue at their modernized positions devoid of any alteration,When $r \geq P_{dp}$, TS_i in secure mode, the position is modernized by,

$$TS_i^{t+1} = T_h^t + d_g \times G_c \times \left(T_h^t - TS_i^t\right) \tag{34}$$

When $r < P_{dp}$, TS_i in dying out mode, and the new-fangled one is computed by,

$$TS_i^{t+1} = TS_i^t \times \left(0.5^{(N(0,1))}\right) \tag{35}$$

$$TS_i^{t+1} = TS_i^t \times \left(0.5 + random\right) \tag{36}$$

1. Input: population
2. Output: population new
3. *when* $S_c^t \geq S_{min}$: Season = winter or else Season = summer
4. End
5. If Season = = winter
6. For P_1=1 to population size
7. When $r \geq P_{dp}$
8. $TS_i^{t+1} = TS_i^t + d_g \times G_c \times \left(T_h^t - TS_i^t\right)$
9. End
10. When $r < P_{dp}$
11. $TS_i^{t+1} = L + random\left(1, D\right) \times \left(U - L\right)$
12. For P_2=1 to nt_1 (whole number of Tree Squirrel under threat)
13. $TS_j^{new} = TS_j^t + d_g \times G_c \times \left(T_{ai}^t - TS_j^t\right) - d_g \times G_c \times \left(TS_i^t - TS_j^t\right)\ i = 1,2,..,N_{fs}$
14. $TS_j^{t+1} = TS_j^{new} \times \left(0.5 + random\right)$
15. End
16. If Season = = summer
17. For P_1=1 to population size

When $r \geq P_{dp}$
$TS_i^{t+1} = T_h^t + d_g \times G_c \times \left(T_h^t - TS_i^t\right)$
End
When $r < P_{dp}$
$TS_i^{t+1} = TS_i^t \times \left(0.5^{(N(0,1))}\right)$
For P_2=1 to nt_2 (whole number of dead Tree Squirrel)
$TS_i^{t+1} = TS_i^t \times \left(0.5 + random\right)$
End
End

Progressive search method is designed to improve the robustness of the algorithm. When $r \geq P_{dp}$ in secure mode, the position is modernized by,

$$TS_i^{t+1} = TS_h^t + Levy(x) \times \left(T_h^t - TS_i^t\right) \tag{37}$$

When $r < P_{dp}$ in dying out mode, choose a dimension arbitrarily and transmute it in the range of L and U.

Input: population
Output: population new
$when\ S_c^t \geq S_{min}$
When $r \geq P_{dp}$
Season = winter
Or elsc
Season = summer
End
When Season = = winter
For P_1=1 to population size
$TS_i^{t+1} = TS_i^t + d_g \times G_c \times \left(T_h^t - TS_i^t\right)$
End
When $r < P_{dp}$
Choose one dimension arbitrarily and alter it between TS_L and TS_H
End
If Season = = summer
For P_1=1 to population size
When $r \geq P_{dp}$
$TS_i^{t+1} = TS_h^t + Levy(x) \times \left(T_h^t - TS_i^t\right)$
End

21. When $r<P_{dp}$
22. Choose a dimension arbitrarily and transmute it in the range of L and U.
23. End
24. End

Outline of Projected Enhanced Tree Squirrel Search Optimization Algorithm (ETSS) for solving optimal reactive power problem

1. Input: pop
2. Output: f_{best}
3. For t = 1 to T (T is the entire generation of the algorithm to be executed) assess the fitness values of the population to modernize the population by jumping exploration method
4. When t == T/n (n -entire sub phase of the entire optimization) compute the equivalent linear regression equations
5. When two or more designed slopes are positive then maintain optimizing by progressive exploration method
6. Else
7. Continue the optimization through the jumping exploration method
8. End

ENHANCED SALP SWARM ALGORITHM

Mechanism of salps when foraging in oceans has been imitated to form the Salp swarm algorithm. In oceans salp chain will be formed and leader salp will be in front and remaining salps are followers. In an s-dimensional exploration space Position of salps is described, "s" indicate is the number of variables. Position of the leader salp will be changed by,

$$SS_n^1 = \begin{cases} Ps_n + r_1\left(\left(U_n - l_n\right)r_2 + l_n\right)r_3 \geq 0 \\ Ps_n - r_1\left(\left(U_n - l_n\right)r_2 + l_n\right)r_3 < 0 \end{cases} \tag{38}$$

$$r_1 = 2e^{-}\left(-\frac{4a}{A}\right)^2 \tag{39}$$

In Salp swarm algorithm "r_1" maintains the equilibrium between the exploration, & exploitation. Followers position are changed by,

$$SS_n^m = \frac{1}{2}ce^2 + v_o e \tag{40}$$

Where $m \geq 2; c = \frac{v_{final}}{v_o}; SS = \frac{SS - SS_o}{e}$

$$SS_n^m = \frac{1}{2}\left(SS_n^m + SS_n^{m-1}\right) \tag{41}$$

In the projected Enhanced Salp Swarm Algorithm (ESS) an inertia weight $\omega \in [0,1]$ used to perk up the convergence rate during exploration. Then Equations (38) and (39) are rewritten by

$$SS_n^1 = \begin{cases} \omega Ps_n + r_1\left(\left(U_n - l_n\right)r_2 + l_n\right) r_3 \geq 0 \\ \omega Ps_n - r_1\left(\left(U_n - l_n\right)r_2 + l_n\right) r_3 < 0 \end{cases} \tag{42}$$

$$SS_n^m = \frac{1}{2}\left(SS_n^m + \omega SS_n^{m-1}\right) \tag{43}$$

$$SS_n^m = \begin{cases} 1 \;\; if \;\; Y_{mn} > 0.5 \\ 0 \;\;\; otherwise \end{cases} \tag{44}$$

Fitness vale calculation done by,

$$Fitness = \rho Err(D) + \varphi \frac{|F|}{|T|} \tag{45}$$

Err(D) - classifier error rate ; |F| - size of identified feature subset,; |T| - entire number of features; $\rho \in [0,1]$; $\varphi = (1-\rho)$; $\rho = 0.9$.

1. Parameters are initialized
2. Population of "S slap's positions are arbitrarily initialized
3. Fitness value of each slap is computed
4. Number of iterations is fixed
5. Modernize the parameter (r_1) through $r_1 = 2e^{-}\left(-\frac{4a}{A}\right)^2$
6. For each salp,
   (a) when m==1, then modernize the position of the principal salp by

$$SS_n^1 = \begin{cases} Ps_n + r_1\left(\left(U_n - l_n\right)r_2 + l_n\right) r_3 \geq 0 \\ Ps_n - r_1\left(\left(U_n - l_n\right)r_2 + l_n\right) r_3 < 0 \end{cases}$$

(b) Or else, modernize the position of the follower salp by $SS_n^m = \frac{1}{2}\left(SS_n^m + SS_n^{m-1}\right)$
(c) Fitness value of every salp will be evaluated.
(d) Modernize Salp- when better solution available.

7. When end criterion is satisfied stop the procedure

SWIM BLADDER OPERATION BASED SHARK ALGORITHM

Swim bladder operation based shark algorithm (SBS) has been modelled based on the – contracted, expand actions of the swim bladder in shark. Exploration to certain areas is done through localization routine search in order to balance the explorative-exploitative of the search. Shark will rise towards the surface, when swim bladder will expand then the density of the shark reduced. By Buoyancy force shark will be pushed in upward direction and by changing the altitude it can touch the surface of the water. When the swim bladder is contracted, then shark will move down towards the base of the water, it escalates the density of the shark and permit the gravity to fetch it closer to the base of the water.

Vertical axis portray the altitude of the Water Column as well as the input parameter of, f(x); Fitness function f(x) can be written as,

$$f(x) = -x \cdot sin\left(\sqrt{|x|}\right) \tag{46}$$

Downward force -F_w ; upward force F_b which operate on the object of mass- "m", volume - "v" submerge in a liquid of density-"ρ" is given by;

$$F_w = m.g \tag{47}$$

$$F_b = v.\rho.g \tag{48}$$

When two forces operate in opposite directions, then the resultant force can be computed by ;

$$F = F_b - F_w \tag{49}$$

Product of mass "m" and acceleration "a" also gives the net force by;

$$m.a = v.\rho.g - m.g \tag{50}$$

When an object submerged in a fluid then the unit of density becomes

$$a = (v-1) \cdot g \tag{51}$$

With respect to acceleration "t" the equation (51) has been written as follows,

$$a \cdot t = (v-1) \cdot g.t \quad (52)$$

$$\text{New\_fangled velocity} = \text{present\_velocity} + a \cdot t \quad (53)$$

Velocity of the shark is updated by,

$$\text{New\_fangled velocity} = \text{present\_velocity} + (v-1) \cdot g.t \quad (54)$$

The value of $g.t$ equal to the present velocity of the object, therefore;

$$\text{New\_fangled velocity} = \text{present\_velocity} + (v-1) \cdot \text{present\_velocity} \quad (55)$$

$$\Rightarrow \text{New\_fangled velocity} = \text{present\_velocity} \cdot v \quad (56)$$

Fittest shark will possess the volume “0” and the flabbiest shark will possess the volume “1”. Scalar value “x” will multiply the velocity of the smallest amount fit shark by “x” and fix the velocity of the fittest shark value as “0” . Volume “v” of shark “j” in the iteration “i” is described as:

$$v_{ji} = x.\frac{maximum\ fitness_i - fitness\ function(position_j)}{maximum\ fitness_i - minimum\ fitness_i} \quad (57)$$

$$V_{ji} = U_{ji} \cdot x.\frac{maximum\ fitness_i - fitness\ function(position_j)}{maximum\ fitness_i - minimum\ fitness_i} \quad (58)$$

V_{ji}, U_{ji} - new-fangled and present velocities of the shark” j”

1. Initialize the parameters
2. For each $j \in N$ do
3. $Fitness_j \leftarrow Fitness\ function(position_j)$
4. End
5. For each iteration do
6. $maximum\ fitness_i \leftarrow maximum(Fitness_1, Fitness_2, .., Fitness_N)$
7. $minimum\ fitness_i \leftarrow minimum(Fitness_1, Fitness_2, .., Fitness_N)$
8. For each $j \in N$ do
9. $V_j \leftarrow V_j \cdot x.\frac{maximum\ fitness_i - fitness\ function(position_j)}{maximum\ fitness_i - minimum\ fitness_i}$

If ($V_j \neq 0$) and $\left(\left|V_j\right| < k\right)$ then

$V_j \leftarrow k.\widehat{V_j}$

End

For each dimension "D" do

$position_{jD} \leftarrow position_{jD} + V_{jD} \cdot trans\ time$

if ($position_{jd} > D_{maximum}$) or ($position_{jD} > D_{minimum}$)

Then

$position_{jD} \leftarrow$ Arbitrary($D_{maximum}, D_{minimum}$)

End

$Fitness_j \leftarrow Fitness\ function(position_j)$

If $maximum\ fitness_i < maximum\left(Fitness_1, Fitness_2, .., Fitness_N\right)$ then

For each $j \in N$ do

If $Fitness_j == maximum\ fitness_i$ then

$V_j \leftarrow k$. Arbitrary velocity

End

SIMULATION RESULTS

In standard IEEE 14 bus system the validity of the projected Enhanced Tree Squirrel Search Optimization Algorithm (ETSS), Enhanced Salp Swarm Algorithm (ESS) and Swim bladder operation based shark algorithm (SBS) has been tested. Table 1 show the limits and comparison of results are presented in Table 2. Then the proposed Enhanced Tree Squirrel Search Optimization Algorithm (ETSS), Enhanced Salp Swarm Algorithm (ESS) and Swim bladder operation based shark algorithm (SBS) simulated in IEEE 30 Bus system. Table 3 shows the comparison results. Finally ETSS, ESS and SBS assessed in IEEE 300 bus system and comparison shown in table 4. Figures 1 to 3 shows the loss comparison. Projected ETSS, ESS and SBS has been compared with standard particle swarm optimization, modified particle swarm optimization, standard Evolutionary programming, Enriched genetic algorithm, faster evolutionary algorithm and standard cuckoo search algorithm. Power loss diminution has been attained.

CONCLUSION

In this chapter Enhanced Tree Squirrel Search Optimization Algorithm (ETSS), Enhanced Salp Swarm Algorithm (ESS) and Swim bladder operation based shark algorithm (SBS) successfully solved the optimal reactive power problem. In ETSS New-fangled winter exploration strategy is implemented in the jumping exploration technique. With reference to the summer modernizing technique, only safe individuals which progress towards Ta are arbitrarily displaced; remaining will continue at their modernized positions devoid of any alteration. Then in Enhanced Salp Swarm Algorithm (ESS) an inertia weight $\omega \in [0,1]$ applied to improve the performance of the algorithm. In the algorithm both the exploration and exploitation are balanced through that optimal solution will be attained. In Swim bladder operation based shark algorithm (SBS) Buoyancy force acting on the shark will push it in upward direction by that shark will change its altitude and touches the surface of the water. Proposed Enhanced Tree Squirrel Search Optimization Algorithm (ETSS), Enhanced Salp Swarm Algorithm (ESS) and Swim bladder operation based shark algorithm (SBS) has been verified in IEEE 14, 30 and 300 bus test systems and actual power loss has been reduced competently. In future proposed algorithms can be applied to large and practical systems.

Table 1. Limits

Parameter	Min (PU)	Max (PU)
Generator Voltage	0.95	1.1
Transformer Tap	0.9	1.1
VAR Source	0	0.20
Bus	Q Min (PU)	Q Max(PU)
1	0	10
2	-40	50
3	0	40
6	-6	24
8	-6	24

Table 2. Analysis of loss reduction

Variables	Base value	Modified PSO (Ali Nasser Hussain et al.,2018)	Standard PSO (Pandya et al., 2015)	Standard EP (Pandya et al., 2015)	SA- Real coded GA (Subbaraj et al., 2009)	ETSS	ESS	SBS
VG1	1.060	1.100	1.100	-	-	1.009	1.010	1.013
VG2	1.045	1.085	1.086	1.029	1.060	1.018	1.009	1.011
VG3	1.010	1.055	1.056	1.016	1.036	1.020	1.022	1.029
VG6	1.070	1.069	1.067	1.097	1.099	1.027	1.030	1.033
VG8	1.090	1.074	1.060	1.053	1.078	1.008	1.010	1.007
T8	0.978	1.018	1.019	1.04	0.95	0.901	0.900	0.905
T9	0.969	0.975	0.988	0.94	0.95	0.904	0.905	0.900
T10	0.932	1.024	1.008	1.03	0.96	0.931	0.929	0.931
QC9	0.19	14.64	0.185	0.18	0.06	0.147	0.145	0.143
PG	272.39	271.32	271.32	-	-	271.80	271.83	271.79
QG	82.44	75.79	76.79	-	-	74.96	74.94	74.90
Diminution in Power Loss (percentage)	0	9.2	9.1	1.5	2.5	20.05	20.25	20.15
Power Loss in MW	13.550	12.293	12.315	13.346	13.216	10.832	10.805	10.819

Figure 1. Assessment of factual power loss

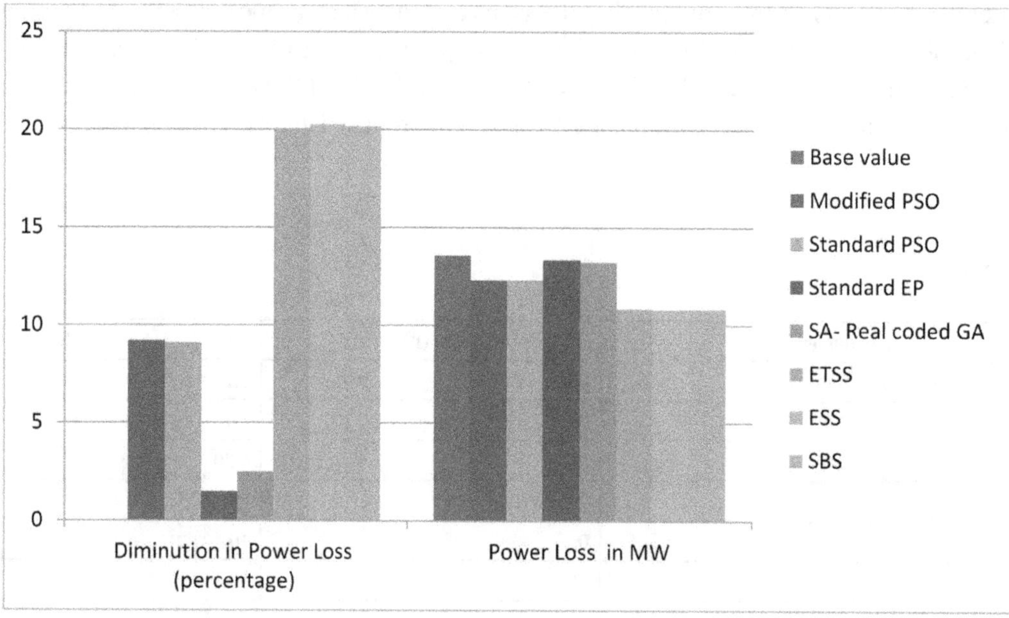

Table 3. Result analysis of power loss

Variables	Base value	Modified PSO (Ali Nasser Hussain et al.,2018)	Standard PSO (Pandya et al., 2015)	Standard EP (Pandya et al., 2015)	SA- Real coded GA (Subbaraj et al., 2009)	ETSS	ESS	SBS
VG1	1.060	1.101	1.100	-	-	1.012	1.013	1.021
VG2	1.045	1.086	1.072	1.097	1.094	1.017	1.019	1.033
VG5	1.010	1.047	1.038	1.049	1.053	1.069	1.062	1.016
VG8	1.010	1.057	1.048	1.033	1.059	1.005	1.008	1.012
VG12	1.082	1.048	1.058	1.092	1.099	1.027	1.029	1.021
VG13	1.071	1.068	1.080	1.091	1.099	1.049	1.041	1.030
T11	0.978	0.983	0.987	1.01	0.99	0.900	0.902	0.929
T12	0.969	1.023	1.015	1.03	1.03	0.918	0.910	0.910
T15	0.932	1.020	1.020	1.07	0.98	0.902	0.907	0.912
T36	0.968	0.988	1.012	0.99	0.96	0.906	0.909	0.909
QC10	0.19	0.077	0.077	0.19	0.19	0.071	0.076	0.090
QC24	0.043	0.119	0.128	0.04	0.04	0.114	0.115	0.112
PG	300.9	299.54	299.54	-	-	298.84	298.88	298.24
QG	133.9	130.83	130.94	-	-	130.96	130.98	130.31
Diminution in Power Loss (percentage)	0	8.4	7.4	6.6	8.3	14.94	14.80	20.12
Power Loss in MW	17.55	16.07	16.25	16.38	16.09	14.928	14.951	14.018

Table 4. Power loss appraisal

Variable	Enriched Genetic algorithm (Reddy 2017)	Faster Evolutionary Algorithm (Reddy2017)	Standard cuckoo search CSA (Reddy 2014)	ETSS	ESS	SBS
Power Loss (MW)	646.2998	650.6027	635.8942	610.2070	611.5131	611.6906

Figure 2. Valuation of factual power loss

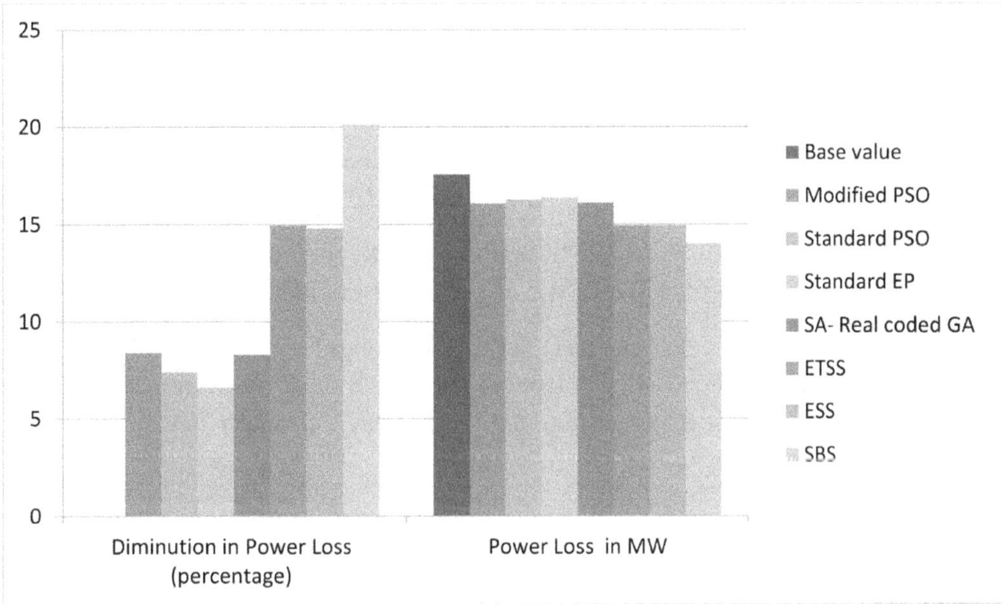

Figure 3. Valuation of factual power loss

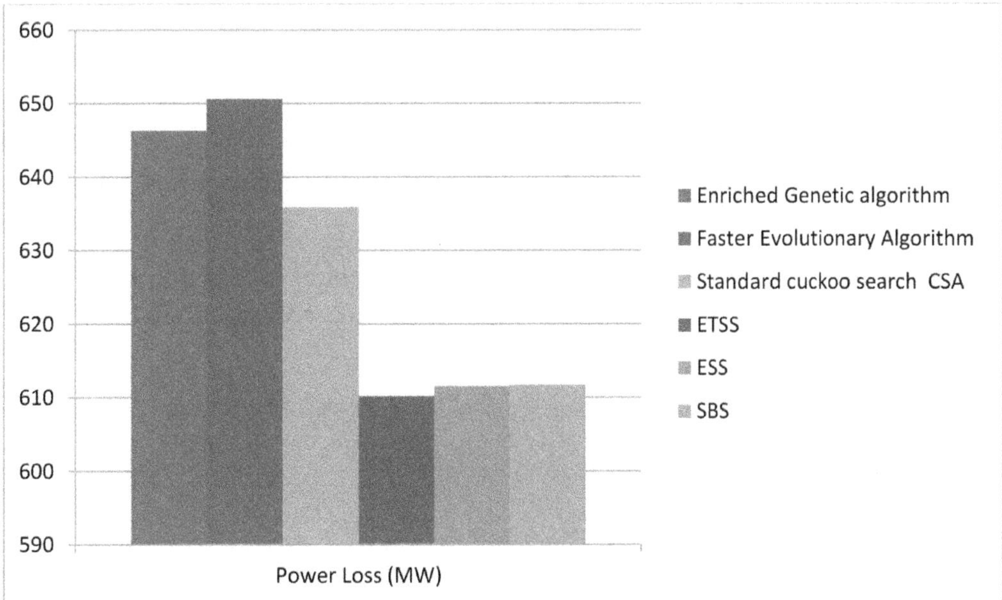

REFERENCES

Antoniewicz, P., & Kazmierkowski, P. (2008). Virtual-flux-based predictive direct power control of AC/DC converters with online inductance estimation. *IEEE Transactions on Industrial Electronics*, *55*(12), 4381–4390. doi:10.1109/TIE.2008.2007519

Azar, A. T. (2010a). *Fuzzy Systems*. IN-TECH.

Azar, A. T. (2010b). Adaptive Neuro-Fuzzy Systems. In A. T. Azar (Ed.), *Fuzzy Systems*. IN-TECH. doi:10.5772/7220

Azar, A. T. (2012). Overview of Type-2 Fuzzy logic systems. *International Journal of Fuzzy System Applications*, *2*(4), 1–28. doi:10.4018/ijfsa.2012100101

Azar, A.T. (2013). Fast neural network learning algorithms for medical applications. *Neural Computing and Applications, 23*(3-4), 1019-1034. . doi:10.1007/s00521-012-1026-y

Azar, A.T., & El-Said, S.A. (2013). Probabilistic neural network for breast cancer classification. *Neural Computing and Applications, 23*(6), 1737-1751. . doi:10.1007/s00521-012-1134-8

Azar, A. T., Kumar, J., Kumar, V., & Rana, K. P. S. (2017d). Control of a Two Link Planar Electrically-Driven Rigid Robotic Manipulator Using Fractional Order SOFC. *Proceedings of the International Conference on Advanced Intelligent Systems and Informatics 2017*. Springer-Verlag.

Azar, A. T., Ouannas, A., & Singh, S. (2017c). Control of New Type of Fractional Chaos Synchronization. *Proceedings of the International Conference on Advanced Intelligent Systems and Informatics 2017*. Springer-Verlag.

Azar, A.T., & Serrano, F.E. (2014). Robust IMC-PID tuning for cascade control systems with gain and phase margin specifications. *Neural Computing and Applications, 25*(5), 983-995. . doi:10.1007/s00521-014-1560-x

Azar, A. T., & Serrano, F. E. (2015a). Stabilization and Control of Mechanical Systems with Backlash. In Advanced Intelligent Control Engineering and Automation. IGI Global. doi:10.4018/978-1-4666-7248-2.ch001

Azar, A. T., & Serrano, F. E. (2015b). Design and Modeling of Anti Wind Up PID Controllers. In Complex system modelling and control through intelligent soft computations. Springer-Verlag. . 1. doi:10.1007/978-3-319-12883-2_1

Azar, A. T., & Serrano, F. E. (2015c). Adaptive Sliding mode control of the Furuta pendulum. In Advances and Applications in Sliding Mode Control systems. Springer-Verlag GmbH. doi:10.1007/978-3-319-11173-5_1

Azar, A. T., & Serrano, F. E. (2015d). Deadbeat Control for Multivariable Systems with Time Varying Delays. In Chaos Modeling and Control Systems Design. Springer-Verlag GmbH. doi:10.1007/978-3-319-13132-0_6

Azar, A. T., & Serrano, F. E. (2016a). Robust control for asynchronous switched nonlinear systems with time varying delays. *Proceedings of the International Conference on Advanced Intelligent Systems and Informatics 2016*, 533, 891-899. 10.1007/978-3-319-48308-5_85

Azar, A. T., & Serrano, F. E. (2016b). Stabilization of Mechanical Systems with Backlash by PI Loop Shaping. *International Journal of System Dynamics Applications*, *5*(3), 20–47. doi:10.4018/IJSDA.2016070102

Azar, A. T., & Vaidyanathan, S. (2015a). Handbook of Research on Advanced Intelligent Control Engineering and Automation. IGI Global. doi:10.4018/978-1-4666-7248-2

Azar, A. T., & Vaidyanathan, S. (2015b). *Computational Intelligence applications in Modeling and Control.* Springer-Verlag.

Azar, A. T., & Vaidyanathan, S. (2015c). *Chaos Modeling and Control Systems Design.* Springer-Verlag.

Azar, A. T., & Vaidyanathan, S. (2016). *Advances in Chaos Theory and Intelligent Control.* Springer-Verlag. doi:10.1007/978-3-319-30340-6

Azar, A. T., Vaidyanathan, S., & Ouannas, A. (2017a). *Fractional Order Control and Synchronization of Chaotic Systems.* Springer-Verlag. doi:10.1007/978-3-319-50249-6

Azar, A. T., Volos, C., Gerodimos, N. A., Tombras, G. S., Pham, V. T., Radwan, A. G., Vaidyanathan, S., Ouannas, A., & Munoz-Pacheco, J. M. (2017b). A novel chaotic system without equilibrium: Dynamics, synchronization and circuit realization. Complexity. doi:10.1155/2017/7871467

Azar, A. T., & Zhu, Q. (2015). *Advances and Applications in Sliding Mode Control systems.* Springer-Verlag.

Baghli, L. (1999). *Contribution à la Commande de la Machine Asynchrone, Utilisation de la Logique Floue, des Réseaux de Neurones et des Algorithmes Génétiques.* Thèse de Doctorat, Département de Génie Electrique Université Henri Poincaré, Nancy-I.

Bimal, & Bos, K. (1994) Expert system, fuzzy logic, and neural network applications in power electronics and motion control. *Proceedings of the IEEE, 82*(8), 1303 – 1323.

Bose, B. (2001) Artificial neural network applications in power electronics. *The 27th Annual Conférence of the IEEE Industrial Electronics Society.* 10.1109/IECON.2001.975533

Bouafia, A., & Krim, F. (2008). A fuzzy-Logic-Based Controller for Three-Phase PWM Rectifier with Unity Power Factor Operation. *J. Electrical Systems*, *4*(1), 36–50.

Bouafia, A., Krim, F., & Gaubert, J. P. (2009). Fuzzy-Logic-Based Switching State Selection for Direct Power Control of Three-Phase PWM Rectifier. *IEEE Transactions on Industrial Electronics*, *56*(6), 1984–1992. doi:10.1109/TIE.2009.2014746

Bouchakour, S. (2005). *Commande Directe de Puissance d'un Convertisseur AC/DC Triphasé Sans Capteurs de Tension.* Memory of Magister Ecole Militaire Polytechnique Algeria.

Boulkroune, A., Bouzeriba, A., Bouden, T., & Azar, A. T. (2016a). Fuzzy Adaptive Synchronization of Uncertain Fractional-order Chaotic Systems. In A. T. Azar & S. Vaidyanathan (Eds.), *Advances in Chaos Theory and Intelligent Control. Studies in Fuzziness and Soft Computing* (Vol. 337). Springer-Verlag. doi:10.1007/978-3-319-30340-6_28

Boulkroune, A., Hamel, S., & Azar, A. T. (2016b). *Fuzzy control-based function synchronization of unknown chaotic systems with dead-zone input.* Springer-Verlag.

Chapuis, Y. A. (1996). *Contrôle directe du couple d'une machine asynchrone par l'orientation de son flux statorique*. Doctoral thesis of INP-Grenoble, France.

Chen, B., & Joos, G. (2008). Direct power control of active filters with averaged switching frequency regulation. *IEEE Transactions on Power Electronics*, *23*(6), 2729–2737. doi:10.1109/TPEL.2008.2004958

Chen, S., & Joos, G. (2001) Direct Power Control of Three Active filer with Minimum Energy Storage Components. *APEC 2001. Sixteenth Annual IEEE Applied Power Electronics Conference and Exposition.* 10.1109/APEC.2001.911703

Cichowlas, M., Malinowski, M., Kazmierkowski, P., Sobczuk, D., & Pou, J. (2005). Active filtering function of three-phase PWM boost rectifier under different line voltage conditions. *IEEE Transactions on Industrial Electronics*, *52*(2), 410–419. doi:10.1109/TIE.2005.843915

Constant, L. (2000). *Modélisation de Dispositifs Electriques par Réseaux de Neurones en Vue de L'émulation Temps Réel.* Thesis Doctorate, Toulouse, France.

Escobar, G., Stankovic, A., Carrasco, M., Galvan, E., & Ortega, R. (2003). Analysis and design of direct power control (DPC) for a three phase synchronous rectifier via output regulation subspaces. *IEEE Transactions on Power Electronics*, *18*(3), 823–830. doi:10.1109/TPEL.2003.810862

Fekik, A., Denoun, H., Benamrouche, N., Benyahia, N., Badji, A., & Zaouia, M. (2016*)* Comparative Analysis of Direct Power Control and Direct power control with space vector modulation of PWM rectifier. *4th IEEE International Conference on Control Engineering & Information Technology (CEIT-2016).* 10.1109/CEIT.2016.7929058

Fekik, A., Denoun, H., Benamrouche, N., Benyahia, N., & Zaouia, M. (2015b). A Fuzzy Logic Based Controller For Three Phase PWM Rectifier With Voltage Oriented Control Strategy. *International Journal of Circuits, Systems and Signal Processing*, *9*, 412–419.

Fekik, A., Denoun, H., Benamrouche, N., Benyahia, N., Zaouia, M., Badji, A., & Vaidyanathan, S. (2017). Improvement of the Performances of the Direct Power Control Using Space Vector Modulation of Three Phases PWM-Rectifier. *International Journal of Control Theory and Applications*, *10*, 125–137.

Fekik, A., Denoun, H., Benamrouche, N., Benyahia, N., Zaouia, M., & Haddad, S. (2015a). Comparative study of PI and FUZZY DC- voltage control for Voltage Oriented Control-PWM rectifier. *The 14th International Conference on Circuits, Systems, Electronics, Control & Signal Processing 2015.*

Fischer, J. R., González, S. A., Carugat, I., Herrán, M. A., Judewicz, M. G., & Carrica, D. O. (2014). Robust predictive control of grid-tied converters based on direct power control. *IEEE Transactions on Power Electronics*, *29*(10), 5634–5643. doi:10.1109/TPEL.2013.2294919

Geyer, T., & Quevedo, E. (2015). Performance of multistep finite control set model predictive control for power electronics. *IEEE Transactions on Power Electronics*, *30*(3), 1633–1644. doi:10.1109/TPEL.2014.2316173

Ghoudelbourk, S., Dib, D., Omeiri, A., & Azar, A. T. (2016). MPPT Control in wind energy conversion systems and the application of fractional control (PIα) in pitch wind turbine. *International Journal of Modelling Identification and Control*, *26*(2), 140–151. doi:10.1504/IJMIC.2016.078329

Grassi, G., Ouannas, A., Azar, A. T., Radwan, A. G., Volos, C., Pham, V. T., Ziar, T., Madian, A., Kyprianidis, I. M., & Stouboulos, I. N. (2017). Chaos Synchronisation Of Continuous Systems Via Scalar Signal. *The 6th International Conference on Modern Circuits and Systems Technologies (MOCAST)*. 10.1109/MOCAST.2017.7937629

Hassanien, A.E., Moftah, H.M., Azar, A.T., & Shoman, M. (2014). MRI Breast cancer diagnosis hybrid approach using adaptive Ant-based segmentation and Multilayer Perceptron neural networks classifier. *Applied Soft Computing, 14*(Part A), 62-71.

Hu, J., Zhu, J., Zhang, Y., Platt, G., Ma, Q., & Dorrell, D. G. (2013). Predictive direct virtual torque and power control of doubly fed induction generators for fast and smooth grid synchronization and flexible power regulation. *IEEE Transactions on Power Electronics*, *28*(7), 3182–3194. doi:10.1109/TPEL.2012.2219321

Hussain, A. N., Abdullah, A. A., & Neda, O. M. (2018). Modified Particle Swarm Optimization for Solution of Reactive Power Dispatch. *Research Journal of Applied Sciences, Engineering and Technology*, *15*(8), 316–327. doi:10.19026/rjaset.15.5917

Jiang, W. (2010) Sliding-mode Control of Single-phase PWM Rectifier for DC Microgrid Applications. *International Conference on Computer and Electrical Engineering*, 53, 82-90. DOI: 10.7763/IPCSIT.2012.V53.No.1.82

Karamanakos, P., Geyer, T., Oikonomou, N., Kieferndorf, F., & Manias, S. (2014). Direct model predictive control: A review of strategies that achieve long prediction intervals for power electronics. *IEEE Industrial Electronics Magazine*, *8*(1), 32–43. doi:10.1109/MIE.2013.2290474

Lamterkati, J., Khafallah, M., & Ouboubker, L. (2014a). A New DPC for Three-phase PWM rectifier with unity power factor operation. *International Journal of Advanced Research in Electrical Electronics and Instrumentation Engineering*, *3*(4), 8273–8285.

Lamterkati, J., Khafallah, M., & Ouboubker, L. (2014b). Comparison of PI and Fuzzy logic DC-Link Voltage Controller for DPC PWM-Rectifier. *International Journal of Enhanced Research in Science Technology and Engineering*, *3*(4), 321–332.

Larrinaga, S., Vidal, M., Oyarbide, E., & Apraiz, J. (2007). Predictive control strategy of DC/AC converters based on direct power control. *IEEE Transactions on Industrial Electronics*, *54*(3), 1261–1271. doi:10.1109/TIE.2007.893162

Larrinaga, S. A., Vidal, M. A. R., Oyarbide, E., & Apraiz, J. R. T. (2007). Predictive control strategy for DC/AC converters based on direct power control. *IEEE Transactions on Industrial Electronics*, *54*(3), 1261–1271. doi:10.1109/TIE.2007.893162

Malinowski, M., Jasinski, M., & Kazmierkowski, P. (2004). Simple Direct power control of three phase PWM rectifier using space vector modulation (DPC-SVM). *IEEE Transactions on Industrial Electronics, 51*(2), 447–454. doi:10.1109/TIE.2004.825278

Malinowski, M., Kaz'mierkowski, M. P., Hansen, S., Blaabjerg, F., & Marques, G. D. (2001). Virtual flux based direct power control of three-phase PWM rectifiers. *IEEE Transactions on Industry Applications, 37*(4), 1019–1027. doi:10.1109/28.936392

Malinowski, M., & Kazmierkowski, M.P. (2000) Simulation Study of Virtual Flux Based Direct Power Control for Three-Phase PWM Rectifiers. *26th Annual Conference of the IEEE Industrial Electronics Society, 2000. IECON 2000.* doi:10.1109/IECON.2000.972411

Malinowski, M., Kazmierkowski, M. P., & Trzynadlowski, A. M. (2003). A Comparative Study of Control Techniques for PWM Rectifiers in AC Adjustable Speed Drives. *IEEE Transactions on Power Electronics, 18*(6), 1390–1396. doi:10.1109/TPEL.2003.818871

Meghni, B., Dib, D., & Azar, A. T. (2017c). A Second-order sliding mode and fuzzy logic control to Optimal Energy Management in PMSG Wind Turbine with Battery Storage. *Neural Computing & Applications, 28*(6), 1417–1434. doi:10.100700521-015-2161-z

Meghni, B., Dib, D., Azar, A. T., Ghoudelbourk, S., & Saadoun, A. (2017a). Robust Adaptive Supervisory Fractional order Controller For optimal Energy Management in Wind Turbine with Battery Storage. *Studies in Computational Intelligence, 688*, 165–202. doi:10.1007/978-3-319-50249-6_6

Meghni, B., Dib, D., Azar, A. T., & Saadoun, A. (2017b). *Effective Supervisory Controller to Extend Optimal Energy Management in Hybrid Wind Turbine under Energy and Reliability Constraints. International Journal of Dynamics and Control.* doi:10.100740435-016-0296-0

Mekki, H., Boukhetala, D., & Azar, A. T. (2015). Sliding Modes for Fault Tolerant Control. In Advances and Applications in Sliding Mode Control systems. Springer-Verlag GmbH. doi:10.1007/978-3-319-11173-5_15

Mezache, A., & Betto, K. (1997). *Estimation et Commande d'un Réacteur de Fabrication de Pâte a Papier par les Réseaux de Neurones Flous. Memory of Engineer.* University of Constantine.

Mondal, S. K., Pinto, J. O. P., & Bose, B. K. (2002). A Neural-Network-Based Space-Vector PWM Controller for a three-Level Voltage-Fed Inverter Induction Motor Drive. *Industry Applications. IEEE Transactions on., 38*(3), 660–669.

Noguchi, T., Tomiki, H., Kondo, S., & Takahashi, I. (1998). Direct power control of PWM converter without power-source voltage sensors. *IEEE Transactions on Industry Applications, 34*(3), 473–479. doi:10.1109/28.673716

Ouannas, A., Azar, A. T., & Abu-Saris, R. (2016a). A new type of hybrid synchronization between arbitrary hyperchaotic maps. *International Journal of Machine Learning and Cybernetics*. Advance online publication. doi:10.100713042-016-0566-3

Ouannas, A., Azar, A. T., & Radwan, A. G. (2016b). *On Inverse Problem of Generalized Synchronization Between Different Dimensional Integer-Order and Fractional-Order Chaotic Systems.* The 28th International Conference on Microelectronics, Cairo, Egypt. 10.1109/ICM.2016.7847942

Ouannas, A., Azar, A. T., & Vaidyanathan, S. (2017f). On A Simple Approach for Q-S Synchronization of Chaotic Dynamical Systems in Continuous-Time. *Int. J. Computing Science and Mathematics*, *8*(1), 20–27.

Ouannas, A., Azar, A. T., & Vaidyanathan, S. (2017g). New Hybrid Synchronization Schemes Based on Coexistence of Various Types of Synchronization Between Master-Slave Hyperchaotic Systems. *International Journal of Computer Applications in Technology*, *55*(2), 112–120. doi:10.1504/IJCAT.2017.082868

Ouannas, A., Azar, A. T., & Vaidyanathan, S. (2017i). A Robust Method for New Fractional Hybrid Chaos Synchronization. *Mathematical Methods in the Applied Sciences*, *40*(5), 1804–1812. doi:10.1002/mma.4099

Ouannas, A., Azar, A. T., & Ziar, T. (2017h). On Inverse Full State Hybrid Function Projective Synchronization for Continuous-time Chaotic Dynamical Systems with Arbitrary Dimensions. *Differential Equations and Dynamical Systems*. doi:10.1007/s12591-017-0362-x

Ouannas, A., Azar, A. T., Ziar, T., & Radwan, A. G. (2017d). Study On Coexistence of Different Types of Synchronization Between Different dimensional Fractional Chaotic Systems. *Studies in Computational Intelligence*, *688*, 637–669. doi:10.1007/978-3-319-50249-6_22

Ouannas, A., Azar, A. T., Ziar, T., & Radwan, A. G. (2017e). Generalized Synchronization of Different Dimensional Integer-order and Fractional Order Chaotic Systems. *Studies in Computational Intelligence*, *688*, 671–697. doi:10.1007/978-3-319-50249-6_23

Ouannas, A., Azar, A. T., Ziar, T., & Vaidyanathan, S. (2017a). On New Fractional Inverse Matrix Projective Synchronization Schemes. *Studies in Computational Intelligence*, *688*, 497–524. doi:10.1007/978-3-319-50249-6_17

Ouannas, A., Azar, A. T., Ziar, T., & Vaidyanathan, S. (2017b). Fractional Inverse Generalized Chaos Synchronization Between Different Dimensional Systems. *Studies in Computational Intelligence*, *688*, 525–551. doi:10.1007/978-3-319-50249-6_18

Ouannas, A., Azar, A. T., Ziar, T., & Vaidyanathan, S. (2017c). A New Method To Synchronize Fractional Chaotic Systems With Different Dimensions. *Studies in Computational Intelligence*, *688*, 581–611. doi:10.1007/978-3-319-50249-6_20

Ouannas, A., Grassi, G., Azar, A. T., Radwan, A. G., Volos, C., Pham, V. T., Ziar, T., Kyprianidis, I. M., & Stouboulos, I. N. (2017j). *Dead-Beat Synchronization Control in Discrete-Time Chaotic Systems.* The 6th International Conference on Modern Circuits and Systems Technologies (MOCAST), Thessaloniki, Greece. 10.1109/MOCAST.2017.7937628

Pandya, S., & Roy, R. (2015). Particle swarm optimization based optimal reactive power dispatch. *Proceeding of the IEEE International Conference on Electrical, Computer and Communication Technologies*, *1*(1), 1-5. 10.1109/ICECCT.2015.7225981

Reddy. (2014). Faster evolutionary algorithm based optimal power flow using incremental variables. *Electrical Power and Energy Systems*, *54*(1), 198-210.

Reddy. (2017). Optimal Reactive Power Scheduling Using Cuckoo Search Algorithm. *International Journal of Electrical and Computer Engineering*, *7*(5), 2349-2356.

Sanjuan, S. (2010). Voltage Oriented Control of Three-Phase Boost PWM Converters Design, simulation and implementation of a 3-phase boost battery Charger. Chalmers University of Technologie.

Singh, S., Azar, A. T., Ouannas, A., Zhu, Q., Zhang, W., & Na, J. (2017). Sliding Mode Control Technique for Multi-switching Synchronization of Chaotic Systems. *9th International Conference on Modelling, Identification and Control (ICMIC 2017)*, Kunming, China. 10.1109/ICMIC.2017.8321579

Soliman, N. S., Said, L. A., Azar, A. T., Madian, A. H., Radwan, A. G., & Ouannas, A. (2017). *Fractional Controllable Multi-Scroll V-Shape Attractor with Parameters Effect*. The 6th International Conference on Modern Circuits and Systems Technologies (MOCAST), Thessaloniki, Greece.

Song, Z., Tian, Y., Chen, W., Zou, Z., & Chen, Z. (2016). Predictive duty cycle control of three-phase active-front-end rectifier. *IEEE Transactions on Power Electronics*, *31*(1), 698–710. doi:10.1109/TPEL.2015.2398872

Subbaraj, P., & Rajnarayan, P. N. (2009). Optimal reactive power dispatch using self-adaptive real coded Genetic algorithm. *Electric Power Systems Research*, *79*(2), 374–38. doi:10.1016/j.epsr.2008.07.008

Tolba, M. F., AbdelAty, A. M., Soliman, N. S., Said, L. A., Madian, A. H., Azar, A. T., & Radwan, A. G. (2017). FPGA implementation of two fractional order chaotic systems. *International Journal of Electronics and Communications*, *28*, 162–172. doi:10.1016/j.aeue.2017.04.028

Vaidyanathan, S., & Azar, A. T. (2015a). Anti-Synchronization of Identical Chaotic Systems using Sliding Mode Control and an Application to Vaidyanathan-Madhavan Chaotic Systems. In Advances and Applications in Sliding Mode Control systems. Springer-Verlag GmbH. doi:10.1007/978-3-319-11173-5_19

Vaidyanathan, S., & Azar, A. T. (2015b). Hybrid Synchronization of Identical Chaotic Systems using Sliding Mode Control and an Application to Vaidyanathan Chaotic Systems. In Advances and Applications in Sliding Mode Control systems. Springer-Verlag GmbH. doi:10.1007/978-3-319-11173-5_20

Vaidyanathan, S., & Azar, A. T. (2015c). Analysis, Control and Synchronization of a Nine-Term 3-D Novel Chaotic System. In Chaos Modeling and Control Systems Design. Springer-Verlag GmbH. doi:10.1007/978-3-319-13132-0_1

Vaidyanathan, S., & Azar, A. T. (2015d). Analysis and Control of a 4-D Novel Hyperchaotic System. In Chaos Modeling and Control Systems Design. Springer-Verlag GmbH. doi:10.1007/978-3-319-13132-0_2

Vaidyanathan, S., & Azar, A. T. (2016a). Takagi-Sugeno Fuzzy Logic Controller for Liu-Chen Four-Scroll Chaotic System. *International Journal of Intelligent Engineering Informatics*, *4*(2), 135–150. doi:10.1504/IJIEI.2016.076699

Vaidyanathan, S., & Azar, A. T. (2016b). *Dynamic Analysis, Adaptive Feedback Control and Synchronization of an Eight-Term 3-D Novel Chaotic System with Three Quadratic Nonlinearities*. Springer-Verlag.

Vaidyanathan, S., & Azar, A. T. (2016c). *Qualitative Study and Adaptive Control of a Novel 4-D Hyperchaotic System with Three Quadratic Nonlinearities.* Springer-Verlag.

Vaidyanathan, S., & Azar, A. T. (2016d). *A Novel 4-D Four-Wing Chaotic System with Four Quadratic Nonlinearities and its Synchronization via Adaptive Control Method.* Springer-Verlag.

Vaidyanathan, S., & Azar, A. T. (2016e). *Adaptive Control and Synchronization of Halvorsen Circulant Chaotic Systems.* Springer-Verlag.

Vaidyanathan, S., & Azar, A. T. (2016f). *Adaptive Backstepping Control and Synchronization of a Novel 3-D Jerk System with an Exponential Nonlinearity.* Springer-Verlag.

Vaidyanathan, S., & Azar, A. T. (2016g). *Generalized Projective Synchronization of a Novel Hyperchaotic Four-Wing System via Adaptive Control Method.* Springer-Verlag.

Vaidyanathan, S., Azar, A. T., & Ouannas, A. (2017a). An Eight-Term 3-D Novel Chaotic System with Three Quadratic Nonlinearities, its Adaptive Feedback Control and Synchronization. *Studies in Computational Intelligence, 688*, 719–746. doi:10.1007/978-3-319-50249-6_25

Vaidyanathan, S., Azar, A. T., & Ouannas, A. (2017c). Hyperchaos and Adaptive Control of a Novel Hyperchaotic System with Two Quadratic Nonlinearities. *Studies in Computational Intelligence, 688*, 773–803. doi:10.1007/978-3-319-50249-6_27

Vaidyanathan, S., Azar, A. T., Rajagopal, K., & Alexander, P. (2015b). Design and SPICE implementation of a 12-term novel hyperchaotic system and its synchronization via active control (2015). *International Journal of Modelling Identification and Control, 23*(3), 267–277. doi:10.1504/IJMIC.2015.069936

Vaidyanathan, S., Idowu, B. A., & Azar, A. T. (2015c). Backstepping Controller Design for the Global Chaos Synchronization of Sprott's Jerk Systems. In Chaos Modeling and Control Systems Design. Springer-Verlag GmbH. doi:10.1007/978-3-319-13132-0_3

Vaidyanathan, S., Sampath, S., & Azar, A. T. (2015a). Global chaos synchronisation of identical chaotic systems via novel sliding mode control method and its application to Zhu system. *International Journal of Modelling Identification and Control, 23*(1), 92–100. doi:10.1504/IJMIC.2015.067495

Vaidyanathan, S., Zhu, Q., & Azar, A. T. (2017b). Adaptive Control of a Novel Nonlinear Double Convection Chaotic System. *Studies in Computational Intelligence, 688*, 357–385. doi:10.1007/978-3-319-50249-6_12

Vazquez, J., & Salmeron, P. (2003). Active power filter control using neural network technologies. *IEE Proceedings. Electric Power Applications, 150*(2), 139–145. doi:10.1049/ip-epa:20030009

Vazquez, S., Leon, J. I., Franquelo, L. G., Rodriguez, J., Young, H., Marquez, A., & Zanchetta, P. (2014). Model predictive control: A review of its applications in power electronics. *IEEE Industrial Electronics Magazine, 8*(1), 16–31. doi:10.1109/MIE.2013.2290138

Vazquez, S., Sanchez, J., Carrasco, J., Leon, J., & Galvan, E. (2008). A model-based direct power control for three-phase power converters. *IEEE Transactions on Industrial Electronics, 55*(4), 1647–11657. doi:10.1109/TIE.2008.917113

Wang, Z., Volos, C., Kingni, S. T., Azar, A. T., & Pham, V. T. (2017). Four-wing attractors in a novel chaotic system with hyperbolic sine nonlinearity. *Optik (Stuttgart)*, *131*, 1071–1078. doi:10.1016/j.ijleo.2016.12.016

Zhang, Y., Xie, W., Li, Z., & Zhang, Y. (2013). Model predictive direct power control of a PWM rectifier with duty cycle optimization. *IEEE Transactions on Power Electronics*, *28*(11), 5343–5351. doi:10.1109/TPEL.2013.2243846

Zhu, Q., & Azar, A. T. (2015). *Complex system modelling and control through intelligent soft computations.* Springer-Verlag.

Chapter 4
Extension of Aspiration Level Model for Optimal Planning of Fast Charging Stations

Manikanta Surya Narayana Suri
Amrita School of Engineering, Bengaluru, India

Deepa Kaliyaperumal
Amrita School of Engineering, Bengaluru, India

ABSTRACT

Electric vehicles will play a dominant role in future transportation due to their friendliness towards the present day environment. The battery which drives the vehicle can be refilled using battery charging and battery swapping techniques. Fast charging stations provide faster service to the customers. Though battery swapping method outperforms battery charging in many ways, the heavy infrastructure requirement of the former requires time in integrating with the real world. Queuing models are used to depict the real-time behavior of service stations. The aspiration level model provides the optimal value of charging piles for the given system capacity in a fast-charging station. The parameters in the aspiration level model can be formulated to an optimization problem. In the present work, the optimal planning for an fast charging station in Beijing is carried out using genetic algorithm. The simulation work is carried out in MATLAB/Simulink.

INTRODUCTION

Internal combustion engine vehicles, though comfortable damages the environment by the release of harmful greenhouse gases (0.3 kg/mile) during the exhaust stroke. Electric vehicles (EVs) are found to be a sustainable solution that mitigates the harm caused by the conventional vehicles. The release of greenhouse gases by the power sector for the fueling of EVs is significantly less (0.16 kg/mile). Many countries have started adopting EVs in their transportation sector as a substitute for internal combustion engine vehicles. Apart from the reduction of greenhouse gases, usage of EVs has other advantages

DOI: 10.4018/978-1-7998-7447-8.ch004

such as regenerative braking, mitigation in noise pollution, smart charging, good running performance, etc(Ban (2019), Xu (2017), Zeng (2019), Shabarish (2020), Zhenhe (2019)).

The power delivered to the wheels of the EV is obtained using a battery that is embedded inside the vehicle. The battery can be charged at the EV battery refilling stations. Battery charging and Battery Swapping are the two modes of refilling EV batteries. In the former, the EV charges in a Battery Charging Station (BCS), and the depleted battery is charged using the standard charging piles. Range anxiety and long charging time(7-15h) cease the development of EVs when batteries are refueled by this technique. This can be overcome by developing a Fast-Charging Station (FCS) where it takes less than 1 hour to charge the battery. In a Battery Swapping Station (BSS), the depleted battery is swapped with a high SOC battery. The swapping is done using swapping robots. The swapping time is within 4-12 minutes. The depleted batteries are then refueled using the inbuilt BCS available inside the BSS. When the EV user is far away from BSS, Battery Swapping Van (BSV) is used to refill the batteries. BSV is the active mode of replenishing energy, where the van contains a stock of fully charged batteries which reaches the EV user and swaps the depleted battery. The battery swapping process has many advantages compared to battery charging method in terms of service time, mileage, easier integration with renewables, grid balance, etc. State of Health (SOH) of the battery can be maintained by appropriately scheduling the charging of the depleted batteries in the inbuilt BCS. BSS also has a stock of Fully Charged Batteries (FCB) which serves the demand during the shortage of FCBs in the inbuilt BCS. However, the growth of BSS is less than FCS owing to the heavy infrastructure requirement of the former. In the future, BSS can replace the FCS due to its outstanding advantages(Neha (2020), Sasikumar (2018), Saxena (2020), Shao (2017)).

There is a requirement for the proper infrastructure of an FCS, to ensure the quality of service for the EV users. The queuing models study the various types of queues, servers, and system capacity concerning to a service station (BCS). These models are derived from the pure birth model and pure death model where there is only the occurrence of arrivals and departures respectively. A system may have a single server or multiple servers based on the incoming arrival demand. Systems with single servers are studied using single sever queuing models, whereas the performance of the system having multiple servers is analyzed through multi server queuing models. If the system can hold infinite arrivals, infinite system capacity queuing models can be used to examine the behavior of the system. However, due to area constraints most of the real time systems cannot accommodate infinite arrivals, In such cases, finite system capacity queuing models are used to evaluate the performance of the system. In this present work, performance of FCS is analyzed using finite system capacity queuing models to account for the practicality. The arrivals in queuing theory assume Poisson probability distribution thus, the service time follow an exponential probability distribution. The reason being the memoryless property associated with these probability distributions. The memory-less property is closely related to the Markov process where for the given present value, the future values do not depend on the past values. The average arrival rate and the average service rate must be known before applying these models. However, some queuing models follow Gaussian probability distribution for both arrival and service times but their practical significance is very low. Various queue measures such as average waiting time of user in system and queue, average queue length in the queue and the system, percentage idleness of servers helps to evaluate the optimal service level of the system. These measures depend upon the queuing model chosen to serve the incoming users (Manikanta (2020), Thiruvonasundari (2020), Mohd (2019), Till(23)).

To obtain the best fit queuing model for a particular service station, either the cost model or the aspiration level model is used. The best fit queuing model guarantee optimal level of service. Cost model optimizes the addition of two conflicting costs: (1) Cost of offering the service and (2) Cost of waiting. The cost model requires prior knowledge on facility costs and waiting time costs. Aspiration level model uses the queue measures of a queuing model to obtain the best fit model. The aspiration level model can be used for planning of FCS in EV developing countries like India, where the accuracy of the data for the cost model is not guaranteed. In the present work, the aspiration level model is carried out to obtain the optimal level of service. It considers two queue measures: (1) Average time spent by EV in the FCS, (2) Percentage Idleness of charging piles in FCS. EVs wait in the queue in FCS as the service time in the FCS is not in a fraction of seconds. Thus, the primary objective of the owner of the FCS is to minimize the waiting time of the EV user in FCS to achieve the users' interest in visiting the charging station again. The secondary objective is to reduce the percentage idleness of the charging piles, thus showing the faith for the money utilized in buying the charging piles. For finite system capacity models, aspiration level model obtains optimum level of service for a fixed system capacity. However, it is worth to solve the model for varying system capacity for better planning of FCS. The present chapter relaxes the assumptions made in the work of Manikanta (2020) by converting the parameters in the aspiration level to an optimization problem. This problem gives the optimal number of charging piles, and the maximum number of EVs that can be accommodated in the FCS. The optimization problem minimizes the average time spent by an EV in FCS and percentage idleness of charging piles in FCS. Since the number of objectives that need to be accomplished is greater than 1, the optimization problem is a multi-objective optimization problem. Multi-objective optimization can be proceeded either by converting the problem to a single objective problem which gives a unique solution or by Pareto optimization which gives multiple non-dominating and diverse solutions. It is worth comparing the solutions obtained using both methods. The pictorial diagram and block diagram of the present system is depicted in figures 1 & 2 respectively (Khalilpourazari (2019), Harifi (2020), Rahnamayan(2020), Linyu (2015)).

Figure 1. Pictorial representation of battery charging process

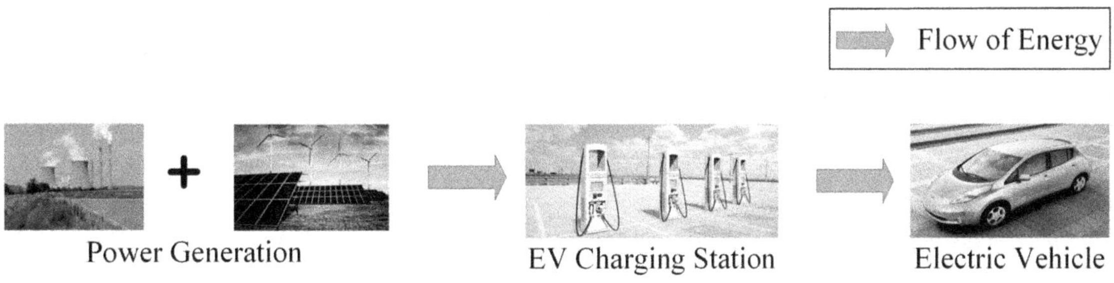

Figure 2. Block diagram of decision process

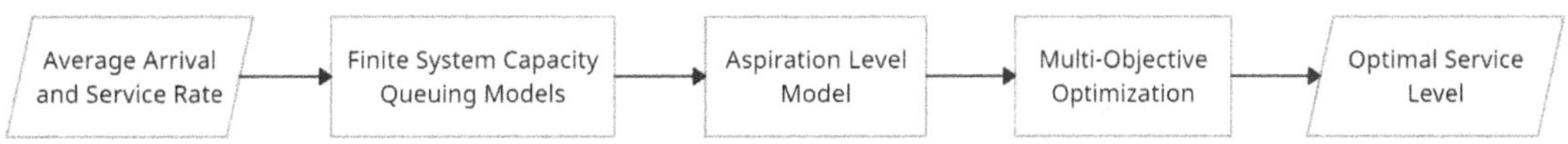

In this chapter, multi objective optimization has been carried out for optimal planning of FCS in Beijing. The optimization problem gives the optimal solutions of number of charging piles in FCS and system capacity of FCS by minimizing the parameters of aspiration level model (average time spent by EV in FCS and percentage idleness of charging piles in FCS). To our best knowledge, multi objective optimization hasn't been carried out for optimal planning of FCS through the aspiration level model. Manikanta (2020) solved the aspiration level model keeping system capacity of FCS as constant. However, while dealing with the finite system capacity queuing models, it is worth to obtain the optimal value of system capacity. Thus, this chapter relaxes the assumptions in the literature through multi objective optimization. The remaining part of the chapter is organized as follows: Section II introduces the finite system capacity queuing models, Section III details the aspiration level model, formulation of the optimization problem is given in Section IV, whereas results are presented in Section V.

FINITE SYSTEM CAPACITY QUEUING MODELS

In general, a queuing model is represented as:

(x/y/z): (a/b/c)

Where,

x = probability distribution of arrivals
y = probability distribution of service time
z = number of parallel servers
a = queue discipline
b = system capacity; finite or infinite
c = size of calling population; finite or infinite

Since the Markovian process is associated with the Poisson and exponential probability distributions, x and y are represented using 'M'. Finite system queuing models are further broken down into single server models and multi-server models. First In First Out (FIFO) rule is commonly used as the queue discipline and the size of the calling population is infinity as any user can reach the FCS. Thus, the finite system capacity queuing model is given by:

(M/M/z): (FIFO/N/∞)

Where N = Maximum number of EVs that can be accommodated in the FCS

The following models give the mathematical formulation of various queue measures of finite system capacity queueing models (Poomrittigul(2020), Yusheng(2019), Marchi Beatrice(2018), Victor Alexander(2015), Taha.H(2011)).

Let

α = Average arrival rate of EVs in the FCS
α_{eff} = Effective average arrival rate of EVs in FCS
β = Average charging time of EV battery in the FCS

σ = Average number of busy charging piles in FCS
L_q = Average number of EVs in the queue
L_s = Average number of EVs in the system
W_q = Average waiting time of an EV in queue
W_s = Average time spent by EV in the FCS
p_n = Probability of finding 'n' EVs in the FCS.
X = Percentage idleness of charging piles in FCS

When the number of EVs in the system reaches N in the FCS, no more EVs are allowed in the system. In such cases, the average arrival rate of EVs is modified to $\alpha_{eff.}$

1. (M/M/1): (FIFO/N/∞): This model consists of a single server for the finite system capacity queuing model. There are N–1 queue slots for EV in the FCS. The queue measures in this model are given by equations (1) – (11):

$$\alpha_n = \begin{Bmatrix} \alpha, & n = 0,1,\ldots,N-1 \\ 0, & n \geq N \end{Bmatrix} \tag{1}$$

$$\beta_n = \beta,\ m = 0,1,\ldots \tag{2}$$

$$\sigma = \frac{\alpha}{\beta} \tag{3}$$

$$p_n = \begin{Bmatrix} \sigma^n p_0, & n \leq N \\ 0, & n > N \end{Bmatrix} \tag{4}$$

$$p_0 = \begin{Bmatrix} \dfrac{1-\sigma}{1-\sigma^{N+1}} & \sigma \neq 1 \\ \dfrac{1}{N+1} & \sigma = 1 \end{Bmatrix} \tag{5}$$

$$\alpha_{eff} = \alpha(1 - p_N) \tag{6}$$

$$L_s = \frac{\sigma\left[1-(N+1)\sigma^N + N\sigma^{N+1}\right]}{(1-\rho)\left(1-\sigma^{N+1}\right)},\ \sigma \neq 1 \tag{7}$$

$$L_s = \frac{N}{2},\ \sigma = 1 \tag{8}$$

$$L_q = L_s - \frac{\alpha_{eff}}{\beta}. \tag{9}$$

$$W_s = \frac{L_s}{\alpha_{eff}} \tag{10}$$

$$W_q = W_s - \frac{1}{\beta} \tag{11}$$

$$X = \left(1 - \frac{\alpha_{eff}}{\beta}\right) * 100 \tag{12}$$

Equations (3), (6), (9-11) are applicable for all finite system capacity queueing models.

2. (M/M/c): (FIFO/N/∞): The system has 'c' parallel servers in this queuing model. Thus, there are N–c queue slots for EV in the FCS. Equations (13) - (18) gives the mathematical model of the queuing measures in this model.

$$\alpha_n = \begin{Bmatrix} \alpha, & n = 0, 1, \ldots, N-1 \\ 0, & n \geq N \end{Bmatrix} \tag{13}$$

$$\beta = \begin{Bmatrix} n\beta, & 0 \leq n \leq c \\ c\beta, & c \leq n \leq N \end{Bmatrix}. \tag{14}$$

$$p_n = \begin{Bmatrix} \dfrac{\sigma^n}{n!} p_0, & 0 \leq n \leq c \\ \dfrac{\sigma^n}{c!c^{n-c}} p_0, & c \leq n \leq N \end{Bmatrix} \tag{15}$$

$$p_0 = \begin{Bmatrix} \left(\sum_{n=0}^{c-1} \dfrac{\sigma^n}{n!} + \dfrac{\sigma^c}{c!} \left(\dfrac{\left(1 - \left(\dfrac{\sigma}{c}\right)\right)^{N-c+1}}{\left(1 - \left(\dfrac{\sigma}{c}\right)\right)} \right) \right)^{-1}, & \left(\dfrac{\sigma}{c}\right) \neq 1 \\ \left(\sum_{n=0}^{c-1} \dfrac{\sigma^n}{n!} + \dfrac{\sigma^c}{c!}(N - c + 1) \right)^{-1}, & \left(\dfrac{\sigma}{c}\right) = 1 \end{Bmatrix} \tag{16}$$

$$g = \frac{\sigma^{c+1}}{(c-1)!(c-\sigma)^2}$$

$$h = \left\{1 - \left(\frac{\sigma}{c}\right)^{N-c+1}\right\}$$

$$r = \left\{(N-c+1)\left(1-\frac{\sigma}{c}\right)\left(\frac{\sigma}{c}\right)^{N-c}\right\}$$

$$L_q = \begin{cases} g*(h-r)*p_0, & \frac{\sigma}{c} \neq 1 \\ \frac{\sigma^c (N-c)(N-c+1)}{2c!} p_0, & \frac{\sigma}{c} = 1 \end{cases} \tag{17}$$

$$X = \left(1 - \frac{\alpha_{eff}}{c\beta}\right)*100 \tag{18}$$

The mathematical formulation of various queuing models helps in analyzing the aspiration level model which is dealt in the next section.

ASPIRATION LEVEL MODEL

Aspiration model is one of the decision-making models which is used to determine the optimum level of service for an FCS. It alleviates the difficulty present in the cost model and can be used for planning of FCS in developing countries like India. Aspiration level model considers two queuing measures:

1. Average time spent by an EV in FCS
2. Percentage idleness of the charging piles in FCS

W_s can be reduced by installing a greater number of charging piles in the system. The desired number of charging piles should map the arrival rate of EVs and the charging time of a pile, otherwise there can be a situation of under usage of fast charging piles which costs more. The optimum number of charging piles can be found out by solving the equations (19) and (20).

$$W_s \leq \varphi \tag{19}$$

$$X \leq \Omega \tag{20}$$

The values of φ and Ω depends on the FCS owner's interest. For the present case study, φ is taken as 45 minutes and Ω as 25%. Solving the aspiration model graphically for different values of charging piles, system capacity is a tedious task. This calls for the formulation of an optimization problem where the objective is to minimize the W_s and X, the decision variables being c and N.

Optimization problem alleviates the difficulty involved in solving the aspiration level model to find the optimal number of charging piles, system capacity associated with the FCS. The next section describes the optimization process of the aspiration level model.

OPTIMIZATION PROBLEM

Optimization is one of the basic tools which is applied in many engineering fields. One can minimize the costs or maximize the profits of a real time problem with the help of optimization. Optimization involves converting the physical model of the problem to its corresponding mathematical model. This model can be solved using different solvers to achieve the final output. The optimization problem consists of objective functions that should be either minimized/maximized, the decision variables on which the objectives are built upon, and the constraints which restrict the decision variables to take up finite values. Figure 3 shows the classification of the optimization problem. For the optimization problem to be linear, the objective functions and the constraints should be linear, the presence of at least one non-linear constraint/objective makes the optimization problem non-linear.

Figure 3. Classification of optimization problem

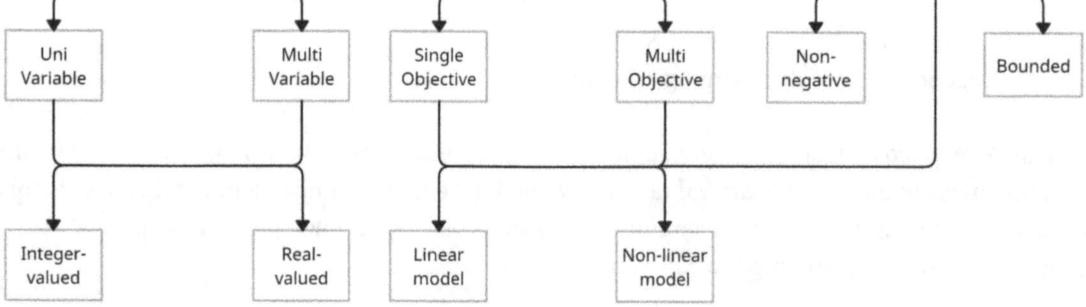

In the present work, the average time spent by an EV in the FCS and the percentage idleness of the charging piles has to be minimized to achieve the best quality of service. The decision variables are the number of charging piles in the FCS and the system capacity of FCS. At the end of the optimization process, the optimal values of c and N are found. It is meaningless to have the number of charging piles greater than system capacity as the excess charging piles will always remain unused. The minimum system capacity is rounded off to the nearest integer of the average arrival rate to account for the worst case. The mathematical model of the optimization problem is given by:

Minimize W_s, X

Decision Variables: c and N
Subjected to:

$$N - c \geq 1 \tag{21}$$

$$N \geq \text{round}(\alpha, 0) \tag{22}$$

$$c \geq 1 \tag{23}$$

$$N, c \in Z^{++} \tag{24}$$

Since the number of decision variables are two and being integer-valued, the number of objective functions are two and non-linear, the optimization is carried out using multi-objective multi-variable constrained non-linear integer programming. Equation (21) shows the presence of a linear inequality constraint. Optimization problems are solved numerically due to the overhead involved in hand calculations. Traditional optimization algorithms are classified into two methods — (1) Direct search methods, where objective function values are used to evaluate the optimum, and (2) Indirect search methods, where the gradient of the objective function is used to track the optimum. Though indirect search methods are computationally expensive, it takes lesser number of iterations to arrive at the optimal solution compared to direct search methods. Most of the practical optimization problems are non-linear and takes significant time to arrive at the optimal value though solved by gradient-based algorithms and the solutions don't guarantee the global optimum. The non-linear constrained problems should necessarily satisfy the Kuhn-Tucker conditions and the objective functions should be convex to guarantee the global optimum. Most of the real-time objective functions can't assure their convex nature. Metaheuristic algorithms overcome the difficulties faced in traditional optimization algorithms by searching in the limited decision space and by generating the solutions quickly. However, the algorithm doesn't guarantee the optimality of the solution obtained. The speed at which the solutions dominate the other disadvantages in these algorithms. Most of the metaheuristic algorithms have biological significance associated with them. They are used in neural network hyperparameter tuning, controller tuning and, in many applications, where optimization is involved. Table 1 depicts different classification of Metaheuristic algorithms and their examples.

Table 1. Classification of metaheuristic algorithms with examples

Sl. No.	Classification	Example
1	Evolutionary Algorithms	Genetic Algorithm (GA), Differential Evolution Algorithm (DE), Memetic Algorithm (ME)
2	Swarm Intelligence Algorithms	Particle Swarm Optimization (PSO), Cuckoo Search (CS), Artificial Bee Colony Algorithm (ABC)
3	Physics Based Algorithms	Gravitational Search (GS), Simulated Annealing (SA), Black-Hole Algorithm (BH)
4	Bio Inspired Algorithms	Firefly Algorithm (FA), Bat Algorithm (BA)
5	Human Based Algorithms	Tabu Search (TS), Soccer League Competition (SLC), Harmony Search (HS)

Metaheuristic algorithms has successfully taken the charge over the traditional optimization algorithms. Unlike single-objective optimization problem, multi-objective optimization problems have multiple optimal solutions which are non-dominating. These solutions are also called as pareto optimal solutions. The pareto front should be diverse so that all pareto solutions are taken into account. There are three approaches to solve the multi-objective optimization problems (*Deb, K.* (2001)):

1. Conversion of Multi-Objective optimization problem to single-objective optimization problem
2. Using classical methods
3. Using pareto optimization

The non-dominating solutions couldn't be found when the problem is solved using the approach (1) as it generates a unique solution. Classical methods of solving multi-objective optimization problems uses weighted sum method approach where weights are added to each objective function before converting to the single-optimization problem (This is the extension of the approach (1) by adapting weights) and the ε constraint method where the other objective functions act as a constraint for the single objective function. The multiple solutions are obtained by changing the weights and constraints for these methods respectively. However, classical methods don't assure the properties of the solutions required for the multi-objective optimization problem. The single objective optimizer has to be run for many times and they require more problem knowledge on choosing weights and the constraints. The objective function values have to be normalized in the range [0,1] if the optimization problem is solved using the approach (1) and (2). Pareto optimization mainly focus on two operations— (1) Non-dominated sorting which is able to identify different levels of pareto solutions, (2) Crowding distance calculation which is able to provide the diversified solutions. Thus, using the pareto optimization one can obtain non-dominated diversified optimal solutions. The non-dominating solutions can be filtered out using the equations (19) and (20) to map with the real time requirement.

The multi-objective optimization problem for the present scenario can be simulated in the software such as MATLAB and the solutions obtained using different approaches of multi-objective optimization problem can be compared.

RESULT AND ANALYSIS

The average EV arrival rate and the average service rate considered is 6.9 EVs/h and 2 EVs/h respectively as dealt by Linyu Wang et.al, FCS in Beijing, China ((Linyu Wang(2015)). It is desired to obtain the optimum number of charging piles and the system capacity of the FCS to achieve the customer's satisfaction and for the optimal operation of the system. The optimization problem formulated in the Section IV is solved using GA tool box in MATLAB. The maximum system capacity is choosen as 20 and the minimum system capacity, according to equation 22, is 7. The minimum and maximum values of charging piles are choosen as 1 and 20 respectively. The optimal solutions in the decision space and the objective space are shown in figure 4.

Figure 4. Optimal solutions of the multi-objective optimization problem in decision space (left) and objective space (right)

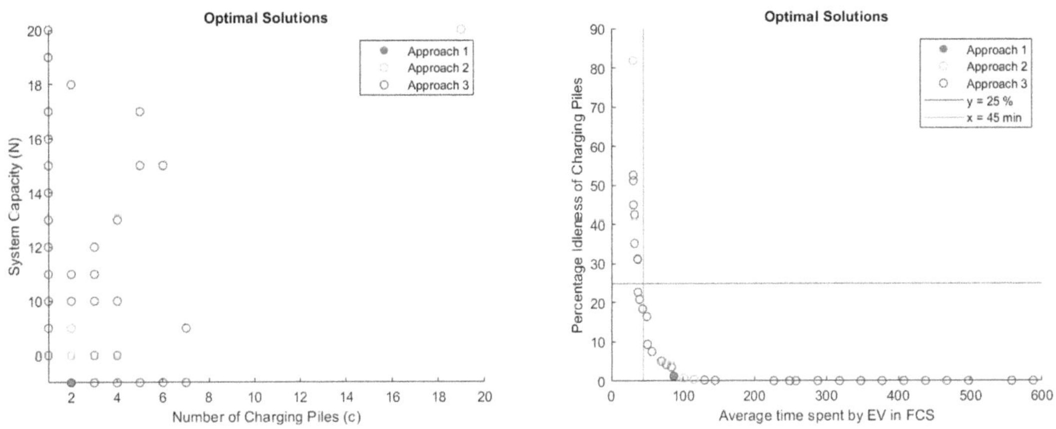

The results obtained using the three approaches are summarized as follows:

Approach 1: The multi-objective optimization problem is converted to single objective optimization problem through addition of the two objective functions. The objective function values are normalized in the range [0,1] so that the orders of both objective functions are same. The mathematical model for the normalization is given by equation (25).

$$i = \frac{j_u - j_{\min}}{j_{\max} - j_{\min}} \qquad (25)$$

Where,

i = Normalized value

j_u = Value corresponding to u^{th} component of j

j_{min} = Minimum value across all j

j_{max} = Maximum value across all j

This approach finds a unique solution which minimizes the sum of W_s and %X. However, this solution doesn't guarantee the optimality of both W_s and %X. The solution finds the optimal value of percentage idleness but it fails to finds the optimal value of W_s. This is due to fact that %X dominates W_s even after normalizing the objective functions. The pareto solutions couldn't be found because the optimization problem is solved as single objective optimization problem. The solution obtained using this approach couldn't satisfy the constraints imposed by aspiration level model. Figure 5 shows the variation of the fitness value with respect to the generation in this approach. Though there is no improvement in the best fitness value after generation 4, it takes nearly 40 iterations for all the population members to reach closer to the best fitness value. The mean fitness value oscillates after generation 40 but it doesn't deviate from the best fitness value.

Figure 5. Evolution of GA in approach 1

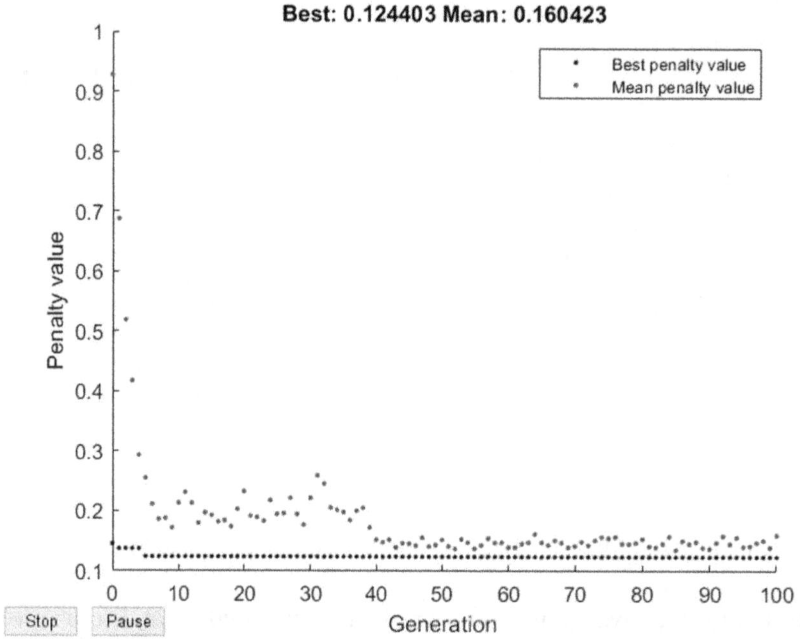

Approach 2*:* This approach generates multiple solutions by simulating the single objective optimization multiple times using weighted sum method. The objective function values are normalized using equation (25) and the weights w1, w2 are added to the objective functions such that w1+ w2 = 1. The single objective optimization problem is simulated 35 times using different values of w1, w2. The objective functions should be convex to guarantee the desired pareto front (Rahnamayan.S(2020)). From the figure 5, it is inferred that though the solver is simulated 35 times, it doesn't provide 35 unique solutions. Most of the times, the same solution is obtained though different weights are used. This is due to non-convexity of the objective functions. Similar to approach 1, this algorithm finds a solution which optimizes %X but fails to optimize of W_s. This approach too fails to satisfy the constraints imposed by aspiration level model.

Approach 3*:* This approach finds the pareto solutions involved in optimizing W_s and %X. Pareto solutions are non-dominating solutions which are optimal for the multi-objective optimization problem. This algorithm unlike the other two approaches, finds multiple optimal solutions which are diversified. The pareto front obtained using this approach closely maps with the desired pareto front (Rahnamayan.S(2020)). This approach doesn't require the convexity of the objective functions to generate the pareto front and the overhead chosen in choosing weights is also reduced using this approach. This approach generates multiple-solutions which satisfies the aspiration level model constraints due to its innate ability to produce the non-dominating diversified solutions. The solutions which solve the aspiration level model are tabulated in Table 2.

Table 2. Optimal solutions satisfying aspiration level model

Sl. No	Decision Space		Objective Space	
	c	N	W_s (min)	%X
1.	4	7	36.72507	22.71987
2.	4	8	39.14562	20.84968
3.	4	10	43.7017	18.41468

The owner of the FCS can choose the one of the solutions listed above according to the preference given to the objective functions. If more priority is given to reduce the average time spent by an EV in FCS, solution (i) is considered as optimal, whereas if the FCS owner is keen to reduce the percentage idleness of charging piles, he/she can choose solution (iii). The pareto optimization outperforms the other two approaches in generating multiple optimal solutions and in solving the aspiration level model. The plots of various queue measures for the optimal solutions are depicted in figures 6 and 7 respectively.

Figure 6. Variation of Lq (left) and Wq (right) with respect to N for c = 4

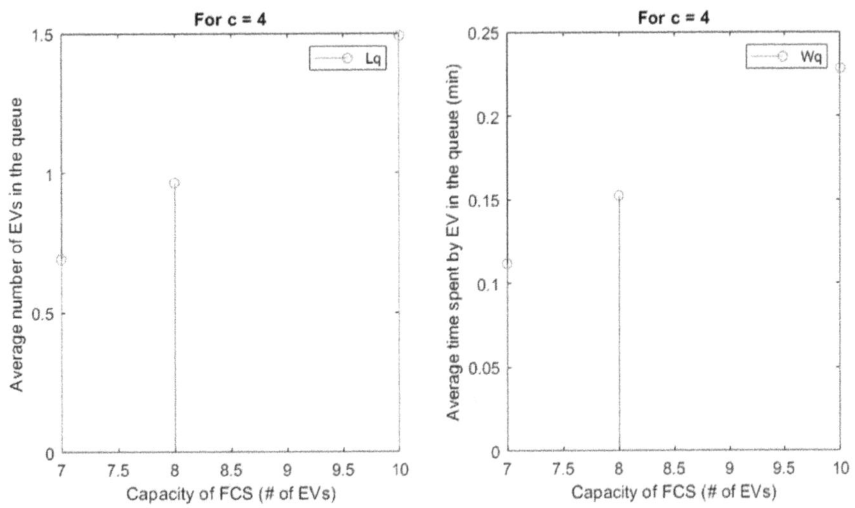

Figure 7. Steady state probabilities of finding EVs in the FCS

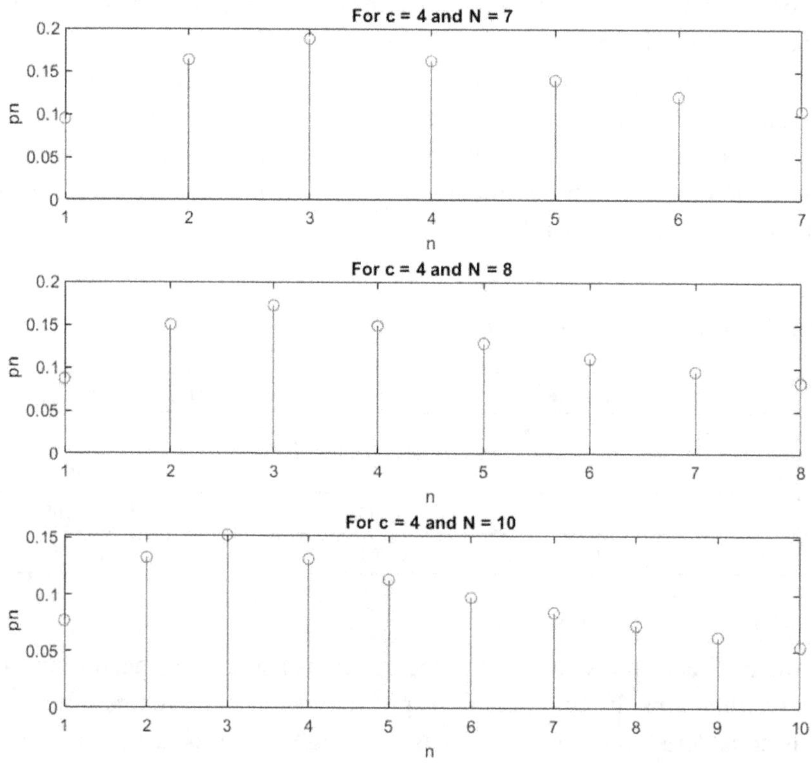

As the system capacity of FCS increases, a greater number of EVs can be accommodated in the station. The queue slots available in the FCS will be N—c for a finite system capacity model. Thus, the number of queue slots increases with increase in N for a given c. Thereby a greater number of EVs can wait in the queue. This increases the average number of EVs waiting in the queue and the corresponding waiting time in queue which is depicted in fig 7. It is more likely to find 3 EVs in the system as the ratio of average arrival rate to the average service rate is closer to 3. This ratio defines the number of busy servers in the FCS.

CONCLUSION

In this chapter, multi-objective optimization is carried out for optimal planning FCS in Beijing (Wang, 2015). This ensures the quality of service to the EV users. Finite system capacity queuing models describes various queue parameters(L_q, L_s, W_q, W_s) of the FCS. Aspiration level model provides the optimum service level for the FCS by considering W_s and %X into account. However, for finite system capacity queuing models it is worth to find the optimal system capacity. Thus, the parameters of the aspiration level model can be converted to the multi-objective optimization problem to obtain all the optimal solutions. The multi objective optimization problem can be solved using three approaches: (1) converting to a single-objective optimization problem, (2) using classical methods, (3) using pareto optimization. The results shows that pareto optimization outperforms the other two approaches due to its innate ability to produce non-dominating and diversified optimal solutions. The pareto optimization which is solved using GA tool box in MATLAB has provided three set of solutions: {(4,7), (4,9), (4,10)}. If more priority is to reduce the average time spent by an EV in FCS, solution (4,7) is considered as optimal, whereas if percentage idleness of charging piles has more importance (4,10) is the optimal solution. However, the pareto front which is solved using genetic algorithm is the local front. In future, the pareto optimization can be carried out using sophisticated techniques such as NSGA-II, NSGA-III or by using modern metaheuristics algorithms.

REFERENCES

Alexander, V. (2015). Application of queuing theory: Analysis of services of commercial banks (A case study of branches of some commercial banks in Abuja, Nigeria). *Journal of Science Teacher Education*, 48.

Ban, M., Yu, J., Li, Z., Guo, D., & Ge, J. (2019). Battery Swapping: An aggressive approach to transportation electrification. *IEEE Electrification Magazine*, *7*(3), 44–54. doi:10.1109/MELE.2019.2925762

Beatrice, Simone, Marco, Lucio &, & Ivan. (2018). *A queuing theory decision support model and discrete event simulations for the smart charging of electric vehicles*. 23rd Summer School F. Turco of Industrial Systems Engineering.

Deb, K. (2001). *Multi-Objective Optimization using Evolutionary Algorithms*. John Wiley & Sons Publications.

Harifi, S., Mohammadzadeh, J., & Khalilian, M. (2020). *Giza Pyramids Construction: an ancient-inspired metaheuristic algorithm for optimization*. Evol. Intel.

Khalilpourazari, S., & Khalilpourazary, S. (2017). An efficient hybrid algorithm based on Water Cycle and Moth-Flame Optimization algorithms for solving numerical and constrained engineering optimization problems. *Journal of Soft Computing*, *23*(5), 1699–1722. doi:10.100700500-017-2894-y

Li, Z., Amir, K., & Song, J. (2019). A comprehensive review of the key technologies for pure electric vehicles. *Energy, Elsevier*, *182*(C), 824–839.

Linyu, W., Wang, X., Sun, Q., & Lai, L. L. (2015). The service modeling of EV charging/swapping station and its cost benefit analysis. *10th International Conference on Advances in Power System Control, Operation & Management*, 1-7.

Manikanta, S., Neha, R., Deepa, K., & Sarada, J. (2020). Application of Aspiration Level Model in determining QoS for an EV battery charging station. *International Conference on Smart Technologies in Computing, Electrical and Electronics*, 71-77.

Mohd, Mohammad, Adil, & Asghar. (2019). A Comprehensive Review on Electric Vehicles Charging Infrastructures and their Impacts on Power-Quality of the Utility Grid. *Journal of eTransportation, 1*, 100006.

Neha, R., Manikanta, S., & Deepa, K. (2020). Integration of Battery Charging and Swapping using Metaheuristics: A review. *2nd International Conference on Machine Learning, Advances in Computing, Renewable Energy and Communication*, 1-10.

Poomrittigul, S., Koomsubsiri, A., Sasithong, P., Deenuch, D., & Wuttisittikulkij, L. (2020). The Simulation of Queuing Model for Bangkok Rapid Transit Train Ticket System Using Python. *34th International Technical Conference on Circuits/Systems, Computers and Communications*, 1-4.

Rahnamayan, S., Mahdavi, S., Deb, K., & Bidgoli, A. (2020). Ranking Multi-Metric Scientific Achievements Using a Concept of Pareto Optimality. *Mathematics*, *8*(6), 956. doi:10.3390/math8060956

Sasikumar, S., & Deepa, K. (2018). LCL topology based single stage boost rectifier topology for wireless EV charging. *Journal of Green Engineering, 8*, 573-596.

Saxena, A., & Deepa, K. (2020). Power quality analysis for electric vehicle charging and its mitigation strategies. *Test Engineering and Management,* 5409 – 5418.

Shabarish, P. R., Krishna, R. P., Prasanth, B., Deepa, K., Manitha, P. V., & Sailaja, V. (2020). Fuzzy based Approach to Enhance the Wireless Charging System in Electric Vehicles. *4th International Conference on Electronics, Communication and Aerospace Technology*, 176-181.

Shao, S., Guo, S., & Qiu, X. (2017). A Mobile battery swapping Service for Electric Vehicles Based on a battery swapping Van. *Energies*.

Taha, H. A. (2011). Operations Research (9th ed.). Pearson Publication.

Thiruvonasundari, D., & Deepa, K. (2020). Electric vehicle battery modelling methods based on state of charge– review. *Journal of Green Engineering, 10*, 24-61.

Till, Simon, Niklas, Patrick, Frances, & Anders. (2018). Fast charging infrastructure for electric vehicles: Today's situation and future needs. *Journal of Transportation Research Part D Transport and Environment, 8*.

Xu, Q., Wang, P., & Tiayang, Z. (2017). Optimal operation of battery swapping-charging systems considering quality-of-service constraints. *IEEE Power & Energy Society General Meeting*, 1-5.

Yusheng, P., Xicheng, Z., Qicheng, X., & Hongjiang, L. (2019). Reducing waiting time at an Airport Security Checkpoint based on the queuing model. *Chinese Control And Decision Conference*, 3860-3865.

Zeng, M., Pan, Y., Zhang, D., Lu, Z., & Li, Y. (2019). Data-Driven Location Selection for Battery Swapping Stations. *IEEE Access: Practical Innovations, Open Solutions*, *7*, 133760–133771. doi:10.1109/ACCESS.2019.2941901

Chapter 5

Power Quality Improvement for Grid-Connected Photovoltaic Panels Using Direct Power Control

Arezki Fekik
Akli Mohand Oulhadj University, Bouira, Algeria

Nashwa Ahmad Kamal
Faculty of Engineering, Cairo University, Egypt

Mohamed Lamine Hamida
Electrical Engineering Advanced Technology Laboratory (LATAGE), Mouloud Mammeri University, Algeria

Hakim Denoun
Electrical Engineering Advanced Technology Laboratory (LATAGE), Mouloud Mammeri University, Algeria

Hamza Houassine
Akli Mohand Oulhadj University, Bouira, Algeria

Sundarapandian Vaidyanathan
Vel Tech University, India

Ahmad Taher Azar
https://orcid.org/0000-0002-7869-6373
Prince Sultan University, Saudi Arabia

Aceng Sambas
Universitas Muhammadiyah Tasikmalaya, Indonesia

ABSTRACT

This chapter displays a control strategy for a photovoltaic system (PV) linked to the network with two phases of a PWM converter, where the first phase is a DC-DC converter linked among the photovoltaic source and the DC-AC converter. The second phase is a DC-AC converter linked to the grid. The maximum power point (MPP) is tracked by DC-DC converter, which increases the DC bus voltage. The P&O (perturbation and observation) technique is utilized as a direct current (DC-DC) converter controller to make the PV arrays work at greatest value of power under changing weather conditions. The DC-AC converter transfers the maximum power extracted from the PV cell into the grid. To improve the energy quality produced by the photovoltaic field other than the performance of the pulse width modulation (PWM) inverter, direct power control (DPC) is used to achieve these improvements. The simulation results showed a good performance of the suggested controller. Decoupled power control is achieved successfully, and a good power quality with low harmonic distortion rate (THD) is obtained.

DOI: 10.4018/978-1-7998-7447-8.ch005

INTRODUCTION

The primary source of energy is considered to be fossil fuel. The combustion of fossil fuels produces a large amount of ozone depleting substances, including carbon dioxide, which is the primary cause of global temperature rise and climate warming, causing significant environmental damage. Due to the limited storage of non-renewable petroleum-based energy, the use of sustainable energies such as PV cells, biomass, wind energy, geothermal, and so on has increased significantly in recent years (Meghni et al. 2017a,b,c; Shalaby et al., 2021; Ghoudelbourk et al., 2016, 2020, 2021; Kamal et al., 2020; Ammar et al., 2019a; Lalouni et al. 2009; Li et al. 2009). Because greenhouse gases are not emitted during renewable energy consumption, it can reduce the risk of the greenhouse effect while also effectively reducing non-renewable energy consumption.

Photovoltaic technology is one of the fastest developing renewable resources due to its abundant availability of solar lighting and lack of negative environmental impacts. Because of advancements in power electronic devices, the demand for energy produced by photovoltaic systems is increasing significantly (Qahouq and Jiang. 2014; Naidu and Singh, 2014; Rajendran et al., 2013; Bouafia et al., 2010). Photovoltaic cells have numerous advantages, including the absence of noise, pollution, and contamination, as well as low maintenance requirements, making them increasingly popular (Melit et al., 2009; Youssef et al., 2017; Amara et al., 2018, 2019a,b; Pilla et al., 2020, 2019). Photovoltaic (PV) technology is now used in a wide range of applications, which can be divided into two categories: autonomous PV systems and network-connected PV systems (Parida et al., 2011).

In remote areas where the association is expensive, autonomous photovoltaic systems with a bank of batteries are used for energy storage (Salas et al., 2006). In contrast, for grid-connected photovoltaic systems, the photovoltaic panels are linked to the utility network without using the battery bank to the extent that the accessible PV control is routed to the electricity grid (Jain and Agarwal. 2007).

The power generated by photovoltaic panels is typically measured using barometric conditions (sunlight and its irradiance or its temperature). Thus, in order to obtain the highest PV production value for a given working situation, a Maximum Power Point Tracker (MPPT) is used to improve control of the PV module's operating point (Ben Smida et al., 2018; Kamal and Ibrahim, 2018). A variety of MPPT algorithms, such as Hill-Climbing, Perturb and Observe (P&O), incremental conductance, intelligent MPPT methods, fractional order MPPT methods, and many others, have been proposed in the literature for improving the competence of the solar system (Joshi and Arora, 2017; Subudhi and Pradhan. 2013; Kamal and Ibrahim, 2018). On the other hand, advancements in power electronics have resulted in the widespread use of non-linear loads, which has resulted in a slew of issues such as harmonics, current disturbance, and so on, which have been introduced into the electrical network. Harmonics reduce power efficiency, power factor, and losses, as well as causing electromagnetic interference with nearby communication correspondence lines and other undesirable outcomes (Li et al., 2009).

The primary goal of modern electrical system research is to improve the quality of energy in modern electrical distribution systems. Finally, after two decades, the majority of loads used by industries and clients were passive and linear, with fewer non-linear loads having less of an impact on the power framework. The introduction of semiconductors and power electronics devices, as well as their ease of control, encourages the use of non-linear loads such as choppers, inverters, rectifiers, and so on. Active power generation is prioritized over power quality improvement. At that point, remuneration for reactive power is required over relief of current harmonics consumed by non-linear loads.

In grid-connected PV systems, the direct (DC) voltage is typically converted to the alternate (AC) voltage using an electronic power converter pulse width modulation (PWM), and system performance is determined by the control technique of a PWM inverter (Sinha et al. 2018; Tareen et al., 2017; Tahir et al., 2018; Elmelegi et al., 2019; Mahmud et al., 2017; Attia, 2019; Malakondareddy et al., 2019; Ahmad & Singh, 2017). There are various energy management configurations for the interface of PV systems connected to the network, such as one-level and two-level structures. The number of steps associated with the PV system that is connected to the network is a critical issue because it determines the overall competence, unwavering quality, and complexity of control in such systems. The elements of the DC-DC and DC-AC control converters are free in a photovoltaic frame coupled to the double-stage network, which influences the comprehensive control to the less demanding assignment than the one-step design.

To achieve highly efficient power generation, the network's photovoltaic inverter is controlled by the injection of a sinusoidal AC current that is synchronized with the supply voltage (Priyadarshi, et al., 2018; Al-Shetwi, et al., 2018; Foti et al., 2018). By selecting an appropriate control strategy, the power interface between the PV system and the grid can become increasingly proficient. Typically, the Voltage Source Inverter (VSI) control structure includes an internal current loop that has a global influence on the system's performance. Some current control systems, such as hysteresis current control, predictive control, dead band control, and voltage control with PWM modulator, have been proposed in the literature (Tsengenes and Adamidis, 2011; Hassaine et al., 2009; Dalapati and Chakraborty, 2009; Blaabjerg et al., 2006). Voltage-Oriented Control (VOC) is the most widely used control method, followed by all others (Fekik et al., 2019, 2015).

To achieve a single rapidly feedback, this strategy employs an external DC connection voltage control loop and an internal current control loop. The execution of the power flow is dependent on the nature of the connected current's control strategy. Another type of high-performance control strategy for PWM converters based on instantaneous power theory is the DPC command (Noguchi et al. 1998; Hu and Zhu. 2011; Atia and Salem, 2015; Salem and Atia, 2015; Fekik et al., 2015a,b; Fekik et al., 2016; Fekik et al., 2017; Hamida et al., 2017 Fekik et al., 2018a,b,c,d,e; Denoun et al., 2018; Hamida et al., 2018; Hamida et al., 2019a,b,c; Fekik et al., 2019a,b; Fekik et al., 2020a,b,c,d). Because the switching states of the inverter are correctly selected by a switching table, the internal current control loop and the pulse width modulator do not exist in that control technique. The latter is affected by errors in the instantaneous active and reactive powers, as well as the angular position of the measured voltages. Direct power control, as opposed to voltage-oriented control, has a simpler calculation, no current loops, no different PWM voltage modulator, no separating active and reactive control, and better performance.

Direct power control's notable shortcomings are the variable and higher exchange rate. Furthermore, the angular data of the grid voltage is required because the inverter's output vectors are primarily determined by these angles. In this case, a PLL (Phase Locked Loop) is required to delete the data, as in the VOC. As a result, in this study, direct power control is used in a PWM inverter powered by a photovoltaic panel and linked to the power grid to improve energy quality.

The chapter is organized as follows. The model of the system to be studied is presented in Section 2. Section 3 depicts the modeling of a photovoltaic panel. The fourth section will go over the modeling of the Boost converter as well as the Perturb & Observer method. In the fifth section, the DPC model is presented. The results of the simulation, as well as a discussion of the results, are displayed in the sixth section. Finally, the conclusion is shown in the seventh section.

PROPOSED METHODOLOGY

Figure 1 depicts the proposed power supply system and control schemes for the photovoltaic side with grid connection. The first phase is the DC / DC boost converter, which raises the relatively low voltage delivered by the solar panel to an appropriate level (300 V) and uses the perturb and observe algorithm to track the maximum power point (MPPT) for the bus on the continuous side. The second phase is the DPC-controlled PWM inverter, which selects switching states from a table based on the errors between the estimated values and the active and reactive power references.

Figure 1. Block diagram of grid connected PV system using DPC

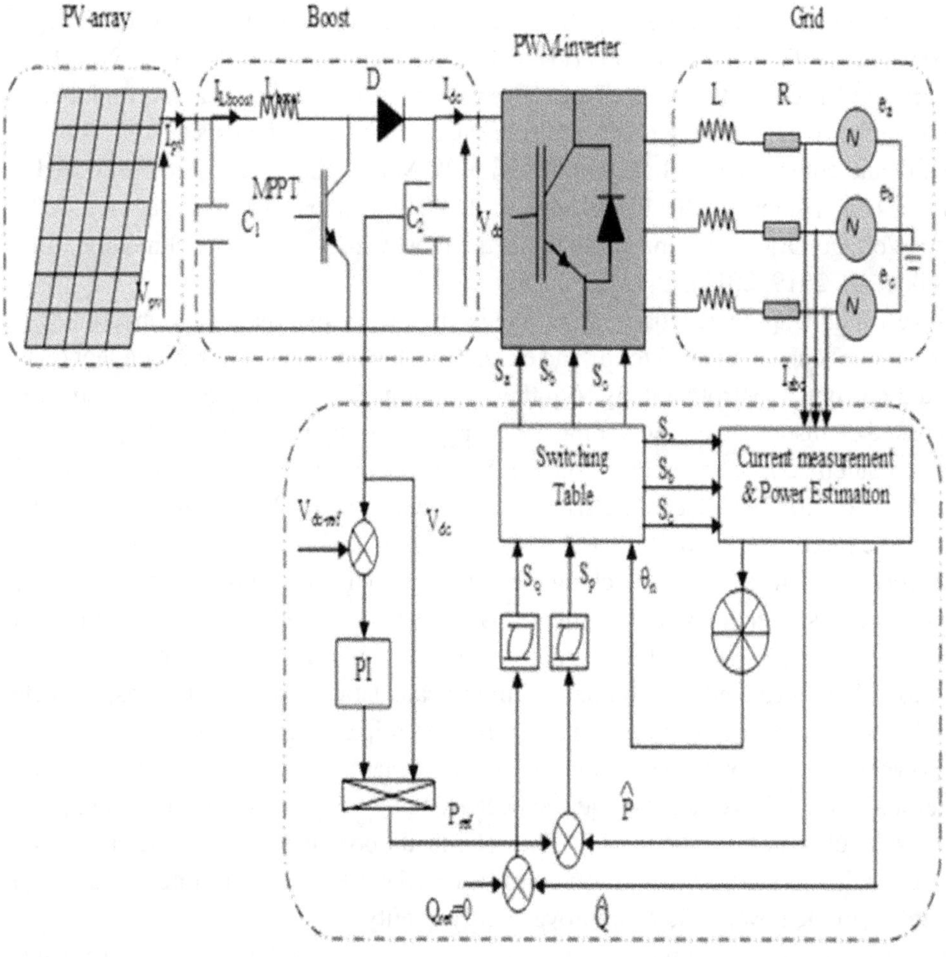

Modelling of the Photovoltaic Panel

Figure 2 depicts the equivalent circuit of a PV system, which consists of a photovoltaic current source, one diode, a shunt resistor, and a series resistor (Lalouni et al., 2009; Li et al., 2009). The five-parameter model shown in Fig. 2(a) can be simplified into the model shown in Fig. 2(b) (b).

Figure 2. Equivalent circuit of PV cell, (a) five parameters model; (b) simplified four parameters model

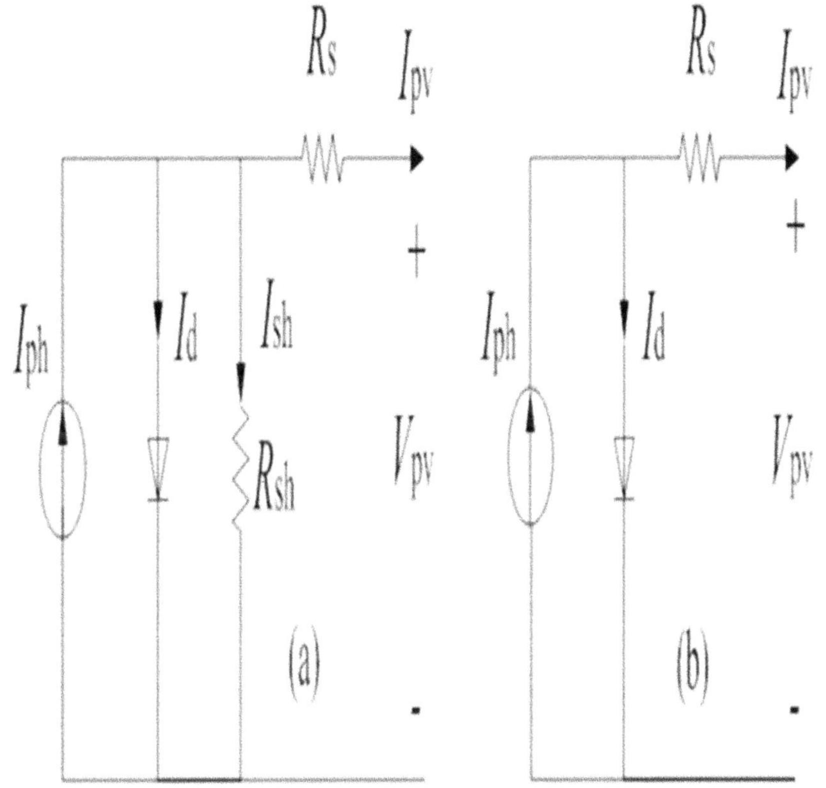

The PV cell model is typically depicted as follows:

$$I_{pv} = I_{ph} - I_d \quad (1)$$

$$I_d = I_0 \left[\exp\left((V_{pv} - R_s I) / V_t \right) - 1 \right] \quad (2)$$

$$I_{ph} = G / G_{ref} \left[I_{ph,ref} + \lambda_I \left(T_c - T_{c,ref} \right) \right] \quad (3)$$

Where I_{ph} is the light current, I_{pv} is current of the load, I_0 is the saturation current, V_{pv} is the output voltage of PV, R_s is the series resistance of PV (Ω), V_t is the thermal voltage, G is the irradiation, T_c is the temperature of cell in Kelvin and λ_I is coefficient of the temperature. When sunlight strikes the surface of the solar panel, the current source generates a photocurrent, and the diode represents the solar cell's p-n transition area. Table 1 shows the specifications of a typical PV module. A MATLAB/Simulink package is used to implement the specifications. Figure 3 depicts the PV power produced for various solar radiation data. Two pairs of ten PV panels are connected in parallel to produce the required power, along with a dc/dc boost converter connected to a dc bus.

Figure 3 depicts the electrical characteristics of the solar cell's output power. Equation can be used to calculate the output power of a PV module (4). The PV array is designed with six series-connected modules per string and ten parallel strings to generate 15 kW of power. In addition, Fig. 3 depicts the output power of a PV module based on standard radiation at 1000 w/m2 and different temperatures (45 and 25 °C)

$$P = V_{pv} . I_{pv} = N_s V_{cell} * N_p I_{cell} \tag{4}$$

Figure 3. Output power, current vs. voltage at different temperature

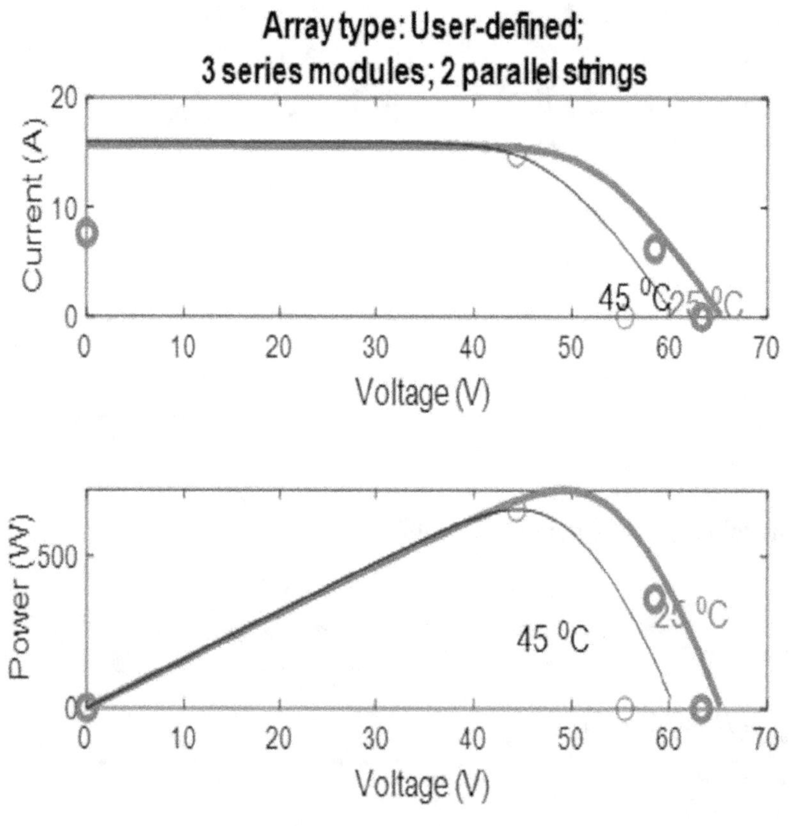

Table 1. PV specification of module data

Parameters	Values
Max power	60.45 W
Voltage at Max Power Point V_{mpp}	19.5 V
Current at Max Power Point I_{mpp}	3.1 A
Open Circuit Voltage V_{oc}	21.1 V
Short Circuit Current I_{sc}	3.8 A
Total Number of Cells per Module (N_{cell})	36
Number of Cells in Parallel	2
Number of Cells in Series	3

Modelling of the DC/DC Converter

The Boost converter can be described using ordinary differential equations as follows:

$$I_{Lboost} = I_{pv} - C_1 \frac{dV_{pv}}{dt} \quad (5)$$

$$I_{dc} = (1-\alpha) I_{LBoost} - C_2 \frac{dV_{dc}}{dt} \quad (6)$$

$$V_{pv} = (1-\alpha) V_{dc} - L_{Boost} \frac{dI_{LBoost}}{dt} \quad (7)$$

Where α is the duty cycle and C_1, L_{boost}, C_2 are the input capacitor, inductor, and the output capacitor respectively. The switch of the Boost Converter is "on" for a period equal to αT, and the switch is "OFF" for a time interval equal to (1 - α) T. I_{dc} and V_{dc} are the output current and the output voltage of the Boost converter.

Maximum Power Point Tracking Techniques

The primary goal of the MPPT control is to extract and track the maximum power point from the PV module. Taking weather conditions into account, the maximum power point corresponding to the best operating point is obtained. The PV module's power fluctuates in response to radiation and temperature. The MPPT of a PV system is typically shown in Fig. 4.

Figure 4. MPPT of PV system

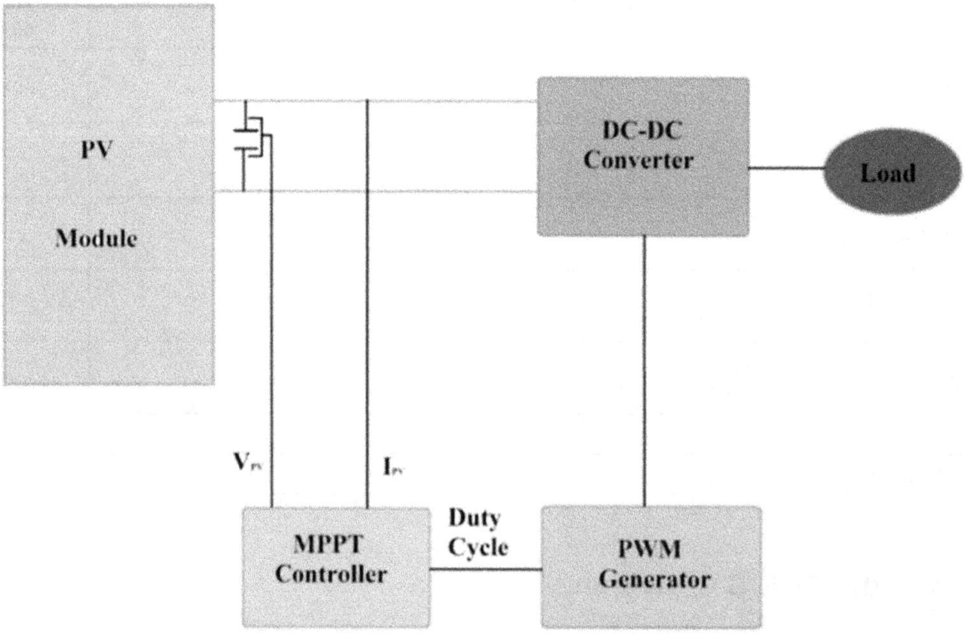

MPPT control strategies play an important role in PV systems because they increase the output power of a PV system, thereby increasing the array's efficiency. There are various methods for tracking the most extreme power point. The traditional technique is the Perturbation and Observation (P&O) strategy. Because its algorithm is simple to implement, the P&O perturbation and observation technique is widely used. This operation is carried out by disrupting the system by increasing or decreasing the operating voltage of the module and then observing the effect on the module's generated power.

The flowchart of the P&O algorithm as it should be applied on the control microprocessor is shown in Fig. 5. According to Fig. 5, the calculation of the current output power P(k) is dependent on the current (I) and voltage measurements (V). The value P (k) is then compared to the prior value P (k -1) produced by the previous measurement. The disturbance will continue in the same direction if the output power is increased. If the power is reduced from one of the most recent measurements, the output voltage disturbance will revert to the previous cycle's direction. The operating voltage is disrupted at each cycle of the MPPT using this technique. Once the MPP is reached, it will oscillate around the ideal operating voltage V. This results in a power loss that is determined by the step width of a single C_p perturbation. When C_p is large, the MPPT algorithm responds quickly to sudden changes in operating conditions, but losses are increased under stable or slightly changing conditions. If C_p is very small, losses will be reduced under stable or changing conditions, but the system will no longer be able to keep up with rapid temperature or insulation changes.

Figure 5. Flowchart of the perturb and observe algorithm (C_p is the perturbation step width)

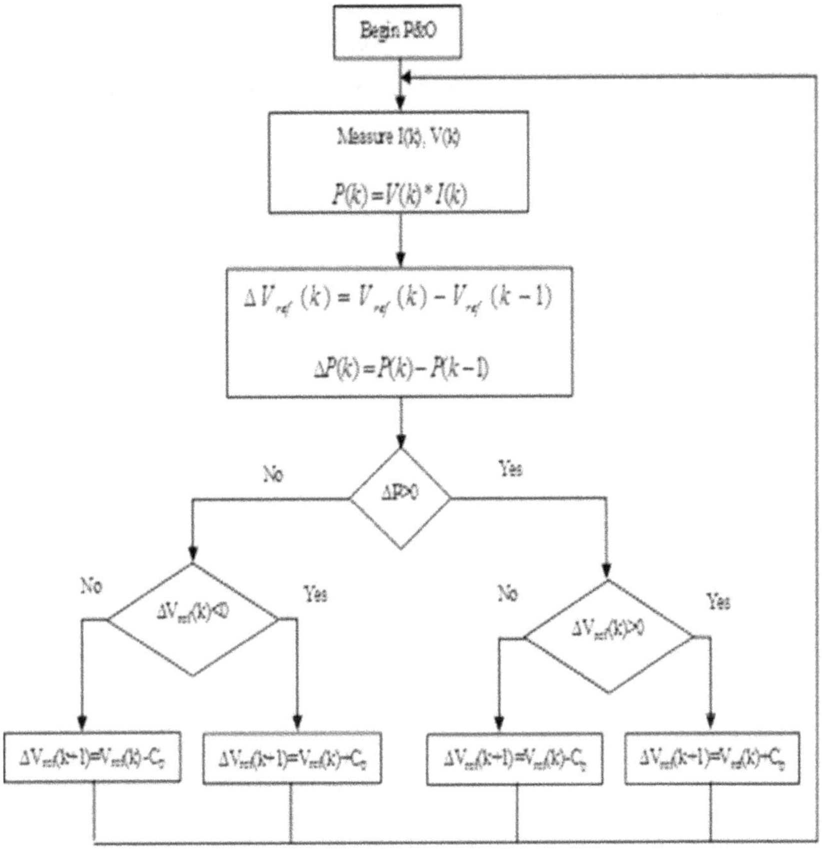

MODELLING OF THE DIRECT POWER CONTROL

Recent decades have witnessed many important developments related to the design of nonlinear systems for many practical applications. Several inspiring approaches have been proposed, such as optimal control, nonlinear feedback control, adaptive control, sliding mode control, nonlinear dynamics, chaos control, chaos synchronization control, fuzzy logic control, fuzzy adaptive control, fractional order control, and robust control and their integrations (Kazim et al., 2021; Ajeil et al., 2020; Koubaa & Azar, 2021; Azar & Kamal, 2021; Abdul-Adheem et al., 2020a,b; Najm et al., 2020; Liu et al., 2020; Humaidi et al., 2020; Elkholy et al., 2020a,b; Mohamed et al., 2020; Ibrahim et al., 2020; Sayed et al., 2020; Barakat et al., 2020; Azar & Vaidyanathan, 2015a,b,c, 2016; Sallam et al., 2020; Ammar et al., 2018; Abdelmalek et al., 2018; Gorripotu et al., 2021; Mekki et al., 2015; Vaidyanathan & Azar, 2020, 2015a,b,c,d, 2016a,b,c,d,e,f,g, 2017a,b,c; Azar & Serrano, 2014, 2015a,b,c,d, 2016a,b, 2017, 2020; Grassi et al., 2017; Ouannas et al., 2016a, Singh et al., 2017; Vaidyanathan et al, 2017a,b,c, 2015a,b,c, 2019; Wang et al., 2017; Giove et al., 2013; Khennaoui et al., 2020a,b; Kammogne et al. 2020; Alain et al., 2020, 2019, 2018a,b; Ammar et al., 2020a,b, 2019a,b; Ammar & Azar, 2020; Ibraheem et al., 2020; Kammogne et al., 2020; Ouannas et al., 2021, 2020a,b,c,d,e, 2019a,b, 2017a,b,c,d,e,f,g, 2016b; Bouakrif et al., 2019; Djeddi et al. 2019; Radwan et al., 2017, 2018; Tolba et al. 2017; Mittal et al. 2021; Singh & Azar 2020;

Singh et al., 2018; Khan et al. 2020a,b; Azar et al., 2021a,b, 2020a,b,c,d,e,f,g,h,i.j,k,l, 2018a,b,c; Azar & Serrano 2018, 2019; Gorripotu et al., 2019; Kumar et al. 2018; AbdelAty et al., 2018; Pham et al., 2018, 2017a,b; Shukla et al., 2018; Khettab et al. 2018; Soliman et al. 2017).

As shown in Fig. 1, the DPC structure consists of a switching table that is applied to a three-phase PWM inverter. It works in the same way as direct torque control (DTC) in induction machines. Instead of torque and stator flux, the controlled quantities are active and reactive instantaneous powers.

The DPC is accomplished by selecting a control vector from a switching table. The latter is determined by the digitized error S_p, S_q of the instantaneous active measurement and reactive powers, as well as the estimated voltage's angular position. To define the working area, the plane (αβ) is divided into twelve sectors. The active power reference is obtained by multiplying the controller output by the DC bus voltage and calculating the difference between the sensed (continuous) voltage and its reference. The reference reactive power is directly imposed equal to zero in order to achieve a unit power factor. The reactive power commands Q_{ref} and the active power P_{ref} are compared with the reactive ($\hat{Q}$) and active ($\hat{P}$) values evaluated in the active and reactive power hysteresis controllers. The output signal S_p of the active power controller is defined as follows:

$$\begin{cases} S_p = 1 \Rightarrow P_{est} < P_{ref} - h_p \\ S_p = 0 \Rightarrow P_{est} > P_{ref} - h_p \end{cases} \tag{8}$$

Similarly, for the output signal S_q of the reactive power controller is defined as follows:

$$\begin{cases} S_q = 1 \Rightarrow Q_{est} \prec Q_{ref} - h_q \\ S_q = 0 \Rightarrow Q_{est} \succ Q_{ref} - h_q \end{cases} \tag{9}$$

The area of the estimated voltage must be known in order to define the optimal switching states. The voltage vector space can be divided into twelve or six sectors, as shown in Fig. 6. These sectors can be mathematically defined as follows (Fekik et al. 2016):

$$(n-2)\frac{\pi}{6} < \theta_n < (n-1)\frac{\pi}{6} \tag{10}$$

Where n = 1,2,3 ... 12 indicating the sector number. It is instantaneously provided by the position of the voltage vector and is calculated as follows:

$$\theta_n = arctg(\frac{v_\beta}{v_\alpha}) \tag{11}$$

Figure 6. Voltage vector position in a fixed frame

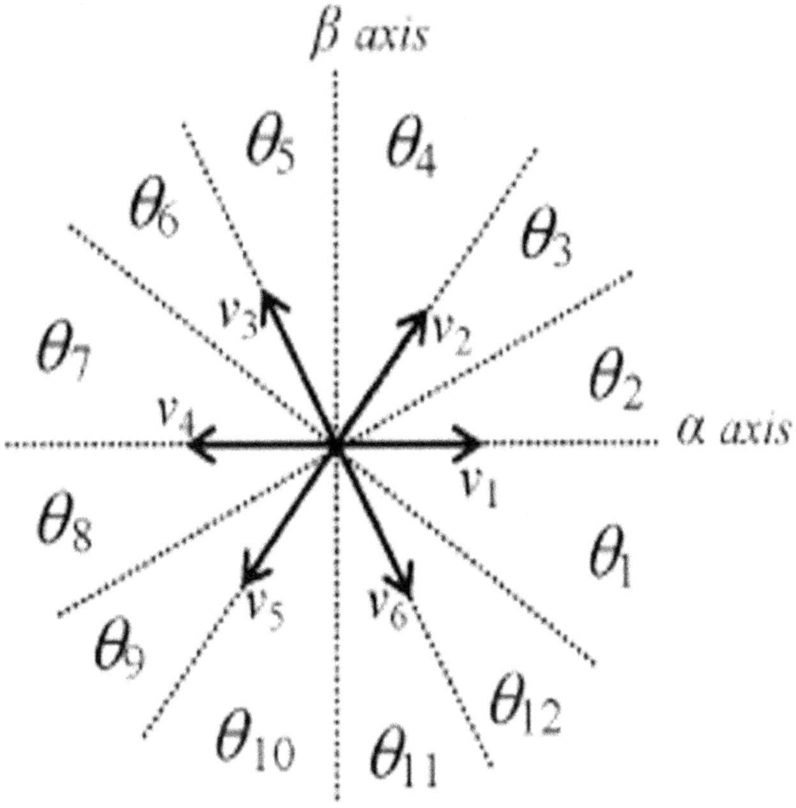

DPC BASED ON VOLTAGE CALCULATION

The sensors should be disconnected from the AC voltage so that an instantaneous active and reactive power as a function of the switching states and the DC voltage can be proposed. This can be expressed as (Fekik et al. 2016):

$$\begin{aligned} \hat{P} &= L\left(\frac{di_a}{dt} + \frac{di_b}{dt} + \frac{di_C}{dt}\right) + V_{dc}(S_a i_a + S_b i_b + S_c i_c) \\ \hat{Q} &= \frac{1}{\sqrt{3}}\left[L\left(\frac{di_a}{dt} i_c - \frac{di_c}{dt} i_a\right) + V_{dc}(S_a(i_b - i_c) + S_b(i_c - i_a) + S_c(i_a - i_b))\right] \end{aligned} \tag{12}$$

The first of the two terms in (12) displays the power in the line inductors, where the internal resistances of the inductors are ignored because the dissipated active power in these resistors is much lower than the involved power. The other terms in the equation represent the power in the converter.

Predicted Voltage

To define the orders, the line voltage method is required. Furthermore, in order to increase the power factor, it is necessary to accurately predict the value of the line voltage in the presence of harmonics. The voltage drop over the inductor can be calculated by inferring the current. As a result, the voltage can be estimated by multiplying the reference voltage by the estimated voltage drop at the converter input (Fekik et al. 2016). This method, on the other hand, has a disadvantage in that the noise is amplified by the current derivative. To avoid this issue, a voltage estimation based on power calculation can be used. The next equation (13) displays the line currents i_a, i_b, i_c in the stationary coordinates α, β.

$$\begin{bmatrix} i_\alpha \\ i_\beta \end{bmatrix} = \sqrt{2/3}\begin{bmatrix} 1 & \frac{-1}{2} & \frac{-1}{2} \\ 0 & \frac{\sqrt{3}}{2} & \frac{-\sqrt{3}}{2} \end{bmatrix}\begin{bmatrix} i_a \\ i_b \\ i_c \end{bmatrix} \tag{13}$$

The equations of the active and reactive powers are expressed as follows:

$$\begin{aligned} \bar{p} &= v_\alpha i_\alpha + v_\beta i_\beta \\ \bar{q} &= v_\alpha i_\beta - v_\beta i_\alpha \end{aligned} \tag{14}$$

This can be written in a matrix from as:

$$\begin{bmatrix} \hat{p} \\ \hat{q} \end{bmatrix} = \begin{bmatrix} v_\alpha & v_\beta \\ -v_\beta & v_\alpha \end{bmatrix}\begin{bmatrix} i_\alpha \\ i_\beta \end{bmatrix} \tag{15}$$

The matrix equation (15) is represented as a function of the measured line current and the estimated power as follows:

$$\begin{bmatrix} \hat{v}_\alpha \\ \hat{v}_\beta \end{bmatrix} = \frac{1}{i_\alpha^2 + i_\beta^2}\begin{bmatrix} i_\alpha & -i_\beta \\ i_\beta & i_\alpha \end{bmatrix}\begin{bmatrix} \hat{P} \\ \hat{Q} \end{bmatrix} \tag{16}$$

Concordia's inverse transform of line voltage is displayed as:

$$\begin{bmatrix} \hat{v}_a \\ \hat{v}_b \\ \hat{v}_c \end{bmatrix} = \sqrt{2/3} \cdot \begin{bmatrix} 1 & 0 \\ -\dfrac{1}{2} & \dfrac{\sqrt{3}}{2} \\ -\dfrac{1}{2} & \dfrac{\sqrt{3}}{2} \end{bmatrix} \cdot \begin{bmatrix} \hat{v}_\alpha \\ \hat{v}_\beta \end{bmatrix} \tag{17}$$

Switching Table

The switchboard for direct power control is shown in Table 2. The initiator of the DTC created the switching table. The voltage vector at the PWM rectifier's input is determined by the switching states Sa, Sb, and Sc. Eight voltage vectors can be applied to the rectifier's input due to the various possible combinations of these three states. The digitized errors S_p, S_q and the working sector θ_n are the inputs of this table where the switching states S_a, S_b and S_c of the converter are stored. The optimum switching state of the converter is selected in each switching state due to the merging of the digital signals S_p, S_q and the number of sectors. Therefore, the choice is made where the error of the active power can be restricted in a band with hysteresis width of $2\Delta h_p$ and the error of the reactive power with a band of $2\Delta h_q$.

Table 2. Switching table

SS_p	SS_q	$\theta\theta_1$	$\theta\theta_2$	$\theta\theta_3$	$\theta\theta_4$	$\theta\theta_5$	$\theta\theta_6$	$\theta\theta_7$	$\theta\theta_8$	$\theta\theta_9$	$\theta\theta_{10}$	$\theta\theta_{11}$	$\theta\theta_{12}$
00	00	VV_6	VV_1	VV_1	VV_2	VV_2	VV_3	VV_3	VV_4	VV_4	VV_5	VV_5	VV_6
0	11	VV_1	VV_2	VV_2	VV_3	VV_3	VV_4	VV_4	VV_5	VV_5	VV_6	VV_6	VV_1
1	00	VV_5	VV_6	VV_6	VV_1	VV_1	VV_2	VV_2	VV_3	VV_3	VV_4	VV_4	VV_5
11	11	VV_3	VV_4	VV_4	VV_5	VV_5	VV_6	VV_6	VV_1	VV_1	VV_2	VV_2	VV_1
$V_1(100)$, $V_2(110)$, $V_3(010)$, $V_4(011)$, $V_5(001)$, $V_6(101)$													

RESULTS AND DISCUSSIONS

The photovoltaic system is connected to the grid, and the overall control systems are simulated with the MATLAB / SIMULINK software. Two simulation tests have been run: keeping the active power constant and varying the reactive power of reference.

Constant Active Power Test

The active and reactive power are kept constant in this test. Figure 7 depicts the DC supply voltage of the PWM-inverter supplied by the Photovoltaic Panels. This voltage is held constant at a predetermined reference voltage.

Figure 7. Input DC voltage for PWM inverter

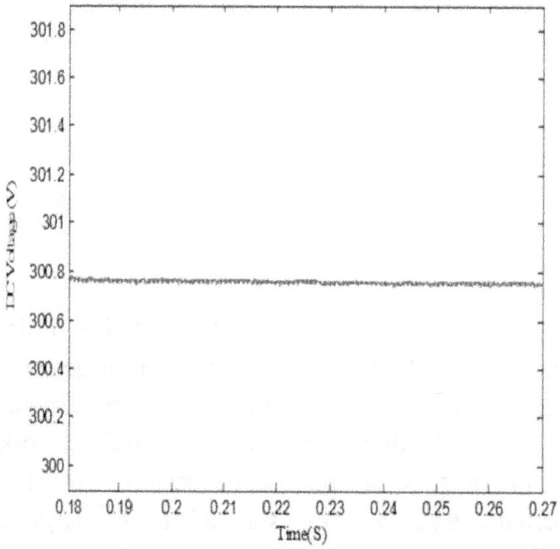

Figure 8 shows the active reference power as well as the estimated value. The estimated active power closely matches the reference calculated by a proportional integral (PI) controller, which regulates the direct voltage delivered by the solar panels.

Figure 8. Active power

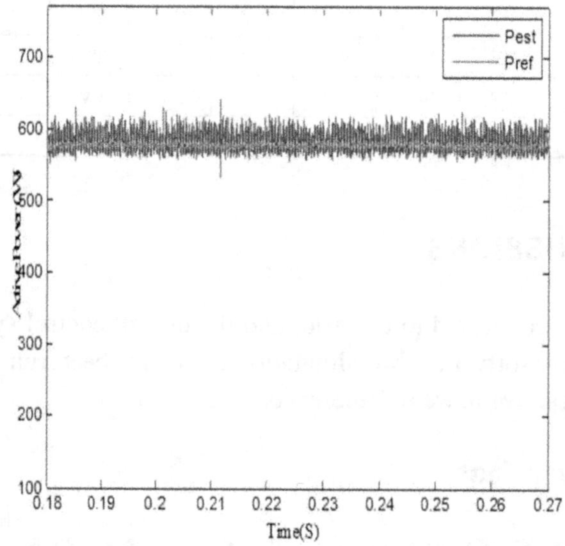

Figure 9 depicts the reactive reference power (red) as well as the predicted value (black). The estimated reactive power follows its reference, which is equal to zero, and thus a unit power factor is possible.

Figure 9. Reactive power

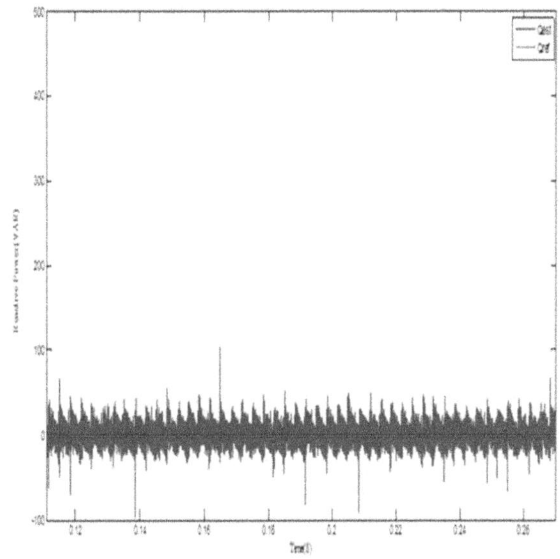

Figure 10 depicts the almost sinusoidal line currents. This results in very high energy quality and a low THD of the current transmitted to the network by the solar panels.

Figure 10. Line current

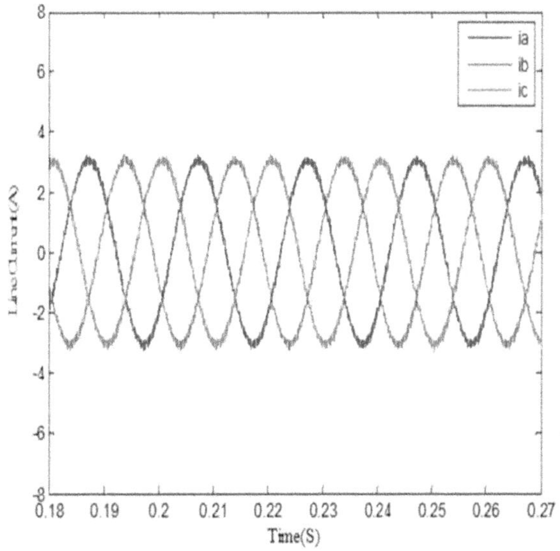

Variation of the Reference Reactive Power Test

In this test, the active power is held constant while the reference reactive power varies. Figure 11 shows that the DC voltage transferred by the photovoltaic panels is kept at the desired reference level.

Figure 11. Input DC voltage for PWM inverter with variation of the reactive power

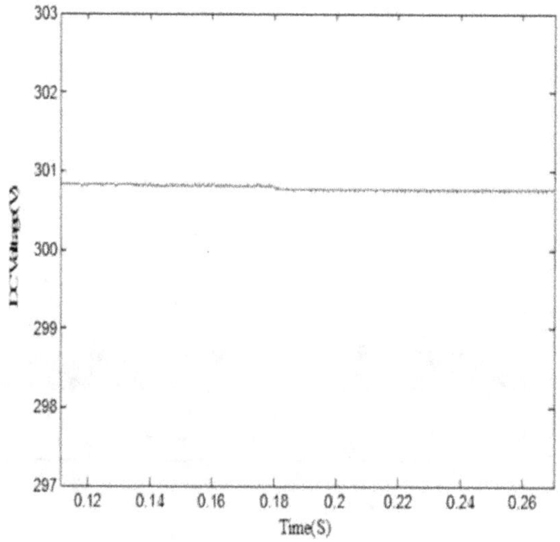

As shown in Fig. 12, the estimated active power follows its reference calculated by regulating the DC voltage of the solar panels even when the reactive power is varied (as shown in Fig. 13), confirming the active and reactive power decoupling control.

Figure 12. Active power with variation of the reactive power

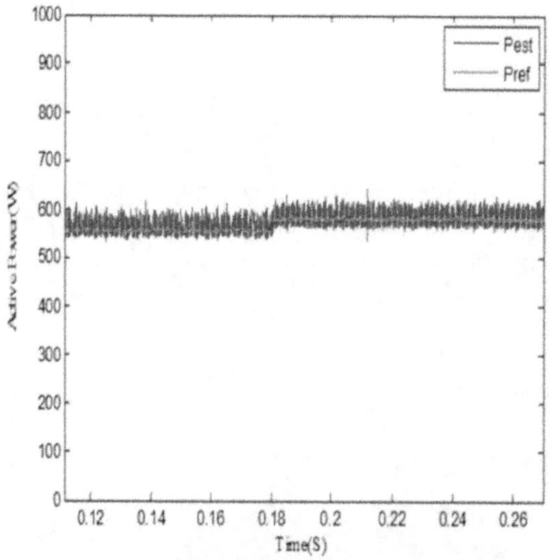

Figure 13. Reactive power with variation of the reactive power

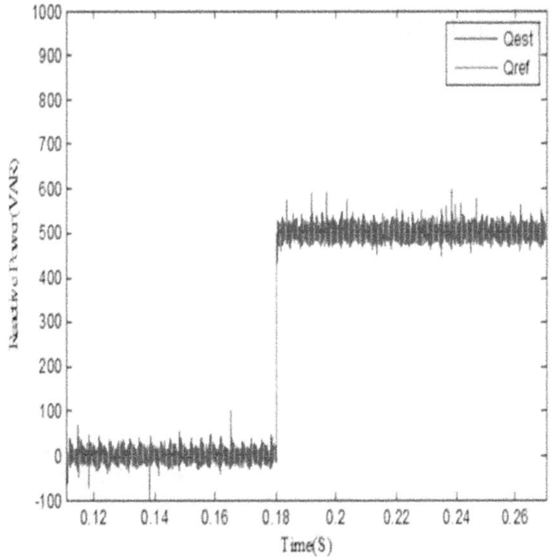

The line currents injected into the PWM-inverter network are almost sinusoidal in shape, as shown in Fig. 14. As a result, a low THD of the current is obtained, and a very high quality of energy is transferred to the grid, confirming a non-polluting energy conversion system.

Figure 14. Line current with variation of the reactive power

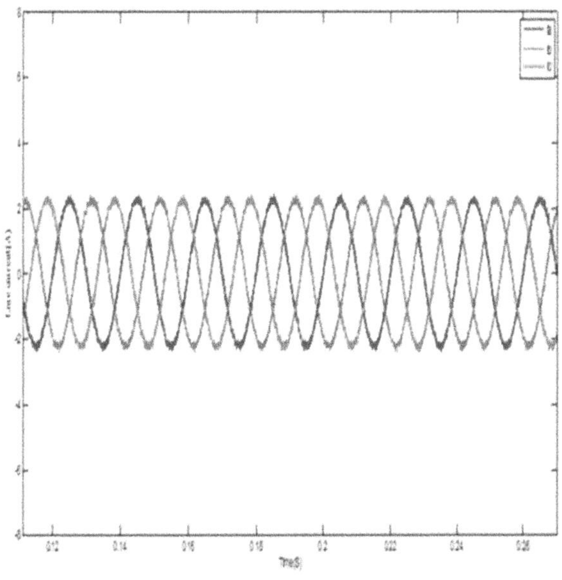

Figure 15 depicts the PWM-inverter output voltage prior to filtering with direct power control. Figure 16 depicts the harmonic spectrum of the current, and the value of THD is less than that required by the IEEE standard, confirming that the DPC-controlled system is non-polluting and ensures a high quality of the current injected into the distribution network.

Figure 15. Output voltage from Phase a of the PWM-inverter before filtering with variation of the reactive power

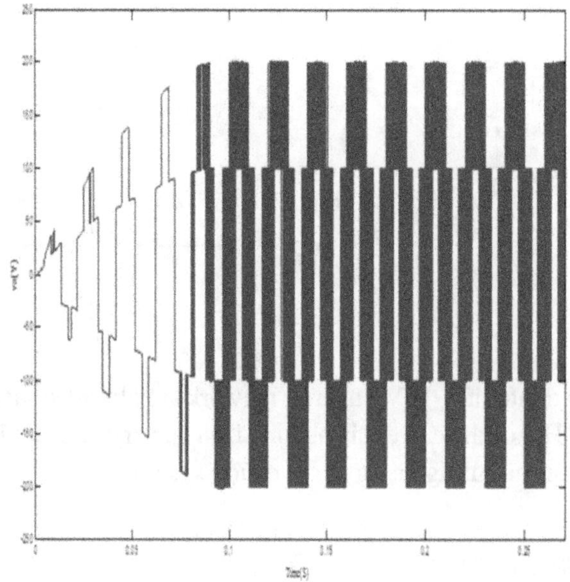

Figure 16. Harmonic Spectrum of line current (THD=1.26%) with variation of the reactive power

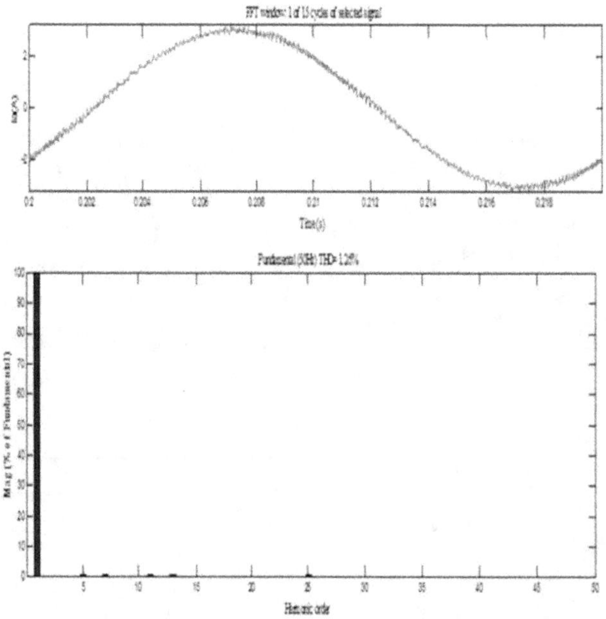

Figures 17 and 18 show the currents and DC voltages delivered by a photovoltaic source from an experimental platform. The load is purely resistive, and the experimental parameters are available at the laboratory level. The DC voltage delivered by the panel is kept at the desired level, and the line currents flowed by the PWM-inverter are sinusoidal.

Figure 17. Experimentation results of DC link Voltage

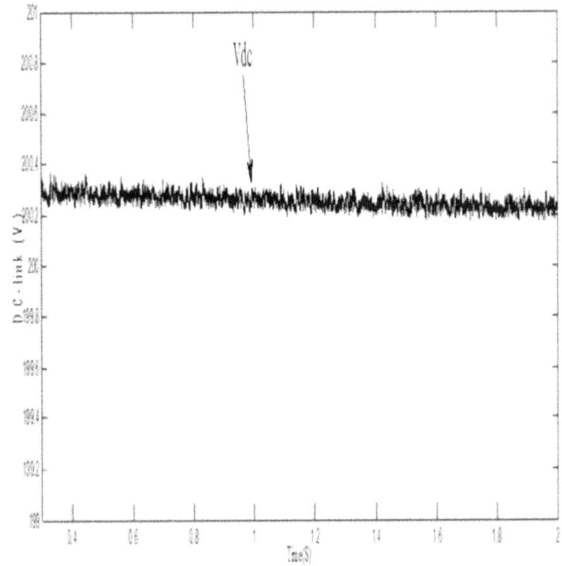

Figure 18. Experimentation results of line currents

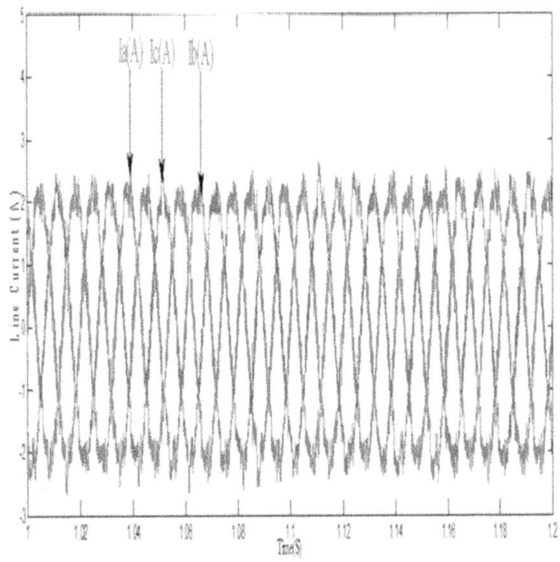

CONCLUSION

The modeling and analysis of a photovoltaic power generation system linked to a three-phase network by a PWM inverter controlled by DPC is presented in this chapter to improve energy quality. As an MPPT algorithm, the Perturb & Observe (P & O) algorithm is used to provide the optimal operating cycle of the boost converter. In comparison to the Voltage Oriented Control (VOC) method, active and reactive power decoupled control is performed without the use of a decoupling system and without changing the coordinates. The DC voltage is adjusted so that it is sensitive to all changing conditions. Large dynamic exposures and decoupled control of active and reactive power were successfully achieved, according to experimental and simulation results. Low THD currents are obtained, confirming the nature of the energy transferred to the grid by photovoltaic panels using the proposed direct power control via the PWM-Inverter control. Replace the two-voltage inverter with a three-voltage inverter is one of the research directions of the proposed study. Another future direction is to design an energy storage system.

REFERENCES

Abdel-Salam, M., El-Mohandes, M. T., & El-Ghazaly, M. (2020). An Efficient Tracking of MPP in PV Systems Using a Newly-Formulated P&O-MPPT Method Under Varying Irradiation Levels. *Journal of Electrical Engineering & Technology, 15*(1), 501–513. doi:10.100742835-019-00283-x

AbdelAty, A. M., Azar, A. T., Vaidyanathan, S., Ouannas, A., & Radwan, A. G. (2018). Applications of continuous-time fractional order chaotic systems. In *Mathematical Techniques of Fractional Order Systems', Advances in Nonlinear Dynamics and Chaos (ANDC)* (pp. 409–449). Elsevier. doi:10.1016/B978-0-12-813592-1.00014-3

Abdelmalek, S., Azar, A. T., & Dib, D. (2018). A novel actuator fault-tolerant control strategy of DFIG-based wind turbines using Takagi-Sugeno Multiple models. *International Journal of Control, Automation, and Systems, 16*(3), 1415–1424. doi:10.100712555-017-0320-y

Abdul-Adheem, W. R., Azar, A. T., Ibraheem, I. K., & Humaidi, A. J. (2020b). Novel Active Disturbance Rejection Control Based on Nested Linear Extended State Observers. *Applied Sciences (Basel, Switzerland), 10*(12), 4069. doi:10.3390/app10124069

Abdul-Adheem, W. R., Ibraheem, I. K., Azar, A. T., & Humaidi, A. J. (2020a). Improved Active Disturbance Rejection-Based Decentralized control for MIMO Nonlinear Systems: Comparison with The Decoupled Control Scheme. *Applied Sciences (Basel, Switzerland), 10*(7), 2515. doi:10.3390/app10072515

Ahmad, Z., & Singh, S. N. (2017). An improved single phase transformerless inverter topology for grid connected PV system with reduce leakage current and reactive power capability. *Solar Energy, 157*, 133–146. doi:10.1016/j.solener.2017.08.007

Ajeil, F., Ibraheem, I. K., Azar, A. T., & Humaidi, A. J. (2020). Autonomous Navigation and Obstacle Avoidance of an Omnidirectional Mobile Robot Using Swarm Optimization and Sensors Deployment. *International Journal of Advanced Robotic Systems, 17*(3), 1–15. doi:10.1177/1729881420929498

Ajeil, F. H., Ibraheem, I. K., Azar, A. T., & Humaidi, A. J. (2020). Grid-Based Mobile Robot Path Planning Using Aging-Based Ant Colony Optimization Algorithm in Static and Dynamic Environments. *Sensors (Basel)*, *20*(7), 1880. doi:10.339020071880 PMID:32231091

Al-Shetwi, A. Q., Sujod, M. Z., & Blaabjerg, F. (2018). Low voltage ride-through capability control for single-stage inverter-based grid-connected photovoltaic power plant. *Solar Energy*, *159*, 665–681. doi:10.1016/j.solener.2017.11.027

Alain, K. S. T., Azar, A. T., Bertrand, F. H., & Romanic, K. (2019). Robust observer-based synchronisation of chaotic oscillators with structural perturbations and input nonlinearity. *International Journal of Automation and Control*, *13*(4), 387–412. doi:10.1504/IJAAC.2019.100467

Alain, K. S. T., Azar, A. T., Kengne, R., & Bertrand, F. H. (2020). Stability analysis and robust synchronisation of fractional-order modified colpitts oscillators. *International Journal of Automation and Control*, *14*(1), 52–79. doi:10.1504/IJAAC.2020.103806

Alain, K. S. T., Romanic, K., Azar, A. T., Vaidyanathan, S., Bertrand, F. H., & Adele, N. M. (2018a), Dynamics analysis and synchronization in relay coupled fractional order colpitts oscillators. In Advances in Systems Analysis, Software Engineering, and High Performance Computing (ASASEHPC). IGI Global. doi:10.4018/978-1-5225-4077-9.ch011

Alain, K. S. T., Romanic, K., Azar, A. T., Vaidyanathan, S., Bertrand, F. H., & Adele, N. M. (2018b). *Dynamics Analysis and Synchronization in Relay Coupled Fractional Order Colpitts Oscillators. Advances in System Dynamics and Control*. IGI-Global. doi:10.4018/978-1-5225-4077-9

Amara, K., Bakir, T., Malek, A., Hocine, D., Bourennane, E. B., Fekik, A., & Zaouia, M. (2019a). An Optimized Steepest Gradient Based Maximum Power Point Tracking for PV Control Systems. *International Journal on Electrical Engineering & Informatics*, *11*(4), 662–683. doi:10.15676/ijeei.2019.11.4.3

Amara, K., Fekik, A., Hocine, D., Bakir, M. L., Bourennane, E. B., Malek, T. A., & Malek, A. (2018, October). Improved performance of a PV solar panel with adaptive neuro fuzzy inference system ANFIS based MPPT. In *2018 7th International Conference on Renewable Energy Research and Applications (ICRERA)* (pp. 1098-1101). IEEE. 10.1109/ICRERA.2018.8566818

Amara, K., Malek, A., Bakir, T., Fekik, A., Azar, A. T., Almustafa, K. M., Bourennane, E. B., & Hocine, D. (2019b). Adaptive neuro-fuzzy inference system based maximum power point tracking for stand-alone photovoltaic system. *International Journal of Modelling. Identification and Control*, *33*(4), 311–321. doi:10.1504/IJMIC.2019.107480

Ammar, H., Ibrahim, M., Azar, A. T., & Shalaby, R. (2020b). Gray Wolf Optimization of Fractional Order Control of 3-Omni Wheels Mobile Robot: Experimental Study. *2020 16th International Computer Engineering Conference (ICENCO)*, 147-152. 10.1109/ICENCO49778.2020.9357384

Ammar, H. H., & Azar, A. T. (2020). Robust Path Tracking of Mobile Robot Using Fractional Order PID Controller. *The International Conference on Advanced Machine Learning Technologies and Applications (AMLTA2019)*, 370-381. 10.1007/978-3-030-14118-9_37

Ammar, H. H., & Azar, A. T. (2020a). Robust path tracking of mobile robot using fractional order pid controller. *The International Conference on Advanced Machine Learning Technologies and Applications (AMLTA2019),* 370–381. 10.1007/978-3-030-14118-9_37

Ammar, H. H., Azar, A. T., Shalaby, R., & Mahmoud, M. I. (2019a). Metaheuristic optimization of fractional order incremental conductance (fo-inc) maximum power point tracking (mppt). *Complexity, 2019*(7687891), 1–13. doi:10.1155/2019/7687891

Ammar, H. H., Azar, A. T., Shalaby, R., & Mahmoud, M. I. (2019b). Metaheuristic Optimization of Fractional Order Incremental Conductance (FO-INC) Maximum Power Point Tracking (MPPT). Complexity. doi:10.1155/2019/7687891

Ammar, H. H., Azar, A. T., Tembi, T. D., Tony, K., & Sosa, A. (2018). Design and Implementation of Fuzzy PID Controller into Multi Agent Smart Library System Prototype. *The International Conference on Advanced Machine Learning Technologies and Applications (AMLTA2018),* 127-137. 10.1007/978-3-319-74690-6_13

Atia, Y., & Salem, M. (2015). Novel deadbeat power control strategy for grid connected systems. *Journal of Electrical Systems and Information Technology, 2*(2), 242–256. doi:10.1016/j.jesit.2015.02.001

Attia, H. (2019). High performance PV system based on artificial neural network MPPT with PI controller for direct current water pump applications. *International Journal of Power Electronics and Drive Systems, 10*(3), 1329. doi:10.11591/ijpeds.v10.i3.pp1329-1338

Azar, A. T., Ammar, H. H., de Brito Silva, G., & Razali, M. S. A. B. (2020b). Self-balancing Robot Modeling and Control Using Two Degree of Freedom PID Controller. *The International Conference on Advanced Machine Learning Technologies and Applications (AMLTA2019),* 64-76.

Azar, A. T., Ammar, H. H., Ibrahim, Z. F., Ibrahim, H. A., Mohamed, N. A., & Taha, M. A. (2020f). Implementation of PID Controller with PSO Tuning for Autonomous Vehicle. *The International Conference on Advanced Intelligent Systems and Informatics AISI 2019,* 288-299. 10.1007/978-3-030-31129-2_27

Azar, A. T., Ammar, H. H., Mayra Beb, M. Y., Garces, S. R., & Boubakarig, A. (2020h). Optimal Design of PID Controller for 2-DOF Drawing Robot using Bat-Inspired Algorithm. *The International Conference on Advanced Intelligent Systems and Informatics AISI 2019,* 175-186.

Azar, A. T., & Kamal, N. A. (2021). Handbook of Research on Modeling, Analysis, and Control of Complex Systems. IGI Global.

Azar, A. T., Kumar, J., Kumar, V., & Rana, K. P. S. (2018c). Control of a Two Link Planar Electrically-Driven Rigid Robotic Manipulator Using Fractional Order SOFC. *Proceedings of the International Conference on Advanced Intelligent Systems and Informatics 2017,* 57-68. 10.1007/978-3-319-64861-3_6

Azar, A. T., Madian, A., Ibrahim, H., Taha, M. A., Mohamed, N. A., & Fathy, F. (2020l). Medical nanorobots: Design, applications and future challenges. In Control Systems Design of Bio-Robotics and Bio-mechatronics with Advanced Applications. Elsevier.

Azar, A. T., Ouannas, A., & Singh, S. (2018b). Control of New Type of Fractional Chaos Synchronization. *Proceedings of the International Conference on Advanced Intelligent Systems and Informatics 2017*, 47-56. 10.1007/978-3-319-64861-3_5

Azar, A. T., Sayed, A. S., Shahin, A. S., Elkholy, H. S., & Ammar, H. H. (2020d). PID Controller for 2-DOFs Twin Rotor MIMO System Tuned with Particle Swarm Optimization. *The International Conference on Advanced Intelligent Systems and Informatics AISI 2019*, 229-242.

Azar, A.T., & Serrano, F.E. (2014). Robust IMC-PID tuning for cascade control systems with gain and phase margin specifications. *Neural Computing and Applications, 25*(5), 983-995. . doi:10.1007/s00521-014-1560-x

Azar, A. T., & Serrano, F. E. (2015a). Stabilization and Control of Mechanical Systems with Backlash. In Advanced Intelligent Control Engineering and Automation. IGI Global. doi:10.4018/978-1-4666-7248-2.ch001

Azar, A. T., & Serrano, F. E. (2015b). Design and Modeling of Anti Wind Up PID Controllers. Complex system modelling and control through intelligent soft computations, 1-44. . 1. doi:10.1007/978-3-319-12883-2_1

Azar, A. T., & Serrano, F. E. (2015c). Adaptive Sliding mode control of the Furuta pendulum. In Advances and Applications in Sliding Mode Control systems. Springer-Verlag GmbH. doi:10.1007/978-3-319-11173-5_1

Azar, A. T., & Serrano, F. E. (2015d). Deadbeat Control for Multivariable Systems with Time Varying Delays. In Chaos Modeling and Control Systems Design. Springer-Verlag GmbH. doi:10.1007/978-3-319-13132-0_6

Azar, A. T., & Serrano, F. E. (2016a). Robust control for asynchronous switched nonlinear systems with time varying delays. *Proceedings of the International Conference on Advanced Intelligent Systems and Informatics 2016*, 533, 891-899. 10.1007/978-3-319-48308-5_85

Azar, A. T., & Serrano, F. E. (2016b). Stabilization of Mechanical Systems with Backlash by PI Loop Shaping. *International Journal of System Dynamics Applications*, *5*(3), 20–47. doi:10.4018/IJSDA.2016070102

Azar, A. T., & Serrano, F. E. (2017). Passivity Based Decoupling of Lagrangian Systems. *Proceedings of the International Conference on Advanced Intelligent Systems and Informatics 2016*, 533, 891-899. 10.1007/978-3-319-48308-5_85

Azar, A. T., & Serrano, F. E. (2018). Fractional Order Sliding Mode PID Controller/Observer for Continuous Nonlinear Switched Systems with PSO Parameter Tuning. *The International Conference on Advanced Machine Learning Technologies and Applications (AMLTA2018)*, 13-22.

Azar, A. T., & Serrano, F. E. (2019). Fractional Order Two Degree of Freedom PID Controller for a Robotic Manipulator with a Fuzzy Type-2 Compensator. In *Proceedings of the International Conference on Advanced Intelligent Systems and Informatics 2018*. Springer. 10.1007/978-3-319-99010-1_7

Azar, A. T., & Serrano, F. E. (2020). Stabilization of Port Hamiltonian Chaotic Systems With Hidden Attractors By Adaptive Terminal Sliding Mode Control. *Entropy (Basel, Switzerland), 22*(1), 122. doi:10.3390/e22010122 PMID:33285897

Azar, A. T., Serrano, F. E., Hameed, I. A., Kamal, N. A., & Vaidyanathan, S. (2020g). Robust H-Infinity Decentralized Control for Industrial Cooperative Robots. *The International Conference on Advanced Intelligent Systems and Informatics AISI 2019*, 254-265.

Azar, A. T., Serrano, F. E., Kamal, N. A., & Koubaa, A. (2021a). Robust Kinematic Control of Unmanned Aerial Vehicles with Non-holonomic Constraints. In *Proceedings of the International Conference on Advanced Intelligent Systems and Informatics 2020*. Springer. 10.1007/978-3-030-58669-0_74

Azar, A. T., Serrano, F. E., Kamal, N. A., & Koubaa, A. (2021b). Leader-Follower Control of Unmanned Aerial Vehicles with State Dependent Switching. In *Proceedings of the International Conference on Advanced Intelligent Systems and Informatics 2020*. Springer. 10.1007/978-3-030-58669-0_76

Azar, A. T., Serrano, F. E., & Koubaa, A. (2020a). Adaptive Fuzzy Type-2 Fractional Order Proportional Integral Derivative Sliding Mode Controller for Trajectory Tracking of Robotic Manipulators. *2020 IEEE International Conference on Autonomous Robot Systems and Competitions (ICARSC)*, 183-187. 10.1109/ICARSC49921.2020.9096163

Azar, A. T., Serrano, F. E., & Koubaa, A. (2020i). Adaptive Fuzzy Type-2 Fractional Order Proportional Integral Derivative Sliding Mode Controller for Trajectory Tracking of Robotic Manipulators. *2020 IEEE International Conference on Autonomous Robot Systems and Competitions (ICARSC),* 183-187. 10.1109/ICARSC49921.2020.9096163

Azar, A. T., Serrano, F. E., & Koubaa, A. (2020j). Adaptive Integral Sliding Mode Force Control of Robotic Manipulators with Parametric Uncertainties and Time-Varying Loads. *2020 IEEE International Conference on Autonomous Robot Systems and Competitions (ICARSC),* 188-193. DOI: 10.1109/ICARSC49921.2020.9096110

Azar, A. T., Serrano, F. E., Koubaa, A., & Kamal, N. A. (2020k). Backstepping H-Infinity Control of Unmanned Aerial Vehicles with Time Varying Disturbances. *The First International Conference of Smart Systems and Emerging Technologies (SMART TECH 2020),* 243 - 248. 10.1109/SMART-TECH49988.2020.00061

Azar, A. T., Serrano, F. E., Koubaa, A., Kamal, N. A., Vaidyanathan, S., & Fekik, A. (2020e). Adaptive Terminal-Integral Sliding Mode Force Control of Elastic Joint Robot Manipulators in the Presence of Hysteresis. *The International Conference on Advanced Intelligent Systems and Informatics AISI 2019,* 266-276. 10.1007/978-3-030-31129-2_25

Azar, A. T., Serrano, F. E., & Vaidyanathan, S. (2018a). Sliding Mode Stabilization and Synchronization of Fractional Order Complex Chaotic and Hyperchaotic Systems. Mathematical Techniques of Fractional Order Systems, 283–317. doi:10.1016/B978-0-12-813592-1.00010-6

Azar, A. T., Serrano, F. E., Vaidyanathan, S., & Albalawi, H. (2020c). Adaptive Higher Order Sliding Mode Control for Robotic Manipulators with Matched and Mismatched Uncertainties. *The International Conference on Advanced Machine Learning Technologies and Applications (AMLTA2019),* 360-369. 10.1007/978-3-030-14118-9_36

Azar, A. T., & Vaidyanathan, S. (2015a). Handbook of Research on Advanced Intelligent Control Engineering and Automation. IGI Global. doi:10.4018/978-1-4666-7248-2

Azar, A. T., & Vaidyanathan, S. (2015b). *Computational Intelligence applications in Modeling and Control.* Springer-Verlag.

Azar, A. T., & Vaidyanathan, S. (2015c). *Chaos Modeling and Control Systems Design.* Springer-Verlag.

Azar, A. T., & Vaidyanathan, S. (2016). *Advances in Chaos Theory and Intelligent Control.* Springer-Verlag. doi:10.1007/978-3-319-30340-6

Barakat, M. H., Azar, A. T., & Ammar, H. H. (2020) Agricultural Service Mobile Robot Modeling and Control Using Artificial Fuzzy Logic and Machine Vision. *The International Conference on Advanced Machine Learning Technologies and Applications (AMLTA2019),* 453-465. 10.1007/978-3-030-14118-9_46

Ben Smida, M., Sakly, A., Vaidyanathan, S., & Azar, A. T. (2018). *Control-Based Maximum Power Point Tracking for a Grid-Connected Hybrid Renewable Energy System Optimized by Particle Swarm Optimization.* IGI-Global. doi:10.4018/978-1-5225-4077-9.ch003

Blaabjerg, F., Teodorescu, R., Liserre, M., & Timbus, A. V. (2006). Overview of control and grid synchronization for distributed power generation systems. *IEEE Transactions on Industrial Electronics, 53*(5), 1398–1409. doi:10.1109/TIE.2006.881997

Bouafia, A., Gaubert, J. P., & Krim, F. (2010). Predictive direct power control of three-phase pulsewidth modulation (PWM) rectifier using space-vector modulation (SVM). *IEEE Transactions on Power Electronics, 25*(1), 228–236. doi:10.1109/TPEL.2009.2028731

Dalapati, S., & Chakraborty, C. (2009). Dynamic performance of a dead-band controlled capacitor charging type inverter. *Simulation Modelling Practice and Theory, 17*(5), 911–934. doi:10.1016/j.simpat.2009.02.010

Denoun, H., Hamida, M. L., Fekik, A., Dyhia, K. A. I. S., & Ghanes, M. (2018, October). Petri Nets Modeling for Two-Cell Chopper Control Using DSPACE 1104. In *2018 6th International Conference on Control Engineering & Information Technology (CEIT)* (pp. 1-6). IEEE.

Denoun, H., Nabil, B., Mustapha, Z., Nacereddine, B., Haddad, S., & Mamar, S. A. (2015). Modelling and Realisation of a Three-Level PWM Inverter Using a DSP Controller to Feed an Asynchronous Machine. In *Handbook of Research on Advanced Intelligent Control Engineering and Automation* (pp. 687-718). IGI Global.

Djeddi, A., Dib, D., Azar, A. T., & Abdelmalek, S. (2019). Fractional Order Unknown Inputs Fuzzy Observer for Takagi–Sugeno Systems with Unmeasurable Premise Variables. *Mathematics, 7*(10), 984. doi:10.3390/math7100984

Elkholy, H. A., Azar, A. T., Magd, A., Marzouk, H., & Ammar, H. H. (2020b) Classifying Upper Limb Activities Using Deep Neural Networks. *Proceedings of the International Conference on Artificial Intelligence and Computer Vision (AICV2020),* 268-282. 10.1007/978-3-030-44289-7_26

Elkholy, H. A., Azar, A. T., Shahin, A. S., Elsharkawy, O. I., & Ammar, H. H. (2020a) Path Planning of a Self Driving Vehicle Using Artificial Intelligence Techniques and Machine Vision. *Proceedings of the International Conference on Artificial Intelligence and Computer Vision (AICV2020),* 532-542. 10.1007/978-3-030-44289-7_50

Elmelegi, A., Aly, M., Ahmed, E. M., & Alharbi, A. G. (2019). A simplified phase-shift PWM-based feedforward distributed MPPT method for grid-connected cascaded PV inverters. *Solar Energy, 187,* 1–12. doi:10.1016/j.solener.2019.05.021

Fadhel, S., Diallo, D., Delpha, C., Migan, A., Bahri, I., Trabelsi, M., & Mimouni, M. F. (2020). Maximum power point analysis for partial shading detection and identification in photovoltaic systems. *Energy Conversion and Management, 224,* 113374. doi:10.1016/j.enconman.2020.113374

Fekik, A., Azar, A. T., Denoun, H., Kamal, N. A., Hamida, M. L., Kais, D., & Amara, K. (2018e). A Backstepping Direct Power Control of Three Phase Pulse Width Modulated Rectifier. In *International Workshop Soft Computing Applications* (pp. 445-456). Springer.

Fekik, A., Azar, A. T., Kamal, N. A., Denoun, H., Almustafa, K. M., Hamida, L., & Zaouia, M. (2020a). Fractional-Order Control of a Fuel Cell–Boost Converter System. In *International Conference on Advanced Machine Learning Technologies and Applications* (pp. 713-724). Springer.

Fekik, A., Azar, A. T., Kamal, N. A., Serrano, F. E., Hamida, M. L., Denoun, H., & Yassa, N. (2020b, October). Maximum Power Extraction from a Photovoltaic Panel Connected to a Multi-cell Converter. In *International Conference on Advanced Intelligent Systems and Informatics* (pp. 873-882). Springer.

Fekik, A., Denoun, H., Azar, A. T., Hamida, M. L., Atig, M., Ghanes, M., . . . Barbot, J. P. (2018c). Direct power control of a three-phase PWM-rectifier based on Petri nets for the selection of switching states. In *2018 7th International Conference on Renewable Energy Research and Applications (ICRERA)* (pp. 1121-1125). IEEE.

Fekik, A., Denoun, H., Azar, A. T., Hamida, M. L., Benamrouche, N., & Vaidyanathan, S. (2017). Direct Power Control of a PWM-Inverter for Grid Connected Photovoltaic System. *5nd international conference on renewable Energy (CIER2017).*

Fekik, A., Denoun, H., Azar, A. T., Hamida, M. L., Zaouia, M., & Benyahia, N. (2018b). Comparative study of two level and three level PWM-rectifier with voltage oriented control. In *International Conference on Advanced Intelligent Systems and Informatics* (pp. 40-51). Springer.

Fekik, A., Denoun, H., Azar, A. T., Kamal, N. A., Zaouia, M., Benyahia, N., & Vaidyanathan, S. (2020d). Direct power control of three-phase PWM-rectifier with backstepping control. In *Backstepping Control of Nonlinear Dynamical Systems* (pp. 215–234). Academic Press.

Fekik, A., Denoun, H., Azar, A. T., Kamal, N. A., Zaouia, M., Yassa, N., & Hamida, M. L. (2019a). Direct Torque Control of Three Phase Asynchronous Motor with Sensorless Speed Estimator. In *International Conference on Advanced Intelligent Systems and Informatics* (pp. 243-253). Springer.

Fekik, A., Denoun, H., Azar, A. T., Koubaa, A., Kamal, N. A., Zaouia, M., ... Yassa, N. (2020c). Adapted Fuzzy Fractional Order proportional-integral controller for DC Motor. In *2020 First International Conference of Smart Systems and Emerging Technologies (SMARTTECH)* (pp. 1-6). IEEE.

Fekik, A., Denoun, H., Azar, A. T., Zaouia, M., Benyahia, N., Hamida, M. L., & Vaidyanathan, S. (2018a). Artificial neural network for PWM rectifier direct power control and DC voltage control. In *Advances in System Dynamics and Control* (pp. 286–316). IGI Global. doi:10.4018/978-1-5225-4077-9.ch010

Fekik, A., Denoun, H., Benamrouche, N., Benyahia, N., Badji, A., & Zaouia, M. (2016). Comparative analysis of direct power control and direct power control with space vector modulation of PWM rectifier. In *2016 4th International Conference on Control Engineering & Information Technology (CEIT)* (pp. 1-6). IEEE. 10.1109/CEIT.2016.7929058

Fekik, A., Denoun, H., Benamrouche, N., Benyahia, N., & Zaouia, M. (2015b). A Fuzzy Logic Based Controller For Three Phase PWM Rectifier With Voltage Oriented Control Strategy. *International Journal of Circuits, Systems and Signal Processing*, *9*, 412–419.

Fekik, A., Denoun, H., Benamrouche, N., Benyahia, N., Zaouia, M., & Haddad, S. (2015a) Comparative study of PI and FUZZY DC- voltage control for Voltage Oriented Control-PWM rectifier. *The 14th International Conference on Circuits, Systems, Electronics, Control & Signal Processing 2015.*

Fekik, A., Denoun, H., Hamida, M. L., Azar, A. T., Atig, M., & Zhu, Q. M. (2018d). Neural Network Based Switching State Selection for Direct Power Control of Three Phase PWM-Rectifier. In *2018 10th International Conference on Modelling, Identification and Control (ICMIC)* (pp. 1-6). IEEE.

Fekik, A., Denoun, H., Zaouia, M., Hamida, M. L., & Vaidyanathan, S. (2019b). A new switching table based neural network for direct power control of three-phase PWM-rectifier. International Journal of Modelling. *Identification and Control, 33*(4), 358–368. doi:10.1504/IJMIC.2019.107484

Foti, S., De Caro, S., Scelba, G., Scimone, T., Testa, A., Cacciato, M., & Scarcella, G. (2018). An optimal current control strategy for asymmetrical hybrid multilevel inverters. *IEEE Transactions on Industry Applications*, *54*(5), 4425–4436. doi:10.1109/TIA.2018.2832024

Ghoudelbourk, S., Azar, A. T., & Dib, D. (2021). Three-level (NPC) Shunt Active Power Filter Based on Fuzzy Logic and Fractional-order PI Controller. *Int. J. Automation and Control*, *15*(2), 149–169. doi:10.1504/IJAAC.2021.113338

Ghoudelbourk, S., Azar, A. T., Dib, D., & Omeiri, A. (2020). Selective Harmonic Elimination Strategy in the Multilevel Inverters for Grid Connected Photovoltaic System. *Int. J. Advanced Intelligence Paradigms*, *15*(3), 317–339. doi:10.1504/IJAIP.2020.105826

Ghoudelbourk, S., Dib, D., Omeiri, A., & Azar, A. T. (2016). MPPT Control in wind energy conversion systems and the application of fractional control (PI^{α}) in pitch wind turbine. *Int. J. Modelling Identification and Control*, *26*(2), 140–151. doi:10.1504/IJMIC.2016.078329

Giove, S., Azar, A. T., & Nordio, M. (2013). Fuzzy Logic Control for Dialysis Application. In Biofeedback Systems and Soft Computing Techniques of Dialysis. Springer-Verlag GmbH. doi:10.1007/978-3-642-27558-6_9

Gorripotu, T. S., Azar, A. T., Pilla, R., & Kamal, N. A. (2021). Impact of Ultra Capacitor on Automatic Load Frequency Control of Nonlinear Power System. In *Intelligent Computing in Control and Communication. Lecture Notes in Electrical Engineering* (Vol. 702, pp. 333–341). Springer. doi:10.1007/978-981-15-8439-8_27

Gorripotu, T. S., Samalla, H., Jagan Mohana Rao, C., Azar, A. T., & Pelusi, D. (2019). TLBO Algorithm Optimized Fractional-Order PID Controller for AGC of Interconnected Power System. In J. Nayak, A. Abraham, B. Krishna, G. Chandra Sekhar, & A. Das (Eds.), *Soft Computing in Data Analytics. Advances in Intelligent Systems and Computing* (Vol. 758). Springer. doi:10.1007/978-981-13-0514-6_80

Grassi, G., Ouannas, A., Azar, A. T., Radwan, A. G., Volos, C., Pham, V. T., Ziar, T., Madian, A., Kyprianidis, I. M., & Stouboulos, I. N. (2017). Chaos Synchronisation Of Continuous Systems Via Scalar Signal. *The 6th International Conference on Modern Circuits and Systems Technologies (MOCAST).* 10.1109/MOCAST.2017.7937629

Hamida, M. L., Denoun, H., Fekik, A., Benyahia, N., Zaouia, M., & Benamrouche, N. (2017). A Comparison Between PWM and Petri Nets Control Applied for Series Multi-cells Chopper. In *International conference on control and signal processing* (*Vol. 26*, pp. 124-128). Academic Press.

Hamida, M. L., Denoun, H., Fekik, A., Kais, D., Benamrouche, N., & Amara, K. (2019b, November). Fuzzy Logic-Feedback Linearization Control for a Three-Cell Flying Capacitor Converter. In *2019 International Conference on Advanced Electrical Engineering (ICAEE)* (pp. 1-5). IEEE. 10.1109/ICAEE47123.2019.9015121

Hamida, M. L., Denoun, H., Fekik, A., & Vaidyanathan, S. (2019a). Control of separately excited DC motor with series multi-cells chopper using PI-Petri nets controller. *Nonlinear Engineering*, *8*(1), 32–38. doi:10.1515/nleng-2017-0174

Hassaine, L., Olias, E., Quintero, J., & Haddadi, M. (2009). Digital power factor control and reactive power regulation for grid-connected photovoltaic inverter. *Renewable Energy*, *34*(1), 315–321. doi:10.1016/j.renene.2008.03.016

Hu, J., & Zhu, Z. Q. (2011). Investigation on switching patterns of direct power control strategies for grid-connected DC-AC converters based on power variation rates. *IEEE Transactions on Power Electronics*, *26*(2), 3582–3598. doi:10.1109/TPEL.2011.2164812

Ibraheem, G. A. R., Azar, A. T., Ibraheem, I. K., & Humaidi, A. J. (2020). A Novel Design of a Neural Network based Fractional PID Controller for Mobile Robots Using Hybridized Fruit Fly and Particle Swarm Optimization. *Complexity*. doi:10.1155/2020/3067024

Ibrahim, H. A., Azar, A. T., Ibrahim, Z. F., & Ammar, H. H. (2020) A Hybrid Deep Learning Based Autonomous Vehicle Navigation and Obstacles Avoidance. *Proceedings of the International Conference on Artificial Intelligence and Computer Vision (AICV2020)*, 296-307. 10.1007/978-3-030-44289-7_28

Jain, S., & Agarwal, V. (2007). Comparison of the performance of maximum power point tracking schemes applied to single-stage grid-connected photovoltaic systems. *IET Electric Power Applications*, *1*(5), 753–762. doi:10.1049/iet-epa:20060475

Joshi, P., & Arora, S. (2017). Maximum power point tracking methodologies for solar PV systems. *RE:view*, *70*, 1154–1177.

Kamal, N. A., Azar, A. T., Elbasuony, G. S., Almustafa, K. A., & Almakhles, D. (2020). PSO-based Adaptive Perturb and Observe MPPT Technique for Photovoltaic Systems. *The International Conference on Advanced Intelligent Systems and Informatics AISI 2019,* 125-135.

Kamal, N. A., & Ibrahim, A. M. (2018). Conventional, Intelligent, and Fractional-Order Control Method for Maximum Power Point Tracking of a Photovoltaic System: A Review. Mathematical Techniques of Fractional Order Systems, 603-671.

Kammogne, A. S. T., Kountchou, M. N., Kengne, R., Azar, A. T., Fotsin, H. B., & Ouagni, S. T. M. (2020). Polynomial Robust Observer Implementation based-passive Synchronization of Nonlinear Fractional-Order Systems with Structural Disturbances. *Frontiers of Information Technology & Electronic Engineering.*, *21*(9), 1369–1386. doi:10.1631/FITEE.1900430

Kazim, M., Azar, A. T., Koubaa, A., & Zaidi, A. (2021). Disturbance Rejection Based Optimized Robust Adaptive Controller for UAVs. *IEEE Systems Journal*, *15*(2), 1–12. doi:10.1109/JSYST.2020.3006059

Khan, A., Singh, S., & Azar, A. T. (2020a). Synchronization between a Novel Integer-Order Hyperchaotic System and a Fractional-Order Hyperchaotic System Using Tracking Control. *The International Conference on Advanced Machine Learning Technologies and Applications (AMLTA2019),* 382-391. 10.1007/978-3-030-14118-9_38

Khan, A., Singh, S., & Azar, A. T. (2020b). Combination-Combination Anti-Synchronization of Four Fractional Order Identical Hyperchaotic Systems. *The International Conference on Advanced Machine Learning Technologies and Applications (AMLTA2019),* 406-414. 10.1007/978-3-030-14118-9_41

Khennaoui, A. A., Ouannas, A., Boulaaras, S., Pham, V. T., & Azar, A. T. (2020a). A fractional map with hidden attractors: Chaos and control. *The European Physical Journal. Special Topics*, *229*(6-7), 1083–1093. doi:10.1140/epjst/e2020-900177-6

Khennaoui, A. A., Ouannas, A., Grassi, G., & Azar, A. T. (2020b) Dynamic Analysis of a Fractional Map with Hidden Attractor. *Proceedings of the International Conference on Artificial Intelligence and Computer Vision (AICV2020),* 731-739. 10.1007/978-3-030-44289-7_68

Khettab, K., Bensafia, Y., Bourouba, B., & Azar, A. T. (2018) Enhanced Fractional Order Indirect Fuzzy Adaptive Synchronization of Uncertain Fractional Chaotic Systems Based on the Variable Structure Control: Robust H∞ Design Approach. Mathematical Techniques of Fractional Order Systems, 559–595.

Koubaa, A., & Azar, A. T. (2021). *Unmanned Aerial Systems: Theoretical Foundation and Applications.* Elsevier.

Kumar, J., Azar, A. T., Kumar, V., & Rana, K. P. S. (2018) Design of Fractional Order Fuzzy Sliding Mode Controller for Nonlinear Complex Systems. Mathematical Techniques of Fractional Order Systems, 249–282. doi:10.1016/B978-0-12-813592-1.00009-X

Lalouni, S., Rekioua, D., Rekioua, T., & Matagne, E. (2009). Fuzzy logic control of stand-alone photovoltaic system with battery storage. *Journal of Power Sources, 193*(2), 899–907. doi:10.1016/j.jpowsour.2009.04.016

Lamine, M., Hakim, D., Arezki, F., Dyhia, K., Nacereddine, B., & Youssef, B. (2019c, July). Control of Three-Cell Inverter with a Fuzzy Logic-Feedback Linearization Strategy to Reduce the Harmonic Content of the Output Current. In *2019 International Conference of Computer Science and Renewable Energies (ICCSRE)* (pp. 1-5). IEEE.

Lamine, M., Hakim, D., Arezki, F., Nabil, B., & Nacerddine, B. (2018, October). Cyclic Reports Modulation Control Strategy for a Five Cells Inverter. In *2018 International Conference on Electrical Sciences and Technologies in Maghreb (CISTEM)* (pp. 1-5). IEEE.

Li, C. H., Zhu, X. J., Cao, G. Y., Sui, S., & Hu, M. R. (2009). Dynamic modeling and sizing optimization of stand-alone photovoltaic power systems using hybrid energy storage technology. *Renewable Energy, 34*(3), 815–826. doi:10.1016/j.renene.2008.04.018

Liu, L., Ma, D., Azar, A. T., & Zhu, Q. (2020). Neural computing Enhanced Parameter Estimation for Multi-Input and Multi-Output Total Nonlinear Dynamic Models. *Entropy (Basel, Switzerland), 22*(5), 510. doi:10.3390/e22050510 PMID:33286282

Mahmud, N., Zahedi, A., & Mahmud, A. (2017). A cooperative operation of novel PV inverter control scheme and storage energy management system based on ANFIS for voltage regulation of grid-tied PV system. *IEEE Transactions on Industrial Informatics, 13*(5), 2657–2668. doi:10.1109/TII.2017.2651111

Malakondareddy, B., Kumar, S. S., Gounden, N. A., & Anand, I. (2019). An adaptive PI control scheme to balance the neutral-point voltage in a solar PV fed grid connected neutral point clamped inverter. *International Journal of Electrical Power & Energy Systems, 110*, 318–331. doi:10.1016/j.ijepes.2019.03.012

Meghni, B., Dib, D., & Azar, A. T. (2017c). A Second-order sliding mode and fuzzy logic control to Optimal Energy Management in PMSG Wind Turbine with Battery Storage. *Neural Computing & Applications, 28*(6), 1417–1434. doi:10.100700521-015-2161-z

Meghni, B., Dib, D., Azar, A. T., Ghoudelbourk, S., & Saadoun, A. (2017a). Robust Adaptive Supervisory Fractional order Controller For optimal Energy Management in Wind Turbine with Battery Storage. *Studies in Computational Intelligence, 688*, 165–202. doi:10.1007/978-3-319-50249-6_6

Meghni, B., Dib, D., Azar, A. T., & Saadoun, A. (2017b). *Effective Supervisory Controller to Extend Optimal Energy Management in Hybrid Wind Turbine under Energy and Reliability Constraints. International Journal of Dynamics and Control.* doi:10.100740435-016-0296-0

Mekki, H., Boukhetala, D., & Azar, A. T. (2015). Sliding Modes for Fault Tolerant Control. Advances and Applications in Sliding Mode Control systems, 407-433. doi:10.1007/978-3-319-11173-5_15

Mellit, A., Kalogirou, S. A., Hontoria, L., & Shaari, S. (2009). Artificial intelligence techniques for sizing photovoltaic systems: A review. *Renewable & Sustainable Energy Reviews, 13*(2), 406–419. doi:10.1016/j.rser.2008.01.006

Mittal, S., Azar, A. T., & Kamal, N. A. (2021). Nonlinear Fractional Order System Synchronization via Combination-Combination Multi-switching. In *Proceedings of the International Conference on Advanced Intelligent Systems and Informatics 2020*. Springer.

Mohamed, N. A., Azar, A. T., Abbas, N. E., Ezzeldin, M. A., & Ammar, H. H. (2020). Experimental Kinematic Modeling of 6-DOF Serial Manipulator Using Hybrid Deep Learning. *Proceedings of the International Conference on Artificial Intelligence and Computer Vision (AICV2020)*, 283-295.

Naidu, N. K. S., & Singh, B. (2014). Sensorless control of single voltage source converter-based doubly fed induction generator for variable speed wind energy conversion system. *IET Power Electronics*, *7*(12), 2996–3006. doi:10.1049/iet-pel.2014.0034

Noguchi, T., Tomiki, H., Kondo, S., & Takahashi, I. (1998). Direct power control of PWM converter without power-source voltage sensors. *IEEE Transactions on Industry Applications*, *34*(3), 473–479. doi:10.1109/28.673716

Ouannas, A., Azar, A. T., & Abu-Saris, R. (2016a). A new type of hybrid synchronization between arbitrary hyperchaotic maps. *International Journal of Machine Learning and Cybernetics*. Advance online publication. doi:10.100713042-016-0566-3

Ouannas, A., Azar, A. T., & Radwan, A. G. (2016b). *On Inverse Problem of Generalized Synchronization Between Different Dimensional Integer–Order and Fractional-Order Chaotic Systems*. The 28th International Conference on Microelectronics, Cairo, Egypt. 10.1109/ICM.2016.7847942

Ouannas, A., Azar, A. T., & Vaidyanathan, S. (2017a). A Robust Method for New Fractional Hybrid Chaos Synchronization. *Mathematical Methods in the Applied Sciences*, *40*(5), 1804–1812. doi:10.1002/mma.4099

Ouannas, A., Azar, A. T., & Vaidyanathan, S. (2017b). A new fractional hybrid chaos synchronization. Int. J. Modelling. *Identification and Control*, *27*(4), 314–322. doi:10.1504/IJMIC.2017.084719

Ouannas, A., Azar, A. T., & Ziar, T. (2020a). Fractional inverse full state hybrid projective synchronisation. *Int. J. Advanced Intelligence Paradigms*, *17*(3-4), 270–298.

Ouannas, A., Azar, A. T., Ziar, T., & Radwan, A. G. (2017f). A Study On Coexistence of Different Types of Synchronization Between Different dimensional Fractional Chaotic Systems. *Studies in Computational Intelligence*, *688*, 637–669. doi:10.1007/978-3-319-50249-6_22

Ouannas, A., Azar, A. T., Ziar, T., & Radwan, A. G. (2017g). Generalized Synchronization of Different Dimensional Integer-order and Fractional Order Chaotic Systems. *Studies in Computational Intelligence*, *688*, 671–697. doi:10.1007/978-3-319-50249-6_23

Ouannas, A., Azar, A. T., Ziar, T., & Vaidyanathan, S. (2017c). On New Fractional Inverse Matrix Projective Synchronization Schemes. *Studies in Computational Intelligence*, *688*, 497–524. doi:10.1007/978-3-319-50249-6_17

Ouannas, A., Azar, A. T., Ziar, T., & Vaidyanathan, S. (2017d). Fractional Inverse Generalized Chaos Synchronization Between Different Dimensional Systems. *Studies in Computational Intelligence*, *688*, 525–551. doi:10.1007/978-3-319-50249-6_18

Ouannas, A., Azar, A. T., Ziar, T., & Vaidyanathan, S. (2017e). A New Method To Synchronize Fractional Chaotic Systems With Different Dimensions. *Studies in Computational Intelligence*, *688*, 581–611. doi:10.1007/978-3-319-50249-6_20

Ouannas, A., Grassi, G., & Azar, A. T. (2020b). A New Generalized Synchronization Scheme to Control Fractional Chaotic Systems with Non-identical Dimensions and Different Orders. *The International Conference on Advanced Machine Learning Technologies and Applications (AMLTA2019),* 415-424. 10.1007/978-3-030-14118-9_42

Ouannas, A., Grassi, G., & Azar, A. T. (2020c). Fractional-Order Control Scheme for Q-S Chaos Synchronization. *The International Conference on Advanced Machine Learning Technologies and Applications (AMLTA2019),* 434-441. 10.1007/978-3-030-14118-9_44

Ouannas, A., Grassi, G., Azar, A. T., & Khennaoui, A. A. (2021). Synchronization Control in Fractional Discrete-Time Systems with Chaotic Hidden Attractors. In A. Hassanien, R. Bhatnagar, & A. Darwish (Eds.), *Advanced Machine Learning Technologies and Applications. AMLTA 2020. Advances in Intelligent Systems and Computing* (Vol. 1141, pp. 661–669). Springer. doi:10.1007/978-981-15-3383-9_59

Ouannas, A., Grassi, G., Azar, A. T., Khennaouia, A. A., & Pham, V. T. (2020d). Chaotic Control in Fractional-Order Discrete-Time Systems. *AISI 2019. The International Conference on Advanced Intelligent Systems and Informatics AISI 2019*, 207-217. 10.1007/978-3-030-31129-2_20

Ouannas, A., Grassi, G., Azar, A. T., Khennaouia, A. A., & Pham, V. T. (2020e). Synchronization of Fractional-Order Discrete-Time Chaotic Systems. *The International Conference on Advanced Intelligent Systems and Informatics AISI 2019*, 218-228. 10.1007/978-3-030-31129-2_21

Ouannas, A., Grassi, G., Azar, A. T., & Singh, S. (2019a). New Control Schemes for Fractional Chaos Synchronization. In *Proceedings of the International Conference on Advanced Intelligent Systems and Informatics 2018*. Springer. 10.1007/978-3-319-99010-1_5

Parida, B., Iniyan, S., & Goic, R. (2011). A review of solar photovoltaic technologies. *Renewable & Sustainable Energy Reviews*, *15*(3), 1625–1636. doi:10.1016/j.rser.2010.11.032

Pham, V. T., Gokul, P. M., Kapitaniak, T., Volos, C., & Azar, A. T. (2018). Dynamics, Synchronization and Fractional Order Form of a Chaotic System With Infinite Equilibria. In Mathematical Techniques of Fractional Order Systems. Elsevier.

Pham, V. T., Vaidyanathan, S., Volos, C. K., Azar, A. T., Hoang, T. M., & Yem, V. V. (2017a). A Three-Dimensional No-Equilibrium Chaotic System: Analysis, Synchronization and Its Fractional Order Form. Studies in Computational Intelligence, 688, 449-470.

Pham, V. T., Volos, C. K., Vaidyanathan, S., & Azar, A. T. (2017b). Dynamics, Synchronization and Fractional Order Form of a Chaotic System without Equilibrium. In C. K. Volos (Ed.), *Nonlinear Systems: Design, Applications and Analysis*. Nova Science Publishers.

Pilla, R., Azar, A. T., & Gorripotu, T. S. (2019). Impact of Flexible AC Transmission System Devices on Automatic Generation Control with a Metaheuristic Based Fuzzy PID Controller. *Energies*, *12*(21), 4193. doi:10.3390/en12214193

Pilla, R., Botcha, N., Gorripotu, T. S., & Azar, A. T. (2020). Fuzzy PID Controller for Automatic Generation Control of Interconnected Power System Tuned by Glow-Worm Swarm Optimization. In *Applications of Robotics in Industry Using Advanced Mechanisms. ARIAM 2019. Learning and Analytics in Intelligent Systems* (Vol. 5, pp. 140–149). Springer. doi:10.1007/978-3-030-30271-9_14

Priyadarshi, N., Padmanaban, S., Bhaskar, M. S., Blaabjerg, F., & Sharma, A. (2018). Fuzzy SVPWM-based inverter control realisation of grid integrated photovoltaic-wind system with fuzzy particle swarm optimisation maximum power point tracking algorithm for a grid-connected PV/wind power generation system: Hardware implementation. *IET Electric Power Applications*, *12*(7), 962–971. doi:10.1049/iet-epa.2017.0804

Qahouq, J. A. A., & Jiang, Y. (2014). Distributed photovoltaic solar system architecture with single-power inductor single-power converter and single-sensor single maximum power point tracking controller. *IET Power Electronics*, *7*(10), 2600–2609. doi:10.1049/iet-pel.2013.0893

Radwan, A. G., Azar, A. T., Vaidyanathan, S., Munoz-Pacheco, J. M., & Ouannas, A. (2017). Fractional-Order and Memristive Nonlinear Systems: Advances and Applications. *Complexity*, *2017*, 3760121. Advance online publication. doi:10.1155/2017/3760121

Radwan, A. G., Emira, A. A., Abdelaty, A., & Azar, A. T. (2018). Modeling and Analysis of Fractional Order DC-DC Converter. *ISA Transactions*, *82*, 184–199. doi:10.1016/j.isatra.2017.06.024 PMID:28709651

Rajendran, S., Govindarajan, U., Reuben, A. B., & Srinivasan, A. (2013). Shunt reactive VAR compensator for grid-connected induction generator in wind energy conversion systems. *IET Power Electronics*, *6*(9), 1872–1883. doi:10.1049/iet-pel.2012.0524

Salas, V., Olias, E., Barrado, A., & Lazaro, A. (2006). Review of the maximum power point tracking algorithms for stand-alone photovoltaic systems. *Solar Energy Materials and Solar Cells*, *90*(11), 1555–1578. doi:10.1016/j.solmat.2005.10.023

Salem, M., & Atia, Y. (2015). Control scheme towards enhancing power quality and operational efficiency of single-phase two-stage grid-connected photovoltaic systems. *Journal of Electrical Systems and Information Technology*, *2*(3), 314–327. doi:10.1016/j.jesit.2015.05.002

Sallam, O. K., Azar, A. T., Guaily, A., & Ammar, H. H. (2020) Tuning of PID controller using Particle Swarm Optimization for cross flow heat Exchanger based on CFD System Identification. *The International Conference on Advanced Intelligent Systems and Informatics AISI 2019*, 300-312. 10.1007/978-3-030-31129-2_28

Sayed, A. S., Azar, A. T., Ibrahim, Z. F., Ibrahim, H. A., Mohamed, N. A., & Ammar, H. H. (2020). Deep Learning Based Kinematic Modeling of 3-RRR Parallel Manipulator. *Proceedings of the International Conference on Artificial Intelligence and Computer Vision (AICV2020)*, 308-321.

Shalaby, R., Ammar, H. H., Azar, A. T., & Mahmoud, M. I. (2021). Optimal Fractional-Order Fuzzy-MPPT for solar water pumping system. *Journal of Intelligent & Fuzzy Systems*, *40*(1), 1175–1190. doi:10.3233/JIFS-201538

Shukla, M. K., Sharma, B. B., & Azar, A. T. (2018). Control and Synchronization of a Fractional Order Hyperchaotic System via Backstepping and Active Backstepping Approach. In Mathematical Techniques of Fractional Order Systems. Elsevier. doi:10.1016/B978-0-12-813592-1.00019-2

Singh, S., & Azar, A. T. (2020). Multi-switching Combination Synchronization of Fractional Order Chaotic Systems. *Proceedings of the International Conference on Artificial Intelligence and Computer Vision (AICV2020),* 655-664. 10.1007/978-3-030-44289-7_61

Singh, S., Azar, A. T., Ouannas, A., Zhu, Q., Zhang, W., & Na, J. (2017). *Sliding Mode Control Technique for Multi-switching Synchronization of Chaotic Systems. 9th International Conference on Modelling, Identification and Control (ICMIC 2017),* Kunming, China. 10.1109/ICMIC.2017.8321579

Singh, S., Azar, A. T., Vaidyanathan, S., Ouannas, A., & Bhat, M. A. (2018). Multiswitching Synchronization of Commensurate Fractional Order Hyperchaotic Systems Via Active Control. In Mathematical Techniques of Fractional Order Systems. Elsevier. doi:10.1016/B978-0-12-813592-1.00011-8

Sinha, A., Jana, K. C., & Das, M. K. (2018). An inclusive review on different multi-level inverter topologies, their modulation and control strategies for a grid connected photo-voltaic system. *Solar Energy, 170*, 633–657. doi:10.1016/j.solener.2018.06.001

Soliman, N. S., Said, L. A., Azar, A. T., Madian, A. H., Radwan, A. G., & Ouannas, A. (2017). *Fractional Controllable Multi-Scroll V-Shape Attractor with Parameters Effect.* The 6th International Conference on Modern Circuits and Systems Technologies (MOCAST), Thessaloniki Greece.

Tahir, S., Wang, J., Baloch, M. H., & Kaloi, G. S. (2018). Digital control techniques based on voltage source inverters in renewable energy applications: A review. *Electronics (Basel)*, *7*(2), 18. doi:10.3390/electronics7020018

Tareen, W. U., Mekhilef, S., Seyedmahmoudian, M., & Horan, B. (2017). Active power filter (APF) for mitigation of power quality issues in grid integration of wind and photovoltaic energy conversion system. *Renewable & Sustainable Energy Reviews*, *70*, 635–655. doi:10.1016/j.rser.2016.11.091

Tolba, M. F., AbdelAty, A. M., Soliman, N. S., Said, L. A., Madian, A. H., Azar, A. T., & Radwan, A. G. (2017). FPGA implementation of two fractional order chaotic systems. *International Journal of Electronics and Communications*, *28*, 162–172. doi:10.1016/j.aeue.2017.04.028

Tsengenes, G., & Adamidis, G. (2011). Investigation of the behavior of a three-phase grid-connected photovoltaic system to control active and reactive power. *Electric Power Systems Research*, *81*(1), 177–184. doi:10.1016/j.epsr.2010.08.008

Vaidyanathan, S., & Azar, A. T. (2015a). Anti-Synchronization of Identical Chaotic Systems using Sliding Mode Control and an Application to Vaidyanathan-Madhavan Chaotic Systems. Advances and Applications in Sliding Mode Control systems, 527-547. doi:10.1007/978-3-319-11173-5_19

Vaidyanathan, S., & Azar, A. T. (2015b). Hybrid Synchronization of Identical Chaotic Systems using Sliding Mode Control and an Application to Vaidyanathan Chaotic Systems. Advances and Applications in Sliding Mode Control systems, 549-569. doi:10.1007/978-3-319-11173-5_20

Vaidyanathan, S., & Azar, A. T. (2015c). Analysis, Control and Synchronization of a Nine-Term 3-D Novel Chaotic System. Chaos Modeling and Control Systems Design, 3-17. doi:10.1007/978-3-319-13132-0_1

Vaidyanathan, S., & Azar, A. T. (2015d). Analysis and Control of a 4-D Novel Hyperchaotic System. Chaos Modeling and Control Systems Design, 19-38. doi:10.1007/978-3-319-13132-0_2

Vaidyanathan, S., & Azar, A. T. (2016a). Takagi-Sugeno Fuzzy Logic Controller for Liu-Chen Four-Scroll Chaotic System. *International Journal of Intelligent Engineering Informatics*, *4*(2), 135–150. doi:10.1504/IJIEI.2016.076699

Vaidyanathan, S., & Azar, A. T. (2016b). Dynamic Analysis, Adaptive Feedback Control and Synchronization of an Eight-Term 3-D Novel Chaotic System with Three Quadratic Nonlinearities. *Studies in Fuzziness and Soft Computing*, *337*, 155–178. doi:10.1007/978-3-319-30340-6_7

Vaidyanathan, S., & Azar, A. T. (2016c). Qualitative Study and Adaptive Control of a Novel 4-D Hyperchaotic System with Three Quadratic Nonlinearities. *Studies in Fuzziness and Soft Computing*, *337*, 179–202. doi:10.1007/978-3-319-30340-6_8

Vaidyanathan, S., & Azar, A. T. (2016d). A Novel 4-D Four-Wing Chaotic System with Four Quadratic Nonlinearities and its Synchronization via Adaptive Control Method. Advances in Chaos Theory and Intelligent Control. *Studies in Fuzziness and Soft Computing*, *337*, 203–224. doi:10.1007/978-3-319-30340-6_9

Vaidyanathan, S., & Azar, A. T. (2016e). Adaptive Control and Synchronization of Halvorsen Circulant Chaotic Systems. *Studies in Fuzziness and Soft Computing*, *337*, 225–247. doi:10.1007/978-3-319-30340-6_10

Vaidyanathan, S., & Azar, A. T. (2016f). Adaptive Backstepping Control and Synchronization of a Novel 3-D Jerk System with an Exponential Nonlinearity. *Studies in Fuzziness and Soft Computing*, *337*, 249–274. doi:10.1007/978-3-319-30340-6_11

Vaidyanathan, S., & Azar, A. T. (2016g). Generalized Projective Synchronization of a Novel Hyperchaotic Four-Wing System via Adaptive Control Method. *Studies in Fuzziness and Soft Computing*, *337*, 275–296. doi:10.1007/978-3-319-30340-6_12

Vaidyanathan, S., & Azar, A. T. (2020). *Backstepping Control of Nonlinear Dynamical Systems. Advances in Nonlinear Dynamics and Chaos (ANDC).* Elsevier.

Vaidyanathan, S., Azar, A. T., Akgul, A., Lien, C. H., Kacar, S., & Cavusoglu, U. (2019). A memristor-based system with hidden hyperchaotic attractors, its circuit design, synchronisation via integral sliding mode control and voice encryption. *Int. J. Automation and Control*, *13*(6), 644–667. doi:10.1504/IJAAC.2019.102665

Vaidyanathan, S., Azar, A. T., & Ouannas, A. (2017a). An Eight-Term 3-D Novel Chaotic System with Three Quadratic Nonlinearities, its Adaptive Feedback Control and Synchronization. *Studies in Computational Intelligence*, *688*, 719–746. doi:10.1007/978-3-319-50249-6_25

Vaidyanathan, S., Azar, A. T., & Ouannas, A. (2017c). Hyperchaos and Adaptive Control of a Novel Hyperchaotic System with Two Quadratic Nonlinearities. *Studies in Computational Intelligence*, *688*, 773–803. doi:10.1007/978-3-319-50249-6_27

Vaidyanathan, S., Azar, A. T., Rajagopal, K., & Alexander, P. (2015b). Design and SPICE implementation of a 12-term novel hyperchaotic system and its synchronization via active control (2015). *International Journal of Modelling Identification and Control*, *23*(3), 267–277. doi:10.1504/IJMIC.2015.069936

Vaidyanathan, S., Idowu, B. A., & Azar, A. T. (2015c) Backstepping Controller Design for the Global Chaos Synchronization of Sprott's Jerk Systems. Chaos Modeling and Control Systems Design, Studies in Computational Intelligence, 39-58. doi:10.1007/978-3-319-13132-0_3

Vaidyanathan, S., Sampath, S., & Azar, A. T. (2015a). Global chaos synchronisation of identical chaotic systems via novel sliding mode control method and its application to Zhu system. *International Journal of Modelling Identification and Control*, *23*(1), 92–100. doi:10.1504/IJMIC.2015.067495

Vaidyanathan, S., Zhu, Q., & Azar, A. T. (2017b). Adaptive Control of a Novel Nonlinear Double Convection Chaotic System. *Studies in Computational Intelligence*, *688*, 357–385. doi:10.1007/978-3-319-50249-6_12

Wang, Z., Volos, C., Kingni, S. T., Azar, A. T., & Pham, V. T. (2017). Four-wing attractors in a novel chaotic system with hyperbolic sine nonlinearity. *Optik (Stuttgart)*, *131*, 1071–1078. doi:10.1016/j.ijleo.2016.12.016

Youssef, A., El-Telbany, M., & Zekry, A. (2017). The role of artificial intelligence in photo-voltaic systems design and control: A review. *Renewable & Sustainable Energy Reviews*, *78*, 72–79. doi:10.1016/j.rser.2017.04.046

Chapter 6
Comparative Analysis of Various Types of Control Techniques for Wind Energy Conversion System

Bibhu Prasad Ganthia
https://orcid.org/0000-0003-2351-7830
IGIT, Sarang, India

Subrat Kumar Barik
KIIT University (Deemed), India

Byamakesh Nayak
KIIT University (Deemed), India

ABSTRACT

The interest towards renewable energy has been enhanced due to zero pollutant emission. Considering the present scenario, wind as a renewable source of energy is highly recommended. As it is freely available and free from pollution, wind can effectively play a role for energy generation. This can produce quality power during grid integrations as the load demands. Due to rapid variations in wind speed, wind energy systems need highly synchronized and powerful controller techniques for power regulations to overcome transients, voltage sags, and swells. A suitable and responsive controller is essential for power generation from wind energy. The controllers for wind energy system are categorized into five controller designs according to their locations to control the demand of the turbine system during grid integrations. In this chapter, various controller designs and implementations are highlighted with reference to previous works and existing studies.

DOI: 10.4018/978-1-7998-7447-8.ch006

INTRODUCTION

Among other renewable energy sources wind energy is a rapidly growing sector (R. Richardson & G. McNerney, 1993). It is because wind power is safe, accessible geographically, inexpensive and useful in rural areas. It is sporadic, however, and needs high investment start-ups, higher transmission costs, and widespread land (H. J. Wagner & J. Mathur, 2009). A physical structure with three main components is in theory a wind power conversion system (WECS). The first is a blade-based rotor. As the wind moves through the blades, it rotates the rotor and thus generates mechanical control. The second is the power transmission from the rotor to the generator. The last is a mechanical generator, which converts mechanical energy into electricity. In fact, wind energy converters are axis graded. (F. D. Bianchi et al. 2007) Two types of wind converters are therefore available. The horizontal converter is the most popular. It requires two or more blades designed for optimal aerodynamics. This machine also provides a mechanism for placing the rotor blades in direction of the wind. (P. Moriarty & S. Butterfield, 2009) The second type is the vertical axis converter consisting of vertically positioned blades.This converter is therefore capable of capturing wind regardless of wind direction. Nonetheless, this converter has collapsed in recent decades due to lower capacity, complicated maintenance and large-scale land usage. Description of wind energy conversion system (WECS) categories and dynamics modeling are given in this chapter. This paper offers an up-to - date description of advanced control methods implemented in WECSs. We picked a number of documents for each technique and studied them primarily in chronological order. (H. Li & Z. Chen, 2008)

The figure 1 shown below illustrates the generation capacity of power from wind energy in India with comparison to other countries for present and future aspects. Wind energy as renewable source is very much effective for both on shore and off shore power generations. Complete hazard free and do not cause any pollution to the environment. This chapter illustrates the various control techniques for generating high efficient powers to compensate the power demands.

Figure 1. Wind power share in generation

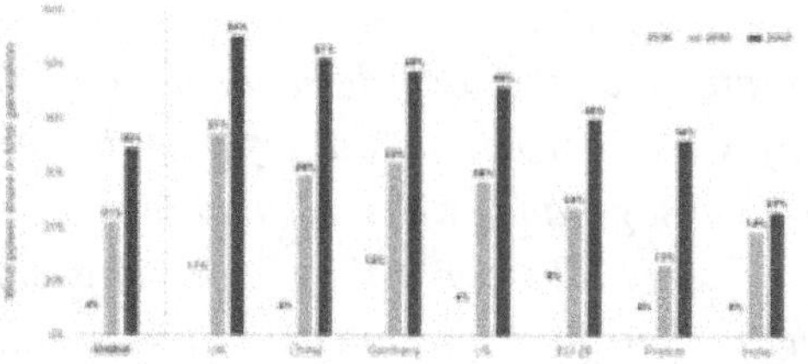

WIND ENERGY CONVERSION SYSTEM

The wind is generated by the intermittent heating of the atmosphere caused by sunlight. As hot and cold air regions form, pressure gradients are created. (Y. zhang Sun et al. 2009) This allows air to escape from the low pressure area. The cinematic energy of the moving air is called wind. The wind conversion system block diagram is shown in Figure 2. The aerodynamically designed blades that result in rotation of the wind turbine are subject to this kinetic power. (J. Laks, L. Pao &A. Wright, 2009) Often the wind speed cannot move the blades at the speed that it is not possible to produce electricity. For some WECS topologies, a gearbox is required for increasing the speed of the rotor shaft.By using generators, the mechanical energy of that same shaft is transformed into electrical energy. The generated electricity is distributed to the interconnected grid. (J. Yan et al., 2009)

Figure 2. Wind energy conversion system

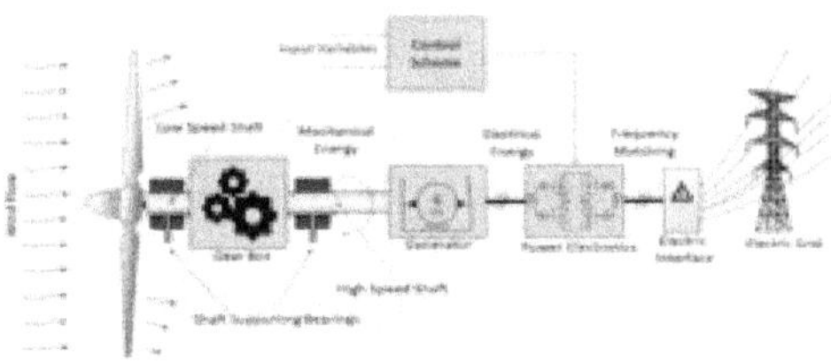

A wind turbine is a mechanical motor that uses aerodynamically shaped blades to absorb power and transform it into usable mechanical power from a fluid flow (wind). (I. Buehring& L. Freris, 1981) The power capacity depends on the wind speed, and to prevent damage to the device, it is important that the power can be controlled and restricted by higher wind speeds. (S. Natarajan, et al., 1991) Power limits can be achieved using certain 3 methods (the blade is fixed so the wind stays in the blade at a lower wind speed), active stall (the blade angle is set to allow a stop on the blades) and/or pitch control (wind is turning off the blades at a low wind speed). The blades can also be managed with an active stall. (I. Tsoumas et al. 2009)

WIND TURBINE SYSTEM

The electrical transformer supplying the winding rotor has a power rating of about 30%, and the DFIG's winding figures are directly related to the grid. (K. Tan and S. Islam, 2004) The third category is a variable speed wind turbine with a full-speed electronic transmission system and a synchronous generator or SCIG. Usually, the two first generators use a multi-stage gearbox. (A. Bratcu et al., 2007) Synchro generators, like a PMSG, can be powered directly by the low-ratio gearbox system; an interesting feature is one or two gearboxes. With the decline in the cost of electronic power converters, the B2B wind turbine is becoming a great option. Direct PMSG-based configurations are an attractive alternative for WECS compared to other wind turbine generators with a variable rpm. (A. Bratcu et al. 2008) In contrast to the DFIG wind turbine system, the WT-PMSG does not necessitate a gearbox and slip-ring to allow high-performance WECS construction, stable operation and low maintenance. In the wind turbines industry, the efficient generation of more power by onshore and offshore wind turbines using PMSGs is continuously being investigated. (L. Jianlin& X. Honghua, 2008) (S. Sanchez et al., 2009)

Figure 3. Wind turbine construction model

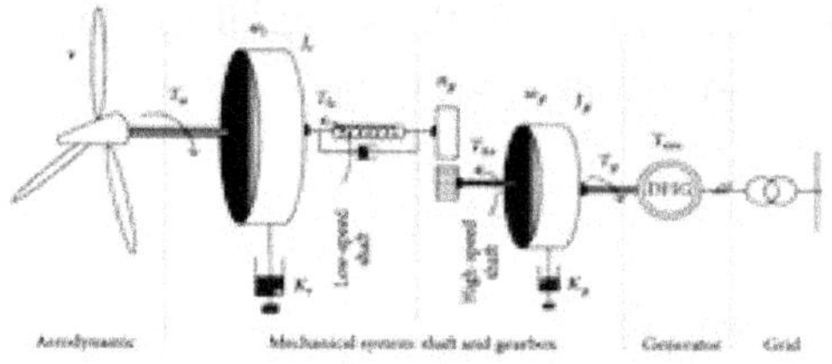

Figure no. 3 shows a detail structure of the wind turbine drive train model. The following equation is used for the calculation of the generated power of WT. (M. Lima et al., 1999)

$$P_m = \frac{1}{2}\rho\pi R^2 C_P(\lambda,\beta)V_w^3 \tag{1}$$

where ρ refers to the density of air, A and V_w denotes the turbine swept area and the speed of the wind respectively. C_p represents the power coefficient, non linear in nature. The value of C_p depends on the tip speed ratio λ and blade pitch angle β given as C_p (λ,β). (R. Rocha et al., 2001)

The tip speed ratio λ can be determined with the help of the equation as given below:

$$\lambda = \frac{\Omega_t R}{V_w} \tag{2}$$

where Ω_t ,R, V_w refers to the rotor speed, the radius and the wind velocity respectively. Torque of the wind turbine T_m produced can be found out using equation as follows:

$$T_m \frac{P_m}{\Omega_t} \tag{3}$$

The C_p can be found out using the equation given below:

$$C_p(\lambda,\beta) = c_1\left(\frac{c_2}{\lambda_i} - c_3\beta - c_4\right)e^{\frac{c_5}{\lambda_i}} + c_6\lambda \tag{4}$$

where,

$$\frac{1}{\lambda_i} = \frac{1}{\lambda + 0.08\beta} - \frac{0.035}{1+\beta^2} \tag{5}$$

CONTROL STRATEGIES

When a wind turbine system is combined with the grid, different control strategies are used for smooth control and operation, as well as improved stability. Maintaining the power system's reliability is critical during integration. As a result, these control strategies are being used to prevent transients that trigger device mismatches. These are divided into four groups based on how different monitoring methods are implemented. They are:

1. Pitch Angle Control Techniques:
    a. Conventional Control
    b. Robust Control
    c. Soft Computing Control
    d. Hydraulic Control
    e. Hybrid Control
2. Converter Control Techniques:
    a. Back-To-BackConverters
    b. ResonantConverters
    c. Multilevel Converters
    d. MatrixConverters
    e. TandemConverters
3. MPPT Control Techniques:
    a. Power Signal Feedback Control
    b. Tip Speed Ratio Control
    c. Hill Climb Control
    d. Optimal Torque Control
    e. Soft Computing Control
4. Grid Side Control Techniques:
    a. Direct Torque Control
    b. Direct Power Control
    c. Stator Voltage Oriented Control
    d. Field Oriented Control
5. FRT and LVRT Control
    a. Series Connected FACTS Devices
    b. Shunt Connected FACTS Devices
    c. Hybrid Connected FACTS Devices

PITCH ANGLE CONTROL TECHNIQUES

When wind speeds are greater than the rated speed and basic control variables such as wind speed, engine rpm, and generator power are chosen, pitch angle adjustment is a very efficient way of adjusting the wind turbine aerodynamic torque. (R. Rocha & L. Filho, 2003) Pitch control is the system for the operation and regulation of the wind turbine angle of the blades. Generally, the machine consists of electric motors, gears or hydraulic cylinders and a supply chain. The PAC technique is shown in figure no. 4 below. (R. Rocha et al., 2005)

Figure 4. Wind energy pitch angle control

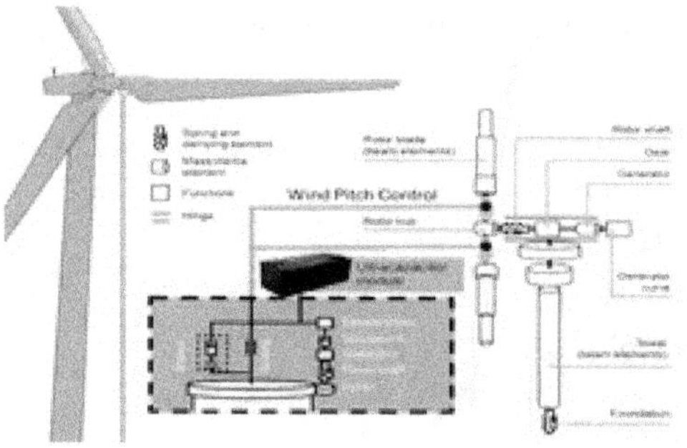

Conventional Control

In figure 5 below, the PI / PID control scheme is the conventional pitch angle control mechanism for controlling rotor speed or power supply. It's usually used to power a small wind turbine. The pitch relation is derived primarily from input parameters such as rotor velocity, generator strength, and wind speed in conventional controllers, as seen in figure 4 below. (B. Wang et al., 2009) The angle reference for wind speed is provided by a sensor, which uses wind speed as data, directly from a wind velocity-angle. Although this approach is easy, it is not possible to accurately calculate the wind speed.Thanks to its reliability, the generator power and rotor speed control is the most developing effective controller. (H. Tien et al., 2007) The inaccuracy signal is received by a PI / PID controller and the wind turbine is referred to by the generated pitch angle. The traditional converter with gain programming is used in order to boost the control efficiency of the non-linear system. Aerodynamic torque's sensitivity to the pitch angle is overcome by the gain scheduling. (D. Bourlis& J. Bleijs, 2009)

Figure 5. Wind energy conventional control (PI controller)

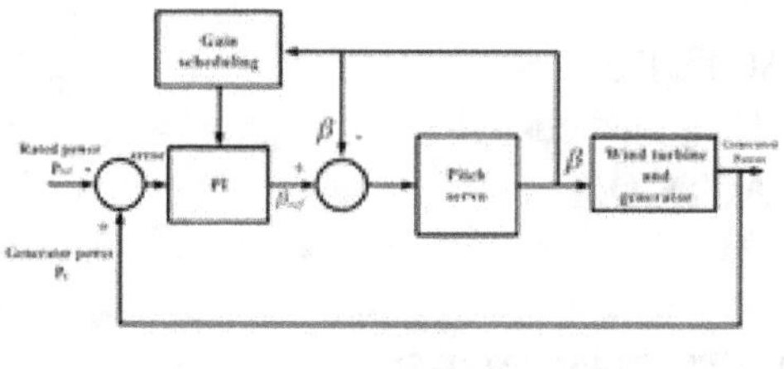

The system's aerodynamic response leads to change of power with the pitch angle respectively. Inversely proportional to the sensitivity of the system and its income. This means that traditional controllers are more robust than the controller without a plan of gains. (J.J. Chen & Z.-C. Ji, 2010)

Robust Control

Figure 6 shows the fundamental configuration of the robust wind energy conversion angle controller. The robust controller provides advance geometry, feedback and feed forward system. To increase the pitch angle controller robust strength, sliding mode control is applied (SMC). The wind turbine's architecture is straightforward and overcomes uncertainties. This method, however, requires previous system knowledge and depends on the wind turbine's mathematical model. The wind turbine system mechanical stress is also increased because the control variables suddenly change.With an H-infinity control, the system sensitivity is reduced. This offers strong resistance to changes in wind speed and turbine parameters. (W. Wang et al., 2010)

Figure 6. Wind energy robust controller

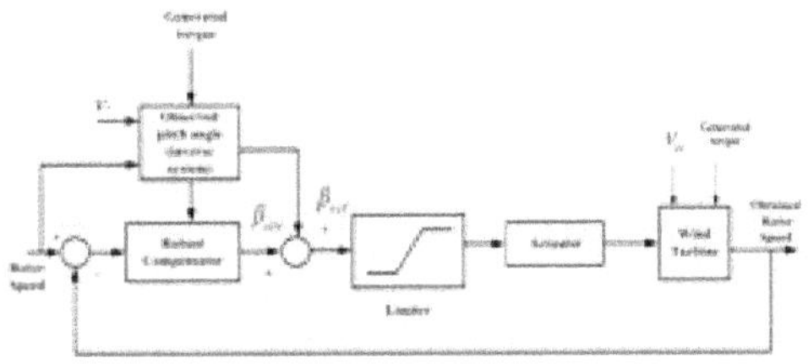

The power output of the turbine is increased as well. This method's only drawback is its difficulty when designing device parameters and weighting task constraints. It proposes an alternative robust controller using linear Gaussian quadratic (LQG) and linear matrix inequalities. (M. Mayosky& I. Cancelo, 1999) Rotor speed trajectory is determined and in a not linear controller the optimal angle of pitch is achieved. In the non-linear controller this device therefore lacks protection for the wind turbines due to mechanical turbulence. (M. Sedighizadeh et al., 2008)

Soft Computing Control

In order to resolve WECS uncertainties due to the fast wind speed variability, soft computer controlled pitch angle controllers are developed. In comparison with another conventional approach, artificial techniques provide accurate, predictive, and rapid responses. (M. Sedighizadehet al., 2007)

Figure 7. Wind energy fuzzy type I controller

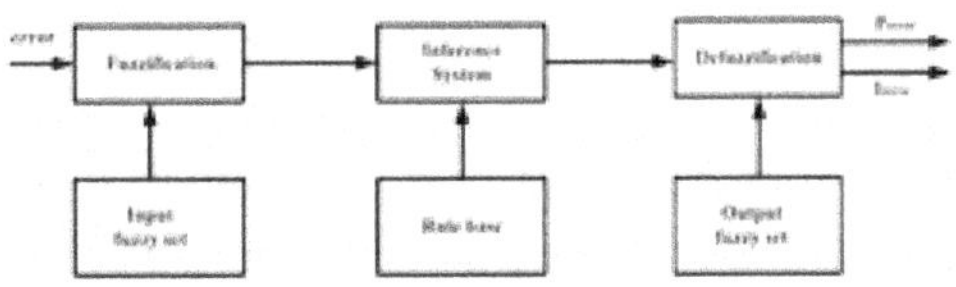

Figure 8. Wind energy fuzzy type II controller

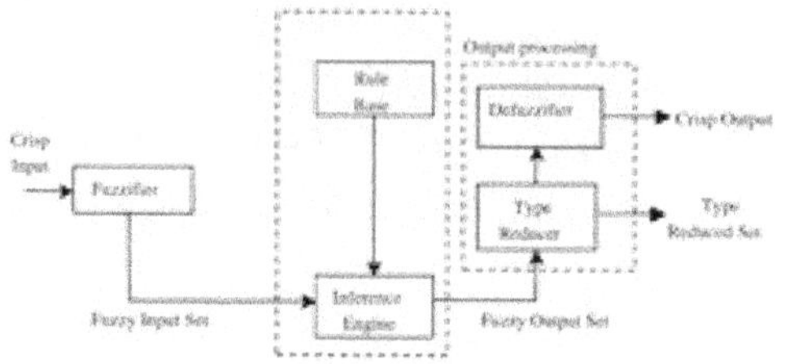

Soft computing aims to alleviate a wide range of issues. Soft computing incorporates the Artificial Network (ANN), Type I in figure 7 &Type II FLC in figure 8, Genetic algorithm based controller and Adaptive neural Fuzzy inference system (ANFIS). Coding based and interlinked models are designed using soft computing control strategies. (J. Hui & A. Bakhshai, 2008)

Hydraulic Control

The speed of the rotor has a major effect on wind turbine absorption. The rotor speed varies with the wind speed, which is determined by an optimum tip speed ratio, in order to obtain full wind power coefficients. A modern closed-loop hydraulic system with hydraulic accumulator is then introduced and used to generate simultaneously variable velocity constants in the wind energy system as shown in Figure 9. As shown in Figure 9. The hydraulic wind turbine with hydraulic accumulator's closing loop mechanism is completely different from the one we are used to. The pump and engine displacements are both variable in this hydraulic wind turbine. (J. Hui & A. Bakhshai, 2008) The rotating joint is an essential feature of the hydraulic wind turbine, which can transmit hydraulic oil without damaging hydraulic lines during ladder operation from the nacelle to the variable motor. (K. Raza et al. 2008)

Figure 9. Wind energy hydraulic control

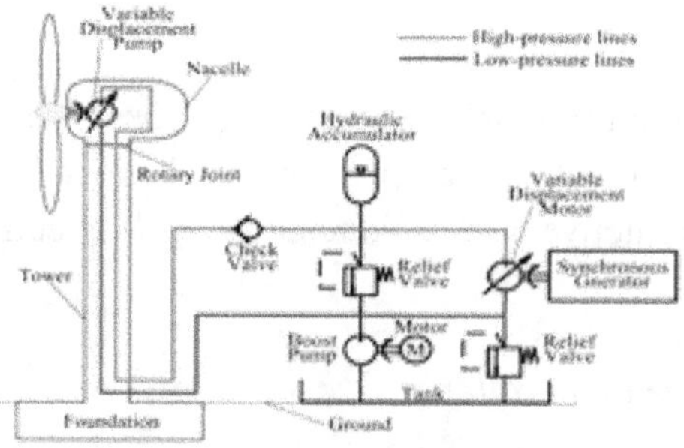

Figure 9 illustrates the conversion and transmission processes for the energy storage system of the hydraulic wind turbine. The pump outlet pressure and input pressures of the motor are calculated by the addition of the hydraulic accumulator to the high-pressure sides. The difference of the wind energy produced by the wind turbine from the generator determines whether the hydraulic accumulator flows into or out of the hydraulic fluid. (K. Raza et al. 2008)

Hybrid Control

Hybrid pitch angle controller was used to address the inconveniences of the above controller. In order to provide effective and full wind power conversion systems, the hybrid controller overcomes all the downsides of the traditional control. In order to estimate rotor speed effectively without a sensor, soft computing technology and a robust controller are hybridized. This system demonstrates good stability performance. In conjunction with traditional controllers SCT technology is often used to reduce system complexity and boost the system's dynamic performance.In order to overcome wind turbine insecurity due to fluctuating wind speeds, ANN mainly radial bias function network is used to establish the integral and proportional gain of the strategies in conventional controller design.(H. K. Khalil et al., 1996)

Figure 10. Wind energy hybrid controller

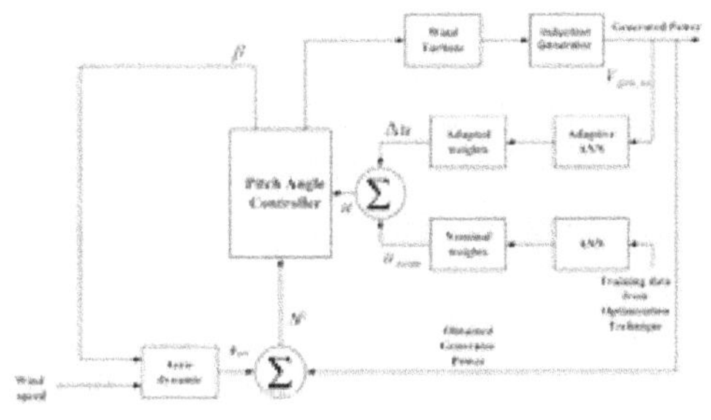

The SCT methodology is used to evaluate the frequency variance and boost the controller response time show in figure 10. A mixture of fuzzy sensor and neural network sensor technology is performed to degree output power for the entire operating area and is based on a various hybrid pitch-angle controllers implemented in WACS. Fuzzy with conventional control techniques, adaptive, genetic algorithm based and sliding modes are comes under hybrid control methods for WECS. (H. De Battista & R. Mantz, 1998)

CONVERTER CONTROL TECHNIQUES

The variable speed wind system includes a power amplification system known as the power converter to change the frequency and voltage of the generator to the grid. In order to increase the power quality and step up the voltage level respectively, most proposed converters require line filters and transformers. (H. De Battista et al., 2000)

Back-To-Back Converters

The topology modules built by wind power electronics manufacturers for energy systems are several forms of power converters shown below in figure 11. The configuration of one of these power converters connected back to back (BTB) on the two sides of the typical DC connection, namely the generator side and the grid side, typically has two identical power converter units. (R. Pena & D. Sbarbaro, 1999) In all commercial WECS as outlined in above figures, the BTB converter configurations can be used.

The two-way power flow from variable voltages / frequency (generator side output) to fixed voltage / frequency (grid side output) and the other way around kept this configuration back to the back of a connected converter. (F. Valenciaga et al., 2004)

Figure 11. Wind energy BTB converter

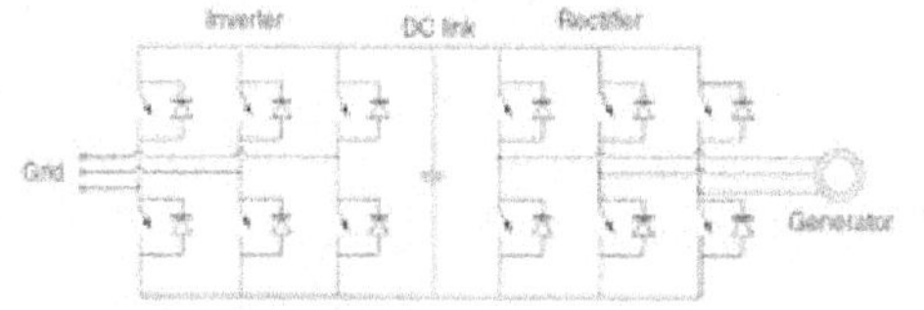

Resonant Converters

The converters may be resonant series, parallel and non-parallel resonant tank combinations shown below figure no. 12. The parallel serial LCC resonant converter gives both series and parallel configurations the desirable features (F. Valenciaga& P. Puleston, 2007). In a WECS, the power output decreases with the wind. The charge decreases in terms of dc-link output, as the CSI actively controls the dc-link voltage. When the wind speed decreases, the device needs higher gain due to a rising input voltage. This WECS requirement is inherently satisfied by the LCC resonant converter feature. (S. F. Pinto et al., 2007)

Figure 12. Wind energy resonant converter

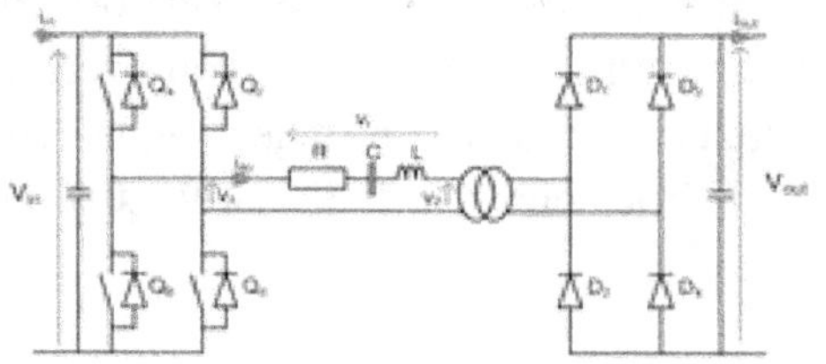

While the WECS works with low power output, low speed and low input voltage, it gives a high return. In comparison, when the service cycle at high speed, the higher fiber voltage convertor and parallel (and LLC) resonant converters lower the dc gain. This unwanted characteristic forces the semiconductor to undergo a large electric stress. (B. Beltran et al., 2008)

Multilevel Converters

The multilevel inverter is based on the assumption that a stepped waveform with several steps can be approximating sine waveforms. The measures are assisted by series of linked batteries or condensers from different DC stages. (I. Munteanu et al., 2008)

Figure 13. Wind energy multilevel converter

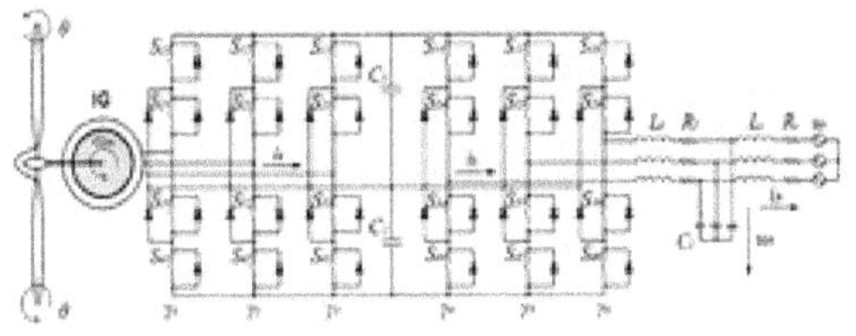

The special multi-level inverter arrangement shown in figure no. 13 makes it possible to achieve high voltages and thus to run a lower voltage rating system. The synthesized output waveform is increased as the number of levels is increasing, producing a fine wave of stairs which approaches the desired sine wave very closely. Multilevel inverters therefore give a better solution at the high power output because with such inverters the high voltage amplifier levels are possible without the problems of high dv / dt and the other associated inverters. (F. Valenciaga& P. Puleston, 2008)

Matrix Converters

The matrix converter is a forced type switched converter which uses a range of regulated two-way switches as key power elements for creating an unregulated frequency variable output voltage system. It has no dc-link circuit and no major energy storage elements are needed. The conventional matrix converter topology consists of a range of nine bi-directional switches that link each input phase to each output phase.The control of the output voltage size, frequency and phase angle, as well as input angle control, can be achieved by correctly operating the switches in the matrix converter. Matrix is a two-directional electricity flow system capable of generating high-quality waveforms for input and output. (X. Zheng et al., 2010)

Figure 14. Wind energy matrix converter (traditional)

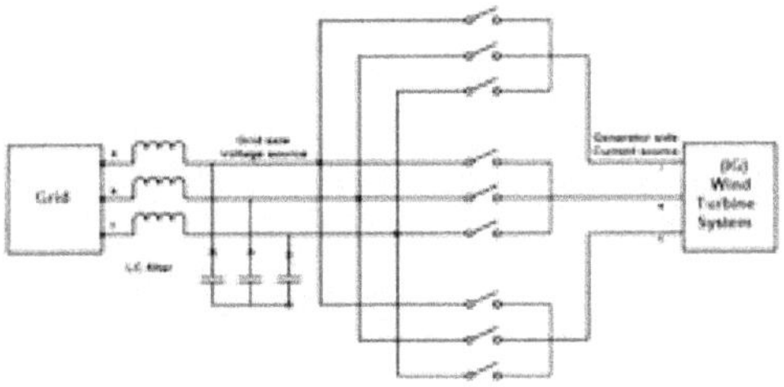

Figure 14 displays the traditional matrix converter schematic diagram. The enhanced matrix converter is based on the "fictional DC link" concept used to control the standard matrix converter. However, between the line-side and load-side converters, there is not a power storage element. Switching problems have been overcome with the switches. (D. Kairous et al., 2010)

Figure 15. Wind energy advanced matrix converter

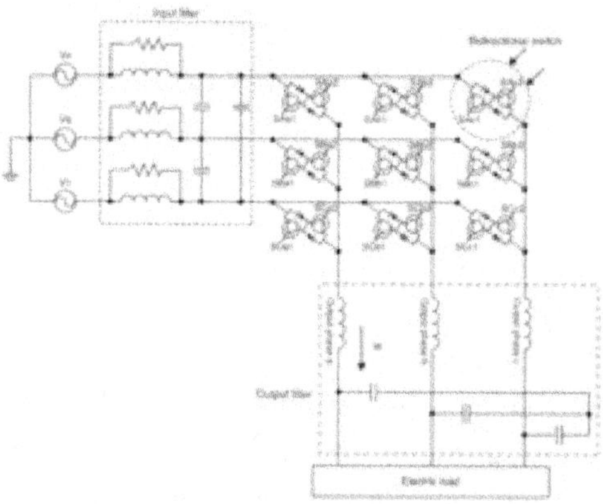

All the turns on and off at zero current on theline-side. As shown in figure no. 15, the matrix converter provides four levels of control for regulating the magnitude, frequency and phase angle of input and output voltage. The above model in Matlab/Simulink setting was implemented and displays the converter and the inverter part in Matrix Converter. (M. Bayata& H. Karegar, 2009)

Tandem Converters

It uses a CSC built on cheap rough thyristors with active shunt filters called the Tandem converter. It uses an active source converter. Using cascaded multilevel inverters, the current waveform is smooth and displayed in the figure below. (D. Dang et al., 2008)

Figure 16. Wind energy tandem converter

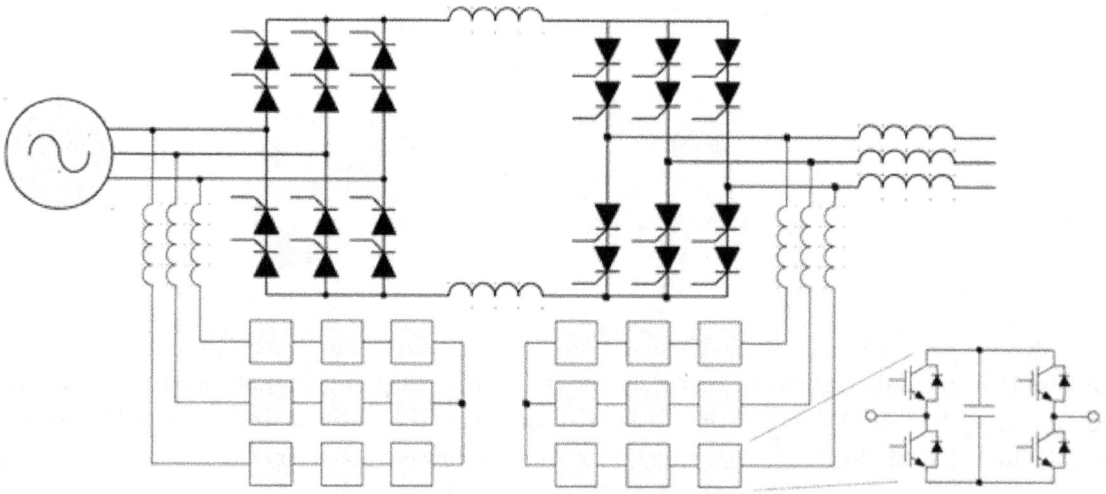

Figure 16 shows that the active filters are typically rated at around 25 percent of the converter power in a tandem converter, resulting in a lower DC-link capacitance requirement. Additionally, the filter current waveform mainly consists of the higher CSC current harmonics, further reducing the capacitance criterion. DC-link energy storage requirements for an MMC and the proposed tandem converter for a 10 percent ripple of DC voltage at a rated power and 2.5Hz AC frequency shown in above figure 16. For example, the DC energy storage for an NPC converter with 1 percent voltage ripple is shown; however the capacitance is likely to be determined by the current specifications for such a converter. (D. Dang et al., 2009)

MPP CONTROL TECHNIQUES

This idea for extracting maximum power from low speed of wind is done by maximum power point technique by adjusting the duty cycle of the converters and by controlling the rotor speed of the wind turbine. This is the effective and highly efficient technique for extracting more power from rapid wind speed variations from low to high. The flow chart shown in figure no. 17below highlights the conditional algorithm for maximum power to achieve. (E. Muhando et al., 2010)

Figure 17. Wind energy MPPT

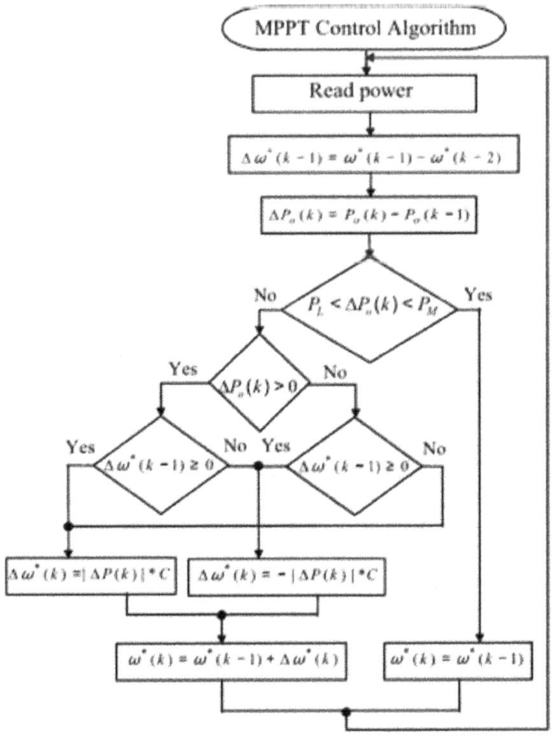

Power Signal Feedback Control

The reference power, i.e. the maximum power produced at a given wind speed, is used for this method. Relevant information of the characteristics of the wind turbine and the WSR is needed for this process. The maximum power is obtained by carrying out the simulation for the wind turbine after obtaining this data. (A. Besheer et al., 2008)

Figure 18. Wind energy power signal feedback control

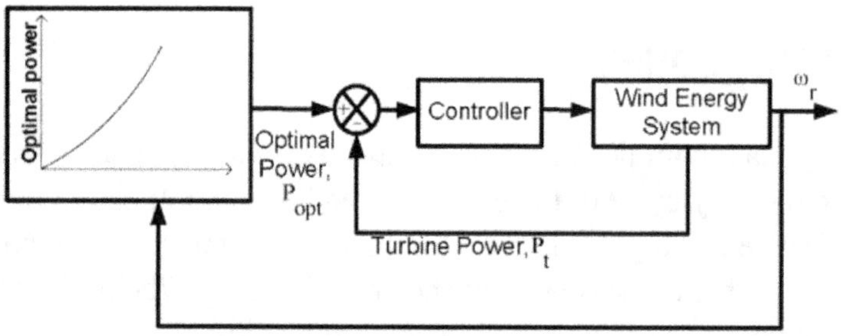

Figure 18 illustrates the characteristic curve for various wind speeds between turbine output torque and turbine rpm. A contrast is made with the present power after receiving the reference power from the power curve. This error is sent to the controller consisting of an integral proportional controller that allows the device to run at its MPP. (X. Zhanget al., 2006) The only downside to using this PSF is that it needs a range of sensors to gain previous reference power information. A drawback of this algorithm is that it uses the properties of the turbine (power coefficient, torque, power curves) to calculate the magnitude of the reference velocity shift compared to the power change.This introduces the algorithm's dependence on the properties of a given turbine. (X. Zhang et al., 2009)

Tip Speed Ratio Control

The goal of this approach is to hold the TSR at a set optimum value such that it is possible to obtain full power. The ideal TSR for a given turbine remains constant, independent of the wind direction. So, if we sustain a constant TSR for that unique turbine, it is guaranteed to extract maximum energy. (X. Jian-jun et al., 2010) We need to evaluate the value of ideal TSR λopt at the beginning itself, either experimentally or theoretically. The error is sent to the controller by comparing the real value λ and the comparison value λopt, which lets the machine run at its optimum position by adjusting the generator speed and reducing the error. The TSR operational block diagram shown below in figure no. 19. (C. Amendola & D. Gonzaga, 2007)

The only drawback of this technique is that prior to setting the reference value of TSR, wind direction and turbine speed (ω) knowledge must be known. The extra sensors would increase the expense of the device to calculate the wind speed, and there are certain frictional errors due to the existence of the anemometer that yield unreliable results. (N. Boonpirom& K. Paithoonwattanakij, 2006)

Figure 19. Wind energy TSR

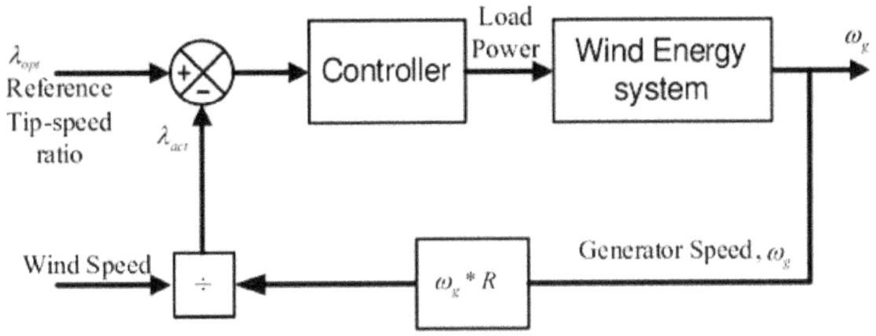

Hill Climb Control

The HCS optimization technique searches constantly for the wind turbine's optimum power. It will solve some of the typical issues usually associated with the other two strategies. The monitoring algorithm relies on the relationship between power and speed shifts, the direction of the operating point, the optimal maximum signal determined to move the device to the highest power point. (Q. Zeng et al., 2008)

Figure 20. Wind Energy hill climb control

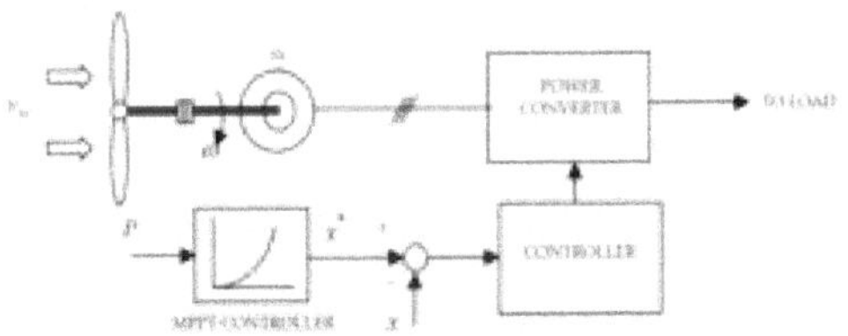

The theory of HCS control and a WECS with HCS controller to monitor maximal power points are seen in Figure No. 20. The wind turbine is attached via a battery and a DC/DC converter to a rectifier. (J. Hui et al., 2010) The proposed MPPT algorithm changes the wind energy conversion system's operating point by explicitly changing the DC/DC converter service cycle on the basis of output power calculation comparisons. The basis of the algorithm is very similar to that of the process of Hill Climb Search (HCS). The unique feature is that by a combination of the change in output power to the task ratio, the changes are introduced. Relationships have been identified between the boost, buck-boost, and buck converters' duty ratios and the shift in output power. Thus, with respect to the service cycle of the converter, the algorithm calculates the operation point modification depending on the shift in power. (X. Yunqi et al., 2010)

Optimal Torque Control

To obtain the maximum extracted power from the WECS, we need to operate the system near to λopt, which makes the system to achieve maximum output power. The main objective of this method is to bring the generator torque close enough to maximum reference torque at particular wind speed. (S. Mishra et al., 2009)

Figure 21. Wind energy optimal torque control

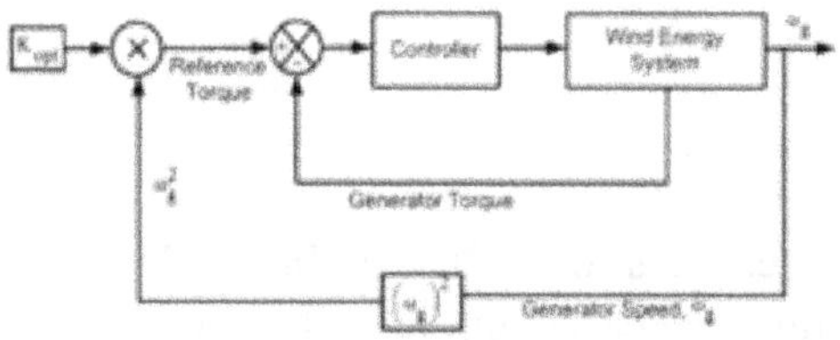

For this technique, the reference power and torque are shown in Figure 21. After making a change in generator torque, this is further related to the controller. When the wind varies, the controller can alter the error and adjust the speed of the wind turbine to achieve the optimal operating position. The benefits of this approach are quick, easy and effective, but because of its indirect calculation of wind speed, it has a lower performance relative to other techniques such as TSR. (M. Ali et al., 2009)

Soft Computing Control

Due to rapid variance in wind direction, soft computing-based pitch angle controllers are designed to resolve the uncertainties of the WECS. In comparison to another conventional approach, artificial techniques provide an effective, predictive and rapid response methodology. The figure no. 22 shown below express the operational block diagram of the soft computing model for wind turbine systems. (Z. Huo et al., 2009)

Figure 22. Wind energy soft computing control

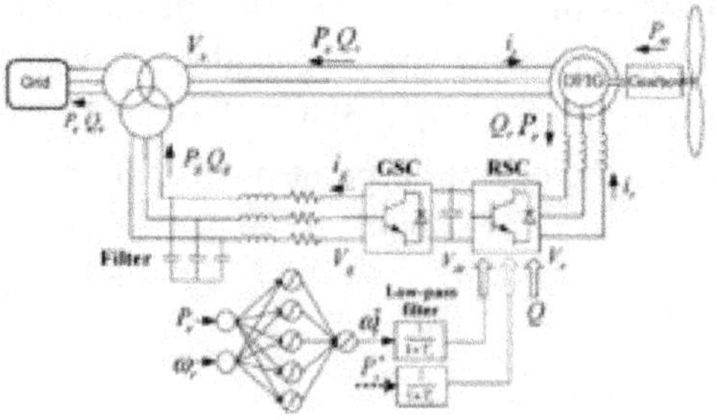

Computational approaches tend to solve a wide range of issues. Fuzzy logic control (FLC), Artificial neural network (ANN), Adaptive neural based fuzzy inference system (ANFIS) and genetic algorithm based controller are included in the soft computing technique. In soft computing based pitch angle controller, many studies have been performed. For pitch angle controllers, the majority of researchers use ANN, FLC and GA based controllers. As shown in figure 22, the ANN controller consists of the input layer, the hidden layer, and the output layer. A user defines the number of nodes in each layer.Generally, the input of the ANN technique is wind speed, power generated, rotor speed, or the combination of these variables. The ANN-based controller is capable of rapid response and can be easily adapted to different conditions. (H. Li et al., 2005)

GRID SIDE CONTROL TECHNIQUES

Direct Torque Control

The DTC principle is known as direct selection of suitable stator voltage vectors attributed to differences between reference and actual stator flux vector magnitude values but mostly electromagnetic torque. The DTC operational block diagram shown below in figure no. 23. (Y. Ren & G. Bao, 2010)

Figure 23. Wind energy DTC

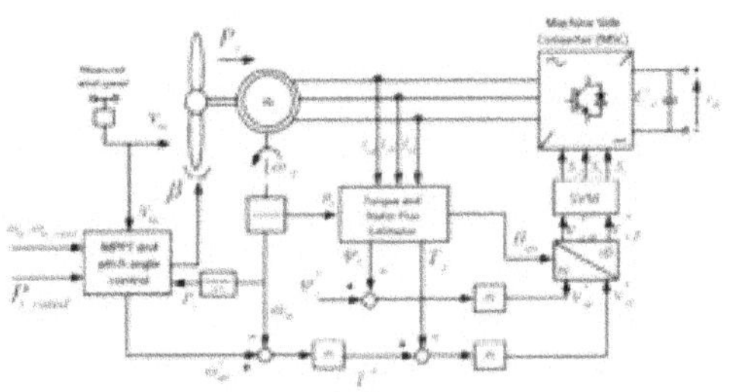

Using the hysteresis flux and torque controllers operated by the switching table, the conventional DTC can be implemented. (J. Brahmi et al., 2009) The variable switching frequency caused by the electromagnetic torque ribs is generated by this conventional DTC system with hysteresis controllers. In addition, the high sampling frequency is necessary for the digital implementation of comparators of hysteresis. (J. Thongam et al., 2009) The alternative to switching table control is to implement enhanced DTC with Space Vector Modulation (DTC-SVM). The DTC-SVM guarantees the stator current of the lower harmonics and enables the electromagnetic torque ripples to be decreased. This method is effective as the SVM modulation technique continuously regulates the switching frequency of the sectors. (L. Zhang et al., 2008)

Direct Power Control

The primary task of the grid side converter (GSC) is to maintain the DC connection voltage and to control the supposed reactive power supplied to the AC grid. (L. Le et al., 2008) The hysteresis controllers and switching table were used for the standard Direct Power Control (DPC). The varied switching frequency of control signals defines this control scheme. Torque and power are indirect control strategies that are estimated and directly controlled, resulting in algorithms that are less complex and faster. (X. Cheng et al., 2009)

Figure 24. Wind energy DPC

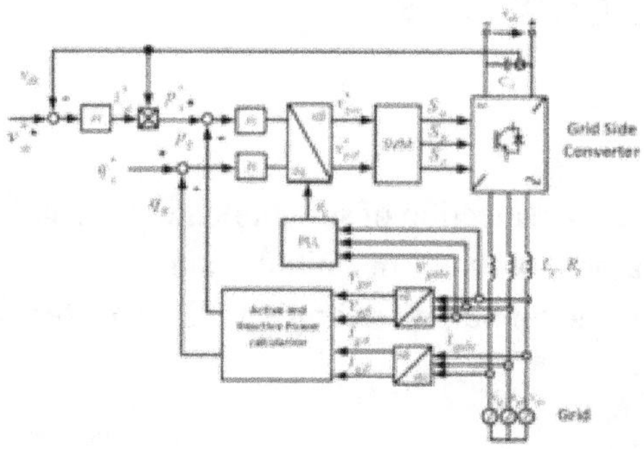

Figure No. 24 shows the complete block diagram of the direct power control technique. The use of the DTC method eliminates internal control loops, and it is also not necessary to transform between reference frames. (J. Qi & Y. Liu, 2010) The performance of generator side control techniques is assessed by incurrent and torque ripples. (F. qing Chen & J. ming Yang, 2009) The primary drawback of DTC is the requirement for differential switching frequency, which is difficult to resolve. (Z. jie Wang et al., 2010) Control implementation is carried out in some advanced control techniques by changing the width of the hysteresis band online or using coordinate transformation, which will take more CPU time in calculation that has lost the benefit of the DPC method. (A. Gargoom et al., 2010)

Stator Voltage Oriented Control

For grid converters, voltage directed control is often used. In VOC, the transformation of coordinates between the stationary and synchronous reference frames is performed to monitor the converter side of the grid. (X. Yao et al., 2007)

The above simulink block diagram figure no. 25 illustrates the grid side converter design using converter, transformation block and conventional controller. (X. Yao et al., 2009)

The above simulink block diagram figure no. 26 illustrates the rotor side converter design using converter, transformation block and conventional controller. (Z. Xing et al., 2009)

The input parameters are translated to dc quantities in the synchronous reference frame, as seen in Figure 26, so filtering and monitoring can easily be achieved. Another value is that, when monitoring dc variables, the PI regulator offers enhanced efficiency. (X. Yao et al., 2009) To achieve grid angle for

Figure 25. Wind energy GSC block

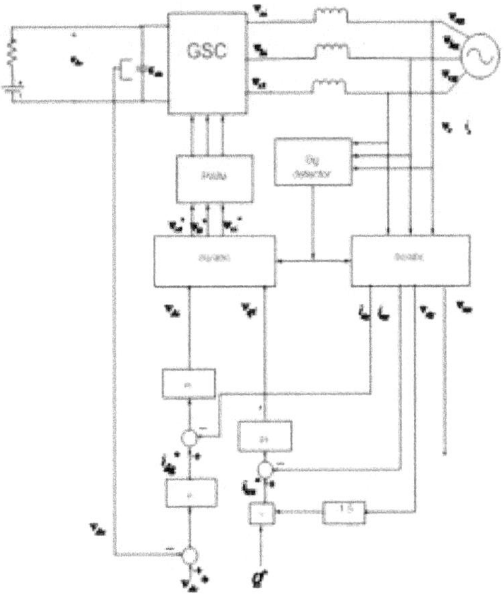

Figure 26. Wind energy RSC block

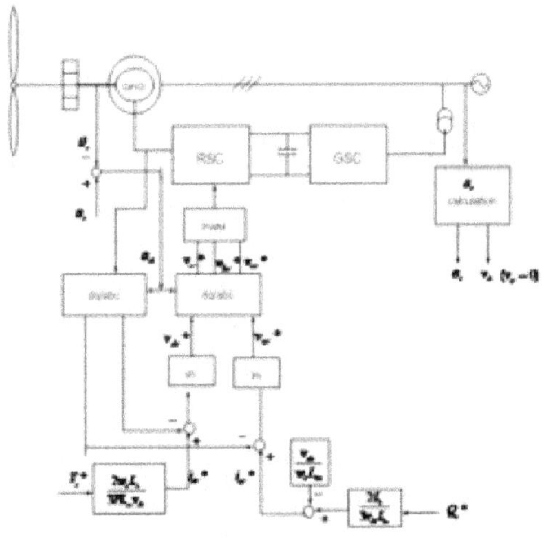

grid synchronisation and coordinate transition, a phase locked loop (PLL) is used. Because of internal control loops, this control technique guarantees fast transient response and high static performance. To provide separate regulation for active and reactive power, grid currents are decomposed into d and q-axis currents. Such regulation allows to achieve high power and sinusoidal grid currents. (M. Kalantar& M. Sedighizadeh, 2005)

Field Oriented Control

Field Oriented Control emerges from the decoupled flux-current and torque-current control of AC drives. In direct current devices, it resembles the concept of decoupled regulation of excitation and armature current. (T. Senjyu et al., 2006) (Y. Wang et al., 2009) As the wound rotor machine is connected to the grid, it regulates the active power and reactive power close-loop and produces the necessary reference flux and torque currents in field-oriented control. (V. Calderaro et al., 2007)

Figure 27. Wind energy field oriented control

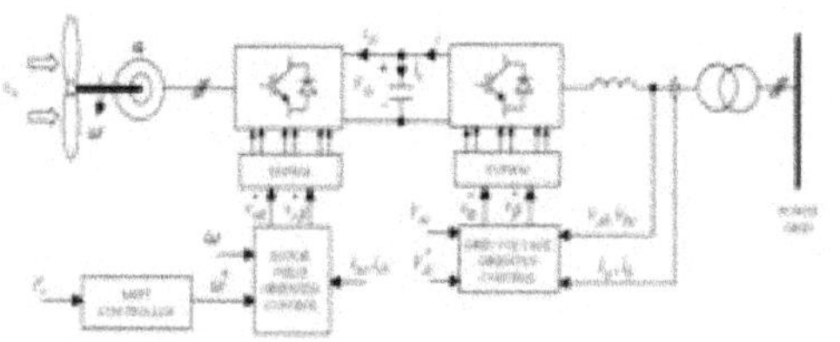

The figure no. 27 shown above highlights the FOC technique implemented for wind energy system under grid integration. (Z. Xin-Fang et al., 2004) (W. Dinghui et al., 2007) The source side converter is connected to the grid with a transformer step-up. Generally, the source-side converters use a power filter to reduce the influx of current harmonics through the power source. (X. Yang et al., 2007) (W.-M. Lin et al., 2010)

FRT and LVRT Control

In the occurrence of a voltage drop, before being allowed to detach, turbines are expected to stay attached for a specific time span. (X. Yao et al., 2007) (X. Yang & X. Liu, 2010) This provision is to ensure that usually cleared faults have no generation failure. Disconnecting a wind generator too quickly, particularly with big wind farms, may have a negative effect on the grid. Grid codes inevitably specify that, for a certain percentage of the nominal voltage and for a given length, massive wind farms have to tolerate voltage sags. (C.M. Hong et al., 2007) (B.P. Ganthia et al., 2021)

These Controls are categorized into three major parts. They are: Series connected Control Techniques, Shunt connected Control Techniques and Hybrid Connected Control Techniques. Further these connected control techniques are use FACTS devices for fault ride through capability and low voltage ride through operational control during faults. (B.P. Ganthia et al., 2020) (Bibhu Prasad Ganthia, et al., 2020)

a. Series Connected Controllers
b. Shunt Connected Controllers
c. Hybrid Connected Controllers

The figure no. 28 shown below illustrates the types of FACTS devices used for these control action.

Figure 28. LVRT techniques

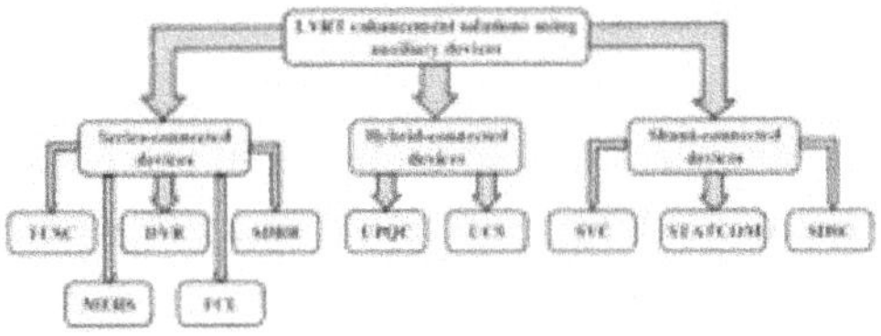

Figure 29. Wind energy FRT and LVRT control

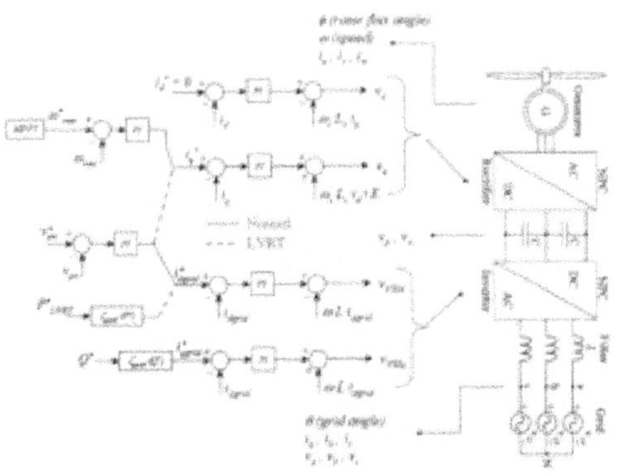

Figure 30. Wind energy FRT and LVRT control

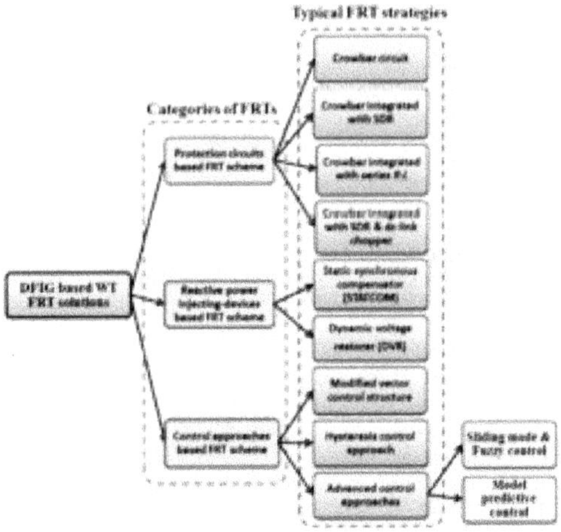

The figure 29 and figure 30 shown above illustrates the type of FRT and function of LVRT under normal and abnormal condition. (Bibhu Prasad Ganthia, et al., 2020) (Bibhu Prasad Ganthia, et al., 2017) These limitations are known as criteria for FRT or LVRT. In the event of network disturbances, the large increase in installed wind capacity in transmission systems requires that wind generation stays in operation. For this reason, the grid codes issued in recent years invariably require wind farms (in particular those connected to HV grids) to withstand voltage drops to a certain percentage of the nominal voltage (in some cases down to 0 percent) and for a specified duration. (Bibhu Prasad Ganthia, et al., 2016) Such requirements are referred to as Fault Ride through (FRT) or Low Voltage Ride through (LVRT) and are described by a characteristic of voltage vs. time that denotes the wind station's minimum required immunity.Fast active and reactive power recovery to before the fault values and after the system voltage returns to normal operating levels are also included in the LVRT requirements. (Bibhu Prasad Ganthia, et al., 2018) In order to provide a voltage support requirement that resembles the behavior of conventional synchronous generators in over-excited operation, some codes impose increased reactive power generation by the wind turbines during the disturbance. The requirements are dependent on the specific characteristics of each power system and the protection used and deviates considerably from each other. (Bibhu Prasad Ganthia, et al., 2021) (A. Pragati et al., 2021)(Bibhu Prasad Ganthia, et al., 2021)

DISCUSSION

The detail analyses on all the control techniques are summarized in below tables.

Table 1. Types of wind turbines

Types	Advantages	Disadvantages
HAWT(*Horizontal Axis Wind Turbine*)	• Considerable growth capacity of wind energy • Link to heavy winds caused by elevated towers	• Installation cost is high initially • Stud tower design required • Wind turbine orientation is needed
VAWT(*Vertical Axis Wind Turbine*)	• Because of the ground level gear box and generator, low unit cost and simpler servicing • No effect on random wind speed	• Efficiency is lower • Torque fluctuates more • Mechanical vibration developed under high speed
FSWT(*Fixed SpeedWind Turbine*)	• Smooth operation • Simpler Design • Robust Controls • Reliable energy source • Low maintenance cost and smooth controls	• Low efficiency • Heavy mechanical Stress • Power fluctuates more under transients • Grid integration is little complex
VSWT(*Variable Speed Wind Turbine*)	• Heavy duty and operation • Effective power generation and control • Negligible stress to turbine system	• More converters are required • Complex control techniques are required • Little costly

Table 2. Summary on pitch angle control strategies

Controller	Rapid wind speed Performance	Efficiency and Effectiveness	Quality and Reliability	Operation and Control	Speed Control	Cost
Soft computing controller	High	Moderate- high	High	Low	Faster	Average
Hybrid controller	Very High	High	High	Moderate	Faster	Average
Robust controller	High	Moderate- high	Moderate- high	High	Moderate	High
Conventional Controller	Low	Low	High	Low	Slow	Low

Table 3. Summary on converter control strategies

Parameter	Back-To-Back Converters	Resonant Converters	Multilevel Converters	Matrix Converters	Tandem Converters
Switching	Bidirectional	Bidirectional	Bidirectional Multiple	Bidirectional Multiple	Bidirectional Multiple
Voltage Mode	Yes	Yes	Yes	Yes	Yes
Current Mode	Yes	Yes	Yes	Yes	Yes
Hysteresis Mode	V-I controlled	V- controlled	V-controlled	V- controlled	V- controlled
Modulation	Required	Required	Required	Required	Required
Power Factor	Normal	Depends	Moderate	Depends	Depends
Response	Fast	Medium	Fast	Faster	Medium
Thermal Stress	Less	Medium	High	High	Medium
Cost	Less	Less	Average	High	Average
THD	Less	Less	Very Less	Very Less	Less
Performance	Fast	Moderate	Fast	Faster	Moderate

Table 4. Summary on MPPT control strategies

Techniques/parameter	PSF	HCS	TSR	OTC	SCT	Hybrid
Efficiency	Moderate	Low	Very High	Moderate- High	High	Very High
Complexity	Simple Required	Simple Not Required	Simple Required	Simple Not Required	Not Required	Not Required
Sensitivity	Yes	No	No	Yes	No	Depends
Convergence speed	Fast	Low	Fast	Fast	Medium	Fast
Memory requirement	Required	Not Required	Not Required	Depends	Not Required	Depends
Prior knowledge	Required	Not Required	Required	Required	Not Required	Depends
Tolerance to rapid variation	Moderate	Low	Moderate- high	Moderate- high	High	Very high
Quality	Normal	More	Moderate-High	Moderate-High	High	High

Table 5. Summary on grid side control strategies

Parameters	DTC	DPC	SVOC	FOC	FRT & LVRT
Implementation	Simple	Simple	Complex	Complex	Complex
Powerquality	Moderate	Moderate	Better	Better	Better
Ripple Values (Power &Current)	More	More	Less	Less	Less
Stability	Moderate	Moderate	Better	Better	Better
Torqueripple	Less	More	Moderate	Moderate	Moderate
DynamicResponsetime	Low	Low	High	High	High
Coordinate transformation	Not Essential	Not Essential	Yes Essential	Yes Essential	Yes Essential
Internal current regulation	Not Essential	Not Essential	Yes Essential	Yes Essential	Yes Essential
ParameterSensitivity	Less	Less	More	More	More
DC linkVoltageripple	No	High	Low	Low	Low
Rotor Position Sensor	Not Required	Not Required	Not Required	Required	Not Required

Table 6. Summary on series connected facts devices control strategies

Methods		Conclusion	Limitation	Summary
Series Connected FACT Devices	**TCSC**	•Capacitive reactance variations •Useful for fault current limitation and voltage unbalance	•High resonance •Heavy harmonic injection	•Offshore wind farm Applicable •Per unit application is medium
	DVR	•Quick voltage recovery •Controlled reactive power	•Faster Phase angle •Faster absorption of Active power	•Compatible and high power usability •More stable
	SDBR	•Quick voltage recovery •Controlled reactive power	•No control on reactive power • High voltage damping • Poor low power factor usage	•Per unit application is better •Less power system stability

Table 7. Summary on shunt connected facts devices control strategies

Methods		Conclusion	Limitation	Summary
Shunt Connected FACT Devices	**SVC**	•Can control reactive power •Stable system under fault • Control over voltage sag and swell	•Depends on the control actions of voltage •Unstable under heavy faults	•Need more reactive power for normal operation during faults
	STATCOM	•Can control reactive power •Responsive towards faults •Reduction in –ve sequence voltage	•Protective system is highly needed •Only operation on reactive powers	•Faster response towards fault clearance and transient action
	SDSC	•High performance under high per units •Less switching losses	•Low voltage operation causes more instability	•Calibrations required •High reactive power is needed for swell and sag operations

Table 8. Summary on hybrid connected facts devices control strategies

<table>
<tr><th colspan="2">Methods</th><th>Conclusion</th><th>Limitation</th><th>Summary</th></tr>
<tr><td rowspan="3">Hybrid Connected FACT Devices</td><td>UPQC</td><td>•Control over active and reactive power
•Compensation is easy in reactive power
•Critical clearance time is more</td><td>•Absorb large active powers
•High capacity DC link capacitor is needed</td><td>•Reliable in control action
•More stability in reactive power compensation
•Smooth operations
•Less switching losses</td></tr>
<tr><td>SFCL</td><td>•Limited fault current
•LVRT curve safety margin is good</td><td>•Unreliable during hybrid connection of UPQC and SFCL combination</td><td>•Less cost but not reliable under stress
•Unstable under heavy rush</td></tr>
<tr><td>UCS</td><td>•Single converter can control both powers during transients</td><td>•Switching losses are high during conduction period</td><td>•Effective for low power mode
•Slow system under high fault conditions</td></tr>
</table>

CONCLUSION

This chapter highlight the different control techniques used in wind turbine systems to get the most out of the wind. Wind energy is mainly efficient for both onshore and offshore power generation as a clean energy source and a non-hazardous source of power generation. The control techniques for better power quality generations are highlighted in this chapter sequentially. This chapter compares all the control techniques, from which the accuracy of implementations can be extracted. The resilient features, optimal control behaviors, predictive designs, re-configured control structures, robust system maintenances and advanced algorithms compared accordingly for wind energy system modeling, operation and control. The grid integrated designs and the power quality balance can achieve from this chapter. This chapter holds a good quality analysis for researchers to enhance their research feasibilities, standardizations and grid regulations.

REFERENCES

Ali, Park, Yu, Murata, & Tamura. (2009). Improvement of wind-generator stability by fuzzy-logic-controlled SMES. *Industry Applications, IEEE Transactions on, 45*, 1045–1051.

Amendola, C., & Gonzaga, D. (2007). Fuzzy-logic control system of a variable-speed variable-pitch wind-turbine and a double-fed induction generator. *Intelligent Systems Design and Applications, 2007. ISDA 2007. Seventh International Conference on*, 252–257.

Bayat, M., & Karegar, H. (2009). Predictive control of wind energy conversion system. *Developments in Renewable Energy Technology (ICDRET), 2009 1st International Conference on the*, 1–5.

Beltran, Ahmed-Ali, & Benbouzid. (2008). Sliding mode power control of variable-speed wind energy conversion systems. *Energy Conversion, IEEE Transactions on, 23*, 551–558.

Besheer, Emara, & Abdel Aziz. (2006). Fuzzy based output feedback control for wind energy conversion system: An LMI approach. *Power Systems Conference and Exposition, 2006. PSCE '06. 2006 IEEE PES*, 2030–2037.

Bianchi, F. D., Battista, H. D., & Mantz, R. J. (2007). *Wind Turbine Control System: Principles, Modelling, and Gain Scheduling Design.* SpringerVerlag.

Boonpirom, N., & Paithoonwattanakij, K. (2006). Wind farm generator control using self-tuning fuzzy controller. *Electrical and Computer Engineering, 2006. ICECE '06. International Conference on*, 221–224.

Bourlis, D., & Bleijs, J. (2009). Gain scheduled controller with wind speed estimation via Kalman filtering for a stall regulated variable speed wind turbine. *Universities Power Engineering Conference (UPEC), 2009 Proceedings of the 44th International*, 1–5.

Brahmi, Krichen, & Ouali. (2009). Sensorless control of PMSG in WECS using artificial neural network. *Systems, Signals and Devices, 2009. SSD '09. 6th International Multi-Conference on*, 1–8.

Bratcu, A., Munteanu, I., & Ceanga, E. (2008). Optimal control of wind energy conversion systems: From energy optimization to multi-purpose criteria - a short survey. *Control and Automation, 2008 16th Mediterranean Conference on*, 759 –766.

Bratcu, A., Munteanu, I., Ceanga, E., & Epure, S. (2007). Energetic optimization of variable speed wind energy conversion systems by extremum seeking control. *EUROCON, 2007. The International Conference on 34;Computer as a Tool, 34*, 2536 –2541.

Buehring, I., & Freris, L. (1981). Control policies for wind-energy conversion systems. *Generation, Transmission and Distribution, IEE Proceedings Conference*, 128, 253 –261.

Calderaro, V., Galdi, V., Piccolo, A., & Siano, P. (2007). Design and implementation of a fuzzy controller for wind generators performance optimization. *Power Electronics and Applications, 2007 European Conference on*, 1–10.

Chen, J.-J., & Ji, Z.-C. (2010). The gain scheduling control for wind energy conversion system based on LPV model. *Networking, Sensing and Control (ICNSC), 2010 International Conference on*, 653–657.

Chen & Yang. (2009). Fuzzy PID controller used in yaw system of wind turbine. *Power Electronics Systems and Applications, 2009. PESA 2009. 3rd International Conference on*, 1–4.

Cheng, X., Lei, Z., & Junqiu, Y. (2009). Fuzzy PID controller for wind turbines. *Intelligent Networks and Intelligent Systems, 2009. ICINIS '09. Second International Conference on*, 74–77.

Dang, D., Wang, Y., & Cai, W. (2008). Nonlinear model predictive control of fixed pitch variable speed wind turbine. *Sustainable Energy Technologies, 2008. ICSET 2008. IEEE International Conference on*, 29–33.

Dang, D., Wang, Y., & Cai, W. (2009). A multi-objective optimal nonlinear control of variable speed wind turbine. *Control and Automation, 2009. ICCA 2009. IEEE International Conference on*, 17–22.

De Battista, Puleston, Mantz, & Christiansen. (2000). Sliding mode control of wind energy systems with DOIG-power efficiency and torsional dynamics optimization. *Power Systems, IEEE Transactions on, 15*, 728–734.

De Battista, H., & Mantz, R. (1998). Sliding mode control of torque ripple in wind energy conversion systems with slip power recovery. *Industrial Electronics Society, 1998. IECON '98. Proceedings of the 24th Annual Conference of the IEEE*, 2, 651 –656.

Dinghui, W., Lili, X., & Zhicheng, J. (2009). Fuzzy adaptive control for wind energy conversion system based on model reference. *Control and Decision Conference, 2009. CCDC '09. Chinese*, 1783–1787.

Ganthia. (2020). Application of Hybrid Facts Devices in DFIG Based Wind Energy System for LVRT Capability Enhancements. *Journal of Mechanics of Continua and Mathematical Sciences. J. Mech. Cont. & Math. Sci., 15*(6), 245-256. doi:10.26782/jmcms.2020.06.00019

Ganthia. (2020). Application of Hybrid Facts Devices in DFIG Based Wind Energy System for LVRT Capability Enhancements. *Journal of Mechanics of Continua and Mathematical Sciences. J. Mech. Cont. & Math. Sci., 15*(6), 139-157. doi:10.26782/jmcms.2020.06.00012

Ganthia, B. P., Agarwal, V., Rout, K., & Pardhe, M. K. (2017). Optimal control study in DFIG based wind energy conversion system using PI & GA. *International Conference on Power and Embedded Drive Control (ICPEDC)*, 343-347.

Ganthia, B. P., & Barik, S. K. (2020). Steady-State and Dynamic Comparative Analysis of PI and Fuzzy Logic Controller in Stator Voltage Oriented Controlled DFIG Fed Wind Energy Conversion System. *J. Inst. Eng. India Ser. B, 101*, 273–286. https://doi.org/10.1007/s40031-020-00455-8

Ganthia, B. P., Barik, S. K., & Nayak, B. (2021). Wind Turbines in Energy Conversion System: Types & Techniques. In V. K. Singh, A. K. Bhoi, A. Saxena, A. F. Zobaa, & S. Biswal (Eds.), Renewable Energy and Future Power Systems. Energy Systems in Electrical Engineering. Springer. https://doi.org/10.1007/978-981-33-6753-1_9

Ganthia, B. P., Barik, S. K., & Nayak, B. (n.d.). Shunt Connected FACTS Devices for LVRT Capability Enhancement in WECS. *Engineering, Technology & Applied Scientific Research, 10*(3), 5819–5823.

Ganthia, B. P., Mohanty, S., Rana, P. K., & Sahu, P. K. (2016). Compensation of voltage sag using DVR with PI controller. *2016 International Conference on Electrical, Electronics, and Optimization Techniques (ICEEOT)*, 2138-2142. doi: 10.1109/ICEEOT.2016.7755068

Ganthia, B. P., Pradhan, R., Sahu, R., & Pati, A. K. (2021). Artificial Ant Colony Optimized Direct Torque Control of Mathematically Modeled Induction Motor Drive Using PI and Sliding Mode Controller. In J. Kumar & P. Jena (Eds.), Recent Advances in Power Electronics and Drives. Lecture Notes in Electrical Engineering (Vol. 707). Springer. https://doi.org/10.1007/978-981-15-8586-9_35.

Ganthia, B. P., Rana, P. K., Patra, T., Pradhan, R., & Sahu, R. (2018). Design and Analysis of Gravitational Search Algorithm Based TCSC Controller in Power System. *Materials Today: Proceedings, 5*(1), 841–847.

Gargoom, A., Haruni, A., Haque, M., & Negnevitsky, M. (2010). Voltage and frequency stabilization using PI-like fuzzy controller for the load side converters of the stand alone wind energy systems. Applied Power Electronics Conference and Exposition (APEC), 2010 Twenty-Fifth Annual IEEE, 2132–2137.

Hong, C.-M., Lin, W.-M., & Cheng, F.-S. (2007). Application of fuzzy neural network sliding mode controller for wind driven induction generator system. *Intelligent Systems Applications to Power Systems, 2007. ISAP 2007. International Conference on*, 1–6.

Hui, J., & Bakhshai, A. (2008). A new adaptive control algorithm for maximum power point tracking for wind energy conversion systems. *Power Electronics Specialists Conference, 2008. PESC 2008*, 4003 –4007.

Hui, J., & Bakhshai, A. (2008). Adaptive algorithm for fast maximum power point tracking in wind energy systems. *Industrial Electronics, 2008. IECON 2008. 34th Annual Conference of IEEE*, 2119–2124.

Hui, J., Bakhshai, A., & Jain, P. K. (2010). A master-slave fuzzy logic control scheme for maximum power point tracking in wind energy systems. *Telecommunications Energy Conference (INTELEC), 32nd International*, 1–6.

Huo, Z., Zheng, Y., Xu, C., & Huang, D. (2009). Robust stability research for wind energy conservation systems based on T-S fuzzy models. *Sustainable Power Generation and Supply, 2009. SUPERGEN '09. International Conference on*, 1–4.

Jian-jun, X., Li-mei, Y., Xiao-ning, Q., Chun-lei, J., & Jian-ren, W. (2010). Study of variable-pitch wind turbine based on fuzzy control. *Future Computer and Communication (ICFCC), 2010 2nd International Conference on,* 1.

Jianlin, L., & Honghua, X. (2008). Research on control system of high power DFIG wind power system. *MultiMedia and Information Technology, 2008. MMIT '08. International Conference on*, 669–672.

Kairous, D., Wamkeue, R., & Belmadani, B. (2010). Advanced control of variable speed wind energy conversion system with DFIG. *Environment and Electrical Engineering (EEEIC), 2010 9th International Conference on*, 41–44.

Kalantar, M., & Sedighizadeh, M. (2005). Adaptive-neural PID control of wind energy conversion systems using wavenets. *Intelligent Control, 2005. Proceedings of the 2005 IEEE International Symposium on, Mediterrean Conference on Control and Automation*, 219–224.

Khalil, H. K. (1996). *Nonlinear Systems* (2nd ed.). Prentice-Hall. Inc.

Laks, J., Pao, L., & Wright, A.(2009). Control of wind turbines: Past, present, and future. *American Control Conference, 2009. ACC '09*, 2096–2103.

Le, L., Ying, L., Tian, P., & Li, P. (2008). Maximum wind capture based on fuzzy PD control strategy. *Electrical Machines and Systems, 2008. ICEMS 2008. International Conference on*, 2542–2546.

Li & Chen. (2008). Overview of different wind generator systems and their comparisons. *Renewable Power Generation, IET, 2*, 123–138.

Li, H., Shi, K., & McLaren, P. (2005, November). Neural-network-based sensorless maximum wind energy capture with compensated power coefficient. *Industry Applications, IEEE Transactions on, 41*, 1548–1556.

Lima, M., Silvino, J., & de Resende, P. (1999). H∞ control for a variablespeed adjustable-pitch wind energy conversion system. *Industrial Electronics, 1999. ISIE '99. Proceedings of the IEEE International Symposium on*, 2, 556–561.

Lin, W.-M., Hong, C.-M., Lee, M.-R., Huang, C.-H., Huang, C.-C., & Wu, B.-L. (2010). Fuzzy sliding mode-based control for PMSG maximum wind energy capture with compensated pitch angle. *Computer Communication Control and Automation (3CA), 2010 International Symposium on, 2*, 397–400.

Mayosky, M., & Cancelo, I. (1999, July). Direct adaptive control of wind energy conversion systems using Gaussian networks. *Neural Networks, IEEE Transactions on, 10*, 898–906.

Mishra, S., Mishra, Y., Li, F., & Dong, Z. (2009). TS-fuzzy controlled DFIG based wind energy conversion systems. *Power Energy Society General Meeting, 2009. PES '09. IEEE*, 1–7.

Moriarty, P., & Butterfield, S. (2009). Wind turbine modeling overview for control engineers. *American Control Conference*, 2090–2095.

Muhando, E., Senjyu, T., Uchida, K., Kinjo, H., & Funabashi, T. (2010, January). Stochastic inequality constrained closed-loop model-based predictive control of MW-class wind generating system in the electric power supply. *Renewable Power Generation, IET, 4*, 23–35.

Munteanu, I., Bacha, S., Bratcu, A., Guiraud, J., & Roye, D. (2008, September). Energy reliability optimization of wind energy conversion systems by sliding mode control. *Energy Conversion, IEEE Transactions on, 23*, 975–985.

Natarajan, Sivakumar, & Sharaf. (1991). Multirate PI control system performance for a wind energy conversion scheme. *American Control Conference*, 2774–2775.

Pena, R., & Sbarbaro, D. (1999). Integral variable structure controllers for small wind energy systems. *Industrial Electronics Society, 1999. IECON '99 Proceedings. The 25th Annual Conference of the IEEE, 3*, 1067–1072.

Pinto, S. F., Aparicio, L., & Esteves, P. (2007). Direct controlled matrix converters in variable speed wind energy generation systems. *Power Engineering, Energy and Electrical Drives, 2007. POWERENG 2007. International Conference on*, 654–659.

Pragati, A., Ganthia, B. P., & Panigrahi, B. P. (2021). Genetic Algorithm Optimized Direct Torque Control of Mathematically Modeled Induction Motor Drive Using PI and Sliding Mode Controller. In J. Kumar & P. Jena (Eds.), Recent Advances in Power Electronics and Drives. Lecture Notes in Electrical Engineering (Vol. 707). Springer. https://doi.org/10.1007/978-981-15-8586-9_32.

Qi, J., & Liu, Y. (2010). PID control in adjustable-pitch wind turbine system based on fuzzy control. *Industrial Mechatronics and Automation (ICIMA), 2010 2nd International Conference on*, 2, 341–344.

Raza, K., Goto, H., Guo, H.-J., & Ichinokura, O. (2008). A novel speedsensorless adaptive hill climbing algorithm for fast and efficient maximum power point tracking of wind energy conversion systems. *Sustainable Energy Technologies, 2008. ICSET 2008. IEEE International Conference on*, 628–633.

Ren, Y., & Bao, G. (2010). Control strategy of maximum wind energy capture of direct-drive wind turbine generator based on neural-network. *Power and Energy Engineering Conference (APPEEC)*, 1–4.

Richardson, R., & McNerney, G. (1993, March). Wind energy systems. *Proceedings of the IEEE, 81*(3), 378–389. doi:10.1109/5.241490

Rocha, R., & Filho, L. (2003). A multivariable H∞ control for wind energy conversion system. *Control Applications, 2003. CCA 2003. Proceedings of 2003 IEEE Conference on*, 1, 206 – 211.

Rocha, R., Filho, L., & Bortolus, M. (2005). Optimal multivariable control for wind energy conversion system - a comparison between H 2 and H∞ controllers. *Decision and Control, 2005 and 2005 European Control Conference. CDC-ECC '05. 44th IEEE Conference on*, 7906 – 7911.

Rocha, R., Resende, P., Silvino, J., & Bortolus, M. (2001). Control of stall regulated wind turbine through H∞: loop shaping method. *Control Applications, 2001. (CCA '01). Proceedings of the 2001 IEEE International Conference on*, 925–929.

Sanchez, S., Bueno, M., Delgado, E., & Giraldo, E. (2009). Optimal PI control of a wind energy conversion system using particles swarm. *Electronics, Robotics and Automotive Mechanics Conference, 2009. CERMA '09*, 332–337.

Sedighizadeh, Arzaghi-Harris, & Kalantar. (2004). Adaptive PID control of wind energy conversion systems using RASP1 mother wavelet basis function networks. *TENCON 2004. 2004 IEEE Region 10 Conference, C*, 524 – 527.

Sedighizadeh, M., Rezazadeh, A., & Khatibi, M. (2008). A self-tuning PID control for a wind energy conversion system based on the Lyapunov approach. *Universities Power Engineering Conference, 2008. UPEC 2008. 43rd International*, 1–4.

Senjyu, T., Sakamoto, R., Urasaki, N., Funabashi, T., & Sekine, H. (2006). Output power leveling of wind farm using pitch angle control with fuzzy neural network. *Power Engineering Society General Meeting.*

Sun, Lin, Li, & Li. (2009). A review on the integration of wind farms with variable speed wind turbine systems into power systems. *Sustainable Power Generation and Supply, 2009. SUPERGEN '09. International Conference on*, 1–6.

Tan, K., & Islam, S. (2004, June). Optimum control strategies in energy conversion of PMSG wind turbine system without mechanical sensors. *Energy Conversion, IEEE Transactions on, 19*, 392–399.

Thongam, J., Bouchard, P., Ezzaidi, H., & Ouhrouche, M. (2009). Artificial neural network-based maximum power point tracking control for variable speed wind energy conversion systems. Control Applications, (CCA) Intelligent Control, (ISIC), 2009 IEEE, 1667–1671.

Tien, H., Scherer, C., & Scherpen, J. (2007). Robust performance of self scheduled LPV control of doubly-fed induction generator in wind energy conversion systems. *Power Electronics and Applications, 2007 European Conference on*, 1–10.

Tsoumas, I., Safacas, A., Tsimplostefanakis, E., & Tatakis, E. (2003). An optimal control strategy of a variable speed wind energy conversion system. *Electrical Machines and Systems, 2003. ICEMS 2003. Sixth International Conference on, 1*, 274 – 277.

Valenciaga, F., & Puleston, P. (2007, June). Variable structure control of a wind energy conversion system based on a brushless doubly fed reluctance generator. *Energy Conversion, IEEE Transactions on*, *22*, 499–506.

Valenciaga, F., & Puleston, P. (2008, September). High-order sliding control for a wind energy conversion system based on a permanent magnet synchronous generator. *Energy Conversion, IEEE Transactions on*, *23*, 860–867.

Valenciaga, F., Puleston, P., & Battaiotto, P. (2004). Variable structure system control design method based on a differential geometric approach: application to a wind energy conversion subsystem. *Control Theory and Applications, IEE Proceedings, 151*, 6 – 12.

Wagner, H. J., & Mathur, J. (2009). *Introduction to Wind Energy Systems:Basics,Technology and Operation*. Springer-Verlag.

Wang, B., Wang, P., Chen, X., & Yuan, X. (2009). Robust control of variable speed adjustable-pitch wind energy conversion system based on Hamiltonian energy theory. *Sustainable Power Generation and Supply, 2009. SUPERGEN '09. International Conference on*, 1 –6.

Wang, W., Wu, D., Wang, Y., & Ji, Z. (2010). H∞ gain scheduling control of PMSG-based wind power conversion system. *Industrial Electronics and Applications (ICIEA), 2010 the 5th IEEE Conference on*, 712–717.

Wang, Y., Fu, Y., & Li, D. (2009). Synthesized power and frequency control strategies based on fuzzy neural networks for wind power generation systems. *Energy and Environment Technology, 2009. ICEET '09. International Conference on, 1*, 869–872.

Wang, Liu, & Liu. (2010). Intelligent controlling of wind turbine based on fuzzy immunity PID algorithmic. *Biomedical Engineering and Computer Science (ICBECS), 2010 International Conference on*, 1–4.

Xin-Fang, Z., Da-Ping, X., & Yi-Bing, L. (2004). Adaptive optimal fuzzy control for variable speed fixed pitch wind turbines. *Intelligent Control and Automation, 2004. WCICA 2004. Fifth World Congress on, 3*, 2481 – 2485.

Xing, Z., Li, Q., Su, X., & Guo, H. (2009). Application of BP neural network for wind turbines. *Intelligent Computation Technology and Automation, 2009. ICICTA '09. Second International Conference on, 1*, 42–44.

Yan, J., Chen, J., Ma, H., Liu, H., & Chen, L. (2009). The survey of electrical control systems of wind turbine generators. *Sustainable Power Generation and Supply, 2009. SUPERGEN 09. International Conference on*, 1 –5.

Yang, X., & Liu, X. (2010). Integral variable structure fuzzy adaptive control for variable speed wind power system. *Logistics Systems and Intelligent Management, 2010 International Conference on, 2*, 1247–1250.

Yang, X., Liu, X., & Wu, Q. (2007). Integral fuzzy sliding mode control for variable speed wind power system. *Automation and Logistics, 2007 IEEE International Conference on*, 1289–1294.

Yao, X., Liu, Y., & Guo, C. (2007). Adaptive fuzzy sliding-mode control in variable speed adjustable pitch wind turbine. *Automation and Logistics, 2007 IEEE International Conference on*, 313–318.

Yao, X., Su, X., & Tian, L. (2009). Wind turbine control strategy at lower wind velocity based on neural network PID control. *Intelligent Systems and Applications, 2009. ISA 2009. International Workshop on*, 1–5.

Yao, X., Su, X., & Tian, L. (2009). Pitch angle control of variable pitch wind turbines based on neural network PID. *Industrial Electronics and Applications, 2009. ICIEA 2009. 4th IEEE Conference on*, 3235 – 3239.

Yao, X., Wen, H., Deng, Y., & Zhang, Z. (2007). Research on rotor excitation neural network PID control of variable speed constant frequency wind turbine. *Electrical Machines and Systems, 2007. ICEMS. International Conference on*, 560–565.

Yunqi, X. (2010). A novel optimum power fuzzy control strategy for doublyfed wind turbine. *Control and Automation (ICCA), 2010 8th IEEE International Conference on*, 165–170.

Zeng, Q., Chang, L., & Shao, R. (2008). Fuzzy-logic-based maximum power point tracking strategy for PMSG variable-speed wind turbine generation systems. *Electrical and Computer Engineering, 2008. CCECE 2008. Canadian Conference on*, 405–410.

Zhang, L., & Li, H. (2008). Pitch control of large scale wind turbine based on fuzzy-PD method. *Electric Utility Deregulation and Restructuring and Power Technologies, 2008. DRPT 2008. Third International Conference on*, 2447 –2452.

Zhang, X., Wang, W., & Liu, Y. (2006). Fuzzy control of variable speed wind turbine. *Intelligent Control and Automation, 2006. WCICA 2006. The Sixth World Congress on, 1*, 3872–3876.

Zhang, X., Zhang, X., Zhou, P., Cheng, J., Wang, W., Jiale, S., & Liu, Y. (2009). Fuzzy control used in variable speed wind turbine. *Automation and Logistics, 2009. ICAL '09. IEEE International Conference on*, 1194 –1198.

Zheng, X., Wei, W., & Xu, D. (2010). Higher-order sliding mode control of DFIG wind energy system under LVRT. *Power and Energy Engineering Conference (APPEEC)*, 1–4.

Chapter 7
Enhanced Symbiotic Organisms Search and Hydrological Cycle Algorithms for Real Power Loss Diminution and Voltage Stability Enhancement

Lenin Kanagasabai
Prasad V. Potluri Siddhartha Institute of Technology, India

ABSTRACT

In this chapter, enhanced symbiotic organisms search (ESOS) algorithm and hydrological cycle (HC) algorithm are projected to solve factual power loss lessening problem. Symbiotic search algorithm is based on the actions between two different organisms in the ecosystem: mutualism, commensalism, and parasitism. Exploration procedure has been initiated arbitrarily, and each organism indicates a solution with fitness value. Quasi-oppositional based learning and chaotic local search have been applied to augment the performance of the algorithm. In this work, hydrological cycle (HC) algorithm has been utilized to solve the optimal reactive power problem. It imitates the circulation of water form land to sky and vice versa. Only definite number of water droplets is chosen for evaporation, and it is done through roulette-wheel selection method. In the condensation stage, water drops move closer, combine, and also collusion occurs as the temperature decreases.

INTRODUCTION

Factual power loss decreasing is foremost objective of the paper. Several orthodox methods and Evolutionary algorithms (Azar & Vaidyanathan, 2015a,b,c, 2016; Azar & Zhu, 2015; Azar & Serrano, 2015a,b,c,d, 2016a,b, 2017; Boulkroune et al, 2016a,b; Ghoudelbourk et al., 2016; Meghni et al, 2017a,b,c; Azar et al., 2017a,b,c,d; Azar 2010a,b, 2012; Mekki et al., 2015; Vaidyanathan & Azar, 2015a,b,c,d, 2016a,b,c,d,e,f,g, 2017a,b,c; Zhu & Azar, 2015; Grassi et al., 2017; Ouannas et al., 2016a,b, 2017a,b,c,d,e,f,g,h,i,j; Singh et

DOI: 10.4018/978-1-7998-7447-8.ch007

al., 2017; Vaidyanathan et al, 2015a,b,c; Wang et al., 2017; Soliman et al., 2017; Tolba et al., 2017).are employed to solve the problem. Major problem is handling the constraints and balancing the exploration, exploitation equally. In this chapter First Enhanced Symbiotic Organisms Search (ESOS) Algorithm is applied to solve the problem. Symbiotic organisms search algorithm is based on the actions between two dissimilar organisms in the ecosystem- mutualism, commensalism and parasitism. In the projected Enhanced Symbiotic Organisms Search (ESOS) Algorithm, Quasi-oppositional based learning and Chaotic local search has been applied to augment the performance of the algorithm. Quasi-oppositional based learning and quasi-opposite numbers is very much successful in discovery of global optimal solution. Chaotic local search has been applied to explore close by present solutions to engender a new-fangled candidate solution. This work presents Hydrological Cycle (HC) algorithm for solving optimal reactive power dispatch problem. It imitates the circulation of water form land to sky and vice versa. Through the process of evaporation, precipitation, condensation water attains different conditions. Projected HC algorithm explores the space, water will move from one point to another point by choosing various nodes. In this HC algorithm quality of the solution depend on quantity of the soil carried by each water drop and a good solution is obtained from the water drop which carries more soil, also the quantity of the soil carried by water drop is utilized to modernize the velocity of the water drops subsequently. Only definite number of water droplets is chosen for evaporation and it done through Roulette-wheel selection method. In the condensation stage water drops are move closer, combine and also collusion will occur as the temperature decreases. ESOS and HC algorithms are corroborated in IEEE 30 bus system. Then ESOS and HC algorithms have been corroborated in IEEE 14, 30, 57 and300 bus test systems deprived of voltage stability index. Projected ESOS and HC algorithms abridged the power loss meritoriously and variables are inside the restrictions.

Problem Formulation

Objective function of the problem is mathematically defined in general mode by,

$$\text{Minimization}\ \tilde{F}(\overline{x},\overline{y}) \tag{1}$$

Subject to

$$E(\overline{x},\overline{y}) = 0 \tag{2}$$

$$I(\overline{x},\overline{y}) = 0 \tag{3}$$

$$x = \left[VG_1,..,VG_{Ng};QC_1,..,QC_{Nc};T_1,..,T_{N_T}\right] \tag{4}$$

$$y = \left[PG_{slack}; VL_1, .., VL_{N_{Load}}; QG_1, .., QG_{Ng}; SL_1, .., SL_{N_T} \right] \tag{5}$$

Then the single objective problem formulation is defined as follows.

The fitness function $\left(OF_1\right)$ is defined to reduce the power loss (MW) in the system is written as,

$$OF_1 = P_{Min} = Min\left[\sum_{m}^{NTL} G_m \left[V_i^2 + V_j^2 - 2*V_i V_j \cos Ø_{ij} \right] \right] \tag{6}$$

Minimization of Voltage deviation fitness function $\left(OF_2\right)$ is given by,

$$OF_2 = Min\left[\sum_{i=1}^{N_{LB}} \left| V_{Lk} - V_{Lk}^{desired} \right|^2 + \sum_{i=1}^{Ng} \left| Q_{GK} - Q_{KG}^{Lim} \right|^2 \right] \tag{7}$$

Then the voltage stability index (L-index) fitness function $\left(OF_3\right)$ is given by,

$$OF_3 = Min\, L_{Max} \tag{8}$$

$$L_{Max} = Max\left[L_j \right]; j = 1; N_{LB} \tag{9}$$

$$\text{And} \begin{cases} L_j = 1 - \sum_{i-1}^{NPV} F_{ji} \dfrac{V_i}{V_j} \\ F_{ji} = -\left[Y_1\right]^1 \left[Y_2\right] \end{cases} \tag{10}$$

Such that

$$L_{Max} = Max\left[1 - \left[Y_1\right]^{-1} \left[Y_2\right] \times \frac{V_i}{V_j} \right] \tag{11}$$

Then the equality constraints are

$$0 = PG_i - PD_i - V_i \sum_{j \in N_B} V_j \left[G_{ij} \cos\left[Ø_i - Ø_j \right] + B_{ij} \sin\left[Ø_i - Ø_j \right] \right] \tag{12}$$

$$0 = QG_i - QD_i - V_i \sum_{j \in N_B} V_j \left[G_{ij} \sin\left[Ø_i - Ø_j \right] + B_{ij} \cos\left[Ø_i - Ø_j \right] \right] \tag{13}$$

Inequality constraints

$$P_{gslack}^{min} \leq P_{gslack} \leq P_{gslack}^{max} \tag{14}$$

$$Q_{gi}^{min} \leq Q_{gi} \leq Q_{gi}^{max}, i \in N_g \tag{15}$$

$$VL_i^{min} \leq VL_i \leq VL_i^{max}, i \in NL \tag{16}$$

$$T_i^{min} \leq T_i \leq T_i^{max}, i \in N_T \tag{17}$$

$$Q_c^{min} \leq Q_c \leq Q_C^{max}, i \in N_C \tag{18}$$

$$\left| SL_i \right| \leq S_{L_i}^{max}, i \in N_{TL} \tag{19}$$

$$VG_i^{min} \leq VG_i \leq VG_i^{max}, i \in N_g \tag{20}$$

Multi objective fitness (MOF) function has been defined by,

$$MOF = OF_1 + x_i OF_2 + y OF_3 = OF_1 + \left[\sum_{i=1}^{NL} x_v \left[VL_i - VL_i^{min} \right]^2 + \sum_{i=1}^{NG} x_g \left[QG_i - QG_i^{min} \right]^2 \right] + x_f OF_3 \tag{21}$$

$$VL_i^{min} = \begin{cases} VL_i^{max}, VL_i > VL_i^{max} \\ VL_i^{min}, VL_i < VL_i^{min} \end{cases} \tag{22}$$

$$QG_i^{min} = \begin{cases} QG_i^{max}, QG_i > QG_i^{max} \\ QG_i^{min}, QG_i < QG_i^{min} \end{cases} \tag{23}$$

Enhanced Symbiotic Organisms Search Algorithm

Symbiotic organisms search algorithm is grounded on the activities between two dissimilar organisms in the ecosystem- mutualism, commensalism and parasitism. Exploration technique has been instigated arbitrarily and each organism designates a solution with fitness value. Three junctures mutualism, commensalism and parasitism will engender new-fangled solutions and it has been rationalised when it is loftier to preceding solution.

In the mutualism juncture or phase two disparate organisms will gain each other. In the algorithm Y_i & Y_j designate the ith & jth organism and nominated haphazardly.

$$Y_i^{new} = Y_i + \text{random}(0,1) \times (Y_{Best} - \text{Mutual vector}(MV) \times \text{benefit factor}(BF_1)) \tag{24}$$

$$Y_j^{new} = Y_j + \text{random}(0,1) \times (Y_{Best} - \text{Mutual vector}(MV) \times \text{benefit factor}(BF_2)) \tag{25}$$

$$\text{Mutual vector}(MV) = \frac{\text{benefit factor}(BF_1) + \text{benefit factor}(BF_2)}{2} \tag{26}$$

In commensalism juncture or phase single organism will be profited but other organism may be debilitated or not abetted. In this period Y_i & Y_j indicate the ith & jth organism and Y_i profited then Y_j is encumbered or not abetted at any genre.

$$Y_i^{new} = Y_i + \text{random}(-1,1) \times (Y_{Best} - Y_j) \tag{27}$$

Ensuing in the parasitism juncture or phase single organism will be profited but other organism will be encumbered.

In the projected Enhanced Symbiotic Organisms Search (ESOS) Algorithm, Quasi-oppositional based learning and Chaotic local search has been applied to amplify the performance of the procedure.

Opposition based learning (OBL) is one of the persuasive optimization tools and it previously protracted to Quasi-oppositional based learning and quasi-opposite numbers is very much efficacious in discovery of global optimal solution.

Let $N(N \in [x, y])$ be a factual number and the N^o (opposite number) can be defined as follows,

$$N^o = x + y - N \tag{28}$$

$$N^{qo} = \text{random}\left(\frac{x+y}{2}, N^{o}\right) \tag{29}$$

In the exploration space it has been protracted as,

$$N_i^{o} = x_i + y_i - N_i \tag{30}$$

$$N_i^{qo} = \text{random}\left(\frac{x_i+y_i}{2}, N_i^{o}\right) \tag{31}$$

Where $(N_1, N_2,..N_d)$ is a point in the dimensional exploration space, $N_i \in [x_i, y_i]$, $i \rightarrow \{1,2,3,..d\}$

Quasi-oppositional based learning amalgamated in the Symbiotic organisms search algorithm for initialization of population and also for generation jump (in each iteration using the generation jumping rate, Jr)

Chaotic local search has been smeared to reconnoitre close by contemporary solutions to engender a new-fangled candidate solution, as follows

$$Y_{Best,k}^{new} = Y_{Best,k} + (Z_k - 0.5) \times (Y_{i,k} - Y_{j,k}) \tag{32}$$

When newfangled solution has superior fitness value than contemporary solution then it will be swapped.

Logistic map has been employed to engender the chaotic number,

$$Z_{k+1} = ¼ \times Z_k \times (1 - Z_k) \tag{33}$$

Start

Fix the control parameters

Set the evaluation counter

In the ecosystem organism population has been arbitrarily engendered through Quasi-oppositional based learning

$$\text{for } i = 1; \text{ecosystem\_size}$$

$$\text{for } j = 1; D$$

$$Y_{i,j}^{o} = \text{Lower bound}_j + \text{Upper bound}_j - Y_{i,j}$$

$$C_{i,j} = \left(\text{Lower bound}_j + \text{Upper bound}_j\right) / 2$$

$$\text{if}\left(Y_{i,j} < C_{i,j}\right)\text{then}$$

$$Y_{i,j}^{qo} = C_{i,j} + \left(Y_{i,j}^{o} - C_{i,j}\right) \times \text{random}$$

Otherwise

$$Y_{i,j}^{qo} = Y_{i,j}^{o} + \left(C_{i,j} - Y_{i,j}^{o}\right) \times \text{random}$$

End

Fitness value of $Y_{i,j}^{qo}$ is computed

Form the set $\left\{Y, Y_{i,j}^{qo}\right\}$ choose most fittest the ecosystem size as preliminary population

$$\text{while}\left(\text{function evaluation number}\left(\text{FE}\right) < \text{maximum FE}\right)$$

$$\text{for } i = 1; \text{ecosystem\_size}$$

Most excellent organism Y_{Best} will be identified

Apply the mutualism phase

Engender arbitrarily one organism Y_j, where $Y_j \neq Y_i$

Compute Y_i and Y_j by

$$Y_i^{new} = Y_i + \text{random}\left(0,1\right) \times \left(Y_{Best} - \text{Mutual vector}\left(\text{MV}\right) \times \text{benefit factor}\left(\text{BF}_1\right)\right)$$

$$Y_j^{new} = Y_j + \text{random}\left(0,1\right) \times \left(Y_{Best} - \text{Mutual vector}\left(\text{MV}\right) \times \text{benefit factor}\left(\text{BF}_2\right)\right)$$

$$\text{Mutual vector}\left(\text{MV}\right) = \frac{\text{benefit factor}\left(\text{BF}_1\right) + \text{benefit factor}\left(\text{BF}_2\right)}{2}$$

Fresh organism's fitness values is computed

New-fangled solution is modernized when it possess superior fitness value

$$\text{FE} = \text{FE} + 2$$

Apply the commensalism phase

Engender arbitrarily one organism Y_j, where $Y_j \neq Y_i$

$$Y_i^{new} = Y_i + random(-1,1) \times (Y_{Best} - Y_j)$$

Fresh organism's fitness values is computed
New-fangled solution is modernized when it possess superior fitness value

$$FE = FE + 1$$

Apply the parasitism phase
Engender arbitrarily one organism Y_j, where $Y_j \neq Y_i$
Compute the $parasite\_vector$
Fresh organism's fitness values is computed
New-fangled solution is modernized when it possess superior fitness value

$$FE = FE + 1$$

End
Quasi-oppositional based learning based generation jumping applied

$$\text{if} \left(random < J_r\right) \text{then apply Quasi-oppositional based learning process}$$

$$\text{for}\, i = 1; ecosystem\_size$$

$$\text{for}\, j = 1; D$$

$$Y_{i,j}^{o} = Lower\,bound_j + Upper\,bound_j - Y_{i,j}$$

$$C_{i,j} = \left(Lower\,bound_j + Upper\,bound_j\right)/2$$

$$\text{if}\left(Y_{i,j} < C_{i,j}\right) \text{then}$$

$$Y_{i,j}^{qo} = C_{i,j} + \left(Y_{i,j}^{o} - C_{i,j}\right) \times random$$

Otherwise

$$Y_{i,j}^{qo} = Y_{i,j}^{o} + \left(C_{i,j} - Y_{i,j}^{o}\right) \times random$$

End
Fitness value of $Y_{i,j}^{qo}$ is computed

Form the set $\left\{Y, Y_{i,j}^{qo}\right\}$ choose most fittest the ecosystem size as present population
Most excellent organism Y_{Best} will be identified
Apply chaotic local search

$\text{for } j = 1 : k$

Chaotic number is modernized by

$$Z_{k+1} = ¼ \times Z_k \times (1 - Z_k)$$

Form the existing population choose two organisms in arbitrary mode
New-fangled organism is computed by

$$Y_{Best,k}^{new} = Y_{Best,k} + (Z_k - 0.5) \times (Y_{i,k} - Y_{j,k})$$

New-fangled organism's fitness value is computed
Modernize the New-fangled solution if it possesses superior fitness value
End

Hydrological Cycle Algorithm

Projected Hydrological Cycle (HC) algorithm emulates the circulation of water form land to sky and vice versa. Through the process of evaporation, precipitation, condensation water conquers altered conditions.

The flow proportion of the water is computed by,

$$\text{flow rate} = \text{velocity} \times \text{cross\_sectional area} \Rightarrow \left[\text{flow rate} = \text{velocity} \times (\text{width} \times \text{depth})\right] \quad (34)$$

In the projected HC methodology Artificial Water Drops (AWD's) are defined by,

$$\text{Hydrological cycle} = \{AWD_1, AWD_2, .., AWD_n\}, n \geq 1 \quad (35)$$

Each water drop possesses the eminence; velocity $\left(\text{velocity}^{AWD}\right)$, amount of the soil carried by each water drop $\left(\text{soil}^{AWD}\right)$ and eminence of the solution with orientation to each water drop $^{\cdot\cdot\ AWD}$

Projected hydrological cycle (HC) algorithm explores the space, water will move from one point to another point by picking countless nodes. By using the state transition rule the next node can be determined by,

$$\text{probability}_{i}^{\text{AWD}}(j) = \frac{f(\text{soil}(i,j))^2 \times g(\text{depth}(i,j))}{\sum_{k \neq vc(\text{AWD})} \left(f(\text{soil}(i,j))^2 \times g(\text{depth}(i,j)) \right)} \tag{36}$$

$$f(\text{soil}(i,j)) = \frac{1}{\mu + \text{soil}(i,j)} \tag{37}$$

$$g(\text{depth}(i,j)) = \frac{1}{\text{depth}(i,j)} \tag{38}$$

$$\text{depth}(i,j) = \frac{\text{Length}(i,j)}{\text{soil}(i,j)} \tag{39}$$

Certainly the velocity for each water droplet varies with orientation to (t+1), it is defined by

$$\text{velocity}_{t+1}^{\text{AWD}} = \left[k \times \text{velocity}_{t}^{\text{AWD}} \right] + \pm \left(\frac{\text{velocity}^{\text{AWD}}}{\text{soil}(i,j)} \right) + \sqrt[2]{\frac{\text{velocity}^{\text{AWD}}}{\text{soil}^{\text{AWD}}}} + \left(\frac{100}{{}^{..}\ \text{AWD}} \right) + \sqrt[2]{\frac{\text{velocity}^{\text{AWD}}}{\text{depth}(i,j)}} \tag{40}$$

Liable on the velocity of water droplet flow soil can be added or jettisoned (sediment transport rate (str)) throughout the crusade and it can be defined by,

$$\text{soil}(i,j) = \begin{cases} str * soil(i,j) - \Delta soil(i,j) - \sqrt[2]{\dfrac{1}{\text{depth}(i,j)}} & \textit{if}\ \text{velocity}^{\text{AWD}} \geq avg\left(\text{entire}_{\text{velocity}}^{\text{AWD's}}\right) erosion \\ str * soil(i,j) + \Delta soil(i,j) + \sqrt[2]{\dfrac{1}{\text{depth}(i,j)}} & or\ else \quad deposition \end{cases} \tag{41}$$

Between node "i" and "j" the quantum of soil swapped and available is given by,

$$\Delta\text{soil}(i,j) = \frac{1}{\text{time}(t)_{i,j}^{\text{AWD}}} \tag{42}$$

$$\mathrm{time}(t)_{i,j}^{AWD} = \frac{\mathrm{distance}(i,j)}{\mathrm{velocity}_{t+1}^{AWD}} \tag{43}$$

In this HC algorithm eminence of the solution be contingent on amount of the soil carried by each water drop and a noble solution is acquired from the water drop which carries more soil, also the amount of the soil carried by water drop is employed to streamline the velocity of the water drops consequently and it described by,

$$\mathrm{soil}^{AWD} = \mathrm{soil}^{AWD} + \frac{\Delta \mathrm{soil}(i,j)}{\cdot\cdot\ ^{AWD}} \tag{44}$$

Then the temperature value has to be get rationalized with orientation to the water drop and it defined by,

$$\mathrm{Temperature}(t+1) = \mathrm{Temperature}(t) + \Delta \mathrm{Temperature} \tag{45}$$

$$\Delta \mathrm{Temperature} = \begin{cases} \beta * \left(\dfrac{\mathrm{Temperature}(t)}{\mathrm{Difference}(d)} \right) \Delta d > 0 \\ \dfrac{\mathrm{Temperature}(t)}{10} \mathit{otherwise} \end{cases} \tag{46}$$

$$\Delta \mathrm{d} = \text{maximum value} - \text{minimum value} \tag{47}$$

$$\text{maximum value} = \text{maximum}\left[\text{quality of solution's}\left(\text{AWD's}\right)\right] \tag{48}$$

$$\text{minimum value} = \text{minimum}\left[\text{quality of solution's}\left(\text{AWD's}\right)\right] \tag{49}$$

Then the evaporation rate of the water drops has been computed haphazardly among one droplet and aggregate number of droplet by,

$$\text{Evaporation rate} = \text{random}\left(1, (\text{number of water drops}(\mathrm{N})\right) \tag{50}$$

Solitary certain number of water droplets is picked for evaporation and it completed through Roulette-wheel selection technique.

In the condensation stage water drops are move closer, combine and also collusion will occur as the temperature decreases. The evaporated water from the overall population is given by,

$$\text{Evaporated water} = \{\text{AWD}_1, \text{AWD}_2, .., \text{AWD}_n\} \tag{51}$$

Certainly two water drops will combine each other or sprung after the collusion, it has been counterfeit in the projected algorithm with orientation to sharing of the information by,

$$\text{Process}(\text{AWD}_1, \text{AWD}_2) = \begin{cases} \mathit{Rebound}(\text{AWD}_1, \text{AWD}_2), \mathit{Similarity} < 50\% \\ \mathit{Combine}(\text{AWD}_1, \text{AWD}_2), \mathit{Similarity} \geq 50\% \end{cases} \tag{52}$$

Throughout combining of water droplets through collusion one of the water droplet will be preponderant then other one will be jettisoned, it described by

$$\text{AWD}_1^{\text{velocity}} = \text{AWD}_1^{\text{velocity}} + \text{AWD}_2^{\text{velocity}} \tag{53}$$

Also throughout the collusion of water droplets there may be leap off between two droplets, through that it will share the information about node eminence and weight of node,

$$\text{weight of the node} = \frac{\text{node}_{\text{occurance}}}{\text{AWD}_{\text{solution}}} \tag{54}$$

In order to dodge the local optima dynamic variables like soil edges of the most excellent solution will be accustomed by,

$$\text{soil}(i, j) = 0.9 * \text{soil}(i, j), \forall (i, j) \text{ most excellent}^{\text{AWD}} \tag{55}$$

Value of node is determined by,

$$Y_{\text{value}} = \sum_{l=1}^{m} n \times 10^{-L} \tag{56}$$

For each variable in the defined problem the values produced by the water droplet is given by,

$$\text{variable}_{\text{value}} = (\text{upper}_{\text{bound}} - \text{lower}_{\text{bound}}) \times \text{value} + \text{Lower}_{\text{bound}} \tag{57}$$

Solution of the water droplet is tranquil of set of nodes visited by droplet which is based on the number of variable and accuracy of the variable,

$$\text{water droplet solution} = \begin{bmatrix} AWD_s^1 12\ldots m_{N\times P} \\ AWD_s^2 12\ldots m_{N\times P} \\ . \\ . \\ AWD_s^k 12\ldots m_{N\times P} \end{bmatrix} \tag{58}$$

Mutation operation is smeared to certain water droplets during collusion in the period of condensation stage to progress the quality of the solution by,

$$\text{variable}_{new} = 1 - ({}^2 \times \text{variable}_{old}) \tag{59}$$

Begin
Parameters of the algorithm initiated
While (end condition is not met)
Cycle = Cycle + 1
Repeat
Iteration = iteration + 1
For every water droplet; next node chosen by,

$$\text{probability}_i^{AWD}(j) = \frac{f(\text{soil}(i,j))^2 \times g(\text{depth}(i,j))}{\sum_{k \neq vc(AWD)} \left(f(\text{soil}(i,j))^2 \times g(\text{depth}(i,j)) \right)}$$

$$f(\text{soil}(i,j)) = \frac{1}{\mu + \text{soil}(i,j)}$$

$$g(\text{depth}(i,j)) = \frac{1}{\text{depth}(i,j)}$$

$$\text{depth}(i,j) = \frac{\text{Length}(i,j)}{\text{soil}(i,j)}$$

Modernization of velocity for each droplet is done by

$$\text{velocity}_{t+1}^{AWD} = \left[k \times \text{velocity}_t^{AWD} \right] + \pm \left(\frac{\text{velocity}^{AWD}}{\text{soil}(i,j)} \right) + \sqrt[2]{\frac{\text{velocity}^{AWD}}{\text{soil}^{AWD}}} + \left(\frac{100}{\cdot\cdot\ AWD} \right) + \sqrt[2]{\frac{\text{velocity}^{AWD}}{\text{depth}(i,j)}}.$$

Modernization of soil and depth for each droplet is done through,

$$\text{soil}(\text{i},\text{j}) = \begin{cases} str * soil(\text{i},\text{j}) - \Delta soil(\text{i},\text{j}) - \sqrt[2]{\dfrac{1}{\text{depth}(\text{i},\text{j})}} & \textit{if } \text{velocity}^{\text{AWD}} \geq avg\left(\text{entire}^{\text{AWD's}}_{\text{velocity}}\right) erosion \\ str * soil(\text{i},\text{j}) + \Delta soil(\text{i},\text{j}) + \sqrt[2]{\dfrac{1}{\text{depth}(\text{i},\text{j})}} & or\ else \quad deposition \end{cases}$$

$$\Delta\text{soil}(\text{i},\text{j}) = \frac{1}{\text{time}(\text{t})^{\text{AWD}}_{\text{i},\text{j}}}$$

Modernization of carrying soil by each droplet is done by,

$$\text{soil}^{\text{AWD}} = \text{soil}^{\text{AWD}} + \frac{\Delta\text{soil}(\text{i},\text{j})}{\text{.. AWD}}$$

End for
Compute the fitness of obtained solutions
Modernize the local optim solution
Modernize the Temperature by

$$\text{Temperature}(\text{t}+1) = \text{Temperature}(\text{t}) + \Delta\text{Temperature}$$

$$\Delta\text{Temperature} = \begin{cases} \beta * \left(\dfrac{\text{Temperature}(\text{t})}{\text{Difference}(\text{d})}\right) & \Delta d > 0 \\ \dfrac{\text{Temperature}(\text{t})}{10} & otherwise \end{cases}$$

$$\Delta\text{d} = \text{maximum value} - \text{minimum value}$$

$$\text{maximum value} = \text{maximum}\left[\text{quality of solution's}(\text{AWD's})\right]$$

$$\text{minimum value} = \text{minimum}\left[\text{quality of solution's}(\text{AWD's})\right]$$

Until (Temperature = Evaporation_ Temperature)
Evaporation method for AWD
Estimate the quality of the solution

Find the Evaporation Rate by,

$$\text{Evaporation rate} = \text{random}\left(1, (\text{number of water drops}\left(\text{N}\right)\right)$$

Choose AWD's for evaporation
End
Condensation method for AWD
Swapping Information between AWD's by

$$\text{Process}\left(\text{AWD}_1, \text{AWD}_2\right) = \begin{cases} \mathit{Rebound}\left(\text{AWD}_1, \text{AWD}_2\right), \mathit{Similarity} < 50\% \\ \mathit{Combine}\left(\text{AWD}_1, \text{AWD}_2\right), \mathit{Similarity} \geq 50\% \end{cases}$$

$$\text{AWD}_1^{\text{velocity}} = \text{AWD}_1^{\text{velocity}} + \text{AWD}_2^{\text{velocity}}$$

$$\text{weight of the node} = \frac{\text{node}_{\text{occurance}}}{\text{AWD}_{\text{solution}}}$$

Find the AWD collector
Modernize the global solution
Modernize the temperature
End
Precipitation method for AWD
Quality of the solution computed
Dynamic parameters of the algorithm are Re-initialized
Modernization of most excellent quality AWD's Global soil through

$$\text{soil}\left(\text{i}, \text{j}\right) = 0.9{*}\text{soil}\left(\text{i}, \text{j}\right), \forall\left(\text{i}, \text{j}\right) \text{most excellent}^{\text{AWD}}$$

Engender new-fangled AWD's population
Allocation of AWD's done arbitrarily
End
End while

Simulation Results

Cogency of Projected Enhanced Symbiotic Organisms Search (ESOS) Algorithm and Hydrological Cycle (HC) algorithm corroborated in in IEEE 30 bus system. On 100 MVA base active and reactive power are 2.8340 (PU) and 1.2620 (PU). Table 1 shows the limits. Tables 2 to 5 give the comparison of values. Projected ESOS and HC is compared with Basic differential evolution algorithm, standard genetic algorithm, Hybrid Artificial Physics–Particle Swarm Optimization, modified particle swarm optimization algorithm, Basic Evolutionary programming algorithm, Self-adaptive real coded Genetic algorithm, canonical genetic algorithm, Adaptive genetic algorithm, Enriched genetic algorithm, Faster evolutionary algorithm, cuckoo search optimization algorithm.

Table 1. Limits

IEEE System	Variables	Minimum (PU)	Maximum (PU)
30 Bus	Generator Voltage	0.95	1.1
	Transformer Tap	0.9	1.1
	VAR Source	0	5
	NB	30	
	NG	6	
	NT	4	
	NQ	9	
	NE	41	
	Base case Power Loss in MW	5.66	

In IEEE 14 Bus system the cogency of the proposed Enhanced Symbiotic Organisms Search (ESOS) Algorithm and Hydrological Cycle (HC) algorithm deprived of voltage stability index has been corroborated. Assessment results are presented in Table 6.

Enhanced Symbiotic Organisms Search (ESOS) Algorithm and Hydrological Cycle (HC) algorithm corroborated in in IEEE 30 Bus system. Evaluation results are presented in Table 7.

Enhanced Symbiotic Organisms Search (ESOS) Algorithm and Hydrological Cycle (HC) algorithm has been corroborated in IEEE 57 Bus system. Evaluation results are presented in Table 8.

Then IEEE 300 bus system is used as an assessment system to corroborate the performance of the Enhanced Symbiotic Organisms Search (ESOS) Algorithm and Hydrological Cycle (HC) algorithm. Table 9 shows the assessment of actual power loss attained after optimization.

CONCLUSION

In this work optimal reactive power dispatch problem has been successfully solved by Enhanced Symbiotic Organisms Search (ESOS) Algorithm and Hydrological Cycle (HC) algorithm. Three stages mutualism, commensalism and parasitism produced new-fangled solutions and it has been updated when it is superior to preceding solution. Quasi-oppositional based learning and quasi-opposite numbers is very much successful in discovery of global optimal solution. Chaotic local search has been applied to explore close by present solutions to engender a new-fangled candidate solution. Projected hydrological cycle (HC) algorithm explores the space, water will move from one point to another point by choosing various nodes. In order to avoid the local optima dynamic variables like soil edges of the most excellent solution are adjusted. Mutation operation is applied to selected water droplets during collusion in the period of condensation stage to improve the quality of the solution. Proposed ESOS, HC algorithms is

Table 2. Appraisal of actual power loss

Parameter	B-DE (ela etal.,2011)	S-GSA (Duman et al., 2012)	APO-PSO (Aljohani et al ., 2019)	ESOS	HC
VlG1	1.1	1.071	1.100	1.090	1.092
VlG2	1.09	1.022	1.084	1.042	1.044
VlG5	1.07	1.040	1.056	1.023	1.025
VlG8	1.07	1.051	1.076	1.044	1.043
VlG1	1.1	0.977	1.091	1.090	1.091
VlG3	5	0.968	1.100	0.982	0.983
QCp10	5	1.653	5.000	4.983	4.985
QCp12	5	4.3722	5.000	5.000	5.000
QCp15	5	0.1199	4.879	4.792	4.791
QCp17	5	2.0876	4.976	4.970	4.968
QCp20	4.41	0.357	3.821	3.701	3.703
QCp21	5	0.2602	4.541	4.660	4.662
QCp23	2.8004	0.0000	2.354	2.402	2.405
QCp24	5	1.3839	4.654	4.504	4.500
QCp29	2.5979	0.0000	2.175	2.160	2.157
Tp11(6-9)	1.04	1.0985	1.029	1.010	1.008
Tp12(6-10)	0.9097	0.9824	0.911	0.901	0.900
Tp15(4-12)	0.98	1.095	0.952	0.942	0.936
Tp36(28-27)	0.9689	1.0593	0.958	0.940	0.939
True Power Loss in MW	4.555	4.5143	4.398	4.250	4.247
Voltage Deviation in PU	1.9589	0.87522	1.047	1.043	1.040
Voltage stability index in PU	0.5513	0.14109	0.1267	0.1205	0.1207

tested in IEEE 30, bus system- real power loss minimization, voltage deviation minimization, and voltage stability index enhancement has been attained. Then the Proposed ESOS, HC algorithms has been tested in standard IEEE 14, 30, 57, 300 bus test systems without considering the voltage stability index. Projected algorithms reduced the power loss effectively and control variables are within the limits.

Table 3. Assessment with reference to voltage stability enhancement

Parameter	B-DE (ela etal.,2011)	S-GSA (Duman et al., 2012)	APO-PSO (Aljohani et al ., 2019)	ESOS	HC
VlG1	1.01	0.983	1.011	1.023	1.025
VlG2	0.99	1.044	1.001	1.014	1.012
VlG5	1.02	1.020	1.014	1.014	1.013
VlG8	1.02	0.999	1.009	1.012	1.016
VlG1	1.01	1.077	0.954	0.944	0.947
VlG3	1.03	1.044	1.000	1.000	1.000
QCp10	4.94	0	4.102	4.103	4.100
QCp12	1.0885	0.4735	2.124	2.124	2.119
QCp15	4.9985	5	4.512	4.495	4.490
QCp17	0.2393	0	0.000	0.000	0.000
QCp20	4.99	5	5.000	5.000	5.000
QCp21	4.90	0	5.000	5.000	5.000
QCp23	4.9863	4.9998	5.000	5.000	5.000
QCp24	4.9663	5	5.000	5.000	5.000
QCp29	2.2325	5	4.120	4.130	4.127
Tp11(6-9)	1.02	0.9	0.998	0.990	0.989
Tp12(6-10)	0.9038	1.1	0.822	0.810	0.809
Tp15(4-12)	1.01	1.051	0.954	0.941	0.939
Tp36(28-27)	0.9635	0.9619	0.958	0.940	0.941
True Power Loss in MW	6.4755	6.9117	5.698	5.432	5.429
Voltage Deviation in PU	0.0911	0.0676	0.087	0.082	0.081
Voltage stability index in PU	0.14352	0.1349	0.1377	0.1314	0.1315

Table 4. Evaluation with reference to voltage deviation

Parameter	B-DE (ela etal.,2011)	S-GSA (Duman et al., 2012)	APO-PSO (Aljohani et al ., 2019)	ESOS	HC
VlG1	0.936	1.0195	0.924	0.920	0.917
VlG2	1.09	1.1	1.043	1.032	1.034
VlG5	1.09	1.1	1.061	1.050	1.049
VlG8	1.09	1.1	1.061	1.059	1.055
VlG1	1.04	1.1	1.057	1.040	1.035
VlG3	1.09	1.1	1.048	1.041	1.038
QCp10	0.95	1.1	1.091	1.080	1.077
QCp12	0.69	5	0.040	0.031	0.029
QCp15	4.7163	5	0.039	0.030	0.026
QCp17	4.4931	5	0.038	0.032	0.029
QCp20	4.51	5	0.040	0.031	0.030
QCp21	4.48	5	0.037	0.030	0.027
QCp23	4.60	5	0.009	0.012	0.010
QCp24	3.8806	5	0.019	0.011	0.009
QCp29	3.8806	5	0.011	0.010	0.007
Tp11(6-9)	3.2541	5	0.001	0.000	0.001
Tp12(6-10)	0.90	0.9	0.919	0.910	0.909
Tp15(4-12)	0.9029	0.9	0.924	0.914	0.908
Tp36(28-27)	0.90	0.9	0.938	0.924	0.919
True Power Loss in MW	7.0733	4.9752	4.478	4.249	4.242
Voltage Deviation in PU	1.419	0.21579	1.8579	1.8207	1.8200
Voltage stability index in PU	0.1246	0.13684	0.1227	0.1174	0.1176

Table 5. Appraisal with reference to multi – objective formulation

Parameter	APO-PSO (Aljohani et al ., 2019)	ESOS	HC
VlG1	1.020	1.012	1.011
VlG2	1.033	1.024	1.020
VlG5	1.000	1.000	1.000
VlG8	1.004	1.000	1.000
VlG1	1.032	1.022	1.020
VlG3	1.028	1.021	1.024
QCp10	0.051	0.042	0.044
QCp12	0.002	0.000	0.001
QCp15	0.044	0.043	0.042
QCp17	0.009	0.004	0.003
QCp20	0.048	0.040	0.039
QCp21	0.041	0.032	0.030
QCp23	0.033	0.021	0.019
QCp24	0.050	0.042	0.040
QCp29	0.015	0.010	0.008
Tp11(6-9)	1.042	1.040	1.039
Tp12(6-10)	0.909	0.902	0.900
Tp15(4-12)	1.023	1.010	1.009
Tp36(28-27)	0.958	0.942	0.940
True Power Loss in MW	4.842	4.741	4.739
Voltage Deviation in PU	1.009	1.003	1.002
Voltage stability index in PU	0.1192	0.1181	0.1183

Table 6. Assessment of loss

Parameters	Base value	Modified PSO (Ali Nasser Hussain et al.,2018)	Standard PSO (Pandya et al., 2015)	Standard EP (Pandya et al., 2015)	SA- Real coded GA (Subbaraj et al., 2009)	ESOS	HC
VlG1	1.060	1.100	1.100	-	-	1.005	1.004
VlG2	1.045	1.085	1.086	1.029	1.060	1.024	1.023
VlG3	1.010	1.055	1.056	1.016	1.036	1.027	1.021
VlG6	1.070	1.069	1.067	1.097	1.099	1.036	1.034
VlG8	1.090	1.074	1.060	1.053	1.078	1.012	1.010
Tp8	0.978	1.018	1.019	1.04	0.95	0.905	0.902
Tp9	0.969	0.975	0.988	0.94	0.95	0.926	0.924
Tp10	0.932	1.024	1.008	1.03	0.96	0.941	0.940
QCp9	0.19	14.64	0.185	0.18	0.06	0.153	0.151
PGr	272.39	271.32	271.32	-	-	270.64	270.62
QGr	82.44	75.79	76.79	-	-	75.04	75.01
Decrease in Power Loss (%)	0	9.2	9.1	1.5	2.5	23.32	23.79
Power Loss (MW)	13.550	12.293	12.315	13.346	13.216	10.390	10.326

Table 7. True power loss appraisal

Parameters	Base value (Ali Nasser Hussain et al.,2018)	Modified PSO (Ali Nasser Hussain et al.,2018)	Standard PSO (Pandya et al., 2015)	Standard EP (Pandya et al., 2015)	SA- Real coded GA (Subbaraj et al., 2009)	ESOS	HC
VlG1	1.060	1.101	1.100	-	-	1.036	1.034
VlG2	1.045	1.086	1.072	1.097	1.094	1.024	1.022
VlG5	1.010	1.047	1.038	1.049	1.053	1.054	1.050
VlG8	1.010	1.057	1.048	1.033	1.059	1.023	1.021
VlG12	1.082	1.048	1.058	1.092	1.099	1.065	1.062
VlG13	1.071	1.068	1.080	1.091	1.099	1.034	1.032
Tp11	0.978	0.983	0.987	1.01	0.99	0.913	0.910
Tp12	0.969	1.023	1.015	1.03	1.03	0.913	0.911
Tp15	0.932	1.020	1.020	1.07	0.98	0.905	0.904
Tp36	0.968	0.988	1.012	0.99	0.96	0.914	0.912
QCp10	0.19	0.077	0.077	0.19	0.19	0.091	0.090
QCp24	0.043	0.119	0.128	0.04	0.04	0.122	0.120
PGr	300.9	299.54	299.54	-	-	297.13	297.11
QGr	133.9	130.83	130.94	-	-	131.90	131.89
Decrease in Power Loss (%)	0	8.4	7.4	6.6	8.3	19.03	19.65
Power Loss (MW)	17.55	16.07	16.25	16.38	16.09	14.21	14.10

Table 8. Loss comparison

Parameters	Base case (Ali Nasser Hussain et al.,2018)	Modified PSO (Ali Nasser Hussain et al.,2018)	Standard PSO (Pandya et al., 2015)	Standard EP (Pandya et al., 2015)	SA- Real coded GA (Subbaraj et al., 2009)	ESOS	HC
VlG1	1.040	1.093	1.083	0.968	1.027	1.023	1.022
VlG2	1.010	1.086	1.071	1.049	1.011	1.014	1.010
VlG3	0.985	1.056	1.055	1.056	1.033	1.031	1.030
VlG6	0.980	1.038	1.036	0.987	1.001	1.012	1.014
VlG8	1.005	1.066	1.059	1.022	1.051	1.033	1.031
VlG9	0.980	1.054	1.048	0.991	1.051	1.012	1.010
VlG12	1.015	1.054	1.046	1.004	1.057	1.040	1.042
Tp19	0.970	0.975	0.987	0.920	1.030	0.952	0.950
Tp20	0.978	0.982	0.983	0.920	1.020	0.934	0.931
Tp31	1.043	0.975	0.981	0.970	1.060	0.926	0.923
Tp35	1.000	1.025	1.003	-	-	1.017	1.014
Tp36	1.000	1.002	0.985	-	-	1.001	1.000
Tp37	1.043	1.007	1.009	0.900	0.990	1.003	1.001
Tp41	0.967	0.994	1.007	0.910	1.100	0.994	0.992
Tp46	0.975	1.013	1.018	1.100	0.980	1.017	1.013
Tp54	0.955	0.988	0.986	0.940	1.010	0.973	0.972
Tp58	0.955	0.979	0.992	0.950	1.080	0.962	0.960
Tp59	0.900	0.983	0.990	1.030	0.940	0.963	0.961
Tp65	0.930	1.015	0.997	1.090	0.950	1.004	1.003
Tp66	0.895	0.975	0.984	0.900	1.050	0.943	0.942
Tp71	0.958	1.020	0.990	0.900	0.950	1.002	1.001
Tp73	0.958	1.001	0.988	1.000	1.010	1.001	1.000
Tp76	0.980	0.979	0.980	0.960	0.940	0.933	0.931
Tp80	0.940	1.002	1.017	1.000	1.000	1.003	1.002
QCp18	0.1	0.179	0.131	0.084	0.016	0.166	0.165
QCp25	0.059	0.176	0.144	0.008	0.015	0.160	0.159
QCp53	0.063	0.141	0.162	0.053	0.038	0.143	0.141
PGr	1278.6	1274.4	1274.8	1276	1275	1270.13	1270.10
QGr	321.08	272.27	276.58	309.1	304.4	272.73	272.70
Decrease in Power Loss (%)	0	15.4	14.1	9.2	11.6	23.70	23.99
Power Loss (MW)	27.8	23.51	23.86	25.24	24.56	21.209	21.130

Table 9. Evaluation of real power loss

Parameter	Enriched Genetic algorithm (Reddy 2017)	Faster Evolutionary Algorithm (Reddy2017)	Standard cuckoo search CSA (Reddy 2014)	ESOS	HC
Power Loss (MW)	646.2998	650.6027	635.8942	611.9257	611.7980

REFERENCES

Aljohani, T. M., Ebrahim, A. F., & Mohammed, O. (2019). Single and Multiobjective Optimal Reactive Power Dispatch Based on Hybrid Artificial Physics–Particle Swarm Optimization. *Energies*, *12*(12), 2333. doi:10.3390/en12122333

Antoniewicz, P., & Kazmierkowski, P. (2008). Virtual-flux-based predictive direct power control of AC/DC converters with online inductance estimation. *IEEE Transactions on Industrial Electronics*, *55*(12), 4381–4390. doi:10.1109/TIE.2008.2007519

Azar, A. T. (2010a). *Fuzzy Systems*. IN-TECH.

Azar, A. T. (2010b). Adaptive Neuro-Fuzzy Systems. In A. T. Azar (Ed.), *Fuzzy Systems*. IN-TECH. doi:10.5772/7220

Azar, A. T. (2012). Overview of Type-2 Fuzzy logic systems. *International Journal of Fuzzy System Applications*, *2*(4), 1–28. doi:10.4018/ijfsa.2012100101

Azar, A.T. (2013). Fast neural network learning algorithms for medical applications. *Neural Computing and Applications, 23*(3-4), 1019-1034. . doi:10.1007/s00521-012-1026-y

Azar, A.T., & El-Said, S.A. (2013). Probabilistic neural network for breast cancer classification. *Neural Computing and Applications, 23*(6), 1737-1751. . doi:10.1007/s00521-012-1134-8

Azar, A. T., Kumar, J., Kumar, V., & Rana, K. P. S. (2017d). Control of a Two Link Planar Electrically-Driven Rigid Robotic Manipulator Using Fractional Order SOFC. *Proceedings of the International Conference on Advanced Intelligent Systems and Informatics 2017*, 47-56.

Azar, A. T., Ouannas, A., & Singh, S. (2017c). Control of New Type of Fractional Chaos Synchronization. *Proceedings of the International Conference on Advanced Intelligent Systems and Informatics 2017*, 47-56.

Azar, A.T., & Serrano, F.E. (2014). Robust IMC-PID tuning for cascade control systems with gain and phase margin specifications. *Neural Computing and Applications, 25*(5), 983-995. doi:10.1007/s00521-014-1560-x

Azar, A. T., & Serrano, F. E. (2015a). Stabilization and Control of Mechanical Systems with Backlash. In Advanced Intelligent Control Engineering and Automation. IGI Global. doi:10.4018/978-1-4666-7248-2.ch001

Azar, A. T., & Serrano, F. E. (2015b). Design and Modeling of Anti Wind Up PID Controllers. In Complex system modelling and control through intelligent soft computations. Springer-Verlag. DOI . 1. doi:10.1007/978-3-319-12883-2_1

Azar, A. T., & Serrano, F. E. (2015c). Adaptive Sliding mode control of the Furuta pendulum. In Advances and Applications in Sliding Mode Control systems. Springer-Verlag GmbH. doi:10.1007/978-3-319-11173-5_1

Azar, A. T., & Serrano, F. E. (2015d). Deadbeat Control for Multivariable Systems with Time Varying Delays. In Chaos Modeling and Control Systems Design. Springer-Verlag GmbH. doi:10.1007/978-3-319-13132-0_6

Azar, A. T., & Serrano, F. E. (2016a). Robust control for asynchronous switched nonlinear systems with time varying delays. *Proceedings of the International Conference on Advanced Intelligent Systems and Informatics 2016*, 533, 891-899. 10.1007/978-3-319-48308-5_85

Azar, A. T., & Serrano, F. E. (2016b). Stabilization of Mechanical Systems with Backlash by PI Loop Shaping. *International Journal of System Dynamics Applications*, *5*(3), 20–47. doi:10.4018/IJSDA.2016070102

Azar, A. T., & Serrano, F. E. (2017). Passivity Based Decoupling of Lagrangian Systems. *Proceedings of the International Conference on Advanced Intelligent Systems and Informatics 2016*, 533, 891-899. 10.1007/978-3-319-48308-5_85

Azar, A. T., & Vaidyanathan, S. (2015a). Handbook of Research on Advanced Intelligent Control Engineering and Automation. IGI Global. doi:10.4018/978-1-4666-7248-2

Azar, A. T., & Vaidyanathan, S. (2015b). *Computational Intelligence applications in Modeling and Control.* Springer-Verlag.

Azar, A. T., & Vaidyanathan, S. (2015c). *Chaos Modeling and Control Systems Design.* Springer-Verlag.

Azar, A. T., & Vaidyanathan, S. (2016). *Advances in Chaos Theory and Intelligent Control.* Springer-Verlag. doi:10.1007/978-3-319-30340-6

Azar, A. T., Vaidyanathan, S., & Ouannas, A. (2017a). *Fractional Order Control and Synchronization of Chaotic Systems.* Springer-Verlag. doi:10.1007/978-3-319-50249-6

Azar, A. T., Volos, C., Gerodimos, N. A., Tombras, G. S., Pham, V. T., Radwan, A. G., Vaidyanathan, S., Ouannas, A., & Munoz-Pacheco, J. M. (2017b). A novel chaotic system without equilibrium: Dynamics, synchronization and circuit realization. Complexity. doi:10.1155/2017/7871467

Azar, A. T., & Zhu, Q. (2015). *Advances and Applications in Sliding Mode Control systems.* Springer-Verlag.

Baghli, L. (1999). *Contribution à la Commande de la Machine Asynchrone, Utilisation de la Logique Floue, des Réseaux de Neurones et des Algorithmes Génétiques.* Thèse de Doctorat, Département de Génie Electrique Université Henri Poincaré, Nancy-I.

Bimal, & Bos, K. (1994) Expert system, fuzzy logic, and neural network applications in power electronics and motion control. *Proceedings of the IEEE, 82*(8), 1303 – 1323.

Bose, B. (2001). Artificial neural network applications in power electronics. *The 27th Annual Conférence of the IEEE Industrial Electronics Society, 2001. IECON '01.* 10.1109/IECON.2001.975533

Bouafia, A., & Krim, F. (2008). A fuzzy-Logic-Based Controller for Three-Phase PWM Rectifier with Unity Power Factor Operation. *J. Electrical Systems*, *4*(1), 36–50.

Bouafia, A., Krim, F., & Gaubert, J. P. (2009). Fuzzy-Logic-Based Switching State Selection for Direct Power Control of Three-Phase PWM Rectifier. *IEEE Transactions on Industrial Electronics*, *56*(6), 1984–1992. doi:10.1109/TIE.2009.2014746

Bouchakour, S. (2005). *Commande Directe de Puissance d'un Convertisseur AC/DC Triphasé Sans Capteurs de Tension.* Memory of Magister Ecole Militaire Polytechnique Algeria.

Boulkroune, A., Bouzeriba, A., Bouden, T., & Azar, A. T. (2016a). Fuzzy Adaptive Synchronization of Uncertain Fractional-order Chaotic Systems. In A. T. Azar & S. Vaidyanathan (Eds.), *Advances in Chaos Theory and Intelligent Control. Studies in Fuzziness and Soft Computing* (Vol. 337). Springer-Verlag. doi:10.1007/978-3-319-30340-6_28

Boulkroune, A., Hamel, S., & Azar, A. T. (2016b). *Fuzzy control-based function synchronization of unknown chaotic systems with dead-zone input.* Springer-Verlag.

Chapuis, Y. A. (1996). *Contrôle directe du couple d'une machine asynchrone par l'orientation de son flux statorique.* Doctoral thesis of INP-Grenoble, France.

Chen, B., & Joos, G. (2008). Direct power control of active filters with averaged switching frequency regulation. *IEEE Transactions on Power Electronics*, *23*(6), 2729–2737. doi:10.1109/TPEL.2008.2004958

Chen, S., & Joos, G. (2001) Direct Power Control of Three Active filer with Minimum Energy Storage Components. *APEC 2001. Sixteenth Annual IEEE Applied Power Electronics Conference and Exposition.* 10.1109/APEC.2001.911703

Cichowlas, M., Malinowski, M., Kazmierkowski, P., Sobczuk, D., & Pou, J. (2005). Active filtering function of three-phase PWM boost rectifier under different line voltage conditions. *IEEE Transactions on Industrial Electronics*, *52*(2), 410–419. doi:10.1109/TIE.2005.843915

Constant, L. (2000). *Modélisation de Dispositifs Electriques par Réseaux de Neurones en Vue de L'émulation Temps Réel.* Thesis Doctorate, Toulouse, France.

Duman, S., Sönmez, Y., Güvenç, U., & Yörükeren, N. (2012). Optimal reactive power dispatch using a gravitational search algorithm. *IET Generation, Transmission & Distribution*, *6*(6), 563–576. doi:10.1049/iet-gtd.2011.0681

El Ela, A. A., Abido, M. A., & Spea, S. R. (2011). Differential evolution algorithm for optimal reactive power dispatch. *Electric Power Systems Research*, *81*(2), 458–464. doi:10.1016/j.epsr.2010.10.005

Escobar, G., Stankovic, A., Carrasco, M., Galvan, E., & Ortega, R. (2003). Analysis and design of direct power control (DPC) for a three phase synchronous rectifier via output regulation subspaces. *IEEE Transactions on Power Electronics*, *18*(3), 823–830. doi:10.1109/TPEL.2003.810862

Fekik, A., Denoun, H., Benamrouche, N., Benyahia, N., Badji, A., & Zaouia, M. (2016) Comparative Analysis of Direct Power Control and Direct power control with space vector modulation of PWM rectifier. *4th IEEE International Conference on Control Engineering & Information Technology (CEIT-2016)*. 10.1109/CEIT.2016.7929058

Fekik, A., Denoun, H., Benamrouche, N., Benyahia, N., & Zaouia, M. (2015b). A Fuzzy Logic Based Controller For Three Phase PWM Rectifier With Voltage Oriented Control Strategy. *International Journal of Circuits, Systems and Signal Processing*, *9*, 412–419.

Fekik, A., Denoun, H., Benamrouche, N., Benyahia, N., Zaouia, M., Badji, A., & Vaidyanathan, S. (2017). Improvement of the Performances of the Direct Power Control Using Space Vector Modulation of Three Phases PWM-Rectifier. *International Journal of Control Theory and Applications*, *10*, 125–137.

Fekik, A., Denoun, H., Benamrouche, N., Benyahia, N., Zaouia, M., & Haddad, S. (2015a). Comparative study of PI and FUZZY DC- voltage control for Voltage Oriented Control-PWM rectifier. *The 14th International Conference on Circuits, Systems, Electronics, Control & Signal Processing 2015.CSECS'15.*

Fischer, J. R., González, S. A., Carugat, I., Herrán, M. A., Judewicz, M. G., & Carrica, D. O. (2014). Robust predictive control of grid-tied converters based on direct power control. *IEEE Transactions on Power Electronics*, *29*(10), 5634–5643. doi:10.1109/TPEL.2013.2294919

Geyer, T., & Quevedo, E. (2015). Performance of multistep finite control set model predictive control for power electronics. *IEEE Transactions on Power Electronics*, *30*(3), 1633–1644. doi:10.1109/TPEL.2014.2316173

Ghoudelbourk, S., Dib, D., Omeiri, A., & Azar, A. T. (2016). MPPT Control in wind energy conversion systems and the application of fractional control (PIα) in pitch wind turbine. *International Journal of Modelling Identification and Control*, *26*(2), 140–151. doi:10.1504/IJMIC.2016.078329

Grassi, G., Ouannas, A., Azar, A. T., Radwan, A. G., Volos, C., Pham, V. T., Ziar, T., Madian, A., Kyprianidis, I. M., & Stouboulos, I. N. (2017). Chaos Synchronisation Of Continuous Systems Via Scalar Signal. *The 6th International Conference on Modern Circuits and Systems Technologies (MOCAST)*. 10.1109/MOCAST.2017.7937629

Hassanien, A.E., Moftah, H.M., Azar, A.T., & Shoman, M. (2014). MRI Breast cancer diagnosis hybrid approach using adaptive Ant-based segmentation and Multilayer Perceptron neural networks classifier. *Applied Soft Computing, 14*(Part A), 62-71.

Hu, J., Zhu, J., Zhang, Y., Platt, G., Ma, Q., & Dorrell, D. G. (2013). Predictive direct virtual torque and power control of doubly fed induction generators for fast and smooth grid synchronization and flexible power regulation. *IEEE Transactions on Power Electronics*, *28*(7), 3182–3194. doi:10.1109/TPEL.2012.2219321

Hussain, A. N., Abdullah, A. A., & Neda, O. M. (2018). Modified Particle Swarm Optimization for Solution of Reactive Power Dispatch. *Research Journal of Applied Sciences, Engineering and Technology*, *15*(8), 316–327. doi:10.19026/rjaset.15.5917

Jiang, W. (2010). Sliding-mode Control of Single-phase PWM Rectifier for DC Microgrid Applications. *International Conference on Computer and Electrical Engineering*, 53, 82-90. DOI: 10.7763/IPCSIT.2012.V53.No.1.82

Karamanakos, P., Geyer, T., Oikonomou, N., Kieferndorf, F., & Manias, S. (2014). Direct model predictive control: A review of strategies that achieve long prediction intervals for power electronics. *IEEE Industrial Electronics Magazine*, *8*(1), 32–43. doi:10.1109/MIE.2013.2290474

Lamterkati, J., Khafallah, M., & Ouboubker, L. (2014a). A New DPC for Three-phase PWM rectifier with unity power factor operation. *International Journal of Advanced Research in Electrical Electronics and Instrumentation Engineering*, *3*(4), 8273–8285.

Lamterkati, J., Khafallah, M., & Ouboubker, L. (2014b). Comparison of PI and Fuzzy logic DC-Link Voltage Controller for DPC PWM-Rectifier. *International Journal of Enhanced Research in Science Technology and Engineering*, *3*(4), 321–332.

Larrinaga, S., Vidal, M., Oyarbide, E., & Apraiz, J. (2007). Predictive control strategy of DC/AC converters based on direct power control. *IEEE Transactions on Industrial Electronics*, *54*(3), 1261–1271. doi:10.1109/TIE.2007.893162

Larrinaga, S. A., Vidal, M. A. R., Oyarbide, E., & Apraiz, J. R. T. (2007). Predictive control strategy for DC/AC converters based on direct power control. *IEEE Transactions on Industrial Electronics*, *54*(3), 1261–1271. doi:10.1109/TIE.2007.893162

Malinowski, M., Jasinski, M., & Kazmierkowski, P. (2004). Simple Direct power control of three phase PWM rectifier using space vector modulation (DPC-SVM). *IEEE Transactions on Industrial Electronics*, *51*(2), 447–454. doi:10.1109/TIE.2004.825278

Malinowski, M., Kaz'mierkowski, M. P., Hansen, S., Blaabjerg, F., & Marques, G. D. (2001). Virtual flux based direct power control of three-phase PWM rectifiers. *IEEE Transactions on Industry Applications*, *37*(4), 1019–1027. doi:10.1109/28.936392

Malinowski, M., & Kazmierkowski, M.P. (2000) Simulation Study of Virtual Flux Based Direct Power Control for Three-Phase PWM Rectifiers. *26th Annual Confierence of the IEEE Industrial Electronics Society, 2000. IECON 2000*. doi:10.1109/IECON.2000.972411

Malinowski, M., Kazmierkowski, M. P., & Trzynadlowski, A. M. (2003). A Comparative Study of Control Techniques for PWM Rectifiers in AC Adjustable Speed Drives. *IEEE Transactions on Power Electronics*, *18*(6), 1390–1396. doi:10.1109/TPEL.2003.818871

Meghni, B., Dib, D., & Azar, A. T. (2017c). A Second-order sliding mode and fuzzy logic control to Optimal Energy Management in PMSG Wind Turbine with Battery Storage. *Neural Computing & Applications*, *28*(6), 1417–1434. doi:10.100700521-015-2161-z

Meghni, B., Dib, D., Azar, A. T., Ghoudelbourk, S., & Saadoun, A. (2017a). Robust Adaptive Supervisory Fractional order Controller For optimal Energy Management in Wind Turbine with Battery Storage. *Studies in Computational Intelligence*, *688*, 165–202. doi:10.1007/978-3-319-50249-6_6

Meghni, B., Dib, D., Azar, A. T., & Saadoun, A. (2017b). *Effective Supervisory Controller to Extend Optimal Energy Management in Hybrid Wind Turbine under Energy and Reliability Constraints. International Journal of Dynamics and Control.* doi:10.100740435-016-0296-0

Mekki, H., Boukhetala, D., & Azar, A. T. (2015). Sliding Modes for Fault Tolerant Control. In Advances and Applications in Sliding Mode Control systems. Springer-Verlag GmbH. doi:10.1007/978-3-319-11173-5_15

Mezache, A., & Betto, K. (1997). *Estimation et Commande d'un Réacteur de Fabrication de Pâte a Papier par les Réseaux de Neurones Flous. Memory of Engineer*. University of Constantine.

Mondal, S. K., Pinto, J. O. P., & Bose, B. K. (2002). A Neural-Network-Based Space-Vector PWM Controller for a three-Level Voltage-Fed Inverter Induction Motor Drive. *Industry Applications. IEEE Transactions on.*, *38*(3), 660–669.

Noguchi, T., Tomiki, H., Kondo, S., & Takahashi, I. (1998). Direct power control of PWM converter without power-source voltage sensors. *IEEE Transactions on Industry Applications*, *34*(3), 473–479. doi:10.1109/28.673716

Ouannas, A., Azar, A. T., & Abu-Saris, R. (2016a). A new type of hybrid synchronization between arbitrary hyperchaotic maps. *International Journal of Machine Learning and Cybernetics*. Advance online publication. doi:10.100713042-016-0566-3

Ouannas, A., Azar, A. T., & Radwan, A. G. (2016b) On Inverse Problem of Generalized Synchronization Between Different Dimensional Integer-Order and Fractional-Order Chaotic Systems. *The 28th International Conference on Microelectronics, IEEE.* 10.1109/ICM.2016.7847942

Ouannas, A., Azar, A. T., & Vaidyanathan, S. (2017f). On A Simple Approach for Q-S Synchronization of Chaotic Dynamical Systems in Continuous-Time. *Int. J. Computing Science and Mathematics*, *8*(1), 20–27.

Ouannas, A., Azar, A. T., & Vaidyanathan, S. (2017g). New Hybrid Synchronization Schemes Based on Coexistence of Various Types of Synchronization Between Master-Slave Hyperchaotic Systems. *International Journal of Computer Applications in Technology*, *55*(2), 112–120. doi:10.1504/IJCAT.2017.082868

Ouannas, A., Azar, A. T., & Vaidyanathan, S. (2017i). A Robust Method for New Fractional Hybrid Chaos Synchronization. *Mathematical Methods in the Applied Sciences*, *40*(5), 1804–1812. doi:10.1002/mma.4099

Ouannas, A., Azar, A. T., & Ziar, T. (2017h). *On Inverse Full State Hybrid Function Projective Synchronization for Continuous-time Chaotic Dynamical Systems with Arbitrary Dimensions.* Differential Equations and Dynamical Systems., doi:10.100712591-017-0362-x

Ouannas, A., Azar, A. T., Ziar, T., & Radwan, A. G. (2017d). Study On Coexistence of Different Types of Synchronization Between Different dimensional Fractional Chaotic Systems. *Studies in Computational Intelligence*, *688*, 637–669. doi:10.1007/978-3-319-50249-6_22

Ouannas, A., Azar, A. T., Ziar, T., & Radwan, A. G. (2017e). Generalized Synchronization of Different Dimensional Integer-order and Fractional Order Chaotic Systems. *Studies in Computational Intelligence*, *688*, 671–697. doi:10.1007/978-3-319-50249-6_23

Ouannas, A., Azar, A. T., Ziar, T., & Vaidyanathan, S. (2017a). On New Fractional Inverse Matrix Projective Synchronization Schemes. *Studies in Computational Intelligence*, *688*, 497–524. doi:10.1007/978-3-319-50249-6_17

Ouannas, A., Azar, A. T., Ziar, T., & Vaidyanathan, S. (2017b). Fractional Inverse Generalized Chaos Synchronization Between Different Dimensional Systems. *Studies in Computational Intelligence*, *688*, 525–551. doi:10.1007/978-3-319-50249-6_18

Ouannas, A., Azar, A. T., Ziar, T., & Vaidyanathan, S. (2017c). A New Method To Synchronize Fractional Chaotic Systems With Different Dimensions. *Studies in Computational Intelligence*, *688*, 581–611. doi:10.1007/978-3-319-50249-6_20

Ouannas, A., Grassi, G., Azar, A. T., Radwan, A. G., Volos, C., Pham, V. T., Ziar, T., Kyprianidis, I. M., & Stouboulos, I. N. (2017j). Dead-Beat Synchronization Control in Discrete-Time Chaotic Systems. *The 6th International Conference on Modern Circuits and Systems Technologies (MOCAST)*. 10.1109/MOCAST.2017.7937628

Pandya, S., & Roy, R. (2015). Particle swarm optimization based optimal reactive power dispatch. *Proceeding of the IEEE International Conference on Electrical, Computer and Communication Technologies*, *1*(1), 1-5. 10.1109/ICECCT.2015.7225981

Reddy. (2014). Faster evolutionary algorithm based optimal power flow using incremental variables. *Electrical Power and Energy Systems, 54*(1), 198-210.

Reddy. (2017). Optimal Reactive Power Scheduling Using Cuckoo Search Algorithm. *International Journal of Electrical and Computer Engineering, 7*(5), 2349-2356.

Sanjuan, S. (2010). Voltage Oriented Control of Three-Phase Boost PWM Converters Design, simulation and implementation of a 3-phase boost battery Charger. Chalmers University of Technologie.

Singh, S., Azar, A. T., Ouannas, A., Zhu, Q., Zhang, W., & Na, J. (2017) Sliding Mode Control Technique for Multi-switching Synchronization of Chaotic Systems. *9th International Conference on Modelling, Identification and Control (ICMIC 2017)*. 10.1109/ICMIC.2017.8321579

Soliman, N. S., Said, L. A., Azar, A. T., Madian, A. H., Radwan, A. G., & Ouannas, A. (2017). Fractional Controllable Multi-Scroll V-Shape Attractor with Parameters Effect. *The 6th International Conference on Modern Circuits and Systems Technologies (MOCAST)*.

Song, Z., Tian, Y., Chen, W., Zou, Z., & Chen, Z. (2016). Predictive duty cycle control of three-phase active-front-end rectifier. *IEEE Transactions on Power Electronics*, *31*(1), 698–710. doi:10.1109/TPEL.2015.2398872

Subbaraj, P., & Rajnarayan, P. N. (2009). Optimal reactive power dispatch using self-adaptive real coded Genetic algorithm. *Electric Power Systems Research*, *79*(2), 374–38. doi:10.1016/j.epsr.2008.07.008

Tolba, M. F., AbdelAty, A. M., Soliman, N. S., Said, L. A., Madian, A. H., Azar, A. T., & Radwan, A. G. (2017). FPGA implementation of two fractional order chaotic systems. *International Journal of Electronics and Communications*, *28*, 162–172. doi:10.1016/j.aeue.2017.04.028

Vaidyanathan, S., & Azar, A. T. (2015a). Anti-Synchronization of Identical Chaotic Systems using Sliding Mode Control and an Application to Vaidyanathan-Madhavan Chaotic Systems. In Advances and Applications in Sliding Mode Control systems. Springer-Verlag GmbH. doi:10.1007/978-3-319-11173-5_19

Vaidyanathan, S., & Azar, A. T. (2015b). Hybrid Synchronization of Identical Chaotic Systems using Sliding Mode Control and an Application to Vaidyanathan Chaotic Systems. In Advances and Applications in Sliding Mode Control systems. Springer-Verlag GmbH. doi:10.1007/978-3-319-11173-5_20

Vaidyanathan, S., & Azar, A. T. (2015c). Analysis, Control and Synchronization of a Nine-Term 3-D Novel Chaotic System. In Chaos Modeling and Control Systems Design. Springer-Verlag GmbH. doi:10.1007/978-3-319-13132-0_1

Vaidyanathan, S., & Azar, A. T. (2015d). Analysis and Control of a 4-D Novel Hyperchaotic System. In Chaos Modeling and Control Systems Design. Springer-Verlag GmbH. doi:10.1007/978-3-319-13132-0_2

Vaidyanathan, S., & Azar, A. T. (2016a). Takagi-Sugeno Fuzzy Logic Controller for Liu-Chen Four-Scroll Chaotic System. *International Journal of Intelligent Engineering Informatics*, *4*(2), 135–150. doi:10.1504/IJIEI.2016.076699

Vaidyanathan, S., & Azar, A. T. (2016b). *Dynamic Analysis, Adaptive Feedback Control and Synchronization of an Eight-Term 3-D Novel Chaotic System with Three Quadratic Nonlinearities.* Springer-Verlag.

Vaidyanathan, S., & Azar, A. T. (2016c). *Qualitative Study and Adaptive Control of a Novel 4-D Hyperchaotic System with Three Quadratic Nonlinearities.* Springer-Verlag.

Vaidyanathan, S., & Azar, A. T. (2016d). *A Novel 4-D Four-Wing Chaotic System with Four Quadratic Nonlinearities and its Synchronization via Adaptive Control Method.* Springer-Verlag.

Vaidyanathan, S., & Azar, A. T. (2016e). *Adaptive Control and Synchronization of Halvorsen Circulant Chaotic Systems. Advances in Chaos Theory and Intelligent Control.* Springer-Verlag.

Vaidyanathan, S., & Azar, A. T. (2016f). *Adaptive Backstepping Control and Synchronization of a Novel 3-D Jerk System with an Exponential Nonlinearity.* Springer-Verlag.

Vaidyanathan, S., & Azar, A. T. (2016g). *Generalized Projective Synchronization of a Novel Hyperchaotic Four-Wing System via Adaptive Control Method.* Springer-Verlag.

Vaidyanathan, S., Azar, A. T., & Ouannas, A. (2017a). An Eight-Term 3-D Novel Chaotic System with Three Quadratic Nonlinearities, its Adaptive Feedback Control and Synchronization. *Studies in Computational Intelligence*, *688*, 719–746. doi:10.1007/978-3-319-50249-6_25

Vaidyanathan, S., Azar, A. T., & Ouannas, A. (2017c). Hyperchaos and Adaptive Control of a Novel Hyperchaotic System with Two Quadratic Nonlinearities. *Studies in Computational Intelligence*, *688*, 773–803. doi:10.1007/978-3-319-50249-6_27

Vaidyanathan, S., Azar, A. T., Rajagopal, K., & Alexander, P. (2015b). Design and SPICE implementation of a 12-term novel hyperchaotic system and its synchronization via active control (2015). *International Journal of Modelling Identification and Control*, *23*(3), 267–277. doi:10.1504/IJMIC.2015.069936

Vaidyanathan, S., Idowu, B. A., & Azar, A. T. (2015c). Backstepping Controller Design for the Global Chaos Synchronization of Sprott's Jerk Systems. In Chaos Modeling and Control Systems Design. Springer-Verlag GmbH. doi:10.1007/978-3-319-13132-0_3

Vaidyanathan, S., Sampath, S., & Azar, A. T. (2015a). Global chaos synchronisation of identical chaotic systems via novel sliding mode control method and its application to Zhu system. *International Journal of Modelling Identification and Control*, *23*(1), 92–100. doi:10.1504/IJMIC.2015.067495

Vaidyanathan, S., Zhu, Q., & Azar, A. T. (2017b). Adaptive Control of a Novel Nonlinear Double Convection Chaotic System. *Studies in Computational Intelligence*, *688*, 357–385. doi:10.1007/978-3-319-50249-6_12

Vazquez, J., & Salmeron, P. (2003). Active power filter control using neural network technologies. *IEE Proceedings. Electric Power Applications*, *150*(2), 139–145. doi:10.1049/ip-epa:20030009

Vazquez, S., Leon, J. I., Franquelo, L. G., Rodriguez, J., Young, H., Marquez, A., & Zanchetta, P. (2014). Model predictive control: A review of its applications in power electronics. *IEEE Industrial Electronics Magazine*, *8*(1), 16–31. doi:10.1109/MIE.2013.2290138

Vazquez, S., Sanchez, J., Carrasco, J., Leon, J., & Galvan, E. (2008). A model-based direct power control for three-phase power converters. *IEEE Transactions on Industrial Electronics*, *55*(4), 1647–11657. doi:10.1109/TIE.2008.917113

Wang, Z., Volos, C., Kingni, S. T., Azar, A. T., & Pham, V. T. (2017). Four-wing attractors in a novel chaotic system with hyperbolic sine nonlinearity. *Optik (Stuttgart)*, *131*, 1071–1078. doi:10.1016/j.ijleo.2016.12.016

Zhang, Y., Xie, W., Li, Z., & Zhang, Y. (2013). Model predictive direct power control of a PWM rectifier with duty cycle optimization. *IEEE Transactions on Power Electronics*, *28*(11), 5343–5351. doi:10.1109/TPEL.2013.2243846

Zhu, Q., & Azar, A. T. (2015). *Complex system modelling and control through intelligent soft computations*. Springer-Verlag.

Chapter 8
Operation and Startup of Three-Phase Grid-Connected PWM Inverter for an Experimental Test Bench With DSPACE Real-Time Implementation of PQ Control:
Active and Reactive Powers Control Strategy of the Grid Side Converter With Experimental Integration

Kamal Elyaalaoui
https://orcid.org/0000-0002-6755-3588
EMI, Mohammed V University in Rabat, Morocco

Moussa Labbadi
EMI, Mohammed V University in Rabat, Morocco

Khalid Chigane
EMI, Mohammed V University in Rabat, Morocco

Mohammed Ouassaid
EMI, Mohammed V University in Rabat, Morocco

Mohamed Cherkaoui
EMI, Mohammed V University in Rabat, Morocco

ABSTRACT

The main objective of this chapter is the experimental validation of active and reactive power control at the connection point for a three-phase grid connected inverter. It gives an overview on the adopted

DOI: 10.4018/978-1-7998-7447-8.ch008

vector control strategy, regulation of the angle of orientation of the blades (pitch control), synchronization grid side converter to the power network using phase closed loop (PLL). Once the experimental test bench is described, the authors devote a first part to the design of the block circuit diagram of the experimental platform and the control strategy implemented in the DSPace DS1104, and they suggest some steps to associate the inverter to the electrical network. Subsequently, they discuss the experimental results validating the proposed power control. The purpose of this experimental results is the DSPACE real-time implementation of PQ control using three-phase inverter and development of a startup algorithm of the experimental test bench.

INTRODUCTION

To overcome the problem of global warming, some countries have turned to new forms of so-called "renewable" energies such as hydro, solar and wind power. These renewable energies provide solutions to climate change ((GWEC), 2018), which ensures the liberalization of the electricity market and the development of decentralized production (production of electricity at the places of use themselves). They give rise, in the field of Electrical Engineering, to several new technical problems. These problems are caused by the impact of new types of energy source on the grid. One of these sources of energy is wind power, which represents a promising and growing technology ((GWEC), 2018).

It is non-polluting, inexhaustible energy (wind energy), reliable in isolated regions and can be complementary to traditional sectors. Its advantages have led to a gradual increase in the injection of wind energy into the electricity grid in many countries.

Most manufacturers build a large-scale wind turbine of megawatts. These wind turbines are based on three types of generators and large-scale back-to-back converters. The permanent magnet synchronous generator (PMSG) (without gearbox), the double feed induction generator (DFIG) and the squirrel cage induction generator (SCIG). The comparison between DFIG, PMSG and SCIG, shows that the SCIG has a simple construction, low cost and more robust in terms of the impact of fault clearance time and disconnection of the wind turbine from the power grid. These generators can be used in fixed speed wind turbine (Duong, et al., 2014) or variable speed wind turbine (Errami, et al., 2015; Elyaalaoui, et al., 2019; Martinez, et al., JUNE 2012; Folly & Sheetekela, 2009). Regarding fixed speed wind turbines, the generator is connected directly to the grid, which allows the operation for only a specific wind speed, which demand the use of variable speed wind turbine equipped with back-to-back converters to achieve this objective (Elyaalaoui, et al., 2014; Elyaalaoui, et al., 2016). The WT is connected into the network through filter and transformer (Elyaalaoui, et al., 2019).

The power produced by the SCIG and PMSG is completely injected into grid through the AC/DC/AC converter, the filter and the transformer. The DC/AC converter can be used for solar energy (Abedi, et al., 2020) and wind energy conversion system (Elyaalaoui, et al., 2020). The control type of DC/AC converter plays an important goal in power quality improvement (Elyaalaoui, et al., 2020; Gui, et al., 2018) and system robustness against grid faults (Elyaalaoui, et al., 2018; Xu, et al., 2007; Ouassaid, et al., 2016; Elyaalaoui, et al., 2021). The control of wind energy conversion system is divided into two parts: the generator side converter control and the grid side converter control (Ouassaid, et al., 2016). The MPPT algorithm is adopted to generate the torque reference in order to extract the maximum active power control from the wind. But the rotor flux control is developed to orient rotor flux on d-axis and ensure decoupling between flux and torque. Mechanical power is limited to nominal power by "pitch control" strategy. In the grid-side converter control system, the active and reactive power, injected to the grid, are regulated to satisfy the requirements imposed by the Transmission System Operator (TSO) and to maintain the DC bus voltage constant. Therefore, a sufficient amount of reactive power should be available to satisfy the new grid code recommendations and reactive power demand (Santos-Martin, et al., 2008). The new grid code is imposed for the WF to produce or consume the reactive power (Ghennam, et al., 2009).

The TSO impose for renewable power system to satisfy the demanded active and reactive power near the PCC because it causes additional losses in the transmitting lines of the power network and cannot be transmitted over long distances (Elyaalaoui, et al., 2019). To solve the problem of regulating reactive power at the point of common coupling (PCC), a PQ control method of the grid side converter must be developed to track its reference generated by the supervision system. This reactive power is also necessary for the LVRT (Low Voltage Ride-Through) control to satisfy the TSO's recommendations (Elyaalaoui, et al., 2019).

To achieve the converter control objectives a sophisticated hardware design and an adequate board should be used. With technological advances in the ðeld of microelectronics and boards design, new hardware design solutions, such as Digital Signal Processing (DSP) (Shireena, et al., 2007; N.Zerroug, et al., 2018), Digital Signal Processing and control engineering (DSPace) (Gui, et al., 2018; Elyaalaoui, et al., 2020; Djoudi, et al., 2020) and Field Programmable Gate Array (FPGA) (Pappu, et al., 2017; Sahu, et al., 2017) and its application Speciðc Integrated Circuit (ASIC) such as ZedBoard, Spartan (Ilyas, et al., 2020), Artix, Xilinx Spartan 3A DSP FPGA (Prathapa, et al., 2018; xilinx, 2021), are available and can be used as digital targets for the implementation of control algorithms.

For experimentation in control engineering, DSPace device has many advantages like easy implementation of algorithms using Matlab/Simulink software. The DSPace Board is an extend and improved Board family based on DSP to solve the design challenges of analog-to-digital converter. It is equipped with ControlDesk software for experiment, visualization and instrumentation of the electronic control unit development. DS1104 R&D board family offers a unique working environment from the start to the end of experiments.

It can be seen that all papers in the literature discuss the converter types and their applications with experimentation without a detailed startup procedure and test bench design. This chapter proposed the required algorithm for grid connected three-phase converter and the equipment protection. It presents the main parts of the system, three-phase grid connected inverter control, experimental validation of the power control for grid side converter. It focusses on the control of active and reactive powers and integration of inverter to power network as well as the design of the test bench to validate the active and reactive control system. The experimental results are carried out to validate the proposed control method and evaluate its performances as well as the detailed start-up of experimental test bench operation.

The converter control laws with the protection system have been implemented using a dSPACE DS1104 R&D board with TMS320F240 DSP. A DC source power is investigated in this experimental test bench, which is coupled to power network through a Semikron three-phase inverter.

The rest of this chapter is organized as follows, three-phase grid connected inverter control is presented in section 2 and its experimental results carried out using a designed test bench to validate the active and reactive power control in section 3. Finally, a conclusion describes the main contributions of this chapter in section 4.

THREE-PHASE GRID CONNECTED INVERTER CONTROL

The literature in the field of electrical engineering (embedded systems and power electronics) shows that the energy losses of the static converters are more important and the efficiency is the lower. In power electronics, the most used converter types are DC-DC, AC-DC and DC-AC converters.

DC-DC converter is divided to three categories, Boost converter, buck converter and Buck-Boost converter. Buck-Boost converter is a converter of DC-to-DC voltage (Spier, et al., August 2019; Li & Liu, April 2019). The magnitude of the output voltage is adjusted through the control of the duty cycle to ensure that the output voltage of the Buck-Boost converter is less than or greater than the input voltage. It is used mostly in power Electronics for Electric Vehicle (EV) Chargers and battery charging using photovoltaic system. Buck-Boost converter operates as Boost and Buck converter in same time (Samavatian & Radan, December 2014). Boost converter is mostly used to increase the output voltage for power supply (Veerachary, 2018; Rosas-Caro, et al., January 2010; GeorgeGkizas, February 2021), while the buck converter is employed for the output voltage decreasing (Bouchama, et al., October 2020; Zerroug, et al., March 2018).

AC-DC converters are divided into controlled rectifier and non-controlled rectifier with diode. It is employed for the DC load supply, DC motor drive and can be integrated in AC-DC-AC for the wind energy conversion system. The reference (Abdelsalam, et al., July 2015) combines AC-DC and boost converter to propose a three-phase ac–dc buck–boost converter.

The DC voltage can be converted into variable AC voltage using DC-AC inverter (Zhi, et al., 2009; Fekik, et al., 2021). This variation can be in the magnitude of voltage, frequency or phase shift. When the input of the inverter is a constant DC voltage source it is called voltage source inverter. There are two famous inverter categories, Stand-alone inverter and Grid-connected inverter.

Stand-alone inverters (or off-grid mode inverters) are employed to supply load with constant voltage magnitude and frequency without ability of synchronization and connection to utility grid (Arricibita, et al., December 2017; Sefa, et al., 2017).

Grid-connected inverter is the most commonly used type for renewable energy conversion (Chen, et al., Feb. 2018; Elyaalaoui, et al., 2020; Zhang, et al., 18 December 2018; Viola, 2018). These kinds of inverters can be connected to the AC grid because of the synchronization possibility, where synchronization process is a technique of making phase and nominal frequency identical of both AC source (inverter) and AC grid. It is adopted to inject the power into three-phase grid connected wind energy or solar power source (Afshari, et al., 31 July 2017). This kind of inverters is also adopted for AC motor drive such as induction motor, control of synchronous and induction generator and grid connected DC power source (Chinmaya & Singh, January 2019; Boulouiha, et al., 2015). The inverters are connected into grid through the filter (RL, LCL, LLCL) for power quality improvement and reduction of harmonics (Sanatkar-Chayjani & Monfared, 27 July 2016; He, et al., 20 May 2020; Ruan, et al., 2018; Zhang, et al., 18 December 2018; Vukosavic, 2018). It can be used also grid tie transformer to increase voltage and for galvanic isolation goal (Hasan, et al., 2018; Greco, et al., September 2018). From the aforementioned analysis we can conclude that the adequate converter for large-scale power conversion system is three-phase grid-connected inverter.

The grid-side inverter, presented in ***Figure 1***, has been controlled to ensure the current injection the into grid. A vector control using Park reference frame yields better control for grid current and synchronization of inverter voltage to grid using Phase Locked Loop (PLL). PLL is developed to orient the voltage vector on d-axis in order to decouple de d-axis and q-axis. The current components (i_d and i_q) are controlled to track the references ($i_{d\text{-}ref}$ and $i_{q\text{-}ref}$) using PI controllers.

Figure 1. Grid side part of the system studied containing the DC bus and the grid side inverter

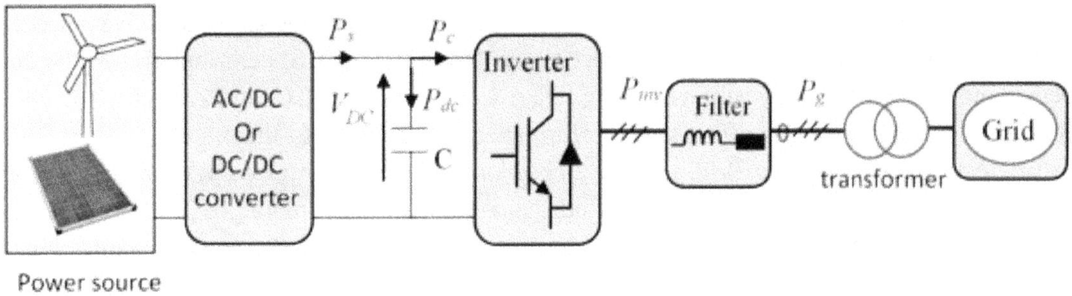

Control of Currents Injected into the Grid

The model of the RL filter expressed in is investigated for the internal loop control system design to generate the voltage control components. The block schemes of the current regulators design are presented in ***Figure 2***. it is assumed that the dynamics of the inverter (internal loop) is very fast than that of power control (external loop).

$$\begin{bmatrix} v_{md} \\ v_{mq} \end{bmatrix} = R_f \begin{bmatrix} i_{fd} \\ i_{fq} \end{bmatrix} + L_f \frac{d}{dt} \begin{bmatrix} i_{fd} \\ i_{fq} \end{bmatrix} + L_f \omega_{res} \frac{d}{dt} \begin{bmatrix} -i_{fq} \\ i_{fd} \end{bmatrix} + \begin{bmatrix} v_{gd} \\ v_{gq} \end{bmatrix}$$

where, v_{md} and v_{mq} are the components of the d and q axis of the voltage at the inverter output. v_{gd} and v_{gq} are the components of the d and q axis of the grid voltage. i_{fd} and i_{fq} are the d and q axis components of the current passing through the filter. ω_{res} is the grid frequency.

The control system has been developed to maintain the DC bus voltage constant and to obtain sinusoidal current with a constant frequency of 50 Hz and a phase shift identical to that of the power network. The block schemes of the current regulators design (it is assumed that the dynamics of the inverter is very fast than that of power control) are presented in Figure 2.

Figure 2.

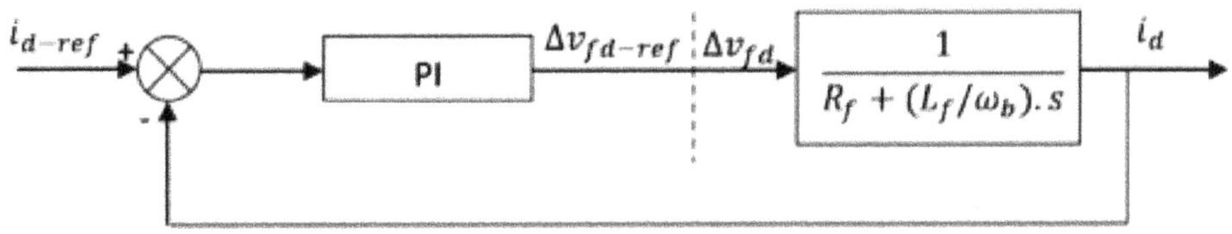

PLL Synchronization in Grid-Connected Inverter

Synchronization is very important issues in the fields of renewable power generation, active power filters and especially in the control of static converters connected to the grid. Synchronization to the desired frequency and phase is typically achieved using Phase-Locked-Loop (PLL) and zero-crossing detector.

The zero-crossing technique offers slow performance too sensitive to line notches because it inaccurately detecting zero crossings. While, Phase locked loop (PLL) is generally sufficient under balanced and unbalanced operating utility conditions (Rodríguez-Valdez & Kerkman, 2010; Ama, et al., 2013). It is one of the most used system for synchronization of the grid-connected converter applications (Rodríguez, et al., MARCH 2007) (Hadjidemetriou, et al., 2013).

For grid-connected inverters, phase-locked loop (PLL) is one of an indispensable part of system control to ensure the synchronization between the power grid and inverter. PLL circuits play an important role in the grid-connected inverter stability and it is designed to ensure a strong robustness under the unbalanced and distorted grid conditions (Chen & Gong, December 2019). Many references in the literature present PLL characteristics and its effect on the system stability (Ali, et al., July 2018) (Chen & Gong, December 2019) (Golestan, et al., Dec. 2017). Three-phase phase-locked loop synchronization algorithms for grid connected renewable energy systems is discussed in (Ali, et al., July 2018). These algorithms for synchronization techniques must guarantee a fast detection and accurately monitoring of phase, amplitude, and frequency of grid voltage (Allouche, et al., 2018). The response of these components is affected by the Low Pass Filter (LPF), which demand the use of an optimal Filter (LPF) for voltage measurement without attenuation and rejecting the required frequency components and degradation of PLL performance (Tan, et al., December 2020; Bo Jiang & Wang, 2014; Wang, et al., 2017; Khazraj, et al., June 2017).

In three phase systems, the PLL usually uses a rotating synchronous reference frame to generate the instantaneous phase angle. Voltage vector is oriented on d-axis (v_d) by controlling the quadrature axis (v_q) voltage to zero, to obtain $\omega_{nom} = \omega_{PLL} = 2.\pi.F_{nom}$, $v_d = V$ and $v_q = 0$. The orientation of voltage vector on d-axis is illustrated in ***Figure 3***. Where F_{nom} is nominal frequency.

Figure 3. Voltage vector is oriented on d-axis

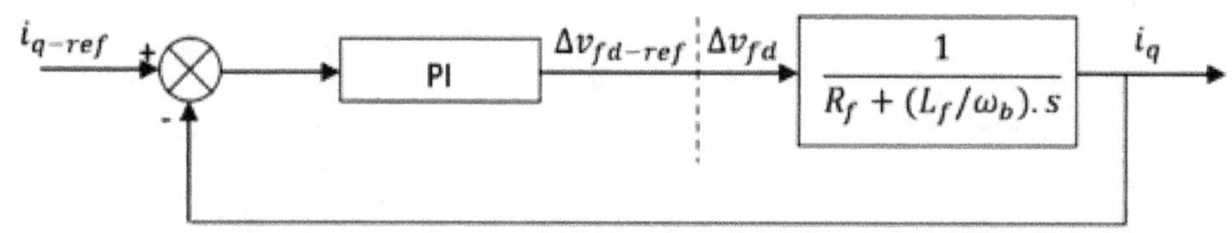

To inject a sinusoidal current with nominal frequency into the power network and to ensure the same phase shift between the inverter and grid voltage, a PLL has been developed. The principle of PLL in (d,q) synchronous reference frame system is illustrated in ***Figure 4***. For a balanced system the grid voltage is given by:

$$\begin{cases} V_a(\theta) = V\sin(\theta) \\ V_b(\theta) = V\sin(\theta - \dfrac{2\pi}{3}) \\ V_c(\theta) = V\sin(\theta + \dfrac{2\pi}{3}) \end{cases}$$

Figure 4. Principle of three-phase PLL in dq reference frame for voltage orientation

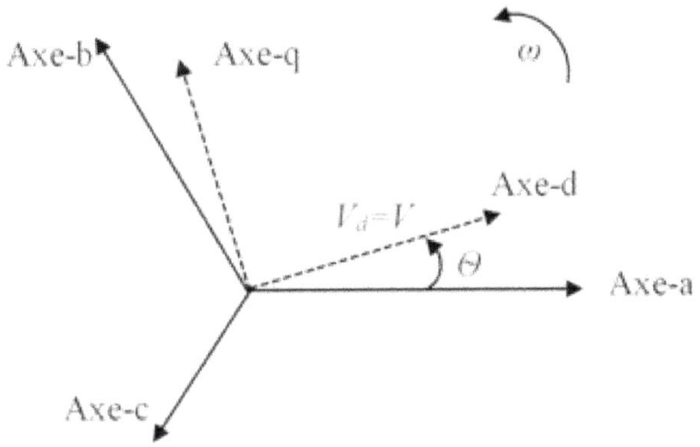

In the design of the closed loop system, it should be considered that the dynamic performance satisfies the fast tracking and good filtering characteristics. However, both requirements cannot be satisfied simultaneously. Therefore, a trade-off is required in designing the PLL and its implementation on DSPace board.

Active Power Control

The required power source for the inverter operation is ensured using the source side converter, which is controlled in most cases to capture the maximal power using Maximum Power Point Tracking (MPPT) algorithm. The reference of active power can be generated using the Maximum Power Point Tracking (MPPT) algorithms.

For the PV system the MPPT algorithm is developed based on Perturbation and Observation (P&O) Method to control (increase or decrease) power output under the conditions of temperature and irradiation condition. The maximum power point tracker, in this case, is an DC/DC converter that connected between the PV solar cells and the inverter (Abdalla, et al., 2019; Mathi & Chinthamalla, 2020; Mohapatra, et al., 26 Feb.-1 March 2019; Fekik, et al., 2020; Djoudi, et al., 2020).

For variable speed wind turbine, the maximum power point tracker is a system based on AC/DC converter (rectifier). This strategy aims to extract maximum power from the wind by controlling generator rotational speed to track the optimum value despite the fluctuation in wind velocity (Hossam H. H. Mousa & E.M.Mohamed, March 2021; Sitharthan, et al., 2020; Apata & Oyedokun, November 2020). In most, cases the MPPT is combined with the pitch control strategy to limit the extracted power at its nominal value and protect wind turbine during high wind speed (Dahbi, et al., 2016).

Active power of wind turbine and PV system can be injected into grid and controlled using the grid side converter (Badihi, et al., 2015; Li, et al., 2012). The control strategy of active power P_g to generate the d-axis current reference is shown in ***Figure 5***. The d-axis current reference ($i_{df\text{-}ref}$) is generated using the active power control loop.

The transfer function from the inverter output voltage to the filter current was approximated as:

$$\left[\frac{i_{fd}}{v_{md}}, \frac{i_{fq}}{v_{mq}}\right]^T = \left[\frac{1}{R_f + sL_f}, \frac{1}{R_s + sL_f}\right]^T$$

Similarly, the transfer function from the filter current to the reactive and active power was approximated as

$$\left[\frac{i_{fd}}{v_{md}}, \frac{i_{fq}}{v_{mq}}\right]^T = \left[\frac{1}{R_f + sL_f}, \frac{1}{R_s + sL_f}\right]^T$$

Figure 5. Block diagram scheme of d-axis grid current and active power control loops

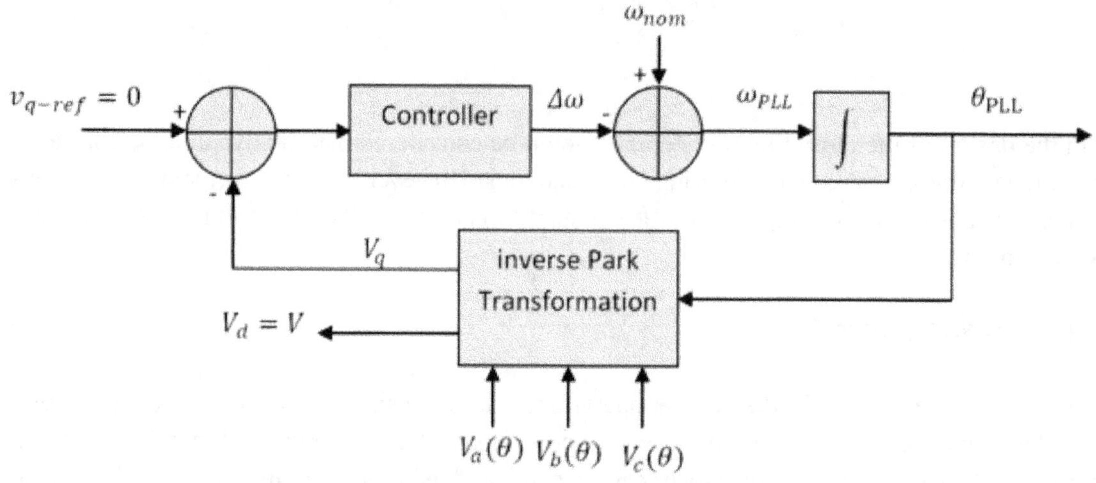

Open transfer function of the system investigated to compute the parameters of active power controllers is given by:

$$FTBO(s) = \frac{\frac{1}{R_f}}{1 + \frac{L_f}{R_f}s} = \frac{k}{1 + \tau s}$$

where gain $k = \frac{1}{R_f}$ and time constant $\tau = \frac{L_f}{R_f}$

The parameters for active power controller are given in appendix (*Table 4*)

Reactive Power Control

The recent research works on power system are oriented towards the control of active and reactive power to participate in voltage and frequency control for Wind Farms and PV system (Tang, et al., 2020) (Anay & Pollitt, 15 July 2020).

The reactive power can be generated to ensure the stability of the power system and participate in volt-var control and local reactive power for voltage regulation. It is required to maintain the grid voltage stability of the power network subjected the voltage drop (Qin, et al., 2011). Witch is required for smart distribution grids to ensure the power exchange across system bus (Khan, et al., March 2021).

In references (Leonardi & Ajjarapu, 2008; Zhai & Liu, 2014; Niazy, et al., 2010; Tian, et al., 2013; Bao, et al., 2003; Elyaalaoui, et al., 2019), the authors show that the reactive power reserve (RPR) is related to the margin of the voltage stability. In a heavily loaded system, a blackout and voltage collapse usually occur when the reactive power reserve of is insufficient. Furthermore, it is needed to support the voltage and transmit the active power demanded by the loads through the grid lines. Therefore, a sufficient amount of reactive power should be available to satisfy the new grid code recommendations and reactive power demand (Santos-Martin, et al., 2008; Elyaalaoui, et al., 2019). The new grid code is imposed for the WF to produce or consume the reactive power (Ghennam, et al., 2009).

The STATCOM proposed in (He, et al., June 2017) for power system under asymmetrical voltage conditions increases the system cost, which demand the use of another strategy to regulate the instantaneous active and reactive powers (Herrera, et al., Jan. 2008) (Hu, et al., 2011). To mitigate the grid faults the inverter should be control the inverter with decoupled active and reactive power control (Talha & Rahim, December 2020) (Zhai & Liu, 2014).

In this part of this work two reactive power regulation strategies are developed in order to generate the q-axis current reference. the first strategy is based on a simple technique where the reference current is calculated directly from the reactive power expression. The second is based on a regulation loop using PI controller. The d-axis current reference is generated through the active power regulation loop (*Figure 5*).

Filter Modeling and Reactive Power Expression

Using the expressions of active power and reactive power, the current references can be generated using the active power P_{ref} and reactive power Q_{ref} as given in (2.25).

$$\begin{cases} i_{d-res} = \dfrac{2}{3}\dfrac{P_{res}v_d + Q_{res}v_q}{v_d^2 + v_q^2} \\ i_{q-res} = \dfrac{2}{3}\dfrac{P_{res}v_q - Q_{res}v_d}{v_d^2 + v_q^2} \end{cases}$$

Adopting the orientation of the grid voltage on d-axis, new expression of filter model and powers are obtained:

$$\begin{bmatrix} v_{md} \\ v_{mq} \end{bmatrix} = R_f \begin{bmatrix} i_{fd} \\ i_{fq} \end{bmatrix} + L_f \frac{d}{dt} \begin{bmatrix} i_{fd} \\ i_{fq} \end{bmatrix} + L_f \omega_{res} \frac{d}{dt} \begin{bmatrix} -i_{fq} \\ i_{fd} \end{bmatrix} + \begin{bmatrix} V \\ 0 \end{bmatrix}$$

$$\begin{cases} P_{res} = \frac{3}{2} i_{fd} V \\ Q_{res} = -\frac{3}{2} i_{fq} V \end{cases}$$

The direct component of the current will be used to control the DC bus voltage. The quadrature component is used to control the reactive power. (Elyaalaoui, et al., 2015).

$$\text{where} \begin{cases} i_{fq\_\max} = \sqrt{i_{nom}^2 - i_{fd\_ref}^2} \\ i_{fq\_\min} = -\sqrt{i_{nom}^2 - i_{fd\_ref}^2} \end{cases}$$

Using the expression of reactive power ensure the reactive power control only at the filter output, not at the PCC, then the reactive power should be controlled using a closed loop control system (Elyaalaoui, et al., 2019).

PCC Reactive Power Control

To solve the problem of reactive power regulation at the PCC, a control strategy based on the PCC power regulation (Qpcc) has been proposed for reactive power to tracks its reference at PCC.

In this part of this section, the proposed reactive power control strategy is called reactive power PI control (RPPIC) (see *Figure 6*). In order to force the reactive power at PCC to follow its reference, the control loop must include the equivalent model of the transformer and transmission line. The reactive power regulation loop includes the transformer and the line (the system between bus 1 and PCC in ***Figure 6***).

Figure 6. Block diagram control loop of q-axis grid current and reactive power

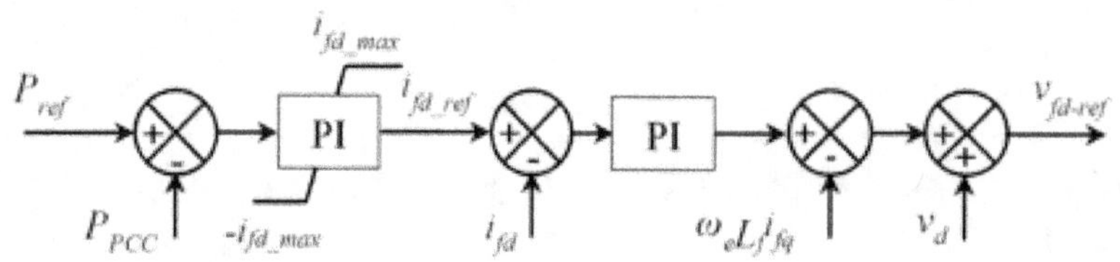

The model of the system connecting the inverter to the power network consists of filter, transformer and transmission line (An, et al., 2012; Li, et al., 2012). For the experimental test the transmission line is neglected compared to transformer. But it is considered for general study and in theorical development.

The mutual components of transformer, noted R_{trm}, L_{trm}, have large impedances, which are neglected in the control system. In this case the modeling expressions of the transformer and modeling of the line are given by:

Transformer model

$$\frac{L_{tr}}{\omega_b}\frac{di_{dl}}{dt} = v_{d2} - v_{d1} - R_{tr}i_{dl} + \omega_e L_{tr} i_{ql}$$

$$L_{tr}\frac{di_{ql}}{dt} = v_{q2} - v_{q1} - R_{tr}i_{ql} - \omega_e L_{tr} i_{dl}$$

The transmission line model is given by:

$$L_{ca}\frac{di_{dl}}{dt} = v_{dg} - v_{d2} - R_{ca}i_{dl} + \omega_e L_{ca} i_{ql}$$

$$\frac{L_{ca}}{\omega_b}\frac{di_{ql}}{dt} = v_{qg} - v_{q2} - R_{ca}i_{ql} - \omega_e L_{ca} i_{dl}$$

The transfer function for reactive power regulation is developed using -. From equation, it can be written:

$$v_{q2} = v_{qg} - \frac{L_{ca}}{\omega_b}\frac{di_{ql}}{dt} - R_{ca}i_{ql} - \omega_e L_{ca} i_{dl}$$

Substitution of v_{q2} in gives:

$$L_{tr}\frac{di_{ql}}{dt} + L_{ca}\frac{di_{ql}}{dt} = v_{qg} - v_{q1} - (R_{tr} + R_{ca})i_{ql} - \omega_e\left(L_{ca} + L_{tr}\right)i_{dl}$$

where $L = L_{tr} + L_{ca}$ and $R = R_{tr} + R_{ca}$.

After simplification, we obtain:

$$L\frac{di_{ql}}{dt} + Ri_{ql} = v_{qg} - v_{q1} - \omega_e L i_{dl}$$

This gives the transfer function for regulation of the q-axis current.

$$\frac{i_{ql}}{\Delta v_{q1}} = \frac{1}{Ls + R}$$

where $\Delta v_q = v_{qg} - v_{q2} - \omega_e L_{ca} i_{dl}$ is voltage across the system.

To calculate the parameters of the PI controller, the Open Loop Transfer Function (FT) is approximated, in unit notation (pu) and in Laplace transformation, by the following equation (An, et al., 2012):

$$TF(s) = \frac{i_{q1}}{Q_{pcc}} = \frac{1}{Ls + R}$$

Where Q_{pcc} the PCC reactive power, $L= L_{ca}+ L_{tr}$ and $R= R_{ca}+ R_{tr}$.

EXPERIMENTAL VALIDATION OF THE POWER CONTROL FOR GRID SIDE CONVERTER

The experimental results are carried out to validate the active and reactive power control for a 3 MW Semikron converter. The image and the electrical scheme of the experimental bench of three phase, connected to the RL filter, are shown in ***Figure 8*** and ***Figure 7***. The test bench is built to demonstrate the effectiveness of the proposed control and filters adopted. The Semikron inverter has been controlled using the DSPace DS1104 card. The environment development mainly consists of two software, MathWorks Matlab / Simulink and ControlDesk. The test bench parameters are given in annexes (Tables 1, 2 and 3).

Data acquisition is achieved using Simulink's ADC blocks connected to the measurement circuit via ADCs of DS1104 panel, in order to process and display in real time all the necessary electrical quantities (voltage, current, power, etc ...) for control system. A sample frequency of 10 kHz and an PWM switching frequency of 8 KHz are chosen.

During the experimental test, two three-phase transformers (127/220 Vac / 1.6 KVA) are used to increase the voltage of the three-phase source and a DC power sources of 3 kW (300 V / 10 A) are connected to the DC bus. The generated power harmonics, transient voltage and current are measured using the C.A 8331 Power Quality Analyzer and PC equipped with "Dataview" software for data analysis.

Description of the Test Bench

The image and the electrical scheme of the realized experimental platform are shown in ***Figure 8*** and ***Figure 7***.

Figure 7.

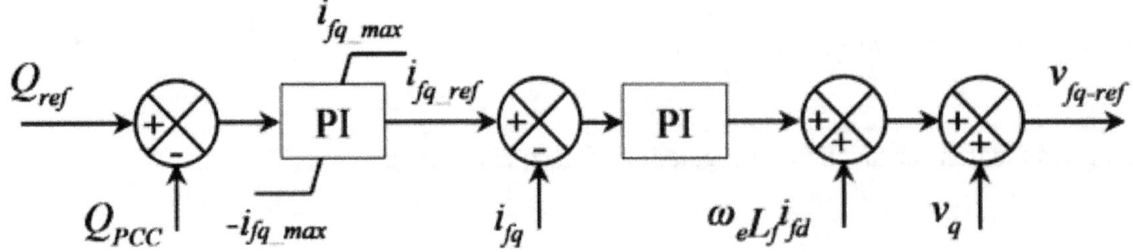

Figure 8.

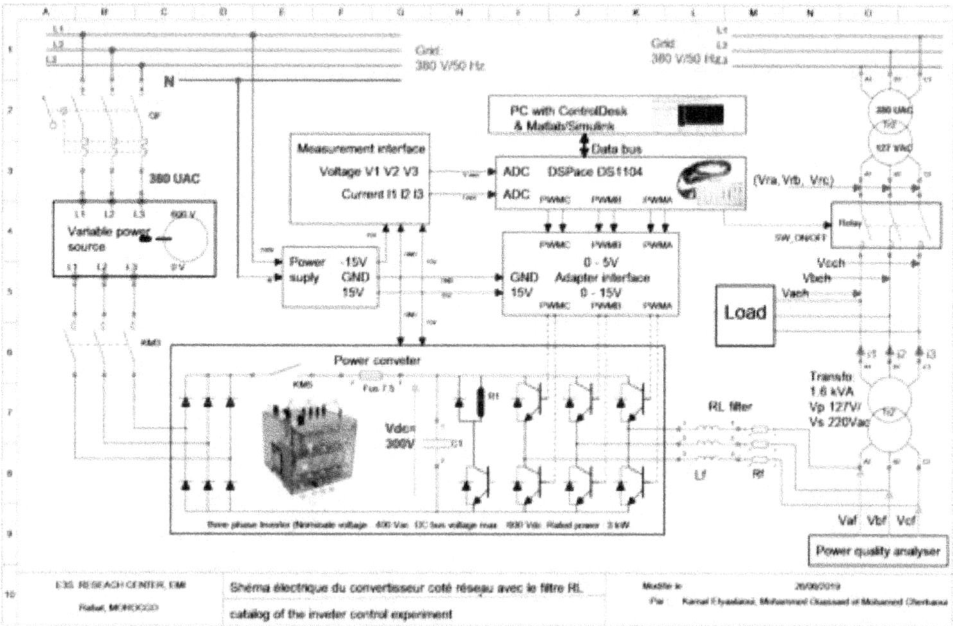

This test bench consists of the following elements:

1. DS1104 DSPace Board.
2. Computer equipped with Matlab / Simulink and ControlDesk.
3. Auxiliary structure of the system:
   a. Power supply circuit of + 15V and -15V
   b. Adaptation circuit (5V-15V)
   c. Circuits for currents and voltages measurement.
4. 4) Three-phase converter (SEMIKRON) consist of rectifier, DC bus and inverter (400 V- 30 A)).
5. 5) Three-phase alternative voltage source to supply the rectifier associated with the three-phase inverter.
6. 6) The power analyzer connected to PC equipped with "DataView" software.

The adaption circuit aims to transform the PWM signal generated from DSPace (0-5V) to the signals required to control the inverter IGBT (0-15V). It is realized using TLP250 optocoupler.

The data acquisition is achieved by the measurement circuits designed based on voltage transducer LV25-P and current transducer LA55-P. The output signal should be within the input range of the dSPACE board [-10V; + 10V]. For a balance three-phase system the three phases of current and voltage can be measured using just two sensors (two phases are measured while the third is estimated).

Experimental Results and Discussion

The purpose of this experimental results is the DSPACE Real-Time Implementation of PQ control using three-phase inverter and development of a startup algorithm of the experimental test bench. The Block circuit diagram of the experimental platform includes the data acquisition part, the control part and power part (inverter, transformars, …) is shown in ***Figure 9***.

Figure 9.

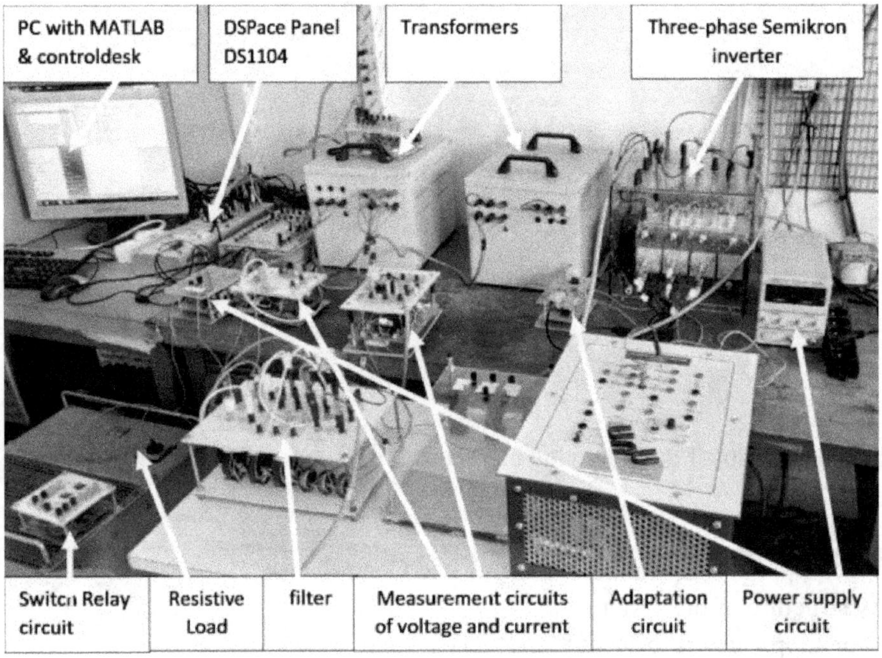

The control strategy implemented in the DSPace DS1104 is presented in, ***Figure 10***. During the first test of the new control types the test bench elements are monitoring and protected using Sequential state diagram illustrated in ***Figure 11***. It is also used to ensure synchronization of inverter connected to grid. In ***Figure 10*** the normalization strategy aims to convert the control voltage limited between the nominal values to the duty cycle required for generation of PWM signal.

Conditions of Synchronization

There are four essential conditions must be satisfied for synchronization of three-phase grid connected power inverter.

1. The inverter must be correctly connected to the grid (phase a, b and c of the inverter connected to phases a, b and c of the grid, respectively);
2. The voltage at the inverter side must be identical to that of the grid;
3. The frequency of inverter and grid must be the same;

Figure 10.

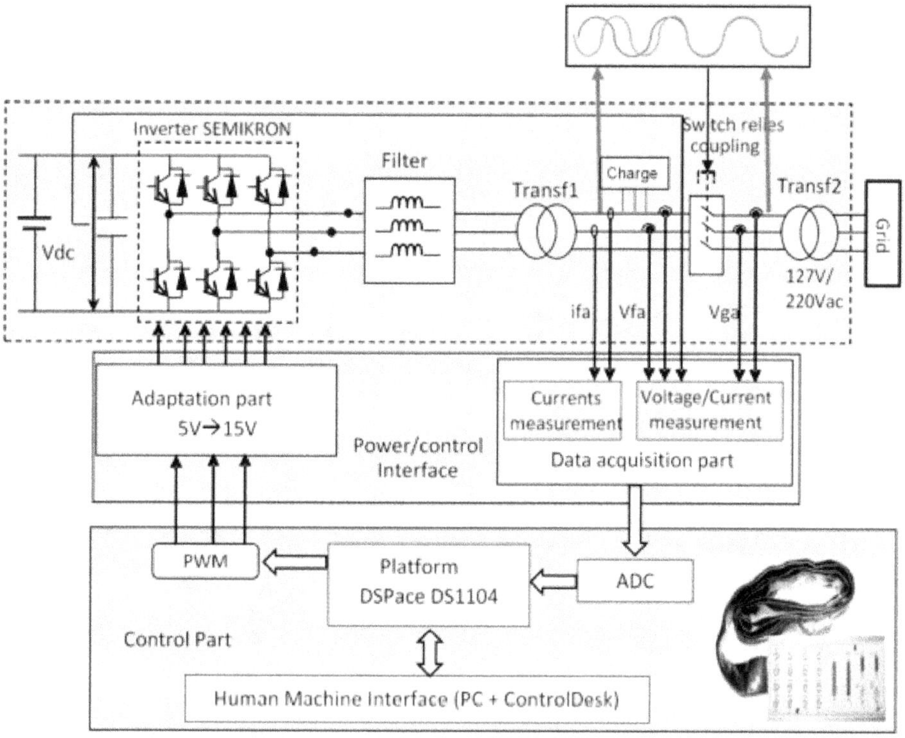

Figure 11.

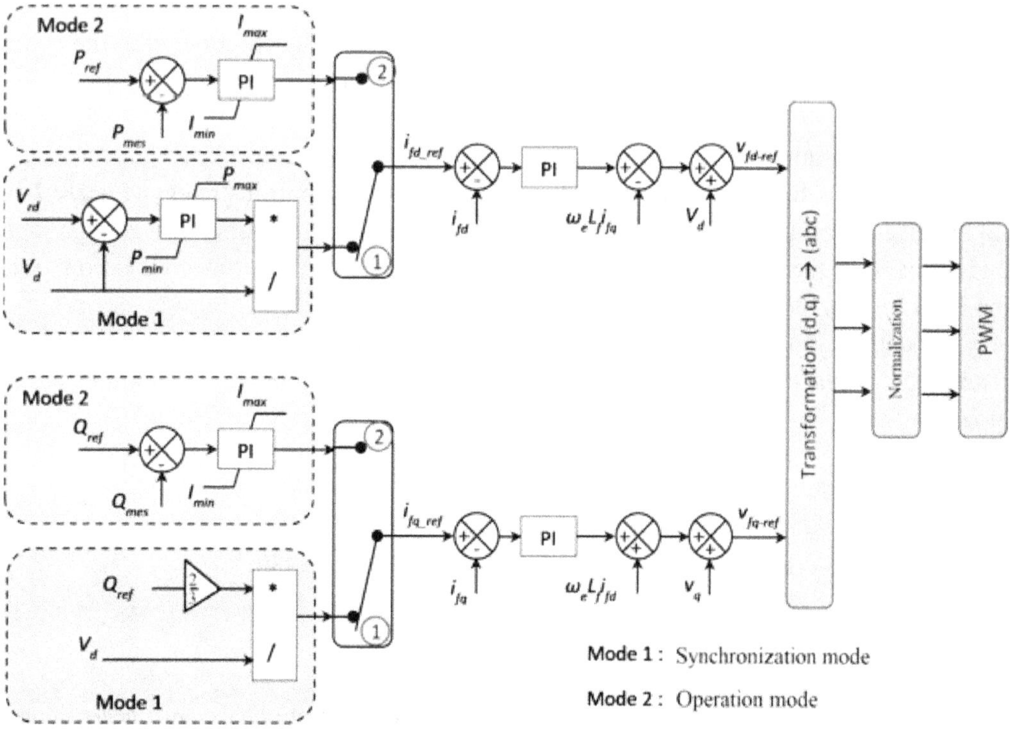

Figure 12.

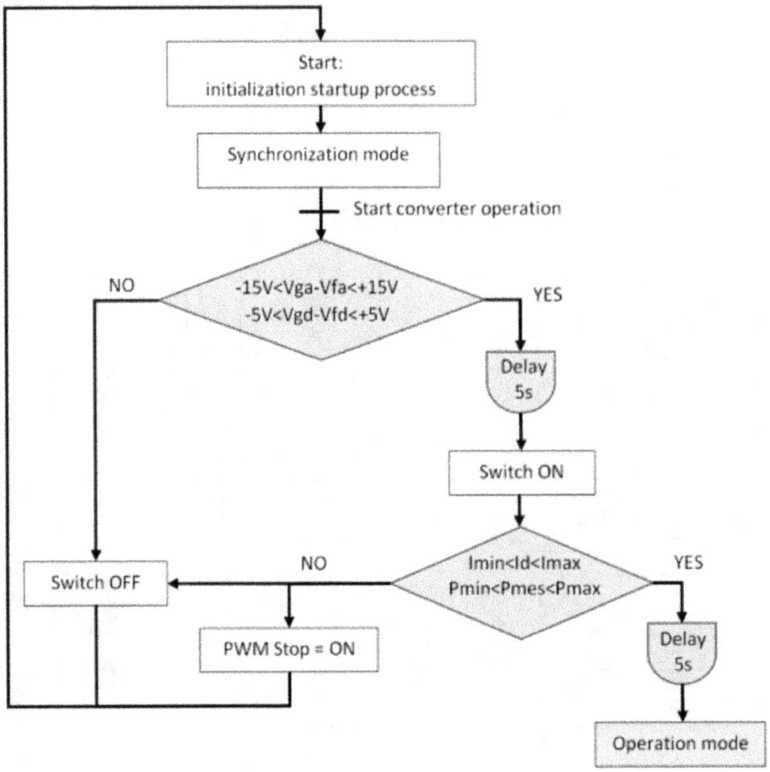

4. The inverter and grid must be having the same phase shift, connection between the inverter and grid should be done when the phase shift difference is minimal.

Sequential state diagram (***Figure 11***) for protection and synchronization of inverter connected to grid is converted to Simulink blocks of (***Figure 12***) associated with the control strategy (***Figure 10***) implemented in DSPace board. The digital phase-locked loop system, for synchronizing a voltage source DC/AC, is one of the main techniques of the three-phase grid connected power inverter used in renewable power source. The PLL for experimental test is presented in ***Figure 12***.

For experimental test of three-phase grid connected inverter, a circuit of 3 relays is used and controlled using a 5V signal generated by the DSPace board when the following conditions are satisfied (see ***Figure 11***):

- -5v<Vsd – Vrd<5v
- -15<Vsa-Vsa<15
- Vdc_min <Vdc< Vdc_max avec Vdc_max=305v et Vdc_min=230v
- Id>Imax=8 A
- P>Pmin=-10 W

Vsd – Vrd: difference between inverter and grid voltages to evaluate phase shift acceptable values.

Vsd – Vrd: difference between d-axis and q-axis of inverter and grid voltages to evaluate voltage magnitude.

Vdc_min and Vdc_max: Minimal and Maximal voltages of DC bus voltage.

Pmin=-10 W: Minimal active power to ensure only the injection into grid.

In order to maintain the grid connected inverter, verify of synchronization and ensure protection of test bench equipment during, sequential state diagram of ***Figure 11*** is converted to Simulink blocks of protection algorithm given in ***Figure 12***. Two Monostable and Bistable RS bascules are used (see ***Figure 12***). The bistable bascule is intended to memorize the fault state. While, the monostable bascule is used to "reset" the fault detection system (reset PWM stop and Grid_trip). The startup steps for three phase grid connected inverter is summarized in the steps:

- Initialized the protection algorithm (set reset to 1 (RESET=1)) to clear the fault memorization cases.
- Disactivate controller (set PI_ON to 0) in order to avoid the converter stuck in its startup before duty cycle generation;
- Activate PWM (put PWM-Stop = ON) to generate the PWM signal;
- Activate the PI regulators (set PI_ON to 1);
- Activate mode 1 (***Figure 10***), to increase gradually the voltage at the load terminal to reach that of the grid;
- Switching from mode 1 to mode 2 for closed loop control of power (***Figure 10***);
- Increase the power reference (P_{ref} < 1.5 KW) to inject power into the grid.

Figure 13.

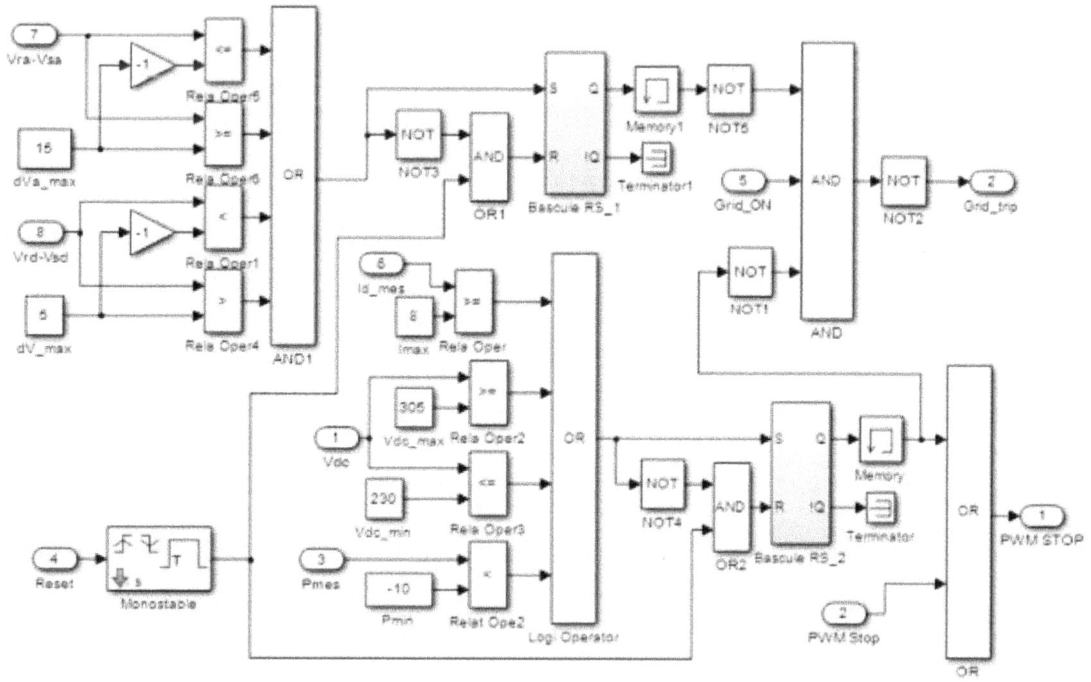

The experiments are performed through the step change of the active and reactive power references. The experimental results of the active and reactive power at the connection bus (PCC) are given in *Figure 14*. The connection bus is located between the transformers as shown in ***Figure** 7*. This bus is equivalent to the PCC for the wind turbine.

Figure 14a shows that the reactive power at connection bus follows its reference and reactive power can be consumed or injected. ***Figure 14b*** shows that the grid voltages at the connection point are sinusoidal. The d-axis voltage $V_d = V_{max}$ and q-axis voltage $V_q = 0$, which confirm the better operation of PLL and the orientation of grid voltage on the d-axis. ***Figure 14c*** and ***Figure 14***d show that the voltages and currents at the connection point are sinusoidal, the data displayed using Dataview software and power quality analyzer (CA.8331). The injected current is shown in ***Figure 14e***.

Figure 14f shows the voltage and current of the 1st phase for zero reactive power injection (Q_{PCC}=0 pu) which ensure the unity power factor. These results are obtained using power quality analyzer (CA.8331).

The THDs of voltage and current injected into grid are given in ***Figure 15***a and ***Figure 15***b, respectively. It can be seen that the harmonics at the output of the filter, using the vector control, are not enough higher (THDV = 5.5% f and THDI = 7.8% f), which confirm that the power quality is acceptable for a small-scale test bench (the transformer nominal power is 1.6 KVA).

From these results, it can be seen that the control strategy of grid connected three-phase inverter based on PI controller is more efficient and allows the injected active and reactive power at PCC to track their references.

Figure 14. Experimental results at the connection bus using vector control: a) active and reactive power responses, b) d-axis and q-axis voltages and c) three-phase voltages during power injection, d) three-phase currents during power injection e) d-axis and q-axes current injected into grid and f) the voltage and current of the 1st phase during unity power factor (zero reactive power)

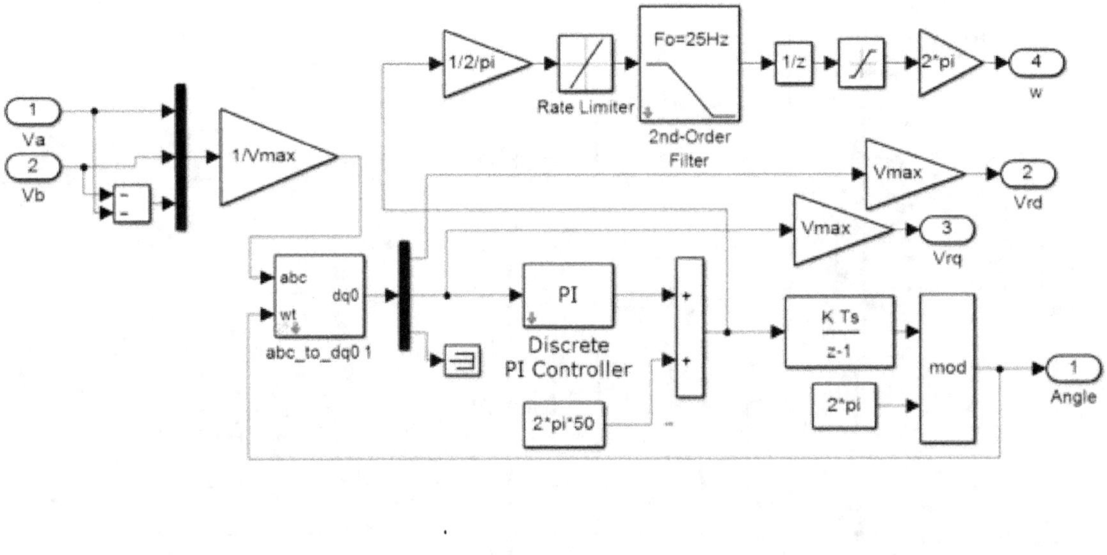

Figure 15. First-phase THDs, a) voltage and b) current injected into grid

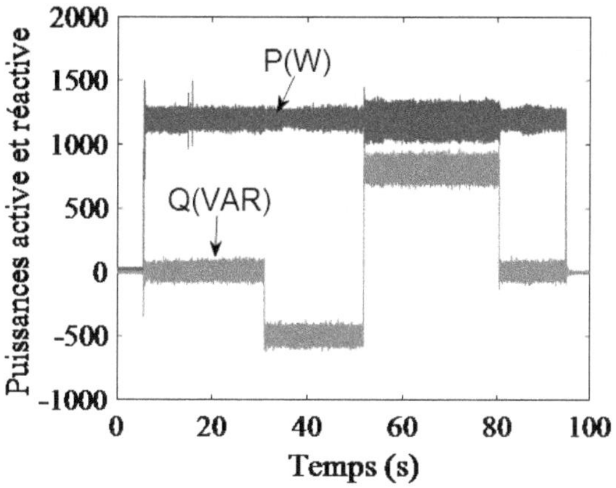

CONCLUSION

The increasing integration of renewable energy conversion systems connected to the electricity distribution network is not only due to technological innovations and lower costs of capturing kinetic energy from the wind and conversion of solar energy. But also due to significant research and development efforts in the field of power electronics. Indeed, the technical performance and reliability of the inverters used for connecting the conversion systems to the electricity distribution network are the important factors that can greatly vary the annual electrical energy production and therefore the economic benefits.

In this chapter, the realized experimental platform is described, then the procedure for integration of three-phase inverter into power network is detailed. The experimental results obtained confirm, on one hand, the DSPACE Real-Time implementation of PQ control using three-phase inverter and, on other hand, the effectiveness of developed start-up algorithm for the converter integration of experimental test bench.

The experimental results of the active and reactive power at the PCC show that the reactive power follows its reference at the PCC. From these figures, it can be seen that the control strategy based on the use of PI controller is more efficient and allows reactive power to the PCC to follow the reference and it can also ensure the unity power factor (Q_{PCC}=0 pu).

Finally, this test bench with its startup can be used also to validate the advanced control system and controllers for a three-phase grid connected inverters.

REFERENCES

Abdalla, O., Rezk, H., & Ahmed, E. M. (2019). Wind driven optimization algorithm based global MPPT for PV system under non-uniform solar irradiance. *Solar Energy*, *180*, 429–444.

Abdelsalam, I., Adam, G. P., Holliday, D., & Williams, B. W. (2015, July). Three-phase ac–dc buck–boost converter with a reduced number of switches. *IET Renewable Power Generation*, *9*(5), 494–502.

Abedi, A., Rezaie, B., Khosravi, A., & Shahabi, M. (2020). DC-bus Voltage Control based on Direct Lyapunov Method for a Converter-based Stand-alone DC Micro-grid. *Electric Power Systems Research, 187.*

Afshari, E. (2017, July 31). Control Strategy for Three-Phase Grid-Connected PV Inverters Enabling Current Limitation Under Unbalanced Faults. *IEEE Transactions on Industrial Electronics, 64*(11), 8908–8918.

Ali, Z. (2018, July). Three-phase phase-locked loop synchronization algorithms for grid-connected renewable energy systems: A review. *Renewable & Sustainable Energy Reviews, 90*, 434–452.

Allouche, A. (2018). A PLL based mechanical faults detection in PMSM at variable speed. *IFAC-PapersOnLine, 51*(24), 1445–1451.

Ama, N. 2013. PLL performance under frequency fluctuation-compliance with standards for distributed generation connected to the grid. *2013 IEEE PES Conference on Innovative Smart Grid Technologies (ISGT Latin America.*

An, H. (2012). Modeling and voltage-control of variable-speed SCAG-based wind farm. *Renewable Energy, 42*, 28–35.

Anay, K. L., & Pollitt, M. G. (2020, July 15). Reactive power procurement: A review of current trends. *Applied Energy, 270.*

Apata, O. & Oyedokun, D. T. O. (2020). An overview of control techniques for wind turbine systems. *Scientific African, 10.*

Arricibita, D., Sanchis, P., González, R., & Marroyo, L. (2017, December). Impedance Emulation for Voltage Harmonic Compensation in PWM Stand-Alone Inverters. *IEEE Transactions on Energy Conversion, 32*(4), 1335–1344.

Badihi, H., Zhang, Y., & Hong, H. (2015). Active power control design for supporting grid frequency regulation in wind farms. *Annual Reviews in Control, 40*, 70–81.

Bao, L., Huang, Z., & Xu, W. (2003). On-line voltage stability monitoring using Var reserves. *IEEE Transactions on Power Systems, 18*(4), 1461–1469.

Bo Jiang, T. X., & Wang, G. (2014). PLL low pass filter design considering unified specification constraints. *Analog Integrated Circuits and Signal Processing, 80*, 113–120.

Bouchama, Z., Khatir, A., Benaggoune, S., & Harmas, M. N. (2020, October). Design and experimental validation of an intelligent controller for DC–DC buck converters. *Journal of the Franklin Institute, 357*(15), 10353–10366.

Boulouiha, H. M. (2015). Direct torque control of multilevel SVPWM inverter in variable speed SCIG-based wind energy conversion system. *Renewable Energy, 80*, 140–152.

Chen, X. (2018, February). Injected Grid Current Quality Improvement for a Voltage-Controlled Grid-Connected Inverter. *IEEE Transactions on Power Electronics, 33*(2), 1247–1258.

Chen, Y. W. X., & Gong, Y. W. C. (2019, December). Analysis of frequency characteristics of phase-locked loops and effects on stability of three-phase grid-connected inverter. *International Journal of Electrical Power & Energy Systems, 113*, 652–663.

Chinmaya, K. A., & Singh, G. K. (2019, January). Modeling and experimental analysis of grid-connected six-phase induction generator for variable speed wind energy conversion system. *Electric Power Systems Research, 166*, 151–162.

Dahbi, A., Nait-Said, N., & Nait-Said, M.-S. (2016). A novel combined MPPT-pitch angle control for wide range variable speed wind turbine based on neural network. *International Journal of Hydrogen Energy, 41*(2), 9427–9442.

Djoudi, H. (2020). Simulation and Experimental Investigation into a Photovoltaic and Fuel Cell Hybrid Integration Power System for a Typical Small House Application. *Electric Power Components and Systems, 48*(14-15), 1598–1613.

Duong, M. Q., Grimaccia, F., Leva, S., & Mussetta, M. (2014). Pitch angle control using hybrid controller for all operating regions of SCIG wind turbine system. *Renewable Energy*, 1–7.

Elyaalaoui, K., Labbadi, M., Ouassaid, M., & Cherkaoui, M. (2021). Optimal Fractional Order Based on Fuzzy Control Scheme for Wind Farm Voltage Control with Reactive Power Compensation. *Mathematical Problems in Engineering, 2021.*

Elyaalaoui, K., Ouassaid, M., & Cherkaoui, M. (2014). Non-linear control and transient stability of SCAG connected to unbalanced AC network. Academic Press.

Elyaalaoui, K., Ouassaid, M., & Cherkaoui, M. (2015). Low Voltage Ride Through for a Wind Turbine Based SCAG. *Third International Conference on Renewable and Sustainable Energy (IRSEC'15)*, 1-6.

Elyaalaoui, K., Ouassaid, M. & Cherkaoui, M. (2016). *Supervision System of a Wind Farm Based On Squirrel Cage Asynchronous Generator.* Academic Press.

Elyaalaoui, K., Ouassaid, M., & Cherkaoui, M. (2018). Primary frequency control using hierarchal fuzzy logic for a wind farm based on SCIG connected to electrical network. *Sustainable Energy. Grids and Networks, 16*, 188–195.

Elyaalaoui, K., Ouassaid, M. & Cherkaoui, M. (2019). Dispatching and Control of Active and Reactive Power for a Wind Farm Considering Fault Ride-Through with a Proposed PI Reactive Power Control. *Renewable Energy Focus, 28*(00), 56-65.

Elyaalaoui, K., Ouassaid, M., & Cherkaoui, M. (2020). Improvement of THD performance of a robust controller for grid-side energy conversion system based on LCL filter without RC sensor. *International Journal of Electrical Power & Energy Systems, 121.*

Errami, Y., Ouassaid, M., & Maaroufi, M. (2015). A performance comparison of a nonlinear and a linear control for grid connected PMSG wind energy conversion system. *International Journal of Electrical Power & Energy Systems, 68*, 180–194.

Fekik, A. (2020). Maximum Power Extraction from a Photovoltaic Panel Connected to a Multi-cell Converter. *International Conference on Advanced Intelligent Systems and Informatics*, 873-882.

Fekik, A. (2021). *Direct power control of three-phase PWM-rectifier with backstepping control.* Advances in Nonlinear Dynamics and Chaos (ANDC).

Folly, K. A., & Sheetekela, S. P. N. (2009). Impact of Fixed and Variable Speed Wind Generators on the Transient Stability of a Power System Network. Academic Press.

Ghennam, T., Francois, B., & Berkouk, E. M. (2009). Local supervisory algorithm for reactive power dispatching of a wind farm. *13th European Conference on Power Electronics and Applications EPE'09*, 1-10.

Gkizas. (2021, February). Optimal robust control of a Cascaded DC–DC boost converter. *Control Engineering Practice, 107.*

Golestan, S., Guerrero, J. M., & Vasquez, J. C. (2017, December). Single-Phase PLLs: A Review of Recent Advances. *IEEE Transactions on Power Electronics, 32*(12), 9013–9030.

Greco, N. (2018, September). Scalable lumped models of integrated transformers for galvanically isolated power transfer systems. *Integration (Amsterdam), 63*, 323–331.

Gui, Y. (2018). Improved Direct Power Control for Grid-Connected Voltage Source Converters. *IEEE Transactions on Industrial Electronics.*

GWEC. (2018). *GWEC Global Wind Report_April 2018.* GWEC.

Hadjidemetriou, L., Kyriakides, E. & Blaabjerg, F. (2013). A New Hybrid PLL for Interconnecting Renewable Energy Systems to the Grid. *IEEE Transactions on Industry Applications, 49*(6), 2709-2719.

Hasan, M. M., Abu-Siada, A., Islam, S. M., & Dahidah, M. S. A. (2018). A New Cascaded Multilevel Inverter Topology with Galvanic Isolation. *IEEE Transactions on Industry Applications, 54*(4), 3463–3472.

He, S., Xiong, J., Wang, Z., & Lin, S. (2020, May 20). Robust AD for LCL-type grid-connected inverter with capacitor current quasi-integral feedback. *IET Power Electronics, 13*(7), 1332–1342.

He, Z., & … . (2017, June). Reactive Power Strategy of Cascaded Delta-Connected STATCOM Under Asymmetrical Voltage Conditions. *IEEE Journal of Emerging and Selected Topics in Power Electronics, 5*(2), 784–795.

Herrera, R. S., SalmerÓn, P., & Kim, H. (2008, January). Instantaneous Reactive Power Theory Applied to Active Power Filter Compensation: Different Approaches, Assessment, and Experimental Results. *IEEE Transactions on Industrial Electronics, 44*(1), 184–196.

Hossam, H. H., Mousa, A.-R. Y., Mohamed, E. M., & March, E. (2021). State of the art perturb and observe MPPT algorithms based wind energy conversion systems: A technology review. *International Journal of Electrical Power & Energy Systems*, 126.

Hu, J., Shang, L., & He, Y. (2011). Direct Active and Reactive Power Regulation of Grid-Connected DC/AC Converters Using Sliding Mode Control Approach. *IEEE Transactions on Power Electronics, 26*(1), 210–222.

Ilyas, A., Khan, M. R., & Ayyub, M. (2020). FPGA based real-time implementation of fuzzy logic controller for maximum power point tracking of solar photovoltaic system. *Optik (Stuttgart), 213.*

Khan, S., Zehetbauer, P., & Schwalbe, R. (2021, March). Evaluation of sensitivity based coordinated volt-var control and local reactive power for voltage regulation and power exchange across system boundaries in smart distribution networks. *Electric Power Systems Research, 192.*

Khazraj, H., da Silva, F. F., Bak, C. L., & Golestan, S. (2017, June). Analysis and design of notch filter-based PLLs for grid-connected applications. *Electric Power Systems Research, 147*, 62–69.

Leonardi, B., & Ajjarapu, V. (2008). *Investigation of various generator reactive power reserve (GRPR) definitions for online voltage stability/security assessment. In Power and Energy Society General Meeting-Conversion and Delivery of Electrical Energy in the 21st Century.* IEEE.

Li, H. (2012). Analysis and estimation of transient stability for a grid-connected wind turbine with induction generator. *Renewable Energy, 36*, 1469–1476.

Li, J., & Liu, J. (2019, April). A Novel Buck–Boost Converter With Low Electric Stress on Components. *IEEE Transactions on Industrial Electronics, 66*(4), 2703–2713.

Martinez, M. I., Tapia, G., Susperregui, A., & Camblong, H. (2012, June). Sliding-Mode Control for DFIG Rotor- and Grid-Side Converters Under Unbalanced and Harmonically Distorted Grid Voltage. *IEEE Transactions on Energy Conversion, 27*(2), 328–339.

Mathi, D. K., & Chinthamalla, R. (2020). Enhanced leader adaptive velocity particle swarm optimisation based global maximum power point tracking technique for a PV string under partially shaded conditions. *IET Renewable Power Generation, 14*(2), 243–253.

Mohapatra, A., Nayak, B., & Saiprakash, C. (2019). Adaptive Perturb & Observe MPPT for PV System with Experimental Validation. *IEEE International Conference on Sustainable Energy Technologies and Systems (ICSETS).*

Niazy, I. 2010. Participation in reactive power market considering generator aging. *IPEC, 2010 Conference Proceedings*, 1067-1072.

Ouassaid, M., Elyaalaoui, K., & Cherkaoui, M. (2016). Sliding mode control of induction generator wind turbine connected to the grid. In Advances and Applications in Nonlinear Control Systems. Springer.

Pappu, H., Kumar, G. S., & Manala, S. (2017). *An FPGA based New Single Phase Five Level Inverter.* Academic Press.

Prathapa, Anandhia, & Sivakumaran. (2018). Xilinx Spartan 3A DSP FPGA based DC Voltage Regulators for PV Systems. *Materials Today Proceedings, 5*(1), 1348-1358.

Qin, W., Wang, P., Han, X. & Du, X. (2011). Reactive Power Aspects in Reliability Assessment of Power Systems. *IEEE Transactions on Power Systems, 26*(1).

Rodríguez, P. (2007, March). Decoupled Double Synchronous Reference Frame PLL for Power Converters Control. *IEEE Transactions on Power Electronics, 22*(2), 584–592.

Rodríguez-Valdez, C. D., & Kerkman, R. J. (2010). Phase Locked Loop for unbalanced utility conditions. *2010 Twenty-Fifth Annual IEEE Applied Power Electronics Conference and Exposition (APEC).*

Rosas-Caro, J., Ramirez, J., Peng, F., & Valderrabano, A. (2010, January). A DC–DC multilevel boost converter. *IET Power Electronics, 3*(1), 129–137.

Ruan, X. (2018). Control Techniques for LCL-Type Grid-Connected Inverters. Springer.

Sahu, N., Londhe, N. D., & Kshirsagar, G. (2017). FPGA applications in inverter and converter circuits: A review on technology, benefits and challenges. *2017 International Conference on Innovations in Information, Embedded and Communication Systems (ICIIECS).*

Samavatian, V., & Radan, A. (2014, December). A novel low-ripple interleaved buck–boost converter with high efficiency and low oscillation for fuel-cell applications. *International Journal of Electrical Power & Energy Systems*, *63*, 446–454.

Sanatkar-Chayjani, M., & Monfared, M. (2016, July 27). Design of LCL and LLCL filters for single-phase grid connected converters. *IET Power Electronics*, *9*(9), 1971–1978.

Santos-Martin, D., Arnaltes, S., & Amenedo, J. R. (2008). Reactive power capability of doubly fed asynchronous generators. *Electric Power Systems Research*, *78*(111), 1837–1840.

Sefa, I. (2017). Three-phase three-level inverter with reduced number of switches for stand-alone PV systems. *2017 IEEE 6th International Conference on Renewable Energy Research and Applications (ICRERA).*

Shireena, W., Vanapallia, S., & Nene, H. (2007). DSP based inverter control for alternate energy systems. *Journal of Power Sources*, *166*(2), 445–449.

Sitharthan, R., Karthikeyan, M., Sundar, D. S., & Rajasekaran, S. (2020). Adaptive hybrid intelligent MPPT controller to approximate effectual wind speed and optimal rotor speed of variable speed wind turbine. *ISA Transactions*, *96*, 479–489. PMID:31202532

Spier, D. W., Oggier, G. G., & Silva, S. A. O. d. (2019, August). Dynamic modeling and analysis of the bidirectional DC-DC boost-buck converter for renewable energy applications. *Sustainable Energy Technologies and Assessments*, *34*, 133–145.

Talha, M., & Rahim, S. N. A. R. (2020, December). PV inverter with decoupled active and reactive power control to mitigate grid faults. *Renewable Energy*, *162*, 877–892.

Tan, L. (2020, December). A novel phase-locked loop applied on frequency fluctuation of EAST power supply. *Energy Reports*, *6*(9), 452–461.

Tang, R., Luo, B., Deng, X., & Shen, Y. (2020). Research on Reactive Power and Voltage Control for Wind Farm Based on coordinate control of DFIGs and SVG. *Procedia Computer Science*, *175*, 460–467.

Tian, J., Su, C., & Chen, Z. (2013). Reactive power capability of the wind turbine with Doubly Fed Induction Generator. *Industrial Electronics Society, IECON 2013-39th Annual Conference of the IEEE*, 5312-5317.

Veerachary, M. (2018). Third-order boost converter. *IET Power Electronics*, *11*(3), 566–575.

Viola, F. (2018). Experimental Evaluation of the Performance of a Three-Phase Five-Level Cascaded H-Bridge Inverter by Means FPGA-Based Control Board for Grid Connected Applications. *Energies*, *11*(12).

Vukosavic, S. N. (2018). *Grid-Side Converters Control and Design: Interfacing Between the AC Grid and Renewable Power Sources*. Academic Press.

Wang, D., Lu, K., Rasmussen, P. O. & Yang, Z. (2017). Comparative study of low-pass filter and phase-locked loop type speed filters for sensorless control of AC drives. *CES Transactions on Electrical Machines and Systems, 1*(2), 207 - 215.

Xilinx. (2021). *XA Spartan-3A DSP FPGA Family.* https://www.xilinx.com/products/silicon-devices/fpga/xa-spartan-3a-dsp.html

Xu, L., Zhi, D., & Yao, L. (2007). Direct Power Control of Grid Connected Voltage Source Converters. *IEEE Power Engineering Society General Meeting.*

Zerroug, N. (2018). DSP-based implementation of fast terminal synergetic control for a DC–DC Buck converter. *Journal of the Franklin Institute*, *355*(5), 2329–2343.

Zerroug, N. (2018, March). DSP-based implementation of fast terminal synergetic control for a DC–DC Buck converter. *Journal of the Franklin Institute*, *355*(5), 2329–2343.

Zhai, J., & Liu, H. (2014). Reactive power control strategy of DFIG wind farms for regulating voltage of power grid. PES General Meeting Conference & Exposition, 2014 IEEE, 1-5.

Zhang, X. (2018, December 18). Direct grid-side current model predictive control for grid-connected inverter with LCL filter. *IET Power Electronics*, *11*(15), 2450–2460.

Zhi, D., Xu, L., & Williams, B. W. (2009). Improved Direct Power Control of Grid-Connected DC/AC Converters. *IEEE Transactions on Power Electronics*, *24*(5), 1280–1292.

APPENDIX

Base values of the test bench are: $\omega_b=2\pi F_b$; $F_b=50$ Hz; $V_b=380$V, $P_b=1.5.10^3$ W, $Z_b=V_b^2/P_b$, $L_b=Z_b/\omega_b$. the parameters of transformers, DC bus and filter are presented in tables 1, 2 and 3.

Table 1. Transformer parameters

Symbol	*Components*	*Value*
Ptr	Rated power	1600 VA
Imax	Nominal current	8 A
V_1	Primary voltage	127 (V)
V_2	Secondary voltage	220 (V)
R_{tr1}	Primary resistance	1.6 (Ω)
L_{tr1}	Primary inductance	7800 (mH)
R_{tr2}	Secondary Resistance	0.6 (Ω)
L_{tr2}	Primary inductance	1400 (mH)

Table 2. DC bus parameters

Symbol	*Components*	*Value*
Vdc	DC bus voltage	300 V
Cf	DC bus capacitor	2200 mF

Table 3. RL filter parameters

Symbol	*Components*	*Value*
Lf	Inductance	3 mH
Rf	Resistance	2 Ω

Table 4. Parameters of the system controllers for experimental test

System Controller	*Parameter*	*Quantity*	*Value*
Current	Kp_i	Proportional gain	1
	Ki_i	Integral gain	641
PLL	Kp_PLL	Proportional gain	225
	Ki_PLL	Integral gain	21
Voltage	Kp_v	Proportional gain	0.0278
	Ki_v	Integral gain	1.3889
Active power	Kp_p	Proportional gain	2.7778e-4
	Ki_p	Integral gain	0.1389
Reactive power	Kp_p	Proportional gain	2.7778e-4
	Ki_p	Integral gain	0.1389

Chapter 9
Multilevel Inverter for Hybrid Fuel Cell/PV Energy Conversion System

Arezki Fekik
https://orcid.org/0000-0001-5053-1764
Akli Mohand Oulhadj University, Bouira, Algeria

Mohamed Lamine Hamida
Electrical Engineering Advanced Technology Laboratory (LATAGE), Mouloud Mammeri University, Algeria

Hakim Denoun
Electrical Engineering Advanced Technology Laboratory (LATAGE), Mouloud Mammeri University, Algeria

Ahmad Taher Azar
https://orcid.org/0000-0002-7869-6373
Prince Sultan University, Riyadh, Saudi Arabia

Nashwa Ahmad Kamal
Faculty of Engineering, Cairo University, Egypt

Sundarapandian Vaidyanathan
Vel Tech University, India

Amar Bousbaine
University of Derby, UK

Nacereddine Benamrouche
Electrical Engineering Technology Laboratory (LATAGE), Mouloud Mammeri University, Algeria

ABSTRACT

Power converters assume a significant part in fuel cell power generation systems and solar power conversion systems which are an alternative to fossil fuel production systems. There is therefore a demand for high quality power conditioning used in PEMFC systems and photovoltaic panels. This chapter proposes a hybrid electric power (FC/PV) production strategy with the use of converter topology as the power interface and also introduces a three-level inverter topology for different operating levels. The converter increases the input voltage to the rated voltage and turns into a DC bus; the multi-level inverter converts the voltage to AC and supplies AC loads. This chapter develops a hybrid electric power generation strategy, which can produce output with positive and zero sequences. Integrating the three-stage inverter into the hybrid renewable energy (FC/PV) production system allows for near sinusoidal current with low THD. The topology of hybrid energy production using the multi-level converter is tested on Matlab.

DOI: 10.4018/978-1-7998-7447-8.ch009

INTRODUCTION

The primary energy source is referred to as fossil fuel. The combustion of fossil fuels produces a large amount of ozone-depleting pollutants, including carbon dioxide, which is the primary cause of rising global temperatures and climate change, both of which are causing significant environmental harm. Because of the limited storage capacity of nonrenewable petroleum-based oil, the use of alternative resources such as PV cells, biomass, wind energy, geothermal, and so on has grown dramatically in the recent years. Because greenhouse gases are not emitted when renewable energy sources are used, the likelihood of the greenhouse effect can be reduced and the use of non-renewable energy can be effectively reduced. Photovoltaic technology is one of the most rapidly evolving technologies due to its abundant supply of solar illumination and lack of negative environmental effects. Significant research progress has been made on the energy generated by photovoltaic systems as a result of the market introduction of power electronic devices (Shalaby et al., 2021; Ghoudelbourk et al., 2016, 2020, 2021; Kamal et al., 2020; Ammar et al., 2019a; Kamal & Ibrahim, 2018; Fekik et al., 2015a,b, 2016, 2018, 2019a,b; Hamida et al., 2017; Amara et al., 2018; Amara et al., 2019a; Amara et al., 2019b; Fekik et al., 2018a,b,c,d,e, 2020a,b,c,d; Denoun et al., 2018; Ben Smida et al., 2018; Hamida et al., 2018, 2019a,b,c).

Photovoltaic cells have many advantages, including being noiseless, pollution-free, and low maintenance. Photovoltaic (PV) technology is now being used in a variety of systems, which can be classified into two types: (1) autonomous PV systems and (2) network-connected PV systems. Autonomous photovoltaic systems are used in conjunction with a battery bank for energy storage at distant intervals where the association is lavish (Dong et al., 2020; Strzelecki, 2008). On the other hand, photovoltaic panels are connected to the power network for grid-connected photovoltaic systems without using the battery bank to the extent that open PV power is redirected to the energy grid. Fuel Cells are another source of renewable energy that has piqued the interest of researchers in recent years. Fuel cells are used to generate electricity by converting hydrogen energy (Edwards et al., 2008; Keshavarzzadeh, Ahmadi & Safaei, 2019; Kumar et al., 2019).

Fuel cells are classified into several types based on their electrolytes. The following are frequently used: (1) lower ambient temperature; (2) lower operating pressure to improve safety; and (3) higher convertibility ratio (Bocci et al., 2014; Cipriani et al., 2014). Because they produce only a low output voltage, fuel cells have a high construction cost if they are required for high power applications. The voltage produced in most renewable energy production systems is continuous, and it is generally converted to alternating voltage (AC) using an electronic power converter with pulse width modulation (PWM). The performance of the system is determined by the control technique of a PWM inverter.

There are various energy management configurations for the interface of renewable energy systems to a source or a hybrid source connected to the grid (Olatomiwa et al., 2016; Etxeberria et al., 2010; Krishna et al., 2015; Krishnamurthy et al., 2019; Murty & Kumar, 2020). The inverter is powered by a hybrid renewable energy source and is connected to the grid or connected to a load. The inverter is controlled by the injection of a sinusoidal alternating current that is synchronized with the supply voltage to achieve very efficient energy production.

By using an adaptive control strategy, the electrical interface between the hybrid system and the grid could become increasingly efficient (Liu et al., 2019; Ye& Li, 2020; Parida & Subudhi, 2020; Mariano-Hernández et al., 2020; Lü et al., 2020; Li et al., 2020). The control structure of a voltage source inverter (VSI) typically includes an internal current loop that influences overall system performance. Many current control techniques, such as hysteresis current control, predictive control, dead-band control, and voltage control with PWM modulator, have been proposed in the control literature.

In this chapter, we present a hybrid electric power (FC / PV) production strategy that employs converter topology as the power interface and a three-level inverter topology for various operating levels. The converter raises the input voltage to the rated voltage and converts it to a DC bus. In contrast, the multi-level inverter converts the voltage to alternating current and supplies alternating current loads. The proposed method of controlling the fuel cell output voltage is based on the design of a fractional controller. The proposed controller regulates this DC-DC converter's output voltage to maintain the voltage delivered by the cell at levels suitable for powering the inverter at three voltage levels. The efficiency of the closed-loop system for photovoltaic (PV) system control, consisting of maximum power extraction followed by a DC-DC converter, is demonstrated, as in the work of Fekik et al. (2020a). The P&O (Perturb and Observe) technique is used as a DC converter controller to ensure that the photovoltaic panels are operating at maximum power. This chapter develops a hybrid electric power generation strategy that can generate output with positive and negative sequences. When a three-stage inverter is integrated into a Hybrid Renewable Energy (FC / PV) production system, it produces near sinusoidal current with low total harmonic distortion (THD). The topology of hybrid energy production using the multi-level converter is tested using MATLAB software.

This chapter of the book is divided into six sections. Section 2 provides a mathematical model of Proton-Exchange Membrane Fuel Cells (PEMFC). Section 3 describes in detail the mathematical model of a PV system. The modeling of the three-level inverter is presented in Section 4. Section 5 presents the results and discussions of MATLAB simulations. Finally, in Section 6, closing remarks are provided.

FUEL CELL SYSTEM

Modeling of Proton-Exchange Membrane Fuel Cells (PEMFC)

As shown in Fig. 1, a simple Proton-Exchange Membrane Fuel Cells (PEMFC) device is a two-electrode divided by a solid membrane serving as an electrolyte (anode and cathode). The anode network draws hydrogen fuel through protons, which effectively pass from the membrane to the cathode. Electrons are captured by an internal circuit that binds the two electrodes by the electric current. In this study, the oxidant (air) travels through a similar network of channels to the cathode, which combines oxygen and electrons in the inner circuit to produce gas (Fekik, 2020a).

Through hydrogen gas ionization, electrons are liberated in the anode electrolyte, and H+ ions are produced (Derbeli et al., 2017; Khan & Mathew, 2019).

$$2H_2 \Leftrightarrow 4H^+ + 4e^-, \tag{1}$$

Water is produced by the interaction of oxygen and electrode electrons, as well as electrode H+ ions from electrolytes, and is given by the equation.

$$O_2 + 4H^+ + 4e^- \Leftrightarrow 2H_2O \tag{2}$$

Figure 1. Operating principle of proton-exchange membrane fuel cells (PEMFC) device

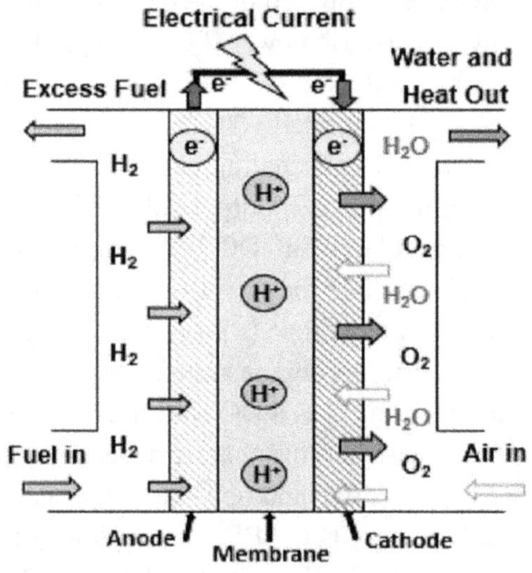

The general cell reaction equation is as follows:

$$2H_2 + O_2 \Leftrightarrow 2H_2O \tag{3}$$

The mathematical model employs a number of parameters. To achieve the best simulation performance, the system parameters must be interpreted. At this point, the output voltage of a single cell should be set, which can be defined as described in the papers (Derbeli et al., 2017; Khan & Mathew, 2019; Fekik et al., 2020).

$$V_{Cell} = E - \eta_{act} - \eta_{ohm} - \eta_{dif} \tag{4}$$

For N cells put in series, creating a stack, the voltage, V_{stack}, can be determined by:

$$V_{Stack} = N_{Cell} * V_{Cell} \tag{5}$$

The electrical power generated by the cell and delivered to the load can be provided by:

$$P_{FC} = V_{FC} * I \tag{6}$$

Using the PEMFC equations given in (Derbeli et al., 2017), V_{FC} is cell voltage for all operating requirements, and the PEMFC system can be modeled as shown in Fig. 2.

DC-DC Converter

Voltage-boosting DC-DC converters generally boost power converters with low voltage input and output at higher voltage (Azizkandi et al., 2019; Hassan et al., 2020). An ideal DC / DC converter circuit is given in Fig. 3.

The duty cycle regulates the mathematical relationship between input and output voltages, as shown by the equation below (Derbeli et al., 2017):

$$V_{Out} = \left(\frac{1}{1-\alpha}\right) * V_{In} \tag{7}$$

with $V_{in}=V_{STACK}$

Figure 3 depicts the DC-DC converter circuit. An inductor, a transistor, a diode, a capacitor filter, and a resistor are all part of the DC-DC converter circuit. It is powered by a DC voltage generator, in this case a PEMFC generator.

When the diode is turned off and the transistor is turned on, the differential equations describing the current dynamic state of the i_L inductor and the dynamic state of the V_{out} output voltage are as follows:

$$\begin{aligned} \frac{di_L}{dt} &= \frac{1}{L} V_{stack} \\ \frac{dV_{Out}}{dt} &= -\frac{1}{C} i_{Out} \end{aligned} \tag{8}$$

When the transistor is turned off and the diode is turned on, the following differential equations characterize the current dynamic state of the i_L inductor and the dynamic state of the output voltage V_{out}:

$$\begin{aligned} \frac{di_L}{dt} &= \frac{1}{L}(V_{stack} - V_{Out}) \\ \frac{dV_{Out}}{dt} &= \frac{1}{C}(i_L - i_{Out}) \end{aligned} \tag{9}$$

Fractional Order PI Controller Design of Fuel Cell

Recent decades have witnessed many important developments related to the design of nonlinear systems for many practical applications. Several inspiring approaches have been proposed, such as optimal control, nonlinear feedback control, adaptive control, sliding mode control, nonlinear dynamics, chaos control, chaos synchronization control, fuzzy logic control, fuzzy adaptive control, fractional order control, and robust control and their integrations (Kazim et al., 2021; Ajeil et al., 2020; Abdul-Adheem et al., 2020a,b; Najm et al., 2020; Liu et al., 2020; Humaidi et al., 2020; Elkholy et al., 2020a,b; Mohamed et al., 2020; Ibrahim et al., 2020; Sayed et al., 2020; Barakat et al., 2020; Azar & Vaidyanathan, 2015a,b,c, 2016; Pilla et al., 2020, 2019; Sallam et al., 2020; Ammar et al., 2018; Abdelmalek et al., 2018; Gorripotu et al.,

Figure 2. Schematic model of PEMFC system

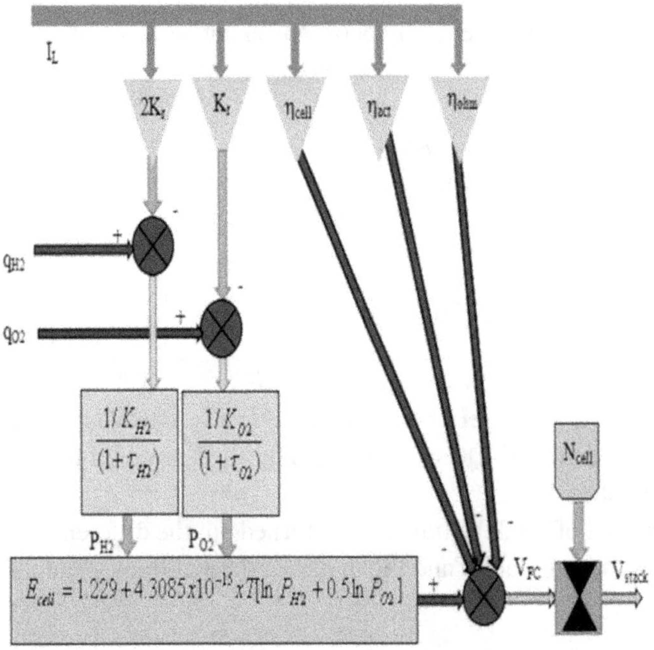

2021; Meghni et al, 2017a,b; Mekki et al., 2015; Vaidyanathan & Azar, 2015a,b,c,d, 2016a,b,c,d,e,f,g, 2017a,b,c; Azar & Serrano, 2014, 2015a,b,c,d, 2016a,b, 2017, 2020; Grassi et al., 2017; Ouannas et al., 2016a, Singh et al., 2017; Vaidyanathan et al, 2017a,b,c, 2015a,b,c, 2019; Wang et al., 2017; Giove et al., 2013).

Fractional order attracted the interest of the community in the physical systems which its applications are progressively extending (Khennaoui et al., 2020a,b; Kammogne et al. 2020; Alain et al., 2020, 2019, 2018a,b; Ammar et al., 2020a,b, 2019a,b; Ammar & Azar, 2020; Ibraheem et al., 2020; Kammogne et al., 2020; Ouannas et al., 2021, 2020a,b,c,d,e, 2019a,b, 2017a,b,c,d,e,f,g, 2016b; Bouakrif et al., 2019; Djeddi et al. 2019; Radwan et al., 2017, 2018; Tolba et al. 2017; Mittal et al. 2021; Singh & Azar 2020; Singh et al., 2018; Khan et al. 2020a,b; Azar et al., 2020, 2018a,b,c; Azar & Serrano 2018, 2019; Gorripotu et al., 2019; Kumar et al. 2018; AbdelAty et al., 2018; Pham et al., 2018, 2017a,b; Shukla et al., 2018; Khettab et al. 2018; Meghni et al. 2017c; Soliman et al. 2017).

Recently, fractional order controllers have been widely used in several articles for these benefits, namely the PID controller's stable efficiency. Podlubny (1994) employs the fractional PID corrector in a fractional control system. Following that, many engineering researchers created fractional PID correctors using various tuning designs and methods (Shah & Agashe, 2016). Fractional order controllers accept traditional PID extensions and are less vulnerable to improvements to core issues in corrector configuration. The fractional PID controller, as opposed to the standard PID controller, has five parameters to be tuned. Thus, the fractional PID controller consists of two extra degrees of freedom when compared to conventional PID controller. The controller displays the transfer function as shown below.

Figure 3. Schematic model of DC-DC converter

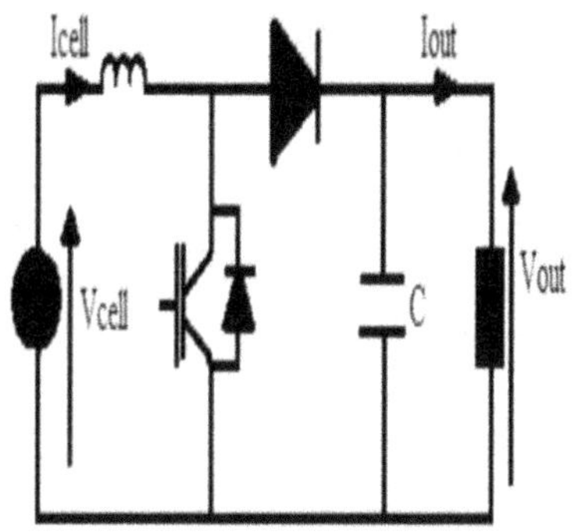

$$G_c(s) = k_p + \frac{k_i}{S^{\lambda}} + S^{\mu} \tag{10}$$

Where λ and μ are the orders of fractional order integration and differentiation, respectively. The type equation shown below can be used to control system feedback (Song et al., 2019; Wu et al., 2019).

$$u(t) - k_p * e(t) + k_i * j^{-\lambda} * e(t) + k_d * D^{\mu} * e(t) \tag{11}$$

Fractional PI controllers are used to stabilize and improve efficiencies in processes such as PEMFC output voltage reaction. Figure 4 depicts the closed loop control system.

In the FOPI regulator, the I operator is typically of fractional orders. Controlling the fractional integrator operator adds one degree of freedom to the k_p and k_i variables. The fractional integrator is represented by this variable. This additional vector independence is provided by the FOPI controller, which has more applications than the Integer order PI controller. The regulation of FOPI signals can be written as (Fekik et al., 2020a):

In the time domain:

$$u(t) = k_p * e(t) + k_i * j^{-\lambda} * e(t) \tag{12}$$

with $e(t) = V_{ref} - V_{load}$

Here D is differential operator. The transfer function is displayed as:

$$C(s) = k_p + \frac{k_i}{S^{\lambda}} \tag{13}$$

The current I_{ref} reference is based on the fractional operation of PI on the V_{load} output voltage and V_{ref}, as shown in Fig. 4, where $I_{ref} \in I$, k_p, and k_i are actual constants that exceed zero and must be chosen carefully because the internal loop requires the first and second I_{ref} derivatives.

PV SYSTEM

PV Modeling

Figure 5 depicts the equivalent circuit of a PV system, which consists of a photovoltaic current source, one diode, a shunt resistor, and a series resistor (Lalouni et al.,2009; Li et al., 2009; Chaibi et al., 2019; Ciocia et al., 2020). The five-parameter model shown in Figure 5(a) can be simplified to that shown in Figure 5(b) (b). In general, the PV cell model looks like this:

$$I_{pv} = I_{ph} - I_d$$
$$I_d = I_0[e^{(\frac{V_{pv} - R_s I}{V_t} - 1)} - 1] \quad (14)$$
$$I_{ph} = \frac{G}{G_{ref}}[I_{phref} + \lambda_t (T_c - T_{cref})]$$

Here, I_{ph} is the light current, Ipv is the load current, I0 is the saturation current, V_{pv} is the PV output voltage, Rs is the PV series resistance (Ω), V_t is the thermal voltage, G is the irradiation, T_c is the temperature of the cell in Kelvin, and I is the temperature coefficient. When sunlight strikes the surface of the solar panel, the current source generates a photocurrent, and the diode represents the solar cell's p-n transition area. Table 1 shows the specifications of a typical PV module. A MATLAB/Simulink package is used to implement the specifications. Figure 3 depicts the PV power produced for various solar radiation data. Two pairs of ten PV panels are connected in parallel to produce the required power, along with a dc/dc boost converter connected to a dc bus.

The output power of PV module can be calculated by Equation (15):

$$P = V_{pv} * I_{pv} = N_a V_{Cell} N_p I_{Cell} \quad (15)$$

MPPT Algorithm

The primary goal of the MPPT control is to extract and track the maximum power point from the PV module. Taking weather conditions into account, the maximum power point corresponding to the best operating point is obtained (Salas et al.,2006; Fadhel et al., 2020; Abe et al., 2020). The PV module's power fluctuates in response to radiation and temperature. Figure 6 depicts the typical MPPT of a PV system.

MPPT control strategies play an important role in PV systems because they increase the output power of a PV system, thereby increasing the array's efficiency. There are various methods for tracking the most extreme power point. The traditional technique is the Perturbation and Observation (P&O) strategy. Because its algorithm is simple to implement, the P&O perturbation and observation technique is widely used (Mohapatra et al. 2019; Jana et al., 2020; Abdel-Salam et al., 2020). This operation is carried out by disrupting the system by increasing or decreasing the operating voltage of the module and then observing the effect on the module's generated power.

Figure 4. Closed loop of PEMFC output voltage

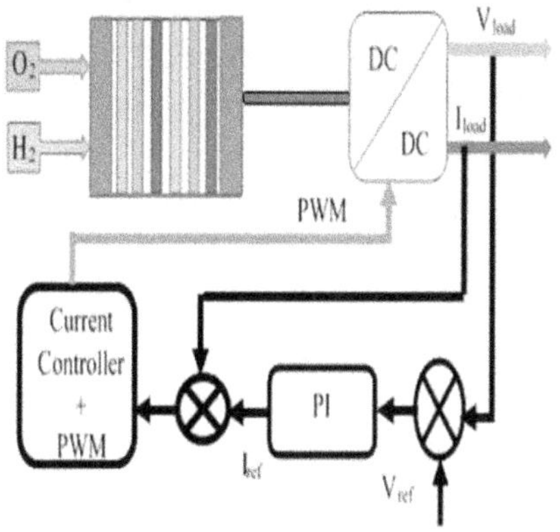

The flowchart of the P&O algorithm as it should be applied to the control microprocessor is shown in Figure 7. According to Figure 7, the calculation of the current output power P(k) is dependent on the current (I) and voltage measurements (V). The value P(k) is then compared to the prior value P(k -1) produced by the previous measurement. The disturbance will continue in the same direction if the output power is increased. If the power is reduced from one of the most recent measurements, the output voltage disturbance will revert to the previous cycle's direction. The operating voltage is disrupted at each cycle of the MPPT using this technique.

Once achieved, the MPPT will oscillate around the ideal voltage V of operation. This results in a power loss that is determined by the step width of a single C_p perturbation. When C_p is large, the MPPT algorithm responds quickly to sudden changes in operating conditions, but losses are increased under stable or slightly changing conditions. If C_p is very small, losses will be reduced under stable or changing conditions, but the system will no longer be able to keep up with rapid temperature or insulation changes (Ali et al, 2020).

THREE LEVEL INVERTERS

The NPC inverter's basic concept is to obtain an output voltage at three levels by superimposing two elementary switches, each supplied by a separate DC voltage source (Zhang et al., 2019).

In this study, a three-phase three-level inverter with three arms and two DC voltage sources was used. The first of these is a photovoltaic panel, and the second is a fuel cell. Each arm has four switches connected in series, as well as two middle diodes. As shown in Fig. 8, each switch is made up of a transistor and a diode mounted head to tail.

Modeling of Three Level Inverter

Each arm's midpoint is linked to a continuous supply. We obtain two secondary sources of direct voltage (E / 2) by using a capacitive voltage divider formed by filtering capacitors C_1 and C_2 of the same capacity (Xu et al., 2019; Wang et al., 2019; Yüksek, & Arifolu, 2020). Being linked to each other at a neutral point o. To avoid load imbalance, the latter are identical ($C_1 = C_2$, i.e. $U_{c1} = U_{c2}$).

1. Different Configurations of the Inverter

To explain the different sequences of the inverter function, we consider the potential conditions of the first leg switches (see Figure 9).

The blocking voltage of a DC-link voltage is equal to the three-level inverter, whereas the maximum dc-link stress of the two-stage inverter is one-fifth.

2. Complementary Command

The inverter leg complementary control is described as follows:

$$\begin{cases} B_{K1} = \overline{B_{K4}} \\ B_{K2} = \overline{B_{K3}} \end{cases} \quad (16)$$

Denoun et al. (2015) demonstrated that the command given by connection (1) best gives the three levels Uc, O, - Uc. Table 2 displays the switch status as well as the corresponding inverter output voltage. Vk (k= a, b, c) are the potentials of the leg.

3. *Model Control of the Three-Level Inverter – Connection Functions*

The status of the switches is determined by these functions. They are equal to 1 if the button is turned on and 0 if the switch is turned off. The following relationship (Berkouk 1995 ; Denoun et al., 2015) defines it:

$$\begin{aligned} F_{K1} &= 1 - F_{K4} \\ F_{K2} &= 1 - F_{K3} \end{aligned} \quad (17)$$

With (K=1,2, 3)

As a consequence, the voltages VAO, VBO, VCO are expressed as follows:

$$\begin{aligned} V_{AO} &= F_{11} * F_{12} * U_{C1} - F_{13} * F_{14} * U_{C2} \\ V_{BO} &= F_{21} * F_{22} * U_{C1} - F_{23} * F_{24} * U_{C2} \\ V_{CO} &= F_{31} * F_{32} * U_{C1} - F_{33} * F_{34} * U_{C2} \end{aligned} \quad (18)$$

However, the inverter's output phase voltage can be calculated using equation (18) as follows:

Figure 5. Equivalent circuit of PV cell, (a) five parameters model; (b) simplified four parameters model

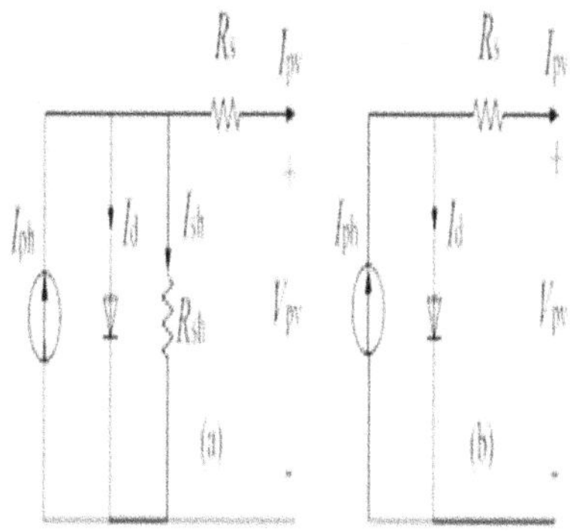

$$V_A = \frac{2V_{AO} - V_{BO} - V_{CO}}{3}$$
$$V_B = \frac{2V_{BO} - V_{CO} - V_{AO}}{3} \quad (19)$$
$$V_C = \frac{2V_{CO} - V_{AO} - V_{BO}}{3}$$

These equations can be rewritten using the voltage branches provided.

$$\begin{bmatrix} V_A \\ V_B \\ V_C \end{bmatrix} = \frac{1}{3}\begin{bmatrix} 2 & -1 & -1 \\ -1 & 2 & -1 \\ -1 & -1 & 2 \end{bmatrix}\begin{bmatrix} F_{11}F_{12} \\ F_{21}F_{22} \\ F_{31}F_{32} \end{bmatrix} U_{C1} - \begin{bmatrix} F_{13}F_{13} \\ F_{23}F_{24} \\ F_{33}F_{34} \end{bmatrix} U_{C2} \quad (20)$$

Space Vector Modulation

1. *Switching States*

As shown in Figure 10, the three-level three-phase inverter has 27 switching states. These are the states that connect to the various DC-link points. When a load is connected to the output of one of these states, the inverter generates the output phase voltage (Tolunay 2012; Zhang et al., 2020; Babaie et al., 2020). Table 3 displays all 27 switching states, 19 voltage vectors, and the generated output voltage.

For the three-level inverter, as with the two-level inverter, the reference vector is generated using three voltage vectors. For the three-level converter, each sector is also divided into four regions, further specifying the output (Tolunay, 2012). The magnitude of the voltage vectors can be defined as follows (Kocalmis & Sunter, 2006):

- Zero Voltage Vectors (ZVV): V=0

Table 1. PV specification of module data

Parameters	Values
Max power	60.45 W
Voltage at Max Power Point V_{mpp}	19.5 V
Current at Max Power Point I_{mpp}	3.1 A
Open Circuit Voltage V_{oc}	21.1 V
Short Circuit Current I_{sc}	3.8 A
Total Number of Cells per Module (N_{cell})	36
Number of Cells in Parallel	2
Number of Cells in Series	3

- Small Voltage Vectors (SVV): V1,4,7,10,13,16
- Medium Voltage Vectors (MVV): V3,6,9,12,15,18
- Large Voltage Vectors (LVV): V2,5,8,11,14,17

Figure 6. MPPT of PV system

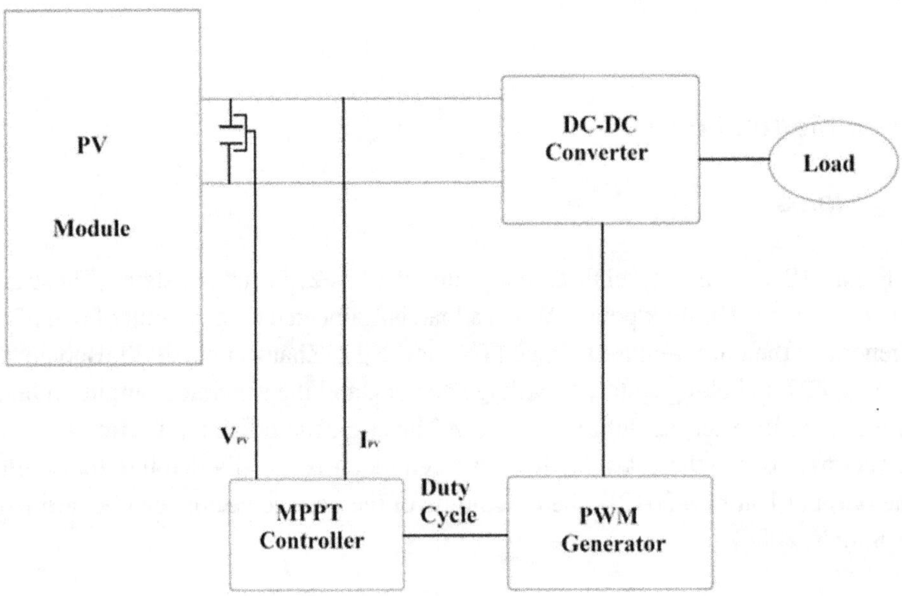

Figure 7. Flowchart of the Perturb and Observe algorithm (C_p is the perturbation step width)

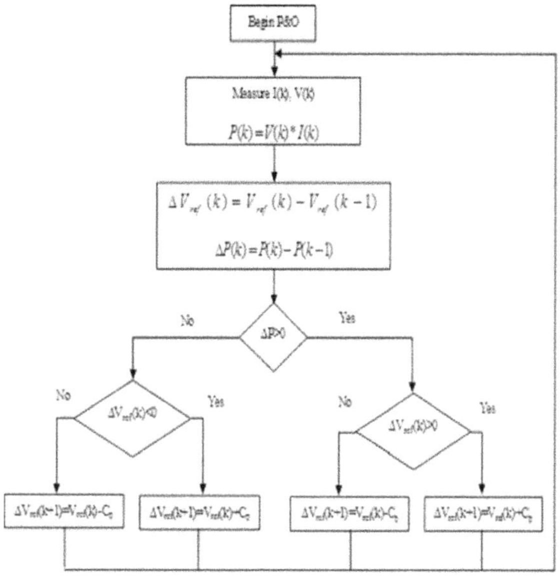

Figure 8. Power circuit of NPC inverter

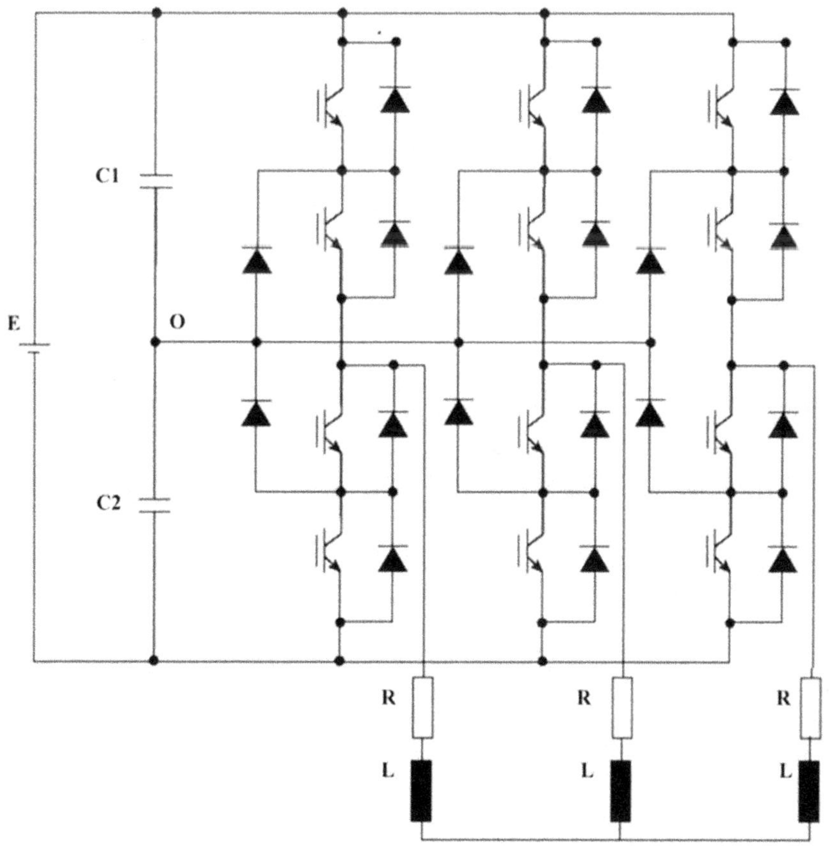

Figure 9. Different configurations of a leg

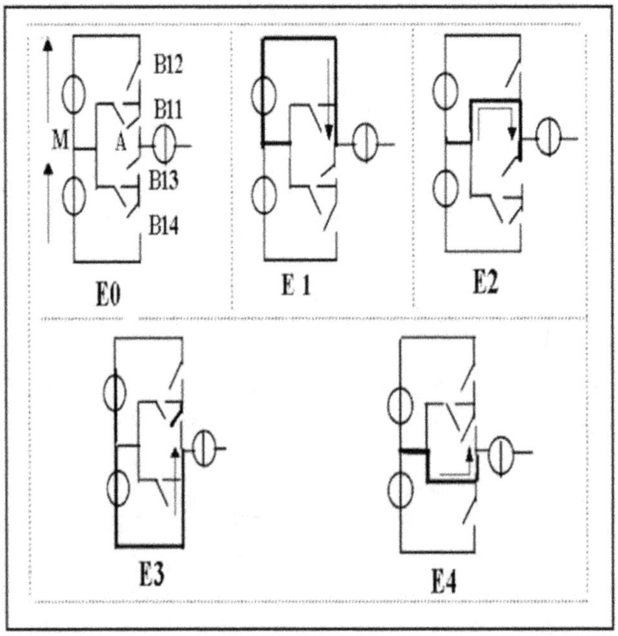

Table 2. Switches states and corresponding voltage

B_{K1}	B_{K2}	B_{K3}	B_{K4}	V_K
0	0	1	1	$-U_{c2}$
0	1	0	1	Unknown
1	0	1	0	0
1	1	0	0	U_{c2}

2. Time Duration

Space vector transformations can be used to describe the reference voltage vector V_{ref}:

$$V_{ref} = \frac{2}{3}(V_{A0} + V_{B0}e^{j\frac{2}{3}\pi} + V_{C0}e^{-j\frac{2}{3}\pi}) \tag{21}$$

The three nearest voltage space vectors can be used to describe V_{ref}. This choice is based on the magnitude and angle of the V_{ref}. For one cycle:

$$V_{ref} = T_1V_x + T_2V_y + T_3V_z \tag{22}$$

Figure 10. Space vector diagram of three-level inverter

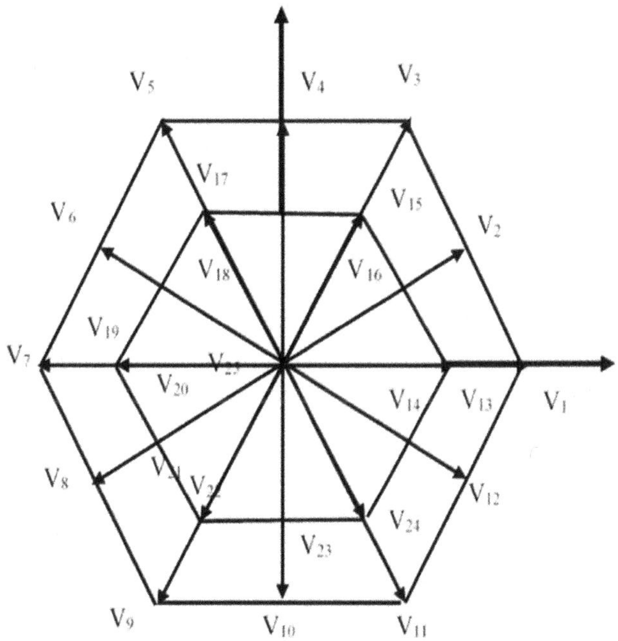

Where T_1, T_2 and T_3 are for Sector 1, Region 3. Figure 11 depicts the space vector diagram for all sectors.When V2 is chosen as the reference axis (maximum unit magnitude), the voltage vectors on the axis are defined as follows:

$$V_x = V_1 = \frac{1}{2}$$
$$V_y = V_3 = \frac{\sqrt{3}}{3}e^{j\frac{\pi}{6}} \tag{23}$$
$$V_z = V_4 = \frac{1}{2}e^{j\frac{\pi}{3}}$$

The reference vector voltage is given by:

$$V_{ref} = \frac{4}{3}E.e^{j\vartheta} \tag{24}$$

V_{ref} has two parts, real part and imaginary part, with the use of equations (22), (23), and (24). The trigonometric relation of V_{ref} is given by:

Table 3. All switches states and corresponding voltage (Magnitude & Angle)

Switching State			Corresponding Voltage Vectors		
S_a	S_b	S_c	Vector	Magnitude	Angle
1	1	1	V_0	0	0
0	0	0			
-1	-1	-1			
1	0	0	V_1	(2/3)*E	0
0	-1	-1			
1	-1	-1	V_2	(4/3)*E	π/3
1	0	-1	V_3	$(2/\sqrt{3})$*E	π/6
1	1	0	V_4	(2/3)*E	π/3
0	0	-1			
1	1	-1	V_5	(4/3)*E	π/3
0	1	-1	V_6	$(2/\sqrt{3})$*E	π/2
0	1	0	V_7	(2/3)*E	2*π/3
-1	0	-1			
-1	1	-1	V_8	(4/3)*E	2*π/3
-1	1	0	V_9	$(2/\sqrt{3})$*E	5*π/6
0	0	0	V_{10}	(2/3)*E	π
-1	0	0			
-1	1	1	V_{11}	(4/3)*E	π
-1	0	1	V_{12}	$(2/\sqrt{3})$*E	-5*π/6
0	0	1	V_{13}	(2/3)*E	-2*π/3
-1	-1	0			
-1	-1	1	V_{14}	(4/3)*E	-2*π/3
0	-1	1	V_{15}	$(2/\sqrt{3})$*E	-π/2
1	0	1	V_{16}	(2/3)*E	-π/3
0	-1	0			
1	-1	1	V_{17}	(4/3)*E	-π/3
1	-1	0	V_{18}	$(2/\sqrt{3})$*E	-π/6

$$\frac{4}{3}E(\cos(\vartheta)+j\sin(\vartheta))=\frac{1}{2}T_1+\frac{\sqrt{3}}{2}[\cos(\frac{\pi}{6})+j\,sin(\frac{\pi}{6})]T_2+\frac{1}{2}[\cos(\frac{\pi}{6})+j\,sin(\frac{\pi}{6})]T_3 \quad (25)$$

The duty cycle equations are simplified by dividing the formula into real and imaginary parts:

Figure 11. Space vector diagram for (a) all sectors (b) sector 1

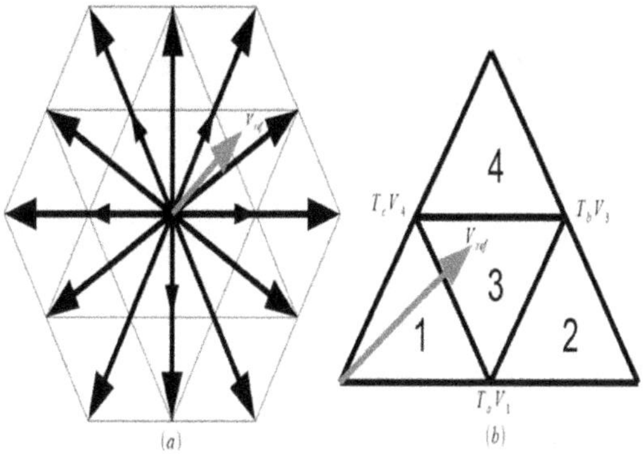

Figure 12. Output voltage of fuel cell V_{FC} and output voltage of DC-DC converter V_{dc}

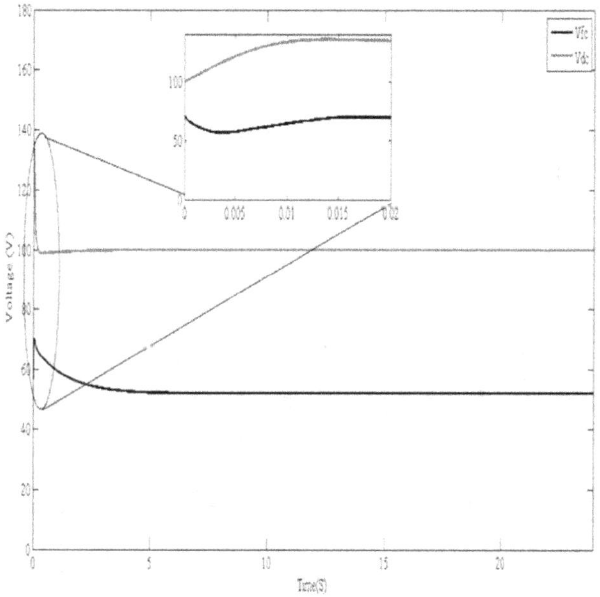

$$\begin{aligned} &Real\_part: \frac{1}{2}T_1 + \frac{\sqrt{3}}{2}\cos(\frac{\pi}{6})T_2 + \frac{1}{2}\cos(\frac{\pi}{6})T_3 = \frac{4}{3}E\cos(\vartheta) \\ &Imaginary\_part: \frac{\sqrt{3}}{2}sin(\frac{\pi}{6})T_2 + \frac{1}{2}sin(\frac{\pi}{6})T_3 = \frac{4}{3}E\sin(\vartheta) \end{aligned} \tag{26}$$

The duty cycles are then given in the following format:

Figure 13. Output voltage of photovoltaic panel V_{pv} and output voltage of DC-DC converter Vdc

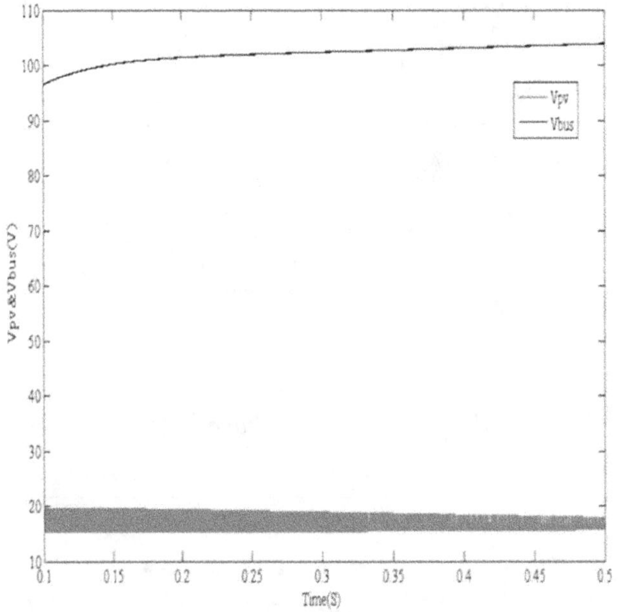

Figure 14. Output Voltage for phase "Va" of three level inverter

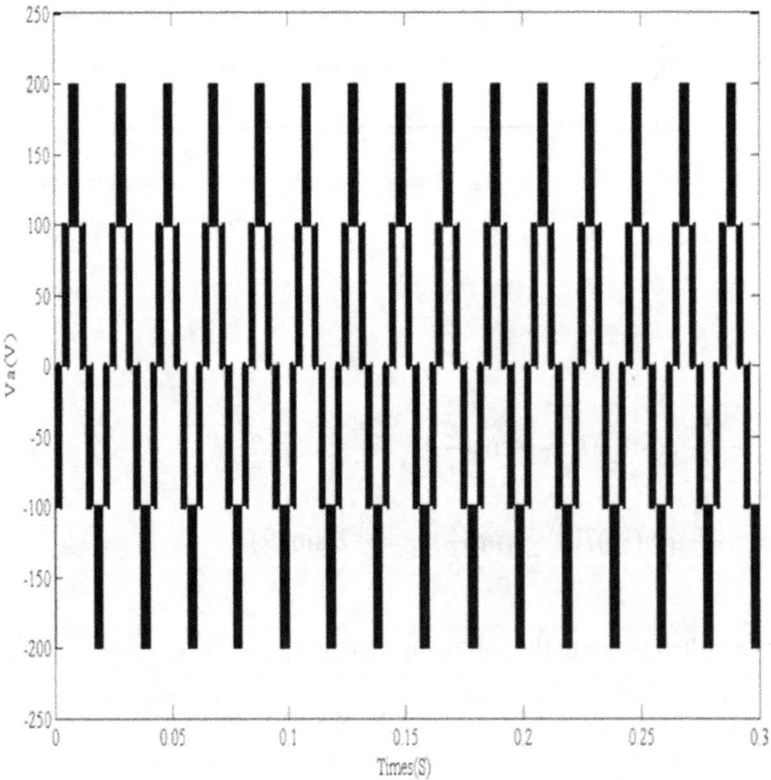

Figure 15. Output Voltage for phase "Vb" of three level inverter

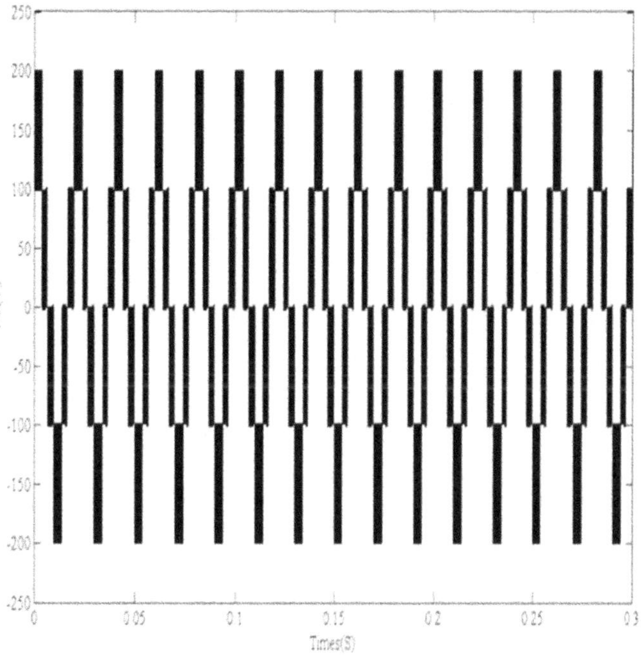

Figure 16. Output Voltage for phase "Vc" of three level inverter

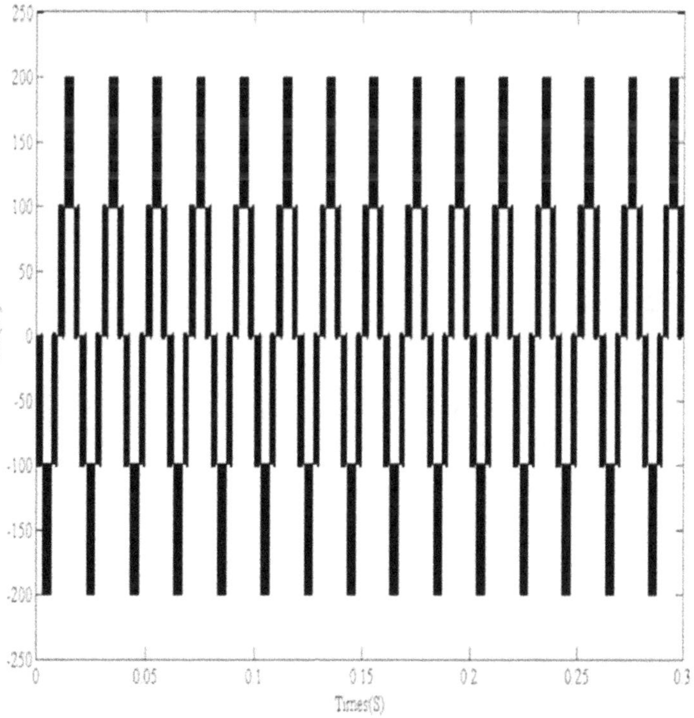

Figure 17. Output current for three phases of three level inverter

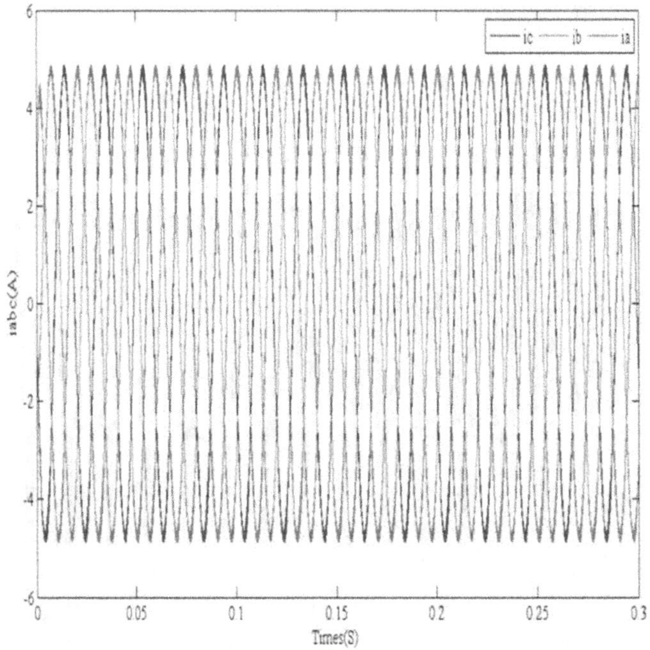

Figure 18. Harmonic spectrum of output current for three phases of three level inverter

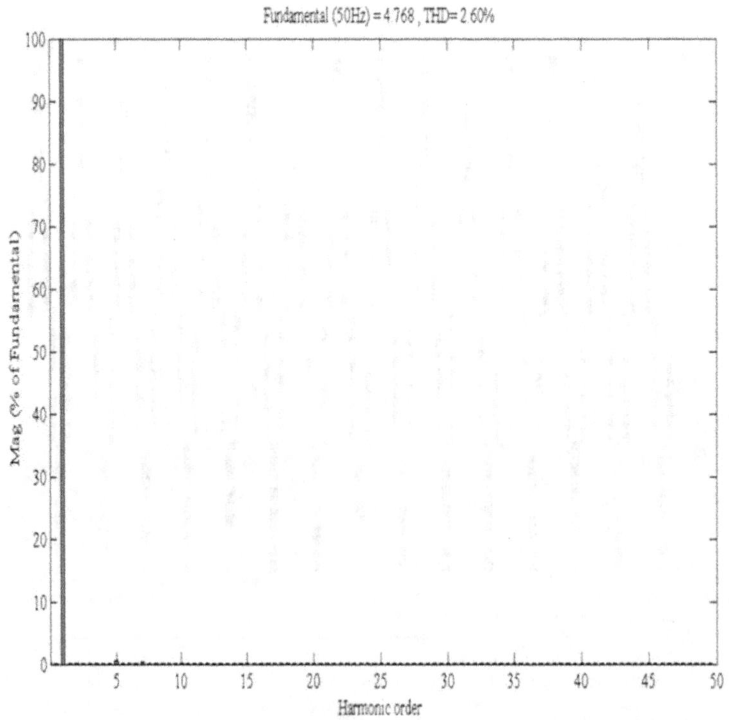

$$T_1 = 1 - 2.a.\sin(\vartheta)$$
$$T_2 = 2.a.\sin(\vartheta + \frac{\pi}{3}) - 1 \qquad (27)$$
$$T_3 = 2.a.\sin(\vartheta - \frac{\pi}{3}) + 1$$

With $a = \frac{8E}{3\sqrt{3}}$ is the index modulation

The equations for each sector and area can be measured in a similar way.

RESULTS AND SIMULATION

The MATLAB / Simulink simulation model consists of two boost converters, one for FC and one for PV using the MPPT technique. The simulation's discrete sample time is set to 50 ms. To obtain the maximum power of the PEMFC, the voltage of the desired DC-DC converter is imposed by $V_{ref} = 100V$, resulting in a DC voltage at the boost output for the two sources FC and PV.

Figure 12 depicts the voltage curves of the fuel cell (black curve) and the dc-dc step-up converter (in red). The voltage produced by the battery is approximately 60 volts, which is the product of the number of cells multiplied by the voltage of a cell after subtracting the cf coefficients, namely (η_{act}; η_{ohm}; η_{dif}). After the transient state, the voltage at the output of the dc-dc step-up converter is maintained at the reference voltage $V_{ref} = 100V$. This voltage control is provided by the PI controller.

Figure 13 depicts the photovoltaic panel's output voltage (red curve), and the voltage curve of the DC-DC step-up converter corresponds to the photovoltaic source (in black). The photovoltaic panel generates approximately 20 volts, which is the product of the number of cells multiplied by the voltage of a cell. After the transient state, the voltage at the output of the DC-DC step-up converter is maintained at the reference voltage $V_{ref} = 100V$. This voltage control is ensured by the MPPT P&O controller.

Figures 14, 15, and 16 show the PWM inverter's output voltage at three different voltage levels. It can be shown that the multi-level converter presents a balanced three-phase voltage system with three levels for each phase that constitutes this inverter that is supplied by a hybrid energy system (PV / FC) that guarantees the load's power.

Figure 17 depicts the PWM inverter's output currents at three voltage levels. We can see that the currents are purely sinusoidal and balanced with an amplitude of 4.78 A to meet the power demand of the loads, which in this case is an R_L load, with the inductance acting as a filter.

Figure 18 depicts the rate of harmonic distortion of the inverter's output current at three levels using the proposed strategy. We can conclude that the quality of the energies transferred by the hybrid source (PV / FC) is successfully ensured because we have a THD of 2.60%.

CONCLUSION

A new hybridization strategy for meeting the energy demand of a three-level inverter is proposed in this chapter. This hybridization technique is based on two sources of renewable energy: a fuel cell and a photovoltaic panel. We presented a control strategy for the inverter at three voltage levels in this paper. This is how the vertical command is used. Two energy sources were considered for the supply of this converter: a fuel cell and a photovoltaic panel. This chapter described a hybrid electric power generation (FC / PV) strategy using converter topology as the power interface, as well as a three-level inverter topology for different operating levels. In the controller design, the converter raises the input voltage to the rated voltage and becomes a DC bus, while the multi-level inverter converts the voltage to alternating current and supplies alternating current loads.

The main contribution is the development of a hybrid electric power generation strategy capable of producing output with both positive and negative sequences. The three-stage inverter's integration into the hybrid renewable energy generation (FC / PV) system enables near sinusoidal current with low THD. The hybrid energy production topology using the multi-level converter was tested in MATLAB, and the simulation results were discussed. Development of an energy management algorithm for the two sources is one of a future direction. Motor control for electric vehicle attractions is another future direction of the the proposed work.

REFERENCES

Abdel-Salam, M., El-Mohandes, M. T., & El-Ghazaly, M. (2020). An Efficient Tracking of MPP in PV Systems Using a Newly-Formulated P&O-MPPT Method Under Varying Irradiation Levels. *Journal of Electrical Engineering & Technology*, *15*(1), 501–513. doi:10.100742835-019-00283-x

AbdelAty, A. M., Azar, A. T., Vaidyanathan, S., Ouannas, A., & Radwan, A. G. (2018). Applications of continuous-time fractional order chaotic systems. In *Mathematical Techniques of Fractional Order Systems', Advances in Nonlinear Dynamics and Chaos (ANDC)* (pp. 409–449). Elsevier. doi:10.1016/B978-0-12-813592-1.00014-3

Abdelmalek, S., Azar, A. T., & Dib, D. (2018). A novel actuator fault-tolerant control strategy of DFIG-based wind turbines using Takagi-Sugeno Multiple models. *International Journal of Control, Automation, and Systems*, *16*(3), 1415–1424. doi:10.100712555-017-0320-y

Abdul-Adheem, W. R., Azar, A. T., Ibraheem, I. K., & Humaidi, A. J. (2020b). Novel Active Disturbance Rejection Control Based on Nested Linear Extended State Observers. *Applied Sciences (Basel, Switzerland)*, *10*(12), 4069. doi:10.3390/app10124069

Abdul-Adheem, W. R., Ibraheem, I. K., Azar, A. T., & Humaidi, A. J. (2020a). Improved Active Disturbance Rejection-Based Decentralized control for MIMO Nonlinear Systems: Comparison with The Decoupled Control Scheme. *Applied Sciences (Basel, Switzerland)*, *10*(7), 2515. doi:10.3390/app10072515

Abe, C. F., Dias, J. B., Notton, G., & Poggi, P. (2020). Computing Solar Irradiance and Average Temperature of Photovoltaic Modules from the Maximum Power Point Coordinates. *IEEE Journal of Photovoltaics*, *10*(2), 655–663. doi:10.1109/JPHOTOV.2020.2966362

Ajeil, F., Ibraheem, I. K., Azar, A. T., & Humaidi, A. J. (2020). Autonomous Navigation and Obstacle Avoidance of an Omnidirectional Mobile Robot Using Swarm Optimization and Sensors Deployment. *International Journal of Advanced Robotic Systems*, *17*(3), 1–15. doi:10.1177/1729881420929498

Ajeil, F. H., Ibraheem, I. K., Azar, A. T., & Humaidi, A. J. (2020). Grid-Based Mobile Robot Path Planning Using Aging-Based Ant Colony Optimization Algorithm in Static and Dynamic Environments. *Sensors (Basel)*, *20*(7), 1880. doi:10.339020071880 PMID:32231091

Alain, K. S. T., Azar, A. T., Bertrand, F. H., & Romanic, K. (2019). Robust observer-based synchronisation of chaotic oscillators with structural perturbations and input nonlinearity. *International Journal of Automation and Control*, *13*(4), 387–412. doi:10.1504/IJAAC.2019.100467

Alain, K. S. T., Azar, A. T., Kengne, R., & Bertrand, F. H. (2020). Stability analysis and robust synchronisation of fractional-order modified colpitts oscillators. *International Journal of Automation and Control*, *14*(1), 52–79. doi:10.1504/IJAAC.2020.103806

Alain, K. S. T., Romanic, K., Azar, A. T., Vaidyanathan, S., Bertrand, F. H., & Adele, N. M. (2018a). Dynamics analysis and synchronization in relay coupled fractional order colpitts oscillators. In Advances in System Dynamics and Control. IGI Global. doi:10.4018/978-1-5225-4077-9.ch011

Alain, K. S. T., Romanic, K., Azar, A. T., Vaidyanathan, S., Bertrand, F. H., & Adele, N. M. (2018b). *Dynamics Analysis and Synchronization in Relay Coupled Fractional Order Colpitts Oscillators.* IGI-Global. doi:10.4018/978-1-5225-4077-9

Ali, M. M., Youssef, A. R., Ali, A. S., & Abdel-Jaber, G. T. (2020). Variable step size PO MPPT algorithm using model reference adaptive control for optimal power extraction. *International Transactions on Electrical Energy Systems*, *30*(1), e12151. doi:10.1002/2050-7038.12151

Amara, K., Bakir, T., Malek, A., Hocine, D., Bourennane, E. B., Fekik, A., & Zaouia, M. (2019a). An Optimized Steepest Gradient Based Maximum Power Point Tracking for PV Control Systems. *International Journal on Electrical Engineering & Informatics*, *11*(4), 662–683. doi:10.15676/ijeei.2019.11.4.3

Amara, K., Fekik, A., Hocine, D., Bakir, M. L., Bourennane, E. B., Malek, T. A., & Malek, A. (2018). Improved performance of a PV solar panel with adaptive neuro fuzzy inference system ANFIS based MPPT. In *2018 7th International Conference on Renewable Energy Research and Applications (ICRERA)* (pp. 1098-1101). IEEE. 10.1109/ICRERA.2018.8566818

Amara, K., Malek, A., Bakir, T., Fekik, A., Azar, A. T., Almustafa, K. M., Bourennane, E. B., & Hocine, D. (2019b). Adaptive neuro-fuzzy inference system based maximum power point tracking for stand-alone photovoltaic system. *International Journal of Modelling. Identification and Control*, *33*(4), 311–321. doi:10.1504/IJMIC.2019.107480

Ammar, H., Ibrahim, M., Azar, A. T., & Shalaby, R. (2020b). Gray Wolf Optimization of Fractional Order Control of 3-Omni Wheels Mobile Robot: Experimental Study. *2020 16th International Computer Engineering Conference (ICENCO)*, 147-152. 10.1109/ICENCO49778.2020.9357384

Ammar, H. H., & Azar, A. T. (2020). Robust Path Tracking of Mobile Robot Using Fractional Order PID Controller. In *The International Conference on Advanced Machine Learning Technologies and Applications (AMLTA2019).* Springer. 10.1007/978-3-030-14118-9_37

Ammar, H. H., & Azar, A. T. (2020a). Robust path tracking of mobile robot using fractional order pid controller. In *The International Conference on Advanced Machine Learning Technologies and Applications (AMLTA2019)*. Springer International Publishing. 10.1007/978-3-030-14118-9_37

Ammar, H. H., Azar, A. T., Shalaby, R., & Mahmoud, M. I. (2019a). Metaheuristic optimization of fractional order incremental conductance (fo-inc) maximum power point tracking (mppt). *Complexity, 2019*(7687891), 1–13. doi:10.1155/2019/7687891

Ammar, H. H., Azar, A. T., Shalaby, R., & Mahmoud, M. I. (2019b). Metaheuristic Optimization of Fractional Order Incremental Conductance (FO-INC) Maximum Power Point Tracking (MPPT). Complexity. doi:10.1155/2019/7687891

Ammar, H. H., Azar, A. T., Tembi, T. D., Tony, K., & Sosa, A. (2018). Design and Implementation of Fuzzy PID Controller into Multi Agent Smart Library System Prototype. In *The International Conference on Advanced Machine Learning Technologies and Applications (AMLTA2018)*. Springer. 10.1007/978-3-319-74690-6_13

Azar, A. T., Kumar, J., Kumar, V., & Rana, K. P. S. (2018c). Control of a Two Link Planar Electrically-Driven Rigid Robotic Manipulator Using Fractional Order SOFC. In *Proceedings of the International Conference on Advanced Intelligent Systems and Informatics 2017*. Springer. 10.1007/978-3-319-64861-3_6

Azar, A. T., Ouannas, A., & Singh, S. (2018b). Control of New Type of Fractional Chaos Synchronization. In *Proceedings of the International Conference on Advanced Intelligent Systems and Informatics 2017*. Springer. 10.1007/978-3-319-64861-3_5

Azar, A.T., & Serrano, F.E. (2014). Robust IMC-PID tuning for cascade control systems with gain and phase margin specifications. *Neural Computing and Applications, 25*(5), 983-995. . doi:10.1007/s00521-014-1560-x

Azar, A. T., & Serrano, F. E. (2015a). Stabilization and Control of Mechanical Systems with Backlash. In Advanced Intelligent Control Engineering and Automation. IGI-Global. doi:10.4018/978-1-4666-7248-2.ch001

Azar, A. T., & Serrano, F. E. (2015b). Design and Modeling of Anti Wind Up PID Controllers. In Complex system modelling and control through intelligent soft computations. Springer-Verlag. . 1. doi:10.1007/978-3-319-12883-2_1

Azar, A. T., & Serrano, F. E. (2015c). Adaptive Sliding mode control of the Furuta pendulum. In Advances and Applications in Sliding Mode Control systems. Springer-Verlag GmbH. doi:10.1007/978-3-319-11173-5_1

Azar, A. T., & Serrano, F. E. (2015d). Deadbeat Control for Multivariable Systems with Time Varying Delays. In Chaos Modeling and Control Systems Design. Springer-Verlag GmbH. doi:10.1007/978-3-319-13132-0_6

Azar, A. T., & Serrano, F. E. (2016a). Robust control for asynchronous switched nonlinear systems with time varying delays. *Proceedings of the International Conference on Advanced Intelligent Systems and Informatics 2016*, 533, 891-899. 10.1007/978-3-319-48308-5_85

Azar, A. T., & Serrano, F. E. (2016b). Stabilization of Mechanical Systems with Backlash by PI Loop Shaping. *International Journal of System Dynamics Applications*, *5*(3), 20–47. doi:10.4018/IJSDA.2016070102

Azar, A. T., & Serrano, F. E. (2018). Fractional Order Sliding Mode PID Controller/Observer for Continuous Nonlinear Switched Systems with PSO Parameter Tuning. In *The International Conference on Advanced Machine Learning Technologies and Applications (AMLTA2018)*. Springer.

Azar, A. T., & Serrano, F. E. (2019). Fractional Order Two Degree of Freedom PID Controller for a Robotic Manipulator with a Fuzzy Type-2 Compensator. In *Proceedings of the International Conference on Advanced Intelligent Systems and Informatics 2018*. Springer. 10.1007/978-3-319-99010-1_7

Azar, A. T., & Serrano, F. E. (2020). Stabilization of Port Hamiltonian Chaotic Systems With Hidden Attractors By Adaptive Terminal Sliding Mode Control. *Entropy (Basel, Switzerland)*, *22*(1), 122. doi:10.3390/e22010122 PMID:33285897

Azar, A. T., Serrano, F. E., & Koubaa, A. (2020). Adaptive Fuzzy Type-2 Fractional Order Proportional Integral Derivative Sliding Mode Controller for Trajectory Tracking of Robotic Manipulators. In *2020 IEEE International Conference on Autonomous Robot Systems and Competitions (ICARSC)*. IEEE. 10.1109/ICARSC49921.2020.9096163

Azar, A. T., Serrano, F. E., & Vaidyanathan, S. (2018a). Sliding Mode Stabilization and Synchronization of Fractional Order Complex Chaotic and Hyperchaotic Systems. In Mathematical Techniques of Fractional Order Systems. Elsevier. doi:10.1016/B978-0-12-813592-1.00010-6

Azar, A. T., & Vaidyanathan, S. (2015a). Handbook of Research on Advanced Intelligent Control Engineering and Automation. IGI Global. doi:10.4018/978-1-4666-7248-2

Azar, A. T., & Vaidyanathan, S. (2015b). *Computational Intelligence applications in Modeling and Control*. Springer-Verlag.

Azar, A. T., & Vaidyanathan, S. (2015c). *Chaos Modeling and Control Systems Design*. Springer-Verlag.

Azar, A. T., & Vaidyanathan, S. (2016). *Advances in Chaos Theory and Intelligent Control*. Springer-Verlag. doi:10.1007/978-3-319-30340-6

Azizkandi, M. E., Sedaghati, F., Shayeghi, H., & Blaabjerg, F. (2019). A high voltage gain DC–DC converter based on three winding coupled inductor and voltage multiplier cell. *IEEE Transactions on Power Electronics*, *35*(5), 4558–4567. doi:10.1109/TPEL.2019.2944518

Babaie, M., Sharifzadeh, M., & Al-Haddad, K. (2020, August). Three-Phase Grid-Connected NPC Inverter Based on a Robust Artificial Neural Network Controller. In *2020 IEEE Power & Energy Society General Meeting (PESGM)* (pp. 1-5). IEEE.

Barakat, M. H., Azar, A. T., & Ammar, H. H. (2020). Agricultural Service Mobile Robot Modeling and Control Using Artificial Fuzzy Logic and Machine Vision. In *The International Conference on Advanced Machine Learning Technologies and Applications (AMLTA2019)*. Springer. 10.1007/978-3-030-14118-9_46

Ben Smida, M., Sakly, A., Vaidyanathan, S., & Azar, A. T. (2018). *Control-Based Maximum Power Point Tracking for a Grid-Connected Hybrid Renewable Energy System Optimized by Particle Swarm Optimization.* IGI-Global. doi:10.4018/978-1-5225-4077-9.ch003

Berkouk, E. M. (1995). *Contribution à la conduite des machines asynchrones monophasée et triphasée alimentées par des convertisseurs directs et indirects. Application aux gradateurs et onduleurs multiniveaux* (Doctoral dissertation). CNAM.

Bocci, E., Di Carlo, A., McPhail, S. J., Gallucci, K., Foscolo, P. U., Moneti, M., Villarini, M., & Carlini, M. (2014). Biomass to fuel cells state of the art: A review of the most innovative technology solutions. *International Journal of Hydrogen Energy*, *39*(36), 21876–21895. doi:10.1016/j.ijhydene.2014.09.022

Bouakrif, F., Azar, A.T., Volos, C.K., Muñoz-Pacheco, J.M., & Pham, V.T. (2019). Iterative Learning and Fractional Order Control for Complex Systems. *Complexity*. . doi:10.1155/2019/7958625

Chaibi, Y., Allouhi, A., Malvoni, M., Salhi, M., & Saadani, R. (2019). Solar irradiance and temperature influence on the photovoltaic cell equivalent-circuit models. *Solar Energy*, *188*, 1102–1110. doi:10.1016/j.solener.2019.07.005

Ciocia, A., Di Leo, P., Fichera, S., Giordano, F., Malgaroli, G., & Spertino, F. (2020, June). A Novel Procedure to Adjust the Equivalent Circuit Parameters of Photovoltaic Modules under Shading. In *2020 International Symposium on Power Electronics, Electrical Drives, Automation and Motion (SPEEDAM)* (pp. 711-715). IEEE. 10.1109/SPEEDAM48782.2020.9161878

Cipriani, G., Di Dio, V., Genduso, F., La Cascia, D., Liga, R., Miceli, R., & Galluzzo, G. R. (2014). Perspective on hydrogen energy carrier and its automotive applications. *International Journal of Hydrogen Energy*, *39*(16), 8482–8494. doi:10.1016/j.ijhydene.2014.03.174

Denoun, H., Hamida, M. L., Fekik, A., Dyhia, K. A. I. S., & Ghanes, M. (2018, October). Petri Nets Modeling for Two-Cell Chopper Control Using DSPACE 1104. In *2018 6th International Conference on Control Engineering & Information Technology (CEIT)* (pp. 1-6). IEEE.

Denoun, H., Nabil, B., Mustapha, Z., Nacereddine, B., Haddad, S., & Mamar, S. A. (2015). Modelling and Realisation of a Three-Level PWM Inverter Using a DSP Controller to Feed an Asynchronous Machine. In *Handbook of Research on Advanced Intelligent Control Engineering and Automation* (pp. 687-718). IGI Global.

Derbeli, M., Farhat, M., Barambones, O., & Sbita, L. (2017). Control of PEM fuel cell power system using sliding mode and super-twisting algorithms. *International Journal of Hydrogen Energy*, *42*(13), 8833–8844. doi:10.1016/j.ijhydene.2016.06.103

Djeddi, A., Dib, D., Azar, A. T., & Abdelmalek, S. (2019). Fractional Order Unknown Inputs Fuzzy Observer for Takagi–Sugeno Systems with Unmeasurable Premise Variables. *Mathematics*, *7*(10), 984. doi:10.3390/math7100984

Dong, S., Kremers, E., Brucoli, M., Rothman, R., & Brown, S. (2020). Techno-enviro-economic assessment of household and community energy storage in the UK. *Energy Conversion and Management*, *205*, 112330. doi:10.1016/j.enconman.2019.112330

Edwards, P. P., Kuznetsov, V. L., David, W. I., & Brandon, N. P. (2008). Hydrogen and fuel cells: Towards a sustainable energy future. *Energy Policy*, *36*(12), 4356–4362. doi:10.1016/j.enpol.2008.09.036

Elkholy, H. A., Azar, A. T., Magd, A., Marzouk, H., & Ammar, H. H. (2020b) Classifying Upper Limb Activities Using Deep Neural Networks. In *Proceedings of the International Conference on Artificial Intelligence and Computer Vision (AICV2020)*. Springer. 10.1007/978-3-030-44289-7_26

Elkholy, H. A., Azar, A. T., Shahin, A. S., Elsharkawy, O. I., & Ammar, H. H. (2020a). Path Planning of a Self Driving Vehicle Using Artificial Intelligence Techniques and Machine Vision. In *Proceedings of the International Conference on Artificial Intelligence and Computer Vision (AICV2020)*. Springer. 10.1007/978-3-030-44289-7_50

Etxeberria, A., Vechiu, I., Camblong, H., & Vinassa, J. M. (2010, October). Hybrid energy storage systems for renewable energy sources integration in microgrids: A review. In *2010 Conference Proceedings IPEC* (pp. 532-537). IEEE. 10.1109/IPECON.2010.5697053

Fadhel, S., Diallo, D., Delpha, C., Migan, A., Bahri, I., Trabelsi, M., & Mimouni, M. F. (2020). Maximum power point analysis for partial shading detection and identification in photovoltaic systems. *Energy Conversion and Management*, *224*, 113374. doi:10.1016/j.enconman.2020.113374

Fekik, A., Azar, A. T., Denoun, H., Kamal, N. A., Hamida, M. L., Kais, D., & Amara, K. (2018e, September). A Backstepping Direct Power Control of Three Phase Pulse Width Modulated Rectifier. In *International Workshop Soft Computing Applications* (pp. 445-456). Springer.

Fekik, A., Azar, A. T., Kamal, N. A., Denoun, H., Almustafa, K. M., Hamida, L., & Zaouia, M. (2020a, February). Fractional-Order Control of a Fuel Cell–Boost Converter System. In *International Conference on Advanced Machine Learning Technologies and Applications* (pp. 713-724). Springer.

Fekik, A., Azar, A. T., Kamal, N. A., Serrano, F. E., Hamida, M. L., Denoun, H., & Yassa, N. (2020b, October). Maximum Power Extraction from a Photovoltaic Panel Connected to a Multi-cell Converter. In *International Conference on Advanced Intelligent Systems and Informatics* (pp. 873-882). Springer.

Fekik, A., Denoun, H., Azar, A. T., Hamida, M. L., Atig, M., Ghanes, M., . . . Barbot, J. P. (2018c, October). Direct power control of a three-phase PWM-rectifier based on Petri nets for the selection of switching states. In *2018 7th International Conference on Renewable Energy Research and Applications (ICRERA)* (pp. 1121-1125). IEEE.

Fekik, A., Denoun, H., Azar, A.T., Hamida, M.L., Benamrouche, N., & Vaidyanathan, S. (2018). Direct Power Control of a PWM-Inverter for Grid Connected Photovoltaic System. *5ème Conférence Internationale des Energies Renouvelables (CIER– 2017) Proceeding of Engineering and Technology –PET, 27*, 1-6.

Fekik, A., Denoun, H., Azar, A. T., Hamida, M. L., Zaouia, M., & Benyahia, N. (2018b, September). Comparative study of two level and three level PWM-rectifier with voltage oriented control. In *International Conference on Advanced Intelligent Systems and Informatics* (pp. 40-51). Springer.

Fekik, A., Denoun, H., Azar, A. T., Kamal, N. A., Zaouia, M., Benyahia, N., & Vaidyanathan, S. (2020d). Direct power control of three-phase PWM-rectifier with backstepping control. In *Backstepping Control of Nonlinear Dynamical Systems* (pp. 215–234). Academic Press.

Fekik, A., Denoun, H., Azar, A. T., Kamal, N. A., Zaouia, M., Yassa, N., & Hamida, M. L. (2019a, October). Direct Torque Control of Three Phase Asynchronous Motor with Sensorless Speed Estimator. In *International Conference on Advanced Intelligent Systems and Informatics* (pp. 243-253). Springer.

Fekik, A., Denoun, H., Azar, A. T., Koubaa, A., Kamal, N. A., Zaouia, M., ... Yassa, N. (2020c, November). Adapted Fuzzy Fractional Order proportional-integral controller for DC Motor. In *2020 First International Conference of Smart Systems and Emerging Technologies (SMARTTECH)* (pp. 1-6). IEEE.

Fekik, A., Denoun, H., Azar, A. T., Zaouia, M., Benyahia, N., Hamida, M. L., & Vaidyanathan, S. (2018a). Artificial neural network for PWM rectifier direct power control and DC voltage control. In *Advances in System Dynamics and Control* (pp. 286–316). IGI Global. doi:10.4018/978-1-5225-4077-9.ch010

Fekik, A., Denoun, H., Benamrouche, N., Benyahia, N., Badji, A., & Zaouia, M. (2016, December). Comparative analysis of direct power control and direct power control with space vector modulation of PWM rectifier. In *2016 4th International Conference on Control Engineering & Information Technology (CEIT)* (pp. 1-6). IEEE. 10.1109/CEIT.2016.7929058

Fekik, A., Denoun, H., Benamrouche, N., Benyahia, N., & Zaouia, M. (2015b). A Fuzzy Logic Based Controller For Three Phase PWM Rectifier With Voltage Oriented Control Strategy. International Journal of Circuits, Systems And. *Signal Processing*, *9*, 412–419.

Fekik, A., Denoun, H., Benamrouche, N., Benyahia, N., Zaouia, M., & Haddad, S. (2015a) Comparative study of PI and FUZZY DC- voltage control for Voltage Oriented Control-PWM rectifier. *The 14th International Conference on Circuits, Systems, Electronics, Control & Signal Processing 2015.CSECS'15.*

Fekik, A., Denoun, H., Hamida, M. L., Azar, A. T., Atig, M., & Zhu, Q. M. (2018d, July). Neural Network Based Switching State Selection for Direct Power Control of Three Phase PWM-Rectifier. In *2018 10th International Conference on Modelling, Identification and Control (ICMIC)* (pp. 1-6). IEEE.

Fekik, A., Denoun, H., Zaouia, M., Hamida, M. L., & Vaidyanathan, S. (2019b). A new switching table based neural network for direct power control of three-phase PWM-rectifier. *International Journal of Modelling. Identification and Control*, *33*(4), 358–368. doi:10.1504/IJMIC.2019.107484

Ghoudelbourk, S., Azar, A. T., & Dib, D. (2021). Three-level (NPC) Shunt Active Power Filter Based on Fuzzy Logic and Fractional-order PI Controller. *Int. J. Automation and Control*, *15*(2), 149–169. doi:10.1504/IJAAC.2021.113338

Ghoudelbourk, S., Azar, A. T., Dib, D., & Omeiri, A. (2020). Selective Harmonic Elimination Strategy in the Multilevel Inverters for Grid Connected Photovoltaic System. *Int. J. Advanced Intelligence Paradigms*, *15*(3), 317–339. doi:10.1504/IJAIP.2020.105826

Ghoudelbourk, S., Dib, D., Omeiri, A., & Azar, A. T. (2016). MPPT Control in wind energy conversion systems and the application of fractional control (PI^{α}) in pitch wind turbine. *Int. J. Modelling Identification and Control*, *26*(2), 140–151. doi:10.1504/IJMIC.2016.078329

Giove, S., Azar, A. T., & Nordio, M. (2013). Fuzzy Logic Control for Dialysis Application. In Biofeedback Systems and Soft Computing Techniques of Dialysis. Springer-Verlag GmbH. doi:10.1007/978-3-642-27558-6_9

Gorripotu, T. S., Azar, A. T., Pilla, R., & Kamal, N. A. (2021). Impact of Ultra Capacitor on Automatic Load Frequency Control of Nonlinear Power System. In *Intelligent Computing in Control and Communication. Lecture Notes in Electrical Engineering* (Vol. 702, pp. 333–341). Springer. doi:10.1007/978-981-15-8439-8_27

Gorripotu, T. S., Samalla, H., Jagan Mohana Rao, C., Azar, A. T., & Pelusi, D. (2019). TLBO Algorithm Optimized Fractional-Order PID Controller for AGC of Interconnected Power System. In J. Nayak, A. Abraham, B. Krishna, G. Chandra Sekhar, & A. Das (Eds.), *Soft Computing in Data Analytics. Advances in Intelligent Systems and Computing* (Vol. 758). Springer. doi:10.1007/978-981-13-0514-6_80

Grassi, G., Ouannas, A., Azar, A. T., Radwan, A. G., Volos, C., Pham, V. T., Ziar, T., Madian, A., Kyprianidis, I. M., & Stouboulos, I. N. (2017). Chaos Synchronisation Of Continuous Systems Via Scalar Signal. *The 6th International Conference on Modern Circuits and Systems Technologies (MOCAST).* 10.1109/MOCAST.2017.7937629

Hamida, M. L., Denoun, H., Fekik, A., Benyahia, N., Zaouia, M., & Benamrouche, N. (2017). A Comparison Between PWM and Petri Nets Control Applied for Series Multi-cells Chopper. In *International conference on control and signal processing* (*Vol. 26*, pp. 124-128). Academic Press.

Hamida, M. L., Denoun, H., Fekik, A., Kais, D., Benamrouche, N., & Amara, K. (2019b, November). Fuzzy Logic-Feedback Linearization Control for a Three-Cell Flying Capacitor Converter. In *2019 International Conference on Advanced Electrical Engineering (ICAEE)* (pp. 1-5). IEEE. 10.1109/ICAEE47123.2019.9015121

Hamida, M. L., Denoun, H., Fekik, A., & Vaidyanathan, S. (2019a). Control of separately excited DC motor with series multi-cells chopper using PI-Petri nets controller. *Nonlinear Engineering*, *8*(1), 32–38. doi:10.1515/nleng-2017-0174

Hassan, W., Lu, Y., Farhangi, M., Lu, D. D. C., & Xiao, W. (2020). Design, analysis and experimental verification of a high voltage gain and high-efficiency DC–DC converter for photovoltaic applications. *IET Renewable Power Generation*, *14*(10), 1699–1709. doi:10.1049/iet-rpg.2019.1020

Humaidi, A. J., Ibraheem, I. K., Azar, A. T., & Sadiq, M. E. (2020). A New Adaptive Synergetic Control Design for Single Link Robot Arm Actuated by Pneumatic Muscles. *Entropy (Basel, Switzerland)*, *22*(7), 723. doi:10.3390/e22070723 PMID:33286496

Ibraheem, G. A. R., Azar, A. T., Ibraheem, I. K., & Humaidi, A. J. (2020). A Novel Design of a Neural Network based Fractional PID Controller for Mobile Robots Using Hybridized Fruit Fly and Particle Swarm Optimization. *Complexity*. doi:10.1155/2020/3067024

Ibrahim, H. A., Azar, A. T., Ibrahim, Z. F., & Ammar, H. H. (2020) A Hybrid Deep Learning Based Autonomous Vehicle Navigation and Obstacles Avoidance. In *Proceedings of the International Conference on Artificial Intelligence and Computer Vision (AICV2020)*. Springer. 10.1007/978-3-030-44289-7_28

Jana, S., Kumar, N., Mishra, R., Sen, D., & Saha, T. K. (2020). Development and implementation of modified MPPT algorithm for boost converter-based PV system under input and load deviation. *International Transactions on Electrical Energy Systems*, *30*(2), e12190. doi:10.1002/2050-7038.12190

Kamal, N. A., Azar, A. T., Elbasuony, G. S., Almustafa, K. A., & Almakhles, D. (2020). PSO-based Adaptive Perturb and Observe MPPT Technique for Photovoltaic Systems. In *The International Conference on Advanced Intelligent Systems and Informatics AISI 2019*. Springer.

Kamal, N. A., & Ibrahim, A. M. (2018). Conventional, intelligent, and fractionalorder control method for maximum power point tracking of a photovoltaic system: A review. In *Fractional Order Systems', Advances in Nonlinear Dynamics and Chaos (ANDC)* (pp. 603–671). Academic Press.

Kammogne, A. S. T., Kountchou, M. N., Kengne, R., Azar, A. T., Fotsin, H. B., & Ouagni, S. T. M. (2020). Polynomial Robust Observer Implementation based-passive Synchronization of Nonlinear Fractional-Order Systems with Structural Disturbances. *Frontiers of Information Technology & Electronic Engineering.*, *21*(9), 1369–1386. doi:10.1631/FITEE.1900430

Kazim, M., Azar, A. T., Koubaa, A., & Zaidi, A. (2021). Disturbance Rejection Based Optimized Robust Adaptive Controller for UAVs. *IEEE Systems Journal*, *15*(2), 1–12. doi:10.1109/JSYST.2020.3006059

Keshavarzzadeh, A. H., Ahmadi, P., & Safaei, M. R. (2019). Assessment and optimization of an integrated energy system with electrolysis and fuel cells for electricity, cooling and hydrogen production using various optimization techniques. *International Journal of Hydrogen Energy*, *44*(39), 21379–21396. doi:10.1016/j.ijhydene.2019.06.127

Khan, A., Singh, S., & Azar, A. T. (2020a). Synchronization between a Novel Integer-Order Hyperchaotic System and a Fractional-Order Hyperchaotic System Using Tracking Control. In *The International Conference on Advanced Machine Learning Technologies and Applications (AMLTA2019)*. Springer. 10.1007/978-3-030-14118-9_38

Khan, A., Singh, S., & Azar, A. T. (2020b). Combination-Combination Anti-Synchronization of Four Fractional Order Identical Hyperchaotic Systems. In *The International Conference on Advanced Machine Learning Technologies and Applications (AMLTA2019)*. Springer. 10.1007/978-3-030-14118-9_41

Khan, M. J., & Mathew, L. (2019). Fuzzy logic controller-based MPPT for hybrid photo-voltaic/wind/fuel cell power system. *Neural Computing & Applications*, *31*(10), 6331–6344. doi:10.100700521-018-3456-7

Khennaoui, A. A., Ouannas, A., Boulaaras, S., Pham, V. T., & Azar, A. T. (2020a). A fractional map with hidden attractors: Chaos and control. *The European Physical Journal. Special Topics*, *229*(6-7), 1083–1093. doi:10.1140/epjst/e2020-900177-6

Khennaoui, A. A., Ouannas, A., Grassi, G., & Azar, A. T. (2020b). Dynamic Analysis of a Fractional Map with Hidden Attractor. In *Proceedings of the International Conference on Artificial Intelligence and Computer Vision (AICV2020)*. Springer. 10.1007/978-3-030-44289-7_68

Khettab, K., Bensafia, Y., Bourouba, B., & Azar, A. T. (2018). Enhanced Fractional Order Indirect Fuzzy Adaptive Synchronization of Uncertain Fractional Chaotic Systems Based on the Variable Structure Control: Robust H∞ Design Approach. In Mathematical Techniques of Fractional Order Systems. Elsevier.

Kocalmis, A., & Sunter, S. (2006, November). Simulation of a space vector PWM controller for a three-level voltage-fed inverter motor drive. In *IECON 2006-32nd Annual Conference On IEEE Industrial Electronics* (pp. 1915-1920). IEEE. 10.1109/IECON.2006.347442

Krishna, K. S., & Kumar, K. S. (2015). A review on hybrid renewable energy systems. *Renewable & Sustainable Energy Reviews*, *52*, 907–916. doi:10.1016/j.rser.2015.07.187

Krishnamurthy, K., Padmanaban, S., Blaabjerg, F., Neelakandan, R. B., & Prabhu, K. R. (2019). Power Electronic Converter Configurations Integration with Hybrid Energy Sources–A Comprehensive Review for State-of the-Art in Research. *Electric Power Components and Systems*, *47*(18), 1623–1650. doi:10.1080/15325008.2019.1689457

Kumar, J., Azar, A. T., Kumar, V., & Rana, K. P. S. (2018). Design of Fractional Order Fuzzy Sliding Mode Controller for Nonlinear Complex Systems. In Mathematical Techniques of Fractional Order Systems. Elsevier. doi:10.1016/B978-0-12-813592-1.00009-X

Kumar, S. S., Kumar, V., Kumar, R., Malyan, S. K., & Pugazhendhi, A. (2019). Microbial fuel cells as a sustainable platform technology for bioenergy, biosensing, environmental monitoring, and other low power device applications. *Fuel*, *255*, 115682. doi:10.1016/j.fuel.2019.115682

Lalouni, S., Rekioua, D., Rekioua, T., & Matagne, E. (2009). Fuzzy logic control of stand-alone photovoltaic system with battery storage. *Journal of Power Sources*, *193*(2), 899–907. doi:10.1016/j.jpowsour.2009.04.016

Lamine, M., Hakim, D., Arezki, F., Dyhia, K., Nacereddine, B., & Youssef, B. (2019c, July). Control of Three-Cell Inverter with a Fuzzy Logic-Feedback Linearization Strategy to Reduce the Harmonic Content of the Output Current. In *2019 International Conference of Computer Science and Renewable Energies (ICCSRE)* (pp. 1-5). IEEE.

Lamine, M., Hakim, D., Arezki, F., Nabil, B., & Nacerddine, B. (2018, October). Cyclic Reports Modulation Control Strategy for a Five Cells Inverter. In *2018 International Conference on Electrical Sciences and Technologies in Maghreb (CISTEM)* (pp. 1-5). IEEE.

Li, C. H., Zhu, X. J., Cao, G. Y., Sui, S., & Hu, M. R. (2009). Dynamic modeling and sizing optimization of stand-alone photovoltaic power systems using hybrid energy storage technology. *Renewable Energy*, *34*(3), 815–826. doi:10.1016/j.renene.2008.04.018

Li, P., Yang, C., Sun, L., Xiang, J., Wen, X., Zhong, J., & Deng, T. (2020). Dynamic characteristics and operation strategy of the discharge process in compressed air energy storage systems for applications in power systems. *International Journal of Energy Research*, *44*(8), 6363–6382. doi:10.1002/er.5362

Liu, C., Wang, Y., Wang, L., & Chen, Z. (2019). Load-adaptive real-time energy management strategy for battery/ultracapacitor hybrid energy storage system using dynamic programming optimization. *Journal of Power Sources*, *438*, 227024. doi:10.1016/j.jpowsour.2019.227024

Liu, L., Ma, D., Azar, A. T., & Zhu, Q. (2020). Neural computing Enhanced Parameter Estimation for Multi-Input and Multi-Output Total Nonlinear Dynamic Models. *Entropy (Basel, Switzerland)*, *22*(5), 510. doi:10.3390/e22050510 PMID:33286282

Lü, X., Wu, Y., Lian, J., Zhang, Y., Chen, C., Wang, P., & Meng, L. (2020). Energy management of hybrid electric vehicles: A review of energy optimization of fuel cell hybrid power system based on genetic algorithm. *Energy Conversion and Management*, *205*, 112474. doi:10.1016/j.enconman.2020.112474

Mariano-Hernández, D., Hernández-Callejo, L., Zorita-Lamadrid, A., Duque-Pérez, O., & García, F. S. (2020). A review of strategies for building energy management system: Model predictive control, demand side management, optimization, and fault detect & diagnosis. *Journal of Building Engineering*.

Meghni, B., Dib, D., & Azar, A. T. (2017b). A Second-order sliding mode and fuzzy logic control to Optimal Energy Management in PMSG Wind Turbine with Battery Storage. *Neural Computing & Applications*, *28*(6), 1417–1434. doi:10.100700521-015-2161-z

Meghni, B., Dib, D., Azar, A. T., Ghoudelbourk, S., & Saadoun, A. (2017). Robust Adaptive Supervisory Fractional order Controller For optimal Energy Management in Wind Turbine with Battery Storage. *Studies in Computational Intelligence*, *688*, 165–202. doi:10.1007/978-3-319-50249-6_6

Meghni, B., Dib, D., Azar, A. T., Ghoudelbourk, S., & Saadoun, A. (2017c). Robust Adaptive Supervisory Fractional order Controller For optimal Energy Management in Wind Turbine with Battery Storage. *Studies in Computational Intelligence*, *688*, 165–202. doi:10.1007/978-3-319-50249-6_6

Meghni, B., Dib, D., Azar, A. T., & Saadoun, A. (2017a). *Effective Supervisory Controller to Extend Optimal Energy Management in Hybrid Wind Turbine under Energy and Reliability Constraints. International Journal of Dynamics and Control*, doi:10.100740435-016-0296-0

Mekki, H., Boukhetala, D., & Azar, A. T. (2015). Sliding Modes for Fault Tolerant Control. In Advances and Applications in Sliding Mode Control systems. Springer-Verlag GmbH. doi:10.1007/978-3-319-11173-5_15

Mittal, S., Azar, A. T., & Kamal, N. A. (2021). Nonlinear Fractional Order System Synchronization via Combination-Combination Multi-switching. In *Proceedings of the International Conference on Advanced Intelligent Systems and Informatics 2020*. Springer. 10.1007/978-3-030-58669-0_75

Mohamed, N. A., Azar, A. T., Abbas, N. E., Ezzeldin, M. A., & Ammar, H. H. (2020). Experimental Kinematic Modeling of 6-DOF Serial Manipulator Using Hybrid Deep Learning. In *Proceedings of the International Conference on Artificial Intelligence and Computer Vision (AICV2020)*. Springer.

Mohapatra, A., Nayak, B., & Saiprakash, C. (2019). Adaptive Perturb & Observe MPPT for PV System with Experimental Validation. In *2019 IEEE International Conference on Sustainable Energy Technologies (ICSET)* (pp. 257-261). IEEE. 10.1109/ICSETS.2019.8744819

Murty, V. V. S. N., & Kumar, A. (2020). Multi-objective energy management in microgrids with hybrid energy sources and battery energy storage systems. *Protection and Control of Modern Power Systems*, *5*(1), 1–20. doi:10.118641601-019-0147-z

Olatomiwa, L., Mekhilef, S., Ismail, M. S., & Moghavvemi, M. (2016). Energy management strategies in hybrid renewable energy systems: A review. *Renewable & Sustainable Energy Reviews*, *62*, 821–835. doi:10.1016/j.rser.2016.05.040

Ouannas, A., Azar, A. T., & Abu-Saris, R. (2016a). A new type of hybrid synchronization between arbitrary hyperchaotic maps. *International Journal of Machine Learning and Cybernetics*. Advance online publication. doi:10.100713042-016-0566-3

Ouannas, A., Azar, A. T., & Radwan, A. G. (2016b). *On Inverse Problem of Generalized Synchronization Between Different Dimensional Integer–Order and Fractional-Order Chaotic Systems*. The 28th International Conference on Microelectronics, Cairo, Egypt. 10.1109/ICM.2016.7847942

Ouannas, A., Azar, A. T., & Vaidyanathan, S. (2017a). A Robust Method for New Fractional Hybrid Chaos Synchronization. *Mathematical Methods in the Applied Sciences*, *40*(5), 1804–1812. doi:10.1002/mma.4099

Ouannas, A., Azar, A. T., & Vaidyanathan, S. (2017b). A new fractional hybrid chaos synchronization. Int. J. Modelling. *Identification and Control*, *27*(4), 314–322. doi:10.1504/IJMIC.2017.084719

Ouannas, A., Azar, A. T., & Ziar, T. (2020a). Fractional inverse full state hybrid projective synchronisation. *Int. J. Advanced Intelligence Paradigms*, *17*(3-4), 270–298.

Ouannas, A., Azar, A. T., Ziar, T., & Radwan, A. G. (2017f). A Study On Coexistence of Different Types of Synchronization Between Different dimensional Fractional Chaotic Systems. *Studies in Computational Intelligence*, *688*, 637–669. doi:10.1007/978-3-319-50249-6_22

Ouannas, A., Azar, A. T., Ziar, T., & Radwan, A. G. (2017g). Generalized Synchronization of Different Dimensional Integer-order and Fractional Order Chaotic Systems. *Studies in Computational Intelligence*, *688*, 671–697. doi:10.1007/978-3-319-50249-6_23

Ouannas, A., Azar, A. T., Ziar, T., & Vaidyanathan, S. (2017c). On New Fractional Inverse Matrix Projective Synchronization Schemes. *Studies in Computational Intelligence*, *688*, 497–524. doi:10.1007/978-3-319-50249-6_17

Ouannas, A., Azar, A. T., Ziar, T., & Vaidyanathan, S. (2017d). Fractional Inverse Generalized Chaos Synchronization Between Different Dimensional Systems. *Studies in Computational Intelligence*, *688*, 525–551. doi:10.1007/978-3-319-50249-6_18

Ouannas, A., Azar, A. T., Ziar, T., & Vaidyanathan, S. (2017e). A New Method To Synchronize Fractional Chaotic Systems With Different Dimensions. *Studies in Computational Intelligence*, *688*, 581–611. doi:10.1007/978-3-319-50249-6_20

Ouannas, A., Grassi, G., & Azar, A. T. (2020b). A New Generalized Synchronization Scheme to Control Fractional Chaotic Systems with Non-identical Dimensions and Different Orders. In *The International Conference on Advanced Machine Learning Technologies and Applications (AMLTA2019). AMLTA 2019*. Springer. 10.1007/978-3-030-14118-9_42

Ouannas, A., Grassi, G., & Azar, A. T. (2020c). Fractional-Order Control Scheme for Q-S Chaos Synchronization. In *The International Conference on Advanced Machine Learning Technologies and Applications (AMLTA2019)*. Springer. 10.1007/978-3-030-14118-9_44

Ouannas, A., Grassi, G., Azar, A. T., & Khennaoui, A. A. (2021). Synchronization Control in Fractional Discrete-Time Systems with Chaotic Hidden Attractors. In A. Hassanien, R. Bhatnagar, & A. Darwish (Eds.), *Advanced Machine Learning Technologies and Applications. AMLTA 2020. Advances in Intelligent Systems and Computing* (Vol. 1141, pp. 661–669). Springer. doi:10.1007/978-981-15-3383-9_59

Ouannas, A., Grassi, G., Azar, A. T., Khennaouia, A. A., & Pham, V. T. (2020d). Chaotic Control in Fractional-Order Discrete-Time Systems. In *AISI 2019. The International Conference on Advanced Intelligent Systems and Informatics AISI 2019*. Springer. 10.1007/978-3-030-31129-2_20

Ouannas, A., Grassi, G., Azar, A. T., Khennaouia, A. A., & Pham, V. T. (2020e). Synchronization of Fractional-Order Discrete-Time Chaotic Systems. In *The International Conference on Advanced Intelligent Systems and Informatics AISI 2019*. Springer. 10.1007/978-3-030-31129-2_21

Ouannas, A., Grassi, G., Azar, A. T., & Singh, S. (2019a). New Control Schemes for Fractional Chaos Synchronization. In *Proceedings of the International Conference on Advanced Intelligent Systems and Informatics 2018. AISI 2018*. Springer. 10.1007/978-3-319-99010-1_5

Parida, A., & Subudhi, B. (2020). Modified leaky LMS-based control strategy for reliable operation of single-stage three-phase grid-tied PV system. *IET Renewable Power Generation*, *14*(9), 1453–1462. doi:10.1049/iet-rpg.2019.0776

Pham, V. T., Gokul, P. M., Kapitaniak, T., Volos, C., & Azar, A. T. (2018) Dynamics, Synchronization and Fractional Order Form of a Chaotic System With Infinite Equilibria. In Mathematical Techniques of Fractional Order Systems. Elsevier.

Pham, V. T., Vaidyanathan, S., Volos, C. K., Azar, A. T., Hoang, T. M., & Yem, V. V. (2017a). A Three-Dimensional No-Equilibrium Chaotic System: Analysis, Synchronization and Its Fractional Order Form. Studies in Computational Intelligence, 688, 449-470.

Pham, V. T., Volos, C. K., Vaidyanathan, S., & Azar, A. T. (2017b). Dynamics, Synchronization and Fractional Order Form of a Chaotic System without Equilibrium. In C. K. Volos (Ed.), *Nonlinear Systems: Design, Applications and Analysis*. Nova Science Publishers.

Pilla, R., Azar, A. T., & Gorripotu, T. S. (2019). Impact of Flexible AC Transmission System Devices on Automatic Generation Control with a Metaheuristic Based Fuzzy PID Controller. *Energies*, *12*(21), 4193. doi:10.3390/en12214193

Pilla, R., Botcha, N., Gorripotu, T. S., & Azar, A. T. (2020). Fuzzy PID Controller for Automatic Generation Control of Interconnected Power System Tuned by Glow-Worm Swarm Optimization. In *Applications of Robotics in Industry Using Advanced Mechanisms. ARIAM 2019. Learning and Analytics in Intelligent Systems* (Vol. 5, pp. 140–149). Springer. doi:10.1007/978-3-030-30271-9_14

Podlubny, I. (1994). Fractional-order systems and fractional-order controllers. Institute of Experimental Physics, Slovak Academy of Sciences.

Radwan, A. G., Azar, A. T., Vaidyanathan, S., Munoz-Pacheco, J. M., & Ouannas, A. (2017). Fractional-Order and Memristive Nonlinear Systems: Advances and Applications. *Complexity*, *2017*, 3760121. Advance online publication. doi:10.1155/2017/3760121

Radwan, A. G., Emira, A. A., Abdelaty, A., & Azar, A. T. (2018). Modeling and Analysis of Fractional Order DC-DC Converter. *ISA Transactions*, *82*, 184–199. doi:10.1016/j.isatra.2017.06.024 PMID:28709651

Salas, V., Olias, E., Barrado, A., & Lazaro, A. (2006). Review of the maximum power point tracking algorithms for stand-alone photovoltaic systems. *Solar Energy Materials and Solar Cells*, *90*(11), 1555–1578. doi:10.1016/j.solmat.2005.10.023

Sallam, O. K., Azar, A. T., Guaily, A., & Ammar, H. H. (2020) Tuning of PID controller using Particle Swarm Optimization for cross flow heat Exchanger based on CFD System Identification. *The International Conference on Advanced Intelligent Systems and Informatics AISI 2019*. Springer. 10.1007/978-3-030-31129-2_28

Sayed, A. S., Azar, A. T., Ibrahim, Z. F., Ibrahim, H. A., Mohamed, N. A., & Ammar, H. H. (2020). Deep Learning Based Kinematic Modeling of 3-RRR Parallel Manipulator. In *Proceedings of the International Conference on Artificial Intelligence and Computer Vision (AICV2020)*. Springer.

Shalaby, R., Ammar, H. H., Azar, A. T., & Mahmoud, M. I. (2021). Optimal Fractional-Order Fuzzy-MPPT for solar water pumping system. *Journal of Intelligent & Fuzzy Systems*, *40*(1), 1175–1190. doi:10.3233/JIFS-201538

Shukla, M. K., Sharma, B. B., & Azar, A. T. (2018). Control and Synchronization of a Fractional Order Hyperchaotic System via Backstepping and Active Backstepping Approach. In Mathematical Techniques of Fractional Order Systems. Elsevier. doi:10.1016/B978-0-12-813592-1.00019-2

Singh, S., & Azar, A. T. (2020). Multi-switching Combination Synchronization of Fractional Order Chaotic Systems. In *Proceedings of the International Conference on Artificial Intelligence and Computer Vision (AICV2020)*. Springer. 10.1007/978-3-030-44289-7_61

Singh, S., Azar, A. T., Ouannas, A., Zhu, Q., Zhang, W., & Na, J. (2017). *Sliding Mode Control Technique for Multi-switching Synchronization of Chaotic Systems. 9th International Conference on Modelling, Identification and Control (ICMIC 2017)*, Kunming, China. 10.1109/ICMIC.2017.8321579

Singh, S., Azar, A. T., Vaidyanathan, S., Ouannas, A., & Bhat, M. A. (2018). Multiswitching Synchronization of Commensurate Fractional Order Hyperchaotic Systems Via Active Control. In Mathematical Techniques of Fractional Order Systems. Elsevier. doi:10.1016/B978-0-12-813592-1.00011-8

Soliman, N. S., Said, L. A., Azar, A. T., Madian, A. H., Radwan, A. G., & Ouannas, A. (2017). *Fractional Controllable Multi-Scroll V-Shape Attractor with Parameters Effect*. The 6th International Conference on Modern Circuits and Systems Technologies (MOCAST), Thessaloniki, Greece.

Song, C., Fei, S., Cao, J., & Huang, C. (2019). Robust synchronization of fractional-order uncertain chaotic systems based on output feedback sliding mode control. *Mathematics*, *7*(7), 599. doi:10.3390/math7070599

Strzelecki, R. M. (Ed.). (2008). *Power electronics in smart electrical energy networks*. Springer Science & Business Media. doi:10.1007/978-1-84800-318-7

Tolba, M. F., AbdelAty, A. M., Soliman, N. S., Said, L. A., Madian, A. H., Azar, A. T., & Radwan, A. G. (2017). FPGA implementation of two fractional order chaotic systems. *International Journal of Electronics and Communications*, *28*, 162–172. doi:10.1016/j.aeue.2017.04.028

Tolunay, B. (2012). *Space Vector Pulse Width Modulation for Three-Level Converters: a LabVIEW Implementation*. Academic Press.

Vaidyanathan, S., & Azar, A. T. (2015a). Anti-Synchronization of Identical Chaotic Systems using Sliding Mode Control and an Application to Vaidyanathan-Madhavan Chaotic Systems. In Advances and Applications in Sliding Mode Control systems. Springer-Verlag GmbH. doi:10.1007/978-3-319-11173-5_19

Vaidyanathan, S., & Azar, A. T. (2015b). Hybrid Synchronization of Identical Chaotic Systems using Sliding Mode Control and an Application to Vaidyanathan Chaotic Systems. In Advances and Applications in Sliding Mode Control systems. Springer-Verlag GmbH. doi:10.1007/978-3-319-11173-5_20

Vaidyanathan, S., & Azar, A. T. (2015c). Analysis, Control and Synchronization of a Nine-Term 3-D Novel Chaotic System. In Chaos Modeling and Control Systems Design. Springer-Verlag GmbH. doi:10.1007/978-3-319-13132-0_1

Vaidyanathan, S., & Azar, A. T. (2015d). Analysis and Control of a 4-D Novel Hyperchaotic System. In Chaos Modeling and Control Systems Design. Springer-Verlag GmbH. doi:10.1007/978-3-319-13132-0_2

Vaidyanathan, S., & Azar, A. T. (2016a). Takagi-Sugeno Fuzzy Logic Controller for Liu-Chen Four-Scroll Chaotic System. *International Journal of Intelligent Engineering Informatics*, *4*(2), 135–150. doi:10.1504/IJIEI.2016.076699

Vaidyanathan, S., & Azar, A. T. (2016b). Dynamic Analysis, Adaptive Feedback Control and Synchronization of an Eight-Term 3-D Novel Chaotic System with Three Quadratic Nonlinearities. *Studies in Fuzziness and Soft Computing*, *337*, 155–178. doi:10.1007/978-3-319-30340-6_7

Vaidyanathan, S., & Azar, A. T. (2016c). Qualitative Study and Adaptive Control of a Novel 4-D Hyperchaotic System with Three Quadratic Nonlinearities. *Studies in Fuzziness and Soft Computing*, *337*, 179–202. doi:10.1007/978-3-319-30340-6_8

Vaidyanathan, S., & Azar, A. T. (2016d). A Novel 4-D Four-Wing Chaotic System with Four Quadratic Nonlinearities and its Synchronization via Adaptive Control Method. *Studies in Fuzziness and Soft Computing*, *337*, 203–224. doi:10.1007/978-3-319-30340-6_9

Vaidyanathan, S., & Azar, A. T. (2016e). Adaptive Control and Synchronization of Halvorsen Circulant Chaotic Systems. *Studies in Fuzziness and Soft Computing*, *337*, 225–247. doi:10.1007/978-3-319-30340-6_10

Vaidyanathan, S., & Azar, A. T. (2016f). Adaptive Backstepping Control and Synchronization of a Novel 3-D Jerk System with an Exponential Nonlinearity. *Studies in Fuzziness and Soft Computing*, *337*, 249–274. doi:10.1007/978-3-319-30340-6_11

Vaidyanathan, S., & Azar, A. T. (2016g). Generalized Projective Synchronization of a Novel Hyperchaotic Four-Wing System via Adaptive Control Method. *Studies in Fuzziness and Soft Computing*, *337*, 275–296. doi:10.1007/978-3-319-30340-6_12

Vaidyanathan, S., Azar, A. T., Akgul, A., Lien, C. H., Kacar, S., & Cavusoglu, U. (2019). A memristor-based system with hidden hyperchaotic attractors, its circuit design, synchronisation via integral sliding mode control and voice encryption. *Int. J. Automation and Control*, *13*(6), 644–667. doi:10.1504/IJAAC.2019.102665

Vaidyanathan, S., Azar, A. T., & Ouannas, A. (2017a). An Eight-Term 3-D Novel Chaotic System with Three Quadratic Nonlinearities, its Adaptive Feedback Control and Synchronization. *Studies in Computational Intelligence*, *688*, 719–746. doi:10.1007/978-3-319-50249-6_25

Vaidyanathan, S., Azar, A. T., & Ouannas, A. (2017c). Hyperchaos and Adaptive Control of a Novel Hyperchaotic System with Two Quadratic Nonlinearities. *Studies in Computational Intelligence*, *688*, 773–803. doi:10.1007/978-3-319-50249-6_27

Vaidyanathan, S., Azar, A. T., Rajagopal, K., & Alexander, P. (2015b). Design and SPICE implementation of a 12-term novel hyperchaotic system and its synchronization via active control (2015). *International Journal of Modelling Identification and Control*, *23*(3), 267–277. doi:10.1504/IJMIC.2015.069936

Vaidyanathan, S., Idowu, B. A., & Azar, A. T. (2015c). Backstepping Controller Design for the Global Chaos Synchronization of Sprott's Jerk Systems. In Chaos Modeling and Control Systems Design. Springer-Verlag GmbH. doi:10.1007/978-3-319-13132-0_3

Vaidyanathan, S., Sampath, S., & Azar, A. T. (2015a). Global chaos synchronisation of identical chaotic systems via novel sliding mode control method and its application to Zhu system. International *Journal of Modelling Identification and Control*, *23*(1), 92–100. doi:10.1504/IJMIC.2015.067495

Vaidyanathan, S., Zhu, Q., & Azar, A. T. (2017b). Adaptive Control of a Novel Nonlinear Double Convection Chaotic System. *Studies in Computational Intelligence*, *688*, 357–385. doi:10.1007/978-3-319-50249-6_12

Wang, Q., He, F., Zhou, X., Geng, C., & Zhang, W. (2019, November). Heat Dissipation System Modeling of Three-level Converter Considering Thermal Coupling Effects. In *2019 IEEE 4th International Future Energy Electronics Conference (IFEEC)* (pp. 1-8). IEEE. 10.1109/IFEEC47410.2019.9014927

Wang, Z., Volos, C., Kingni, S. T., Azar, A. T., & Pham, V. T. (2017). Four-wing attractors in a novel chaotic system with hyperbolic sine nonlinearity. *Optik (Stuttgart)*, *131*, 1071–1078. doi:10.1016/j.ijleo.2016.12.016

Wu, G. C., Deng, Z. G., Baleanu, D., & Zeng, D. Q. (2019). New variable-order fractional chaotic systems for fast image encryption. *Chaos (Woodbury, N.Y.)*, *29*(8), 083103. doi:10.1063/1.5096645 PMID:31472503

Xu, S., Zhang, J., Huang, Y., & Jatskevich, J. (2019). Dynamic average-value modeling of three-level T-type grid-connected converter system. *IEEE Journal of Emerging and Selected Topics in Power Electronics*, *7*(4), 2428–2442. doi:10.1109/JESTPE.2019.2894057

Ye, K., & Li, P. (2020). A new adaptive PSO-PID control strategy of hybrid energy storage system for electric vehicles. *Advances in Mechanical Engineering*, *12*(9), 1687814020958574. doi:10.1177/1687814020958574

Yüksek, H. I., & Arifoğlu, U. (2020). Modeling of three-phase three-level rectifier with space vector pulse width modulation method in matlab/simulink program. *Sigma: Journal of Engineering & Natural Sciences/Mühendislik ve Fen Bilimleri Dergisi, 38*(1).

Zhang, H., Zheng, W., Yu, Y., & Li, H. (2019, August). A Hybrid Discontinuous PWM Neutral Point Balance Method Based on Zero-Sequence Component for Three-Level NPC Converter. In *2019 22nd International Conference on Electrical Machines and Systems (ICEMS)* (pp. 1-5). IEEE. 10.1109/ICEMS.2019.8922291

Zhang, Z., Zhu, Z., Hu, B., Tang, X., Liu, C., & Wang, Z. (2020). Uniform carrier-based PWM method for three-phase three-level three-wire and four-wire converter system with neutral-point balancing. *IET Power Electronics*, *13*(14), 3081–3089. doi:10.1049/iet-pel.2020.0160

Chapter 10
Power Analysis Using Various Types of Wind Turbines

Bibhu Prasad Ganthia
https://orcid.org/0000-0003-2351-7830
Indira Gandhi Institute of Technology, Sarang, India

Monalisa Mohanty
ITER, SIksha 'O' Anusandhan (Deemed), India

Jai Kumar Maherchandani
College of Technology and Engineering, Maharana Pratap University of Agriculture and Technology, India

ABSTRACT

This chapter highlights on the design, operation, and comparative analysis of different types of wind turbine systems with respect to steady state and transient phenomenal activities under rapid wind speed variations. Here, Type I, which is fixed speed induction generator based, and Type II, which is DFIG based, variable speed operated systems are initially compared. In the next part, Type III wind turbine system is presented, which uses DFIG; later, it is compared with the Type IV WT system, which uses permanent magnet synchronous generator. This chapter provides a comparative overview on existing wind power systems including an analytic discussion of key principles and innovations for wind turbines. In this energy conversion system, various designs of wind turbines, pitch angle controlled based variable speed wind turbines governed by help of electronic power converters, were preferred. This scope of dynamic simulation-based study is implemented using MATLAB Simulink to convey the feasibility of the proposed wind turbine models.

DOI: 10.4018/978-1-7998-7447-8.ch010

INTRODUCTION

The demand for global energy led to a reduction in fossil fuel supply and an increase in CO_2 levels. A great many countries have changed their energy policies to include more efficiency, due to the harmful effects of the carbon emissions on the environment. Because of the ubiquity, and because wind energy is an choice for many countries to contribute and to increase the dependence on an electric power network. The 22,446 MW installed offshore wind power generation in 2014 accounted for a fraction of the world's total installed wind power capacity. In 2050, the wind turbine could generate 25-30 percent of the global grid. The wind power production capacity installed in India is 37693.75 MW and is expected to rise to 60,000 MW by 2022. (B.P. Ganthia & S.K. Barik, 2020) (B.P. Ganthia et al., 2020)

WIND ENERGY CONVERSION SYSTEM

Wind is created by irregular atmospheric heating of sunlight. The creation of hot and cold air regions results in a pressure gradient. The air flow from a region of high pressure to the region of low pressure cause the wind flow. Energy in motion is called the kinetic energy of water. The wind power conversion network block diagram with all parts is shown in figure 1. The aerodynamically engineered blades, which results in the wind turbine rotation, are influenced by this cinematic energy. Often the wind velocity is not enough to move the blades at a rate that cannot produce electricity. In some of WECS topologies, gearbox is helps to raise the speed of the rotor shaft. The induction generators used to convert mechanical energy of the shaft into electrical energy. This electrical energy is delivered to the grid. (B.P. Ganthia & S.K. Barik, 2020)

Figure 1. Wind energy conversion system

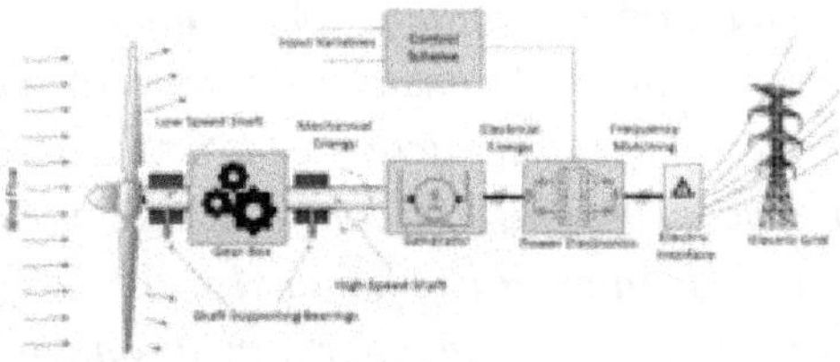

Every wind turbine is a mechanical motor that includes aerodynamic blades to extract and transform power from a turbulent motion, the wind. The available power will rely upon the wind speed so it is important that the power at different wind speeds is controlled and restricted in order to avoid grid disruption. In general they are the three most common forms of conventional wind generator systems. One type is a continuous wind turbine system with a regular squirrel cage induction generator (SCIG), which is connected to the power supply network explicitly. The second type is a wind turbine variable speed system with Double Fed Induction Generator (DFIG). (B.P. Ganthia et al., 2020) (Bibhu Prasad Ganthia et al., 2020) The power rate of the winding rotor electric converter is approximately 30%; the DFIG winding stator is connected directly to the grid. The third type is a variable wind turbo with a fully-rated, synchronous generator or SCIG electronic conversion system. Usually, two generators use a multi-stage transmission. Synchronous generators can be driven directly by low ratio gear box systems, such as permanent synchronous magnet generator (PMSG). An interesting alternative is one or two-phase gear box. (Bibhu Prasad Ganthia et al., 2020)

WIND TURBINE SYSTEM

The power level for the electrical converter supplying the winding rotor is around 30 percent; the winding statistics for the DFIG are connected directly to the grid. The third category is a variable speed wind turbine with a full-speed electronic transmission system and a synchronous generator or SCIG. Usually, the two first generators use a multi-stage gearbox. Synchro generators, like a PMSG, can be powered directly by the low-ratio gearbox system; an interesting feature is one or two gearboxes. With the decline in the cost of electronic power converters, the B2B wind turbine is becoming a great option. Direct PMSG-based configurations are an attractive alternative for WECS compared to other wind turbine generators with a variable rpm. In contrast to the DFIG wind turbine system, the WT-PMSG does not necessitate a gearbox and slip-ring to allow high-performance WECS construction, stable operation and low maintenance. In the wind turbines industry, the efficient generation of more power by onshore and offshore wind turbines using PMSGs is continuously being investigated. (A. Honrubia-Escribano et al. 2018) (L. Sun et al., 2018)

Figure 2. Power curve for wind turbine system

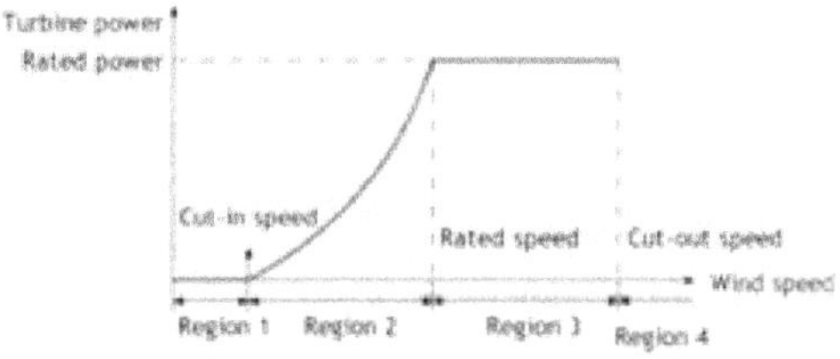

Figure 2 illustrates the power curve into four areas. Region I: Standby, Region II: full power regulation, Region III: regulation of power rating and Region 4: shut-down of turbines. Region I: Standby type. Considering that Region II has low wind speeds and its blades below turbine capacity, maximum power can be extracted at the highest output. Area 3 is a control region that mainly conserves low noise and rotor torque. The pitch angle in Region III is isolated from its maximum power conversion angle by power measurement. (E. Guest et al. 2018) (J. Hu et al. 2017)

The model is revised to reflect changes with the evolution of the technology. Each sort of generic WTG has one to four models, depending on the topology. These models are linked together for the simulation of events in the standard transient stability study, such as 20 to 30 seconds. The table no. 1 summarizes the popular WTG versions which are now available for industrial applications at onshore and offshore areas.

Table 1. Models of wind turbine system available

Model	Description
Generic Wind Generator Model	
WT1G1	Type I generator system direct connected
WT2G1	Type- II Induction generator based external rotor resistor
WT3G1	Type-III WTs with Doubly-fed induction generator
WT3G2U	Type-III and Version II, Doubly-fed induction generator
WT4G1	Type-IV and Version II, Type-IV WTs with power converter
W4G2U	Power Converter built in Wind Turbine Systems
Generic Wind Electrical Model	
WT2E1	Type-II Wind Turbine with Rotor resistance control
WT3E1	Type III wind turbine with electrical controllers
WT4E1	Type IV wind turbine with electrical controllers
W4E2U	Type IV version II wind turbine with electrical controllers
Generic Wind Mechanical Model	
WT12T1	Type I and Type II WTs with 2-Mass Model
WT3T1	Type III WTs with Mechanical Model
Generic Wind Pitch Control Model	
WT3P1	Type III Wind Turbine with Pitch control model
Generic Wind Aerodynamic Model	
WT12A1	Type I and Type II with Pseudo-governor model
Voltage and Frequency Protection Model	
FRQDCA	Under/Over frequency generator with normal relay WT system
FRQTPA	Under/Over frequency generator with trip relay WT system
VGTDCA	Class I WTs with Under/Over frequency generator with disconnection relay
VTGTPA	Class II WTs with Under/Over frequency generator with trip relay

TYPE I WIND TURBINE

Type-I wind turbine system shown in figure 3 is basically common and based on fixed or constant speed operated. This can be designed in stand alone or grid connected mode using a step up high rated power transformer. In this wind turbine system an squirrel cage induction generator (SCIG) which is operated in asynchronous mode is directly connected with the grid with constant speed controlled devices. This type of induction generator draws reactive power when it is connected to the AC grid system with a capacitor used as condenser which helps in minimizing the reactive power effect in the grid system. This condenser of capacitor bank switched continuously for the reactive power compensation and act as a soft starter.

Fixed speed or constant speed operated wind turbine system mechanisms have many useful activities as it is very lost cost, availability in the market is nominal and maintenance free. The disadvantage of Type-I wind turbine system is it is unable to withstand for long duration with affixed speed of rotations. This instability due to the speed control makes the grid unstable. This method cannot maintain

Figure 3. Type I wind turbine system (squirrel cage induction generator with fixed speed operation)

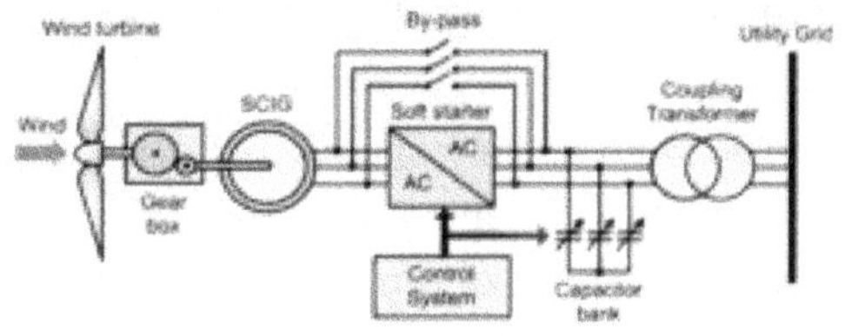

the constant speed for the grid system. So the Type-II wind turbine system coming into the feature with variable speed and constant voltage and frequency for withstanding the high mechanical strength and stress. (J. Jallad et al. 2017) (Hee-Sang Ko et al. 2008)

TYPE II WIND TURBINE SYSTEM

This Type-II wind turbine system coming into feature due to no control over the speed in Type-I wind turbine system. In this type of turbine system wound rotor induction generators (WRIG) are used partial variable wind speed rotating system is present with use of variable rotor resistance. In the figure 4 shown below the model is used type-II turbine system with grid integrations is presented. Here in this model a rotor resistance in connected for the rotor speed control during wind speed variations and can control the speed of the rotor. A capacitor bank as condenser helps in reactive power compensation and acts as a soft starter for the stability. The converter control system is implemented in the model for the rotor side converter control on speed. (Hee-Sang Ko et al. 2008)

Figure 4. Type II wind turbine system (wound rotor induction generator operated by variable rotor resistance with Partial variable speed)

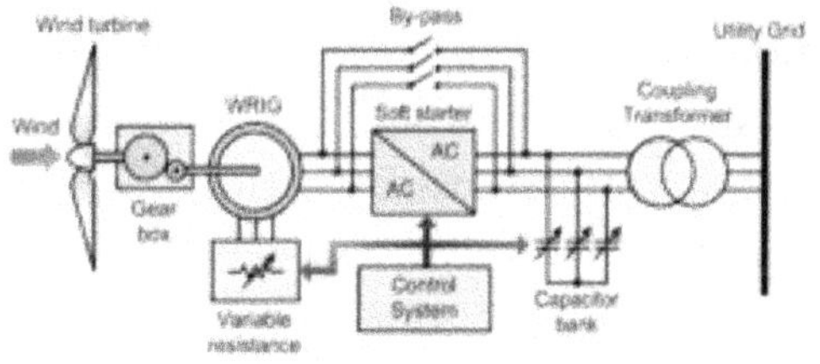

This featured win turbine system uses the converter control mechanism with the help of rotor resistance to control the rotor speed during grid integrations. This means that the rotor resistance can be controlled, but that the optically patented system prevents the need for costly ring-slipping that requires breasts and maintenance. The slip and thus the wind turbine power output can be operated by varying the rotor resistance. With the size of the variable rotor resistance, the dynamic speed control range is usually up to 10 percent higher than the synchronic speed. The heat loss energy from the external power converter is dumped. The disadvantage of this model is partial control on rotor speed hence the Type-III wind turbine system is coming into features. (J. Hu et al. 2017) (Hee-Sang Ko et al. 2008)

TYPE III WIND TURBINE SYSTEM WITH VARIABLE SPEED AND PARTIAL SCALE POWER CONVERTER CONTROL

The DFIG rule is based on a specific wind turbine operated by speed and a WRIG and a partial power converter (rated approximately 30 percent of nominal value) on the rotor circuit shown in figure 5. Using electronic electricity converters, wind turbines run at variable (or adjustable) speeds, allowing more effective power recovery than fixed speed system. Other important benefits include the reduction of mechanical losses which make for lighter mechanical structures and a more active power output; cost is very efficient, smooth pitch control, power quality enhancement and faster performance, reduced high noise level and can be operated in islanding mode.

Figure 5. Type III wind turbine system (DFIG operated by partial-scale power converter)

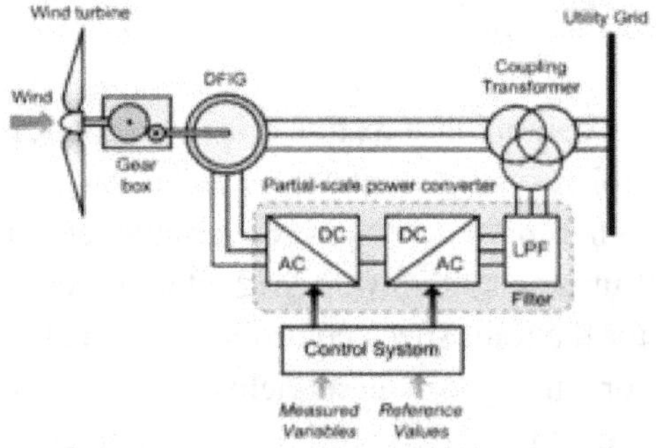

As shown in the figure 6 the rotor control and stator control is linked to the grid directly and the rotor frequency and thus the rotor speed are controlled through a partial power converter. In addition, this converter helps in the compensation of reactive power into the grid and the power flow in the grid fully controlled by Type-III wind turbine system. This method is economically desirable because of the smaller power converter with faster response. The main disadvantage of Type-III wind turbine system is absence of brushes and it uses slip rings which need heavy maintenance and this system needs high protective systems. With this drawbacks of DFIG based wind turbine system Type-IV wind turbine system is coming into features. But the Type-III DFIG based wind turbine system effective and used for both onshore and offshore applications. The grid protection system also taken care in Type-III wind turbine system using various control techniques. (J. Jallad et al. 2017) (Hee-Sang Ko et al. 2008)

TYPE IV WIND TURBINE SYSTEM

This type of wind turbine system is advanced and is fully controlled over rotor side and grid side. In this model permanent magnet synchronous generator is used for reducing maintenance like in Type-III slip rings. This model also can be operated in standalone as well as grid integrated mode. The advantages of this model are PMSG has brushes which demand less maintenance. The rotor side controller can control the pitch angle and the grid side converter controller can helps in controlling the reactive power using the adaptive techniques. The direct on line system helps to run the system in both synchronous and asynchronous modes. The generator system can excite by PMSG or WRSG which develop maximum efficiency in the wind turbine system during grid integrations. (D. Thiruvonasundari & K. Deepa, 2021)

Figure 6. Type IV wind turbine system (PMSG and full-scale power converter)

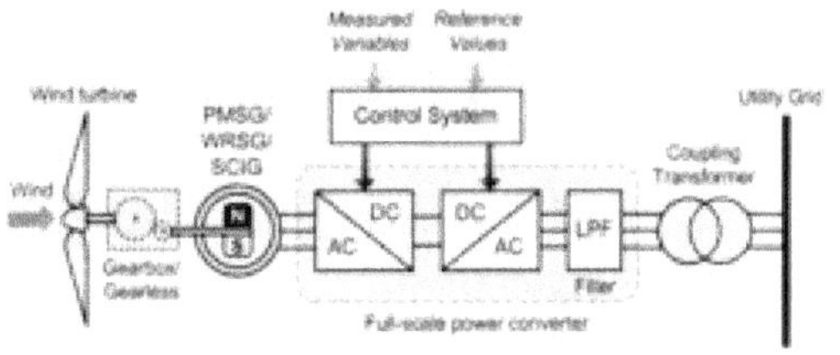

The model is shown in the figure 6 where the all controllers are inserted in the system for faster response and higher control over variable speeds. This wind turbine system needs larger permanent magnets which bare very much costly and for the full converter control this demands faster protective system with hybrid systems that can control as well as maintain the safety measures for weak grid system. (Hee-Sang Ko et al. 2008) (B. P. Ganthia et al. 2017)

DFIG MODELING

The model in Figure 8 illustrates a DFIG phasor diagram based on GSC (grid side control technique) functions of the voltage control (stator) Mechanism on the RSC (rotor side control). This control uses a PI control conventional system for the steady state analysis and transient analysis with settling times for the removal of faults during the grid integrated sudden wind speed variations. (B.P. Ganthia & S.K. Barik, 2020) (B. P. Ganthia et al. 2016)

Figure 7. Equivalent circuit of DFIG in: (a) d-axis and (b) q-axis

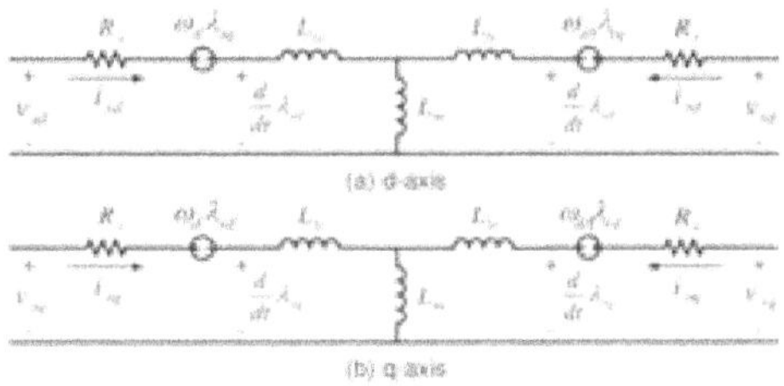

$$v_{ds} = -R_s x i_{ds} + (x_s + x_m) \text{x}\, i_{qs} + x_m x i_{qr} \tag{1}$$

$$v_{qs} = -R_s x i_{qs} - (x_s + x_m) x i_{ds} + x_m x i_{dr} \tag{2}$$

$$v_{dr} = -R_r x i_{dr} + (1-\omega)x((x_r + x_m)x i_{qr} + x_m x i_{qs} \tag{3}$$

$$v_{qr} = -R_r x i_{qr} + (1-\omega)x((x_r + x_m)x i_{dr} + x_m x i_{ds} \tag{4}$$

Where: υ_{ds}, υ_{qs} are d and q axes of the stator voltages; υ_{dr}, υ_{qr}, d and q axes of the rotor voltages; ids, iqs, d and q axes of the stator currents; i_{dr}, i_{qr}, d and q axes of the rotor currents; R_a, R_r, stator and rotor resistance; xs, stator self-reaction; x_r, rotor self-reaction; x_m, rotor speed. The inserted active and reactive forces in the network can be expressed as given blow:

$$P = v_{ds} x i_{ds} + v_{qs} x i_{qs} + v_{dr} x i_{dr} + v_{qr} x i_{qr} \tag{5}$$

$$Q = -\frac{x_m v i_{dr}}{x_s + x_m} - \frac{v^2}{x_m} \tag{6}$$

Here v is the magnitude of grid voltage.

$$\left.\begin{array}{l} J\dfrac{dw_m}{dt} = T_e - T_m \\ T_e = \dfrac{3P}{2}\mathrm{Re}(j\vec{\lambda}_s \vec{i}_s^{\,*}) = -\dfrac{3P}{2}\mathrm{Re}(j\vec{\lambda}_s \vec{i}_r^{\,*}) \end{array}\right\} \tag{7}$$

Here J = MI-moment of inertial of the rotor (Kgm^2)
P = number of poles
Tm = Mechanical Torque (N.m)
Te = Electromagnetic Torque(N.m)
w_m = Rotor Mechanical Speed (rad/s)
The wind turbine system mechanical torque T_m is:

$$T_m = \frac{p_w}{\omega} \tag{14}$$

The wind turbine system power as P_w is:

$$p_w = \frac{1}{2}\rho c_p A * V_w^3 \tag{15}$$

Where: c_p - power coefficient, ω - rotor speed, ρ - air density (kg / m^2), A - rotor area (m^2) and V_w (m/s) - wind velocity. Thus the electrical torque T_e and the mathematical link between T_m and T_e results:

$$T_e = x_m (i_{qr} i_{ds} - i_{dr} i_{qs}) \tag{16}$$

$$\frac{d\omega}{dt} = \frac{1}{2H}(T_m - T_e) \tag{17}$$

Here H is denotes rotor inertia. (B.P. Ganthia & S.K. Barik, 2020) (B.P. Ganthia et al., 2018)

PMSG MODELING

A standard circuit diagram with an armature function along with the back electromotive forces (emf) can be used to electrically characterize the permanent magnet synchronously with normal condition. This model predicts equilibrium is deficient, the emf generated are sinusoidal so hysteresis and eddy current losses can be neglected with current dynamics. Hence the voltage equations are:

$$\begin{bmatrix} Uam \\ Ubm \\ Ucm \end{bmatrix} - \begin{bmatrix} Ua \\ Ub \\ Uc \end{bmatrix} = (Rm + sL)\begin{bmatrix} iam \\ ibm \\ icm \end{bmatrix} \tag{18}$$

$$Rm = \begin{bmatrix} Rm & 0 & 0 \\ 0 & Rm & 0 \\ 0 & 0 & Rm \end{bmatrix} \qquad L = \begin{bmatrix} Laa & Lab & Lac \\ Lab & Lbb & Lbc \\ Lac & Lbc & Lcc \end{bmatrix} \tag{19}$$

Here s is the laplace transform for t > 0, U is the voltage notation and R is the resistance. L is the mutual reactance with all combinations of the matrix representations.

Figure 8. D axis and q axis phasor diagram of PMSG

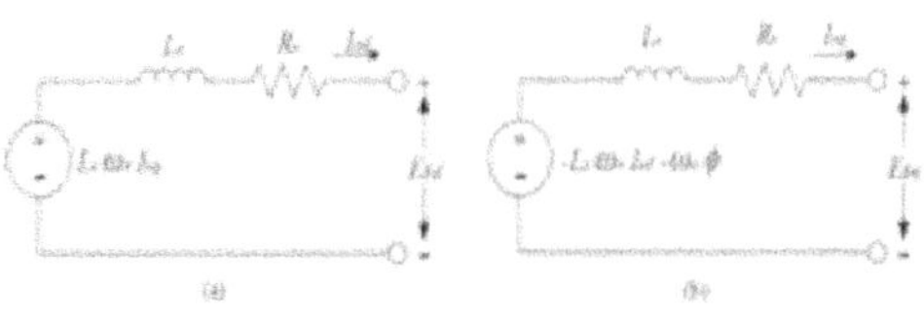

PMSG uses permanent magnets which are helps in developing fluxes and it converts aerodynamic operated mechanical power to the electrical power in the system and transfer the power to the grid during grid integrations. This system uses converter models at both the rotor and grid side for the faster switching with a dc link as filter. The equivalent circuit diagram presented above in the figure no. 8. The voltage at the stator with respect to the rotating frame shown below as:

$$E_{sd}= -R_sI_{sd}-L_s(d/dt(I_{sd}))+ \omega L_sI_{sq} \tag{20}$$

$$E_{sq}= -R_sI_{sq}-L_s(d/dt(I_{sq}))-\omega L_sI_{sd}+ \omega_e \varphi \tag{21}$$

where E_{sd},E_{sq}, are d and q axes voltages, I_{sd}, I_{sq} are currents. R_s is resistance, L_s is self-inductance, φ is flux linkage and ω_e is rotational speed of PMSG.

The PMSG electromagnetic torque is:

$$T_e =3/2\{pI_{sd}((L_d-L_q)I_{sd} + \phi)\} \tag{22}$$

where L_d, L_q are inductance, p is pole pairs. (A. Moghadasi et al., 2016) (Siraj et al., 2014) (Hong-Woo Kim et al., 2009)

SIMULINK RESULTS

A constant wind speed (13 m / s) was used to run the wind turbine. The valued value was selected for this wind speed. Voltage sags were simulated on the grid and the wind turbine's true and reactive power response observed. Net tension is reduced by 1 p.u. 0.8 p.u. 0.8 p. The sag lasts 18 cycles (0.3 seconds) at t=15s. The test aims to show that the model is effectively responsive to events in the dynamic time frame and that the machine's response to this event is realistic. The changes in grid voltage when the sag begins and ends are expected to result in an immediate answer. Note that as predicted from a fixed-speed wind turbine, the velocity does not change by much (roughly 2%). The actual power and reactive power outputs are disturbed, and the outputs show that after the ends of the sag, a mechanical oscillation happens and eventually the oscillation dampens.

Type-I FSIG WTs Simulink Model and Results

The wind turbine model with mathematical parameter calculation block diagram represented in the figure 9.

Figure 9. Mathematical Modeling of Wind Turbine system

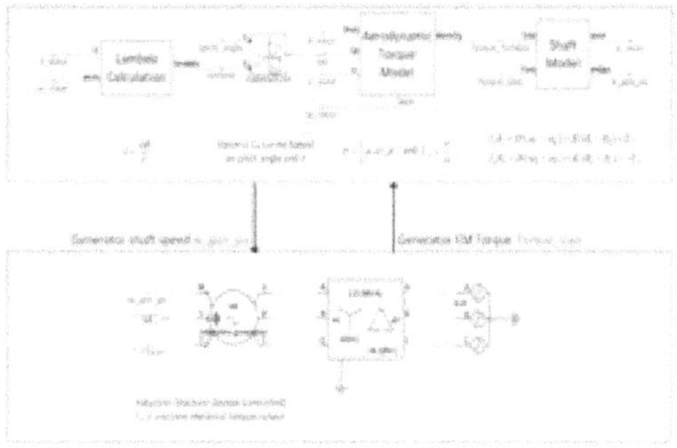

Figure 10. Type-I FSIG simulink results

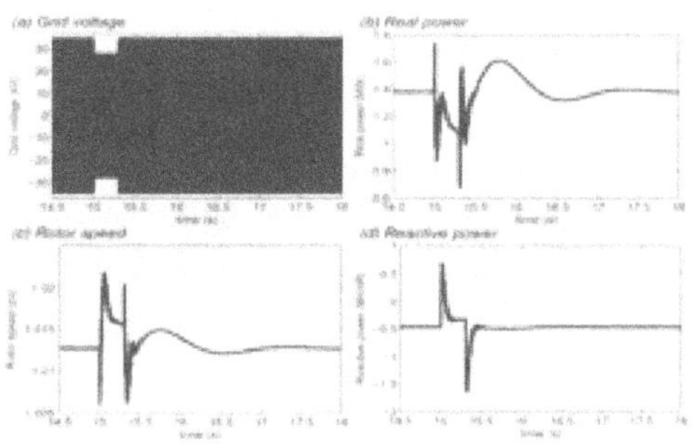

Figure 11. Type-II DFIG simulink results (grid voltage, output current, real power, reactive power, rotor speed and pitch angle)

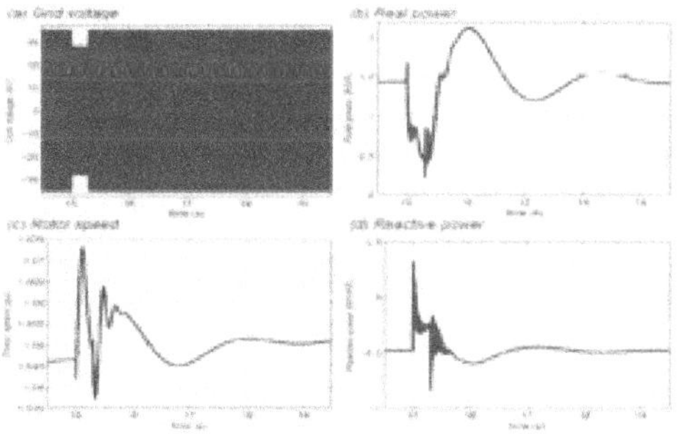

Type-II WTs Simulink Results

The changes in grid voltage when the sag begins and ends are expected to result in an immediate answer. The output is greater (about 5 percent) than that of the wind turbine with a set speed. The real power and the reactive power generated by the wind turbines is also disturbed; however, because of the rotor control, the disturbance again differs qualitatively and quantitatively from the response of the fixed wind turbine. The findings also show that after the sag stops a mechanical oscillation occurs, and that the oscillation damps out.

Type-III DFIG WTs Simulink Model and Results

Figure 12. Type-III DFIG simulink block diagram

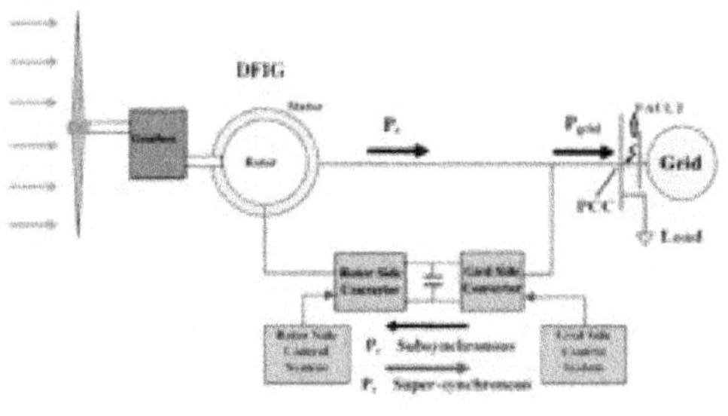

Figure 13. Type-III DFIG simulink results (grid voltage, output current, real power, reactive power, rotor speed and pitch angle)

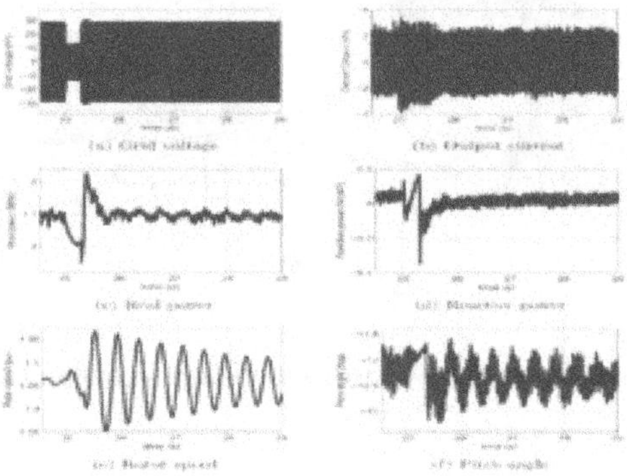

Figure 14. Type-IV PMSG Control Design

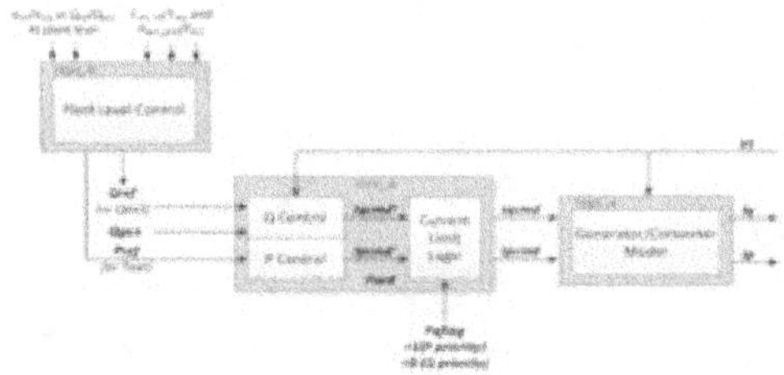

Type-IV DFIG WTs Simulink Model and Results

The above simulink results shown in figure 17 are comparing the power output during cut-in speed (4m/s) and cut-out speed (12m/s).

Figure 15. Type-IV PMSG simulink results

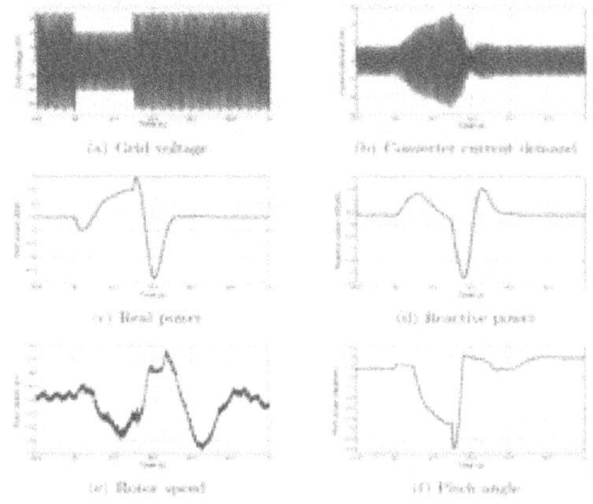

Figure 16. Comparison based on Simulink Results (a) Pitch Angle generation (b) Power Coefficient generation (c) Pitch Angle generation (pu) (d) Pitch Angle variation (e) Generated Active Power during 4m/sec (f) Generated Reactive Power during 4m/sec (g) Generated Active Power during 12m/sec (h) Generated Reactive Power during 12m/sec

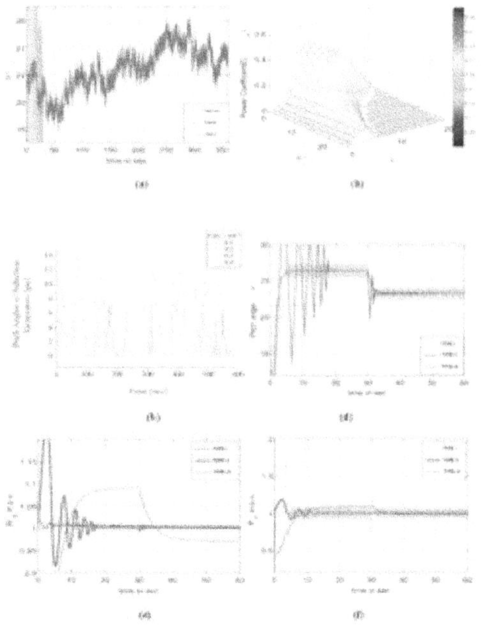

DISCUSSION

Table 2. Types of wind turbine system comparison (type based)

Indices	Type I Wind Turbine	Type II Wind Turbine	Type III Wind Turbine	Type IV Wind Turbine
WT Mode	Fixed Speed	Variable Speed Partial control	Variable speed Full control	Variable speed Full control
Generator	SCIG	WRIG	DFIG	PMSG
Grid Link	Soft starter Capacitor bank	Partial scale converter	Full scale converter	Full scale converter
Mechanical Drive	Gear	Gear	Gear	Gear less
Speed Range	~2% Slip N_{synch}	-30 ± 30% N_{synch}	0-1 N_{rated}	0-1 N_{rated}
PAC (Pitch Angle)	Fixed	PAC control	PAC control	PAC control
Controller	Rotor	RSC and PAC	PAC, RSC, GSC	PAC, RSC, GSC
Power Curve	Minimal	Average	Fully controlled	Fully controlled

Table 3. Types of wind turbine system comparison (technique based)

Performance	Machine Types and Techniques	
	Type-III DFIG based WTs	Type-IV PMSG based WTs
Manufacturing	Complicated Rotor, Simple Stator	Simple Rotor, Simple Stator
Stator	Radial, Coreless	Radial, Coreless
Rotor	Coil	Permanent Magnet
Carbon Brush	Inbuilt	Not Required
Slip Ring	Inbuilt	Not Required
Converter	30-35% Output	100% Output
LVRT	Needed	Needed
Reactive Power	Demand Controller	Fully Controlled
Grid Support	Average/Good	Very Good
Maintenance	Highly Needed	Very Less
Cost	High Maintenance Cost	High Initial and Installation Cost
Advantages	• Low initial investment • Fast Response • Heavy load with maintenance • Medium/High efficiency • High speed needs a gearbox • Cheaper than PMSG + Full converter	• No coil and No slip rings • No rotor maintenance • Low, Medium and High speed use • High Efficiency • Small and Lighter generator • High Annual Power
Disadvantages	• Difficult to service the rotor hub • No Low and medium speed operation • Low efficiency • Needs more controller for smooth and high efficiency	• High initial investment • Demand high standard software • Sophisticated Process

CONCLUSION

From the simulink design and electro-mechanical transformation conclude the DFIG and PMSG can be use according to the demand of the grid and cost of the farm shown in table 2 and 3. Wind power is being used worldwide very quickly and wind turbines are using many different technologies. This paper evaluates the three main types of wind turbines used over the last 20 years. At the load end of a transmission line connected to the wind generators demands the faster response towards the wind speed variations and full control over active and reactive powers. Hence DFIG and PMSG based Type-III and Type-IV wind turbine system installed in the grid integrations for onshore and offshore applications to facilitate the load demands.

REFERENCES

Ganthia. (2020a). Application of Hybrid Facts Devices in DFIG Based Wind Energy System for LVRT Capability Enhancements. *Journal of Mechanics of Continua and Mathematical Sciences, 15*(6), 245-256. . doi:10.26782/jmcms.2020.06.00019

Ganthia, B. P., Agarwal, V., Rout, K., & Pardhe, M. K. (2017). Optimal control study in DFIG based wind energy conversion system using PI & GA. *International Conference on Power and Embedded Drive Control (ICPEDC)*, 343-347. 10.1109/ICPEDC.2017.8081112

Ganthia, B. P., & Barik, S. K. (2020). Steady-State and Dynamic Comparative Analysis of PI and Fuzzy Logic Controller in Stator Voltage Oriented Controlled DFIG Fed Wind Energy Conversion System. *J. Inst. Eng. India Ser. B, 101*(3), 273–286. doi:10.100740031-020-00455-8

Ganthia, B. P., Barik, S. K., & Nayak, B. (n.d.). Shunt Connected FACTS Devices for LVRT Capability Enhancement in WECS. *Engineering, Technology & Applied Scientific Research, 10*(3), 5819–5823.

Ganthia, B. P., Mohanty, S., Rana, P. K., & Sahu, P. K. (2016). Compensation of voltage sag using DVR with PI controller. *2016 International Conference on Electrical, Electronics, and Optimization Techniques (ICEEOT)*, 2138-2142. 10.1109/ICEEOT.2016.7755068

Ganthia, B. P., Rana, P. K., Patra, T., Pradhan, R., & Sahu, R. (2018). Design and Analysis of Gravitational Search Algorithm Based TCSC Controller in Power System. *Materials Today: Proceedings, 5*(1), 841–847. doi:10.1016/j.matpr.2017.11.155

Ganthia. (2020b). Application of Hybrid Facts Devices in DFIG Based Wind Energy System for LVRT Capability Enhancements. *Journal of Mechanics of Continua and Mathematical Sciences, 15*(6), 139-157. . doi:10.26782/jmcms.2020.06.00012

Guest, E., Jensen, K. H., & Rasmussen, T. W. (2018, June). Sequence Domain Harmonic Modeling of Type-IV Wind Turbines. *IEEE Transactions on Power Electronics, 33*(6), 4934–4943. doi:10.1109/TPEL.2017.2735303

Honrubia-Escribano, A., Jiménez-Buendía, F., Gómez-Lázaro, E., & Fortmann, J. (2018, March). Field Validation of a Standard Type 3 Wind Turbine Model for Power System Stability, According to the Requirements Imposed by IEC 61400-27-1. *IEEE Transactions on Energy Conversion*, *33*(1), 137–145. doi:10.1109/TEC.2017.2737703

Hu, J., Sun, L., Yuan, X., Wang, S., & Chi, Y. (2017, July). Modeling of Type 3 Wind Turbines With df/dt Inertia Control for System Frequency Response Study. *IEEE Transactions on Power Systems*, *32*(4), 2799–2809. doi:10.1109/TPWRS.2016.2615631

Jallad, J., Mekhilef, S., & Mokhlis, H. (2017). Frequency Regulation Strategies in Grid Integrated Offshore Wind Turbines via VSC-HVDC Technology: A Review. *Energies*, *10*(9), 1244. doi:10.3390/en10091244

Kim, Kim, & Ko. (2010). Modeling and control of PMSG-based variable-speed wind turbine. *Electric Power Systems Research, 80*(1), 46-52. doi:10.1016/j.epsr.2009.08.003

Ko, Yoon, Kyung, & Hong. (2008). Modeling and control of DFIG-based variable-speed wind-turbine. *Electric Power Systems Research, 78*(11), 1841-1849. doi:10.1016/j.epsr.2008.02.018

Siraj, K., Siraj, H., & Nasir, M. (2014). Modeling and control of a doubly fed induction generator for grid integrated wind turbine. Power Electronics and Motion Control Conference and Exposition (PEMC). IEEE. doi:10.1109/EPEPEMC.2014.6980613

Sun, L., Peng, C., Hu, J., & Hou, Y. (2018, May). Application of Type 3 Wind Turbines for System Restoration. *IEEE Transactions on Power Systems*, *33*(3), 3040–3051. doi:10.1109/TPWRS.2017.2762009

Thiruvonasundari, D., & Deepa, K. (2021). Optimized Passive Cell Balancing for Fast Charging in Electric Vehicle. *Journal of the Institution of Electronics and Telecommunication Engineers*, 1–9. Advance online publication. doi:10.1080/03772063.2021.1886604

Chapter 11
Flying Capacitor Inverter Integration in a Renewable Energy System

Mohamed Lamine Hamida
LATAGE Laboratory, Mouloud Mammeri University, Tizi Ouzou, Algeria

Arezki Fekik
Akli Mohand Oulhadj University, Bouira, Algeria

Hakim Denoun
LATAGE Laboratory, Mouloud Mammeri University, Tizi Ouzou, Algeria

Aghiles Ardjal
L2CSP Laboratory, Mouloud Mammeri University, Tizi Ouzou, Algeria

Aicha Aissa Bokhtache
GEER Laboratory, Hassiba Benbouali University, Chlef, Algeria

ABSTRACT

This chapter presents a three-cell flying capacitor converter photovoltaic (PV) system. This system consists of a DC-DC boost power converter connected in series with a multicell inverter. The perturb and observe MPPT technique has been used to extract the maximum power from the solar panel and generate the duty signal to control the switch of the DC-DC converter. The three-cell flying capacitor inverter ensures the conversion of the output voltage of the boost chopper to the alternative voltage. This topology is made up of hybrid association of commutation cells, which makes it possible to share the voltage constraint on several switches. A closed loop control based on PWM has been proposed to control the capacitor voltages of the inverter. The output current is controlled using a PI regulator. The aim of the proposed three cell inverter is to produce an approximate sinusoidal output current with a very low THD. The simulation results assess the effectiveness of the control.

DOI: 10.4018/978-1-7998-7447-8.ch011

INTRODUCTION

Renewable energy sources have arisen as a realistic alternative to significantly reduce carbon emissions and limit the rate of global warming (Ameur et al., 2018). The major drawbacks of renewable energies reside in its changing conditions, abundance in the region, and cost. The choice of one source or more, to meet the system requirement, depends on these criteria. Among the various sustainable energy sources, solar energy is the most elegant with zero emissions (Knez & Jereb, 2013).

For both domestic and industrial users, it has become a popular source of electricity (Knez & Jereb, 2013). This includes different applications for solar electric cars, charging stations for vehicles, a significant range of water pumping systems and stand-alone systems for places where there is a lack of reliable access to the grid (Hassan et al., 2017).

Photovoltaic power generation contains solar PV panels, where a DC-DC converter supplies the output of a solar panel to the DC connection. The voltage from the DC connection is supplied to the DC-AC inverter and is supplied to the load(Amara et al., 2018). The performance of the solar PV is not constant and varies according to the temperature and solar irradiation(Mosa et al., 2017).

Therefore, even under varying climatic changes, optimum power must be derived from the PV module for the efficient operation of the PV panel. To achieve these goals, the maximum power point tracking (MPPT) based techniques are used to extract the maximum power from the panel using a DC-DC converter (Manna & Akella, 2021). The use of MPPT in the PV system improves the solar module's performance and life span (Bahgat et al., 2005). Before connecting a PV system to the AC load, the output voltage of the solar modules must be converted into an AC voltage, hence the need to introduce a DC-AC converter (Ciobotaru et al., 2006).

Nowadays, using low voltage strength semiconductors, we can synthesize high voltage multilevel waveforms(Othman et al., 2020). This discipline affects many areas of use, ranging from a few Watts to several hundred megawatts, such as electrical machine control (Hamida et al., 2019), railway traction (Guo et al., 2019), high voltage direct current HVDC electrical energy transport (Bakas et al., 2020) and parallel active filtering (Rao et al., 2020).

Harmonics, which are the source of tension wave distortion, circulating in non-linear loads such as motors, induce circulating currents in conducting torque decreasing materials, when exposed to a variable magnetic flux. The International Electrotechnical Commission (IEC) specifies the measurement of appropriate values and methods of correction (Fujita et al., 1998). In order to limit the rate of harmonic disturbance, a variety of methods to reduce harmonics have been suggested, which are based on passive or active filters (Green & Marks, 2005; Marzouki et al., 2015). Power electronic solution is one of these reduction methods. It has shown a very important technological development: PWM rectifiers (Fekik et al., 2015; Fekik et al., 2016; Fekik et al., 2017 ; Fekik et al., 2018; Fekik et al., 2019 ; Fekik et al., 2021). This is accomplished by the development of semiconductors, power components and new energy conversion systems (Kesarwani & Stauth, 2015; Rajagopal et al., 2019). Flying capacitor inverters occur among these systems (Escalante & Erdmann, 2021).

The flying capacitor converter structure is relatively recent, it was introduced at the beginning of the 1990's (Meynard, Foch, Thomas, et al., 2002), this topology has acquired a fundamental interest in the high-energy systems (Meynard, Foch, Forest, et al., 2002). This topology ensures equal stress distribution on the different low voltage semiconductor devices connected in series and improves the output waveforms and harmonic content when an adequate phase control strategy is applied (Boulouiha et al., 2019).

The flying capacitor converter is integrated in the various technological applications, such as renewable energy systems ((Arezki Fekik et al., 2020); Sahoo et al., 2019; Amghar et al., 2014; di Benedetto et al., 2021) active power filters ((Yofo & Djondiné, 2019), motors drive ((Hamida, Denoun, Fekik, & Vaidyanathan, 2019), etc.

A standalone PV system is presented in this chapter using a flying capacitor inverter incorporated with a DC-DC boost converter. To take advantage of the large potential of the flying capacitor inverter structure, an appropriate distribution of the voltages crossing each cell is needed. In order to control this converter, we must wisely choose the adequate model to allow the control synthesis and the precision to achieve the desired behavior. In this paper the cyclic reports modulation control has been proposed for a three-cell inverter to control the capacitor voltages. The conventional PI regulator has been used to improve the harmonic content of the output current. To extract the peak energy from the solar panel, the P&O powered MPPT technique is used.

The organization of the paper is as follows: Modeling of the PV and the DC-DC boost converter has been represented in section 2, the three-cell inverter modeling and its control have been presented in section 3. The simulation results have been explained in section 4. Finally, conclusions followed with future scope are drawn in section 5.

THE OVERALL SYSTEM MODELING

Figure. 1 shows the proposed system schematic/block diagram. Output terminal of the PV array is connected to the AC load through two stages of power conversion. The first stage is a boost converter which is used to track maximum power point of the PV array and to boost the PV output voltage. The flying capacitor converter is utilized in the second stage in order to invert DC voltage to the AC voltage and control the load current. The control system has three main features. These ones are maximum power point tracking of the PV array; in this chapter, perturbation and observation (P&O) method has been used. For load current control; the conventional proportional and integral (PI) regulator is used. The last feature is the PWM based strategy for switching the flying capacitor inverter. This control ensures the regulation of the capacitor's voltages.

PV Modeling

The modeling of a solar cell is an important step of analyzing a solar PV system. To study the behavior of PV cells, under the variation of irradiance and temperature, electrical equivalent circuits and mathematical models are needed. Figure.2 gives the equivalent electric circuit for a PV panel.

Since an individual cell produces only about 0.5 V, a single cell is practically of no use, apart from in extra low power appliances. Instead, the basic building block for PV applications is a module consisting of several prewired cells in series, all encased in tough, weather-resistant packages(Rosa-Clot & Tina, 2017).

A simple equivalent circuit model for a PV cell consists of a real diode in parallel with an ideal current source. The ideal current source delivers current in proportion to the solar flux to which it is exposed. There are two conditions of interest for the actual PV and for its equivalent circuit, which are:

1. the current that flows when the terminals are shorted together (the short-circuit current, I_{sc});
2. the voltage across the terminals when the leads are left open (the open-circuit voltage, V_{oc}).

Figure 1. Solar PV- flying capacitor inverter system

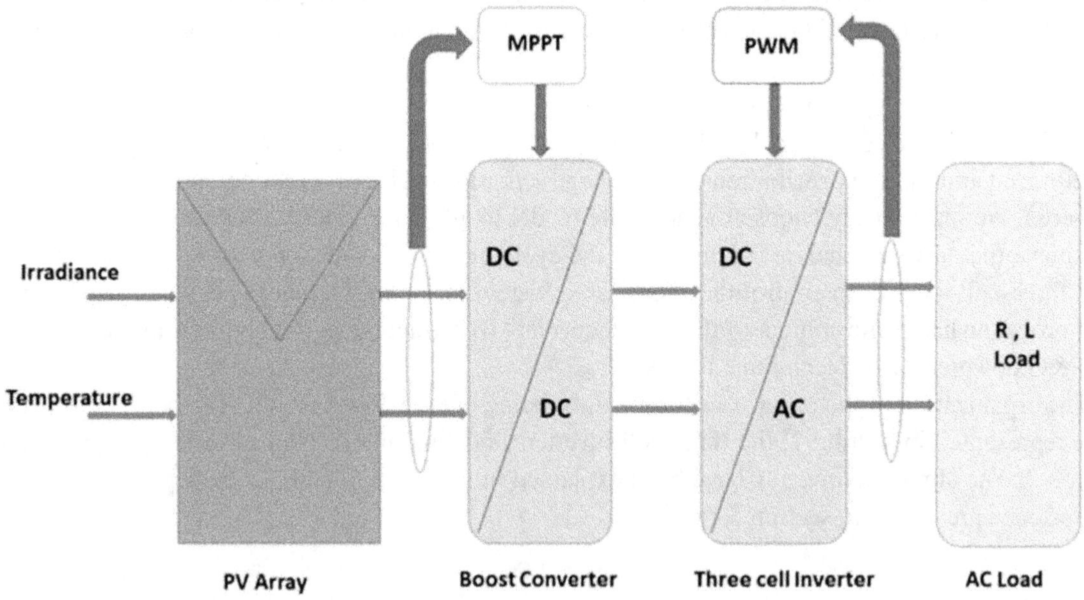

When the leads of the equivalent circuit for the PV cell are shorted together, no current flows in the (real) diode since V_d=0, so the whole current from the ideal source flows through the shorted leads.

Figure 2. Single-diode electrical PV module equivalent circuit

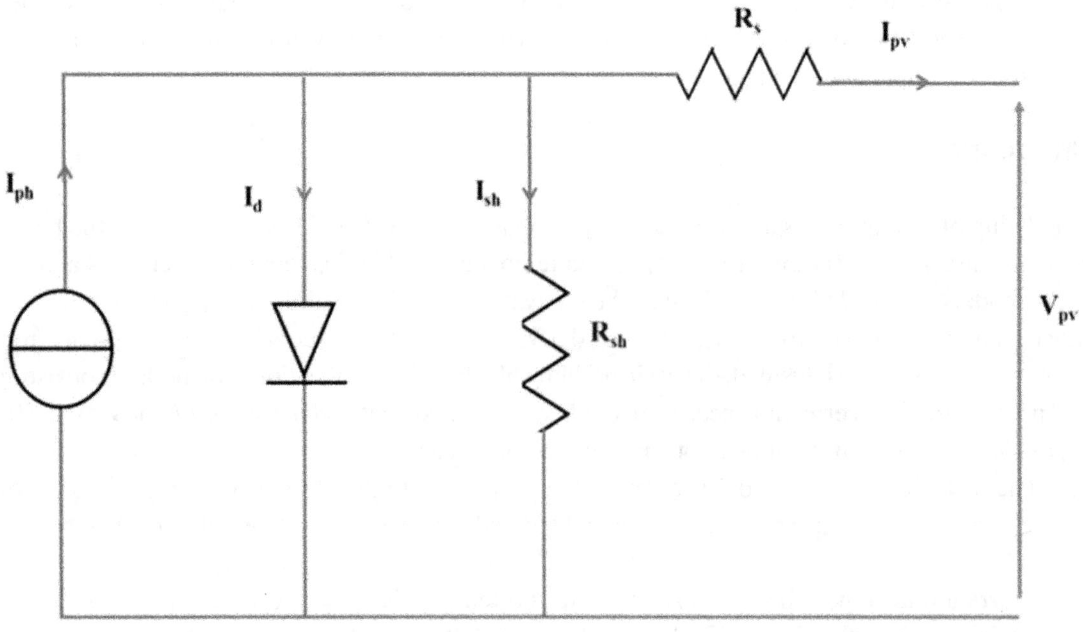

Since the short-circuit current must equal I_{sc}, the magnitude of the ideal current source itself must be equal to I_{sc}. When the leads from the PV cell are left open, the load current, I, is null and the V on the load is equal to $V_{oc} = V_d$. (Rosa-Clot & Tina, 2017)

From this circuit, and applying Kirchhoff's law, we shall derive the following equations:

$$I_{pv} = I_{ph} - I_d - I_{sh} \tag{1}$$

$$I_{pv} = I_{ph} - I_0\left(\frac{V_{pv} + R_s I_{pv}}{e^{n_s \cdot V_t \cdot Q_d} - 1}\right) - \frac{V_{pv} + R_s I_{pv}}{R_{sh}} \tag{2}$$

Where:

- V_{PV} : PV module voltage (V)
- I_{PV} : PV module current (A)
- I_{ph}: light current (A)
- I_0: diode reverse saturation current (A)
- Q_d : diode ideality factor
- n_s: number of cells in series
- R_s: series resistance (Ω)
- R_{sh}: shunt resistance (Ω)
- $V_t = kT_c/q$ is the thermal voltage (V), k is Boltzmann's constant, T_c is the cell temperature, and q is the charge of an electron.

In this chapter, a solar panel (KC200GT) has been used; the parameters of the module at standard test condition have been given in Table 1.

Table 1. Parameters of the adjusted model of the KC200GT solar array at nominal operating conditions

Maximum power	200 W (+10% / -5%)
The voltage at maximum power point (V_{MPP})	26.3V
Open circuit voltage (V_{oc})	32.9 V
Current at maximum power point (I_{MPP})	7.61A
Short circuit current (I_{sc})	8.21 A

Figure 3 shows the *Ipv-Vpv* and *Ppv-Vpv* characteristics, respectively for different solar insolation conditions at constant cell temperature $T_c = 25°C$. With the increase of the solar insolation, the short circuit current of the PV module increases and the maximum power output is also increased.

Figure 4 shows the *Ipv-Vpv* and *Ppv-Vpv* characteristics, respectively for different cell operating temperatures where solar insolation is equal to 1000W/m^2. With the increase of the temperature, the open

Figure 3. I_{pv}-V_{pv} and P_{pv}-V_{pv} characteristics of KC200GT module, (varying insolation) and Tc=25°C

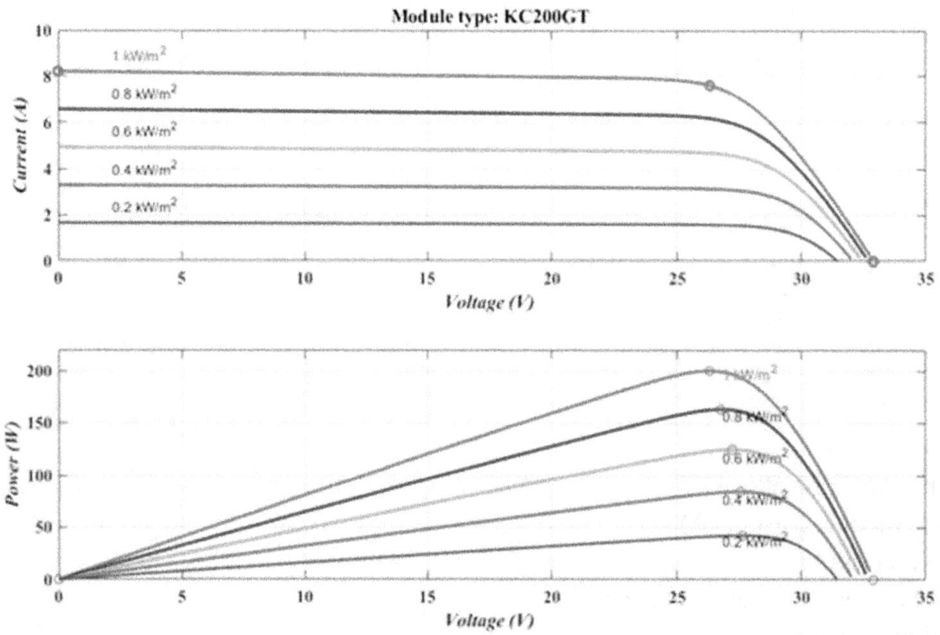

circuit voltage has decreased; however, the short circuit current has increased to some extent, hence the maximum power output has decreased.

Maximum Power Point Tracking (MPPT)

The maximum power delivered by a PV module is heavily dependent on meteorological conditions. To make an optimal use of PV sources, it is necessary to operate at the MPP. This can happen by varying the duty cycle of dc-dc converter. In fact, the search for the MPP must be carried out automatically. This is quite possible by adopting one of the adaptation approaches known as Maximum Power Point Tracking (MPPT) techniques (Yadav et al., 2020).

There are many such methods to track the maximum power such as incremental conductance, perturb and observe, genetic algorithm, fractional open-circuit voltage, etc. In this paper, due to its many advantages the perturb and observe algorithm is adopted. In fact, it is easy to implement using various controllers such as Arduino, microcontroller, etc. The maximum power point determination speed can be controlled by varying the perturbation value(Mousa et al., 2021). Figure 5 depicts the P&O algorithm.

DC-DC Boost Converter Modeling

The dc-dc converter is a power electronics circuit, which uses an inductor, a transformer, or a capacitor as an energy-storage element to convert electrical power from one voltage level into another voltage level by switching action (Vangari et al., 2015).

Figure 4. I_{pv}-V_{pv} and P_{pv}-V_{pv} characteristics of KC200GT module, (varying temperature) and insolation=1000w/m²

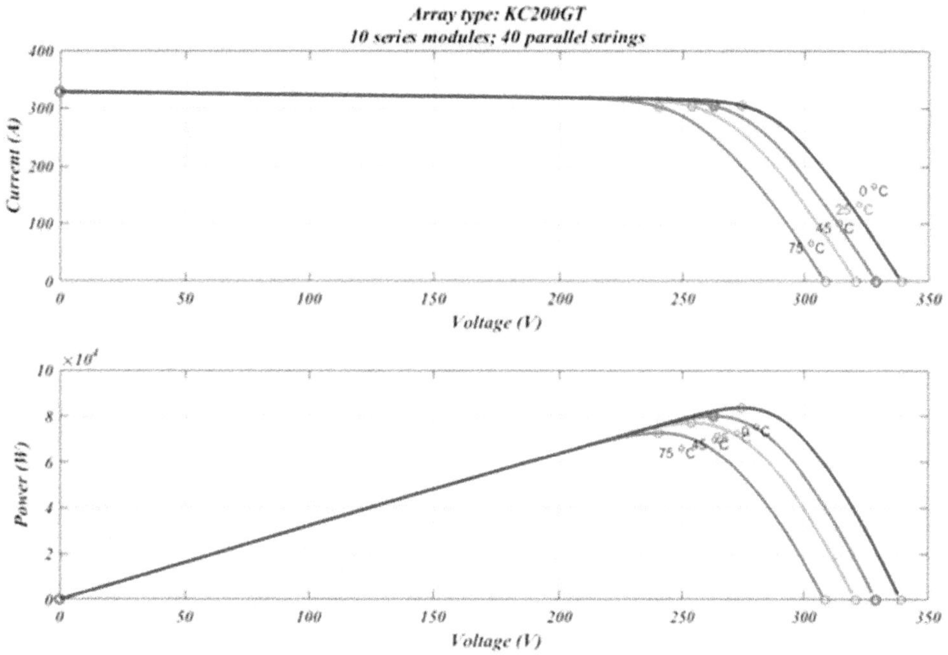

Figure 5. P&O algorithm representation

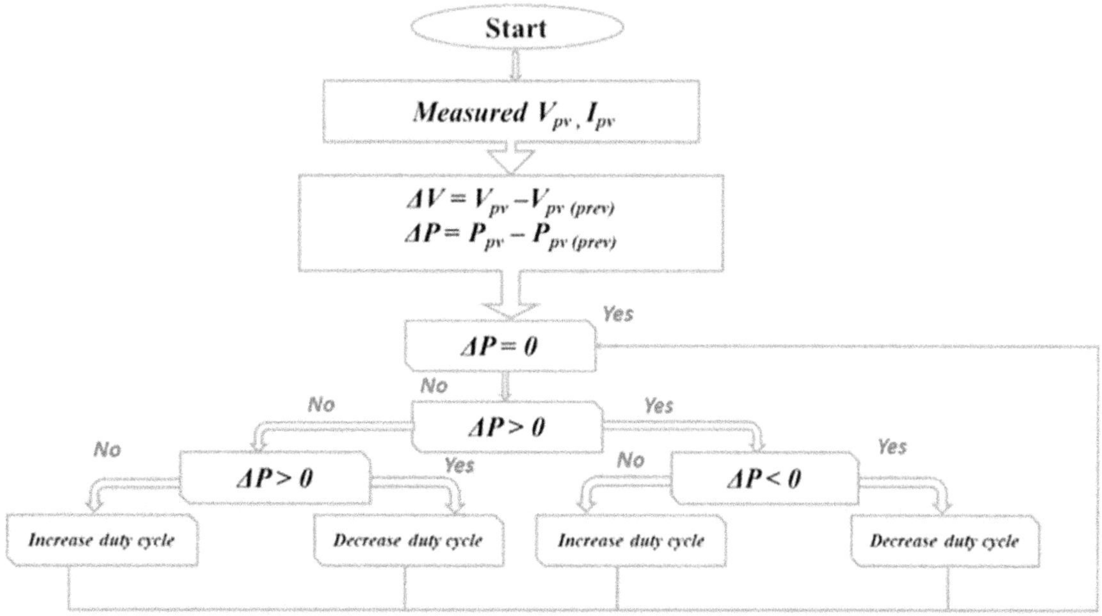

The advantage of using a dc-dc converter is two fold: First is the efficiency. In ideal conditions, all the elements in the circuit, such as inductor, capacitor, switch and diode, don't consume any energy. So the ideal efficiency is 100%. In practice, the efficiency of dc/dc converter can exceed 90%, which is very high for power transformation (Rosas-Caro et al., 2010).

The other advantage is the ease to control and integrate. The voltage transformation in dc/dc converter is achieved by changing the duty cycle of the pulse train, which is called PWM (pulse width modulation). The power needed for the pulse train is very small and can be neglected. Moreover the pulse train can be easily constructed (Rex & Praba, 2020).

Because of the existence of the switches, the boost converter works in two modes. When the switch is closed, the inductor stores energy and the capacitor releases energy. When the switch is open, the inductor releases energy and the capacitor stores energy (Tan & Hoo, 2015).

Figure 6. DC-DC boost converter topology

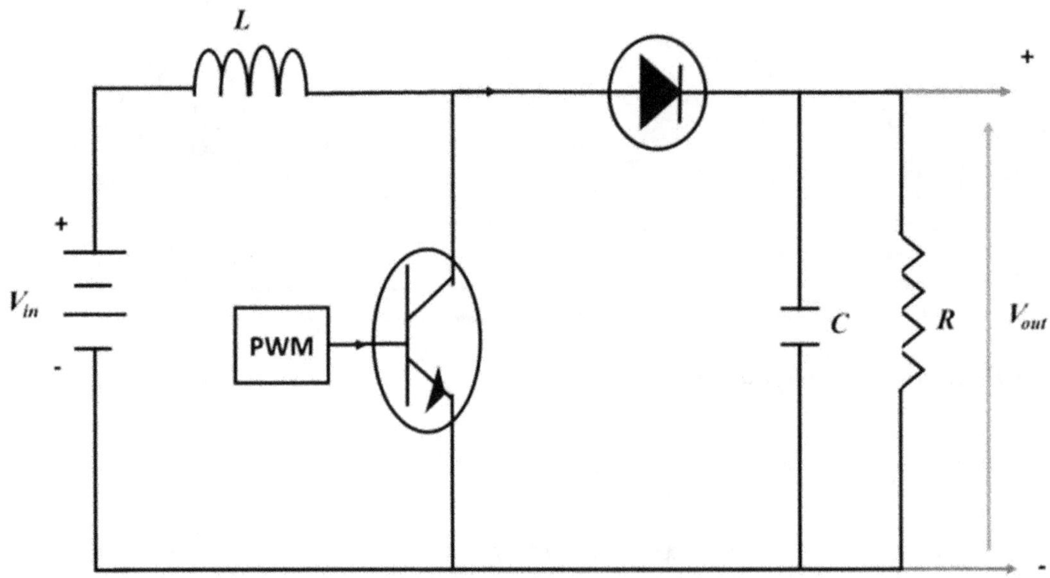

The first law involves the energy balance (3), which requires that the input energy equals the output energy:

$$P_{in} = P_{out} \Rightarrow I_{in}V_{in} = I_{out}V_{out} \tag{3}$$

The second law is the charge balance (4), which means the input charge is equal to the output charge. Because of the switch, the input current can only provide charge to the output side when the switch is open, and the time is (1-α) T in one period (T).

$$Q_{in} = Q_{out} \Rightarrow I_{in}(1-\alpha)T = I_{out}T \tag{4}$$

Using the two equations we can after some algebra, derive the basic relationship between the input voltage and output voltage.

$$V_{out} = \frac{V_{in}}{1-\alpha} \tag{5}$$

Where α is the duty cycle, which is a positive number less than 1. From the relationship we can see clearly that $V_{out} > V_{in}$.

The model equations of this converter are the following:

- When the switch is closed:

$$L\frac{di_L}{dt} = V_{in} \tag{6}$$

$$C\frac{dV_C}{dt} = -\frac{V_C}{R} \tag{7}$$

- When the switch is open:

$$L\frac{di_L}{dt} = V_{in} - V_C \tag{8}$$

$$C\frac{dV_C}{dt} = i_L - \frac{V_C}{R} \tag{9}$$

THREE -CELL INVERTER MODEL

The power that exits the PV array is in direct current (DC) so it needs to be converted into alternate current (AC) to be used locally or injected into the public distribution grid. This is the main function of the inverter.

The principle scheme of a three-cell inverter connected in series with an *R,L* load is shown in Figure 7. Each cell is controlled by a binary function s_{ck} (t) є {0, 1}, *k*=1,2,3 where s_{ck} =1 indicates that the upper switch is closed and the lower switch is open.

The system of equations (1) shows the average model of the three-cell inverter.

Figure 7. Three-cell inverter topology

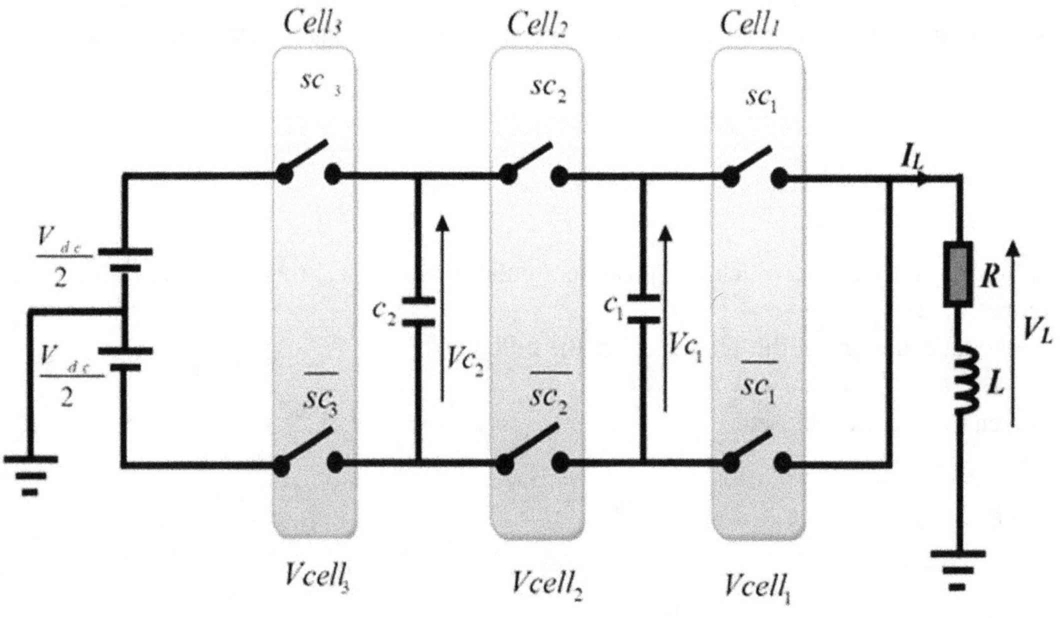

$$\begin{cases} \dot{V}_{C_1} = \dfrac{\alpha_2 - \alpha_1}{C_1} I_L = \dot{x}_1 \\ \dot{V}_{C_2} = \dfrac{\alpha_3 - \alpha_2}{C_2} I_L = \dot{x}_2 \\ \dot{I}_L = \dfrac{1}{L}(\alpha_1 - \alpha_2).V_{C1} + \dfrac{1}{L}(\alpha_2 - \alpha_3).V_{C2} + \dfrac{\alpha_3}{L} V_{dc} - \dfrac{R}{L} I_L - \dfrac{1}{2L} V_{dc} = \dot{x}_3 \end{cases} \tag{10}$$

From the equation (1) we can define the state equation of this system:

$$\left\{ \dot{X} = A.X + B(X).u \right. \tag{11}$$

With

A= (0, 0, R/L) ^{t} (12)

$$B(X)=\begin{bmatrix} -\frac{x_3}{C_1} & \frac{x_3}{C_1} & 0 \\ 0 & -\frac{x_3}{C_2} & \frac{x_3}{C_2} \\ \frac{x_1}{L} & \frac{x_2-x_1}{L} & \frac{V_{dc}-x_2}{L} \end{bmatrix} \tag{13}$$

$X=(x_1, x_2, x_3)^t$ and $u=(\alpha_1, \alpha_2, \alpha_3)$ t represent the cyclic reports vector.Where Vc_1, Vc_2 and I_L represent the average values of the capacitor's voltages and the load current respectively.

The equilibrium state of a multicellular inverter is reached when the cell's voltage assumes the same value as $V_{Cellk}=V_{dc}/p, k=1,\ldots,p$. At this desired situation, the voltage across the floating capacitors should be maintained at the following references (Hamida, et al., 2019):

$$V_{ck}=k\frac{V_{dc}}{3},\ k=1,2,3. \tag{14}$$

Capacitor's Voltages and Load Current Control Strategy

The duty cycle modulation method is used to control the capacitor voltage of the three-cell inverter. The principle of this method is based on a modification of the cyclic reports: if the cyclic reports of the cell_{k+1} and cell_i are different, the average current I_{ck} in the capacitor C_k is not equal to zero, the voltage V_{ck} evolves (Hamida et al., 2018). When an imbalance occurs on a capacitor voltage V_{ck}, the average current I_{ck} which crosses the capacitor for a cutting period T_c is such as the voltage V_{ck} tends towards its balance value. The variation of the capacitor voltage ΔV_{ck} is obtained by equation (15) (Hanafi et al., 2014).

$$\Delta V_{ck}(t+T_c)-V_{ck}(t)=\frac{k.V_{dc}}{p}-V_{ck}=\frac{I_{ck}T_c}{C_k} \tag{15}$$

The current I_{ck} can be expressed according of the average load current I_L and the cyclic reports α_{k+1} and α_k. Replacing it in equation (15), we obtain:

$$\Delta V_{ck}=\frac{T_c}{C_k}I_L[\alpha_{k+1}-\alpha_k]=\frac{T_c}{C_k}I_L M_k \tag{16}$$

Where:

$$M_k=\alpha_{k+1}-\alpha_k \text{ and } I_{ck}=I_L[\alpha_{k+1}-\alpha_k] \tag{17}$$

Based on equations (15) and (16) M_k can be written as below:

$$M_k = \frac{C_k}{T_c I_L}[\frac{kV_{dc}}{p} V_{ck}] \tag{18}$$

We introduce into the equation (18) the gain G which will be chosen to avoid any saturation of the cyclic reports and ensure an important balancing dynamic in the starting phase, taking into account the sign of the charging current which is sinusoidal.

$$M_k = sign(i_L) G \frac{C_k}{T_c I_L}[\frac{kV_{dc}}{p} V_{ck}] \tag{19}$$

The control law can be synthesized by equations (20)

$$\begin{cases} \alpha_3 = \alpha_{ref} \\ \alpha_2 = \alpha_3 \text{-} sign\ (i_L)\ G.\frac{C_2}{T_c I_L}\ [\frac{2V_{dc}}{3} \text{-} V_{c2}] \\ \alpha_1 = \alpha_2 \text{-} sign\ (i_L)\ G.\frac{C_1}{T_c I_L}\ [\frac{V_{dc}}{3} \text{-} V_{c1}] \end{cases} \tag{20}$$

The control signal of each cell is generated by the intersection between carrying triangular signal with frequency f_c and the sinusoidal modulating signal with frequency 50 Hz.

The Proportional Integral (PI) controller is used to control the load current and generate the cyclic duty reference. Figure 8 shows the basic control strategy.

RESULTS AND DISCUSSION

The simulation results are obtained using MATLAB/SIMULINK. The specifications used in implementing the flying capacitor inverter are the following:

- Input voltage V_{dc}= 381.3 V
- Floating capacitors C_1=C_2= 47 µF
- Resistive load R= 33 Ω
- Inductive load L= 50 mH
- Cutting frequency f_c = 16 kHz

The PV and the DC-DC converter output voltage are shown in the Figure 9 and Figure 10.

Figure 9 shows the PV output voltage, we can see that the voltage value stabilizes around of V_{oc} *3.

Figure 8. The basic scheme of the proposed PI- cyclic reports modulation control

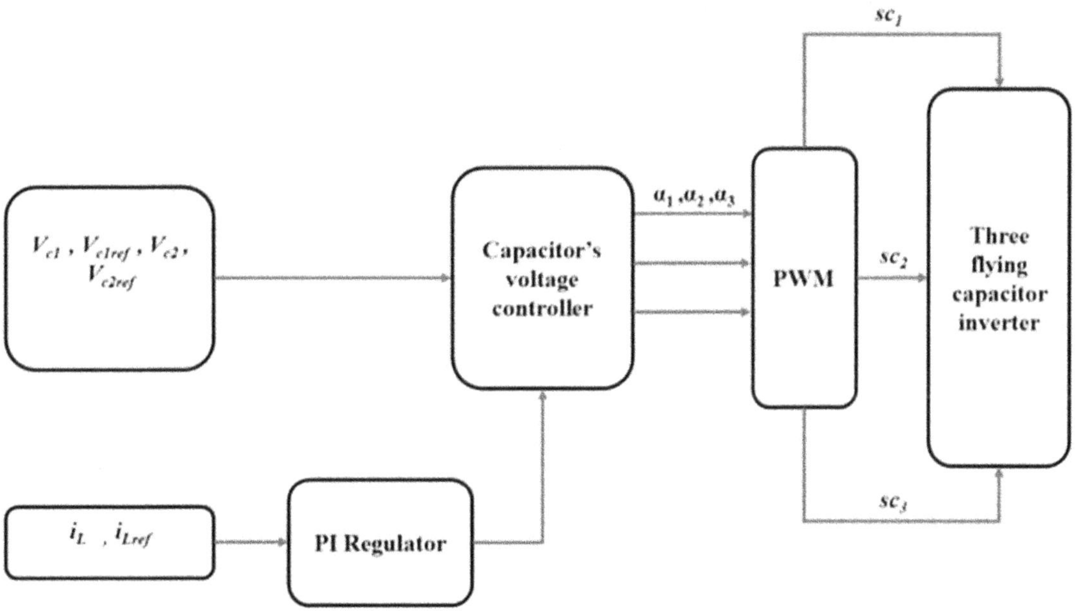

Figure 9. PV output voltage with irradiance =1000 W/m2 at Tc = 25 ◦C

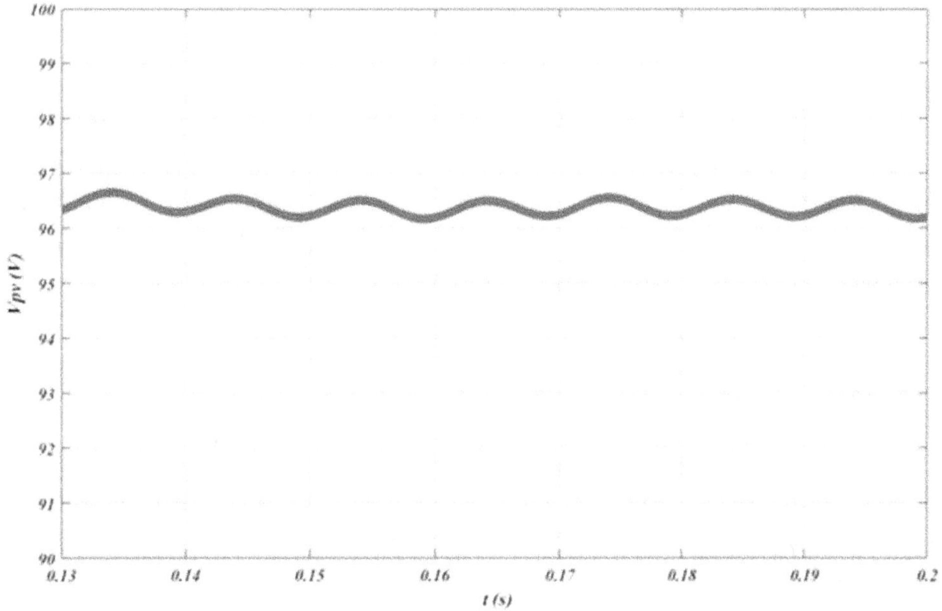

Figure 10 shows the DC supply voltage of the three-cell inverter delivered by the dc-dc boost converter; this voltage is kept at a desired reference voltage.

Figure 10. DC-DC boost converter output voltage

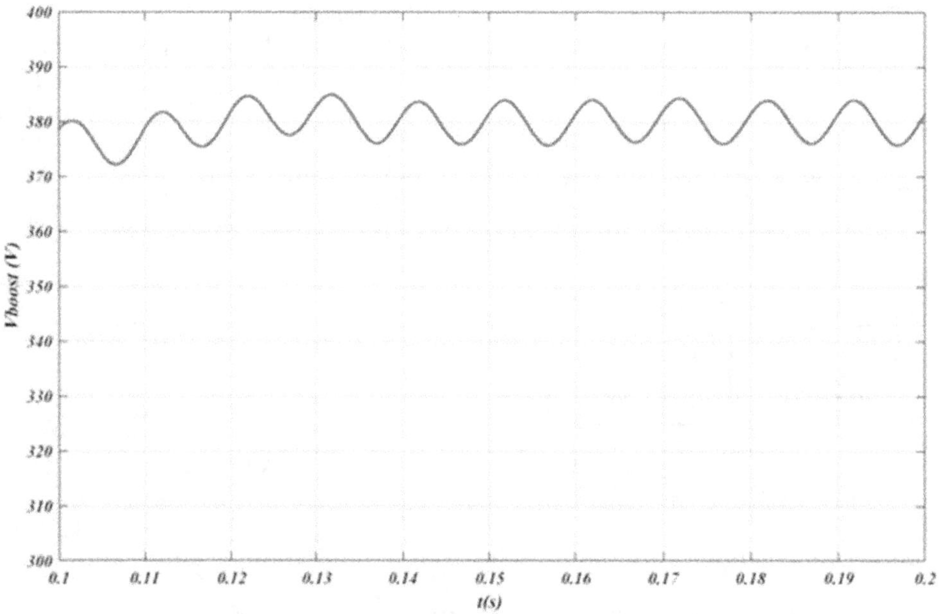

Figure 11 and Figure 12 show the evolution of the floating voltages and the load current, we find that the floating voltages join their references respectively ($V_{dc}/3$ and $2V_{dc}/3$) and the load current remains equal to the desired value.

Figure 13 shows that the group of harmonics is centered around 48 KHz ($3{*}f_c$) which corresponding to the output frequency of the converter; this represents a very good property compared to the classical converters, since it is the frequency seen by the load. The p proposed control gives a low THD of 0.13%.

As shown in Figure 14, the output voltage takes its four possible values. We notice that the output voltage is computed between ± 190.65 V (± $V_{dc}/2$).

CONCLUSION

In this work, a three-cell inverter topology has been proposed for the solar PV energy system. To extract the maximum power from the PV panel the P&O algorithm based MPPT technique has been used. To achieve the performances of the flying capacitor inverter, a cyclic report modulation control has been implemented. Simulation results show the convergence of the capacitor's voltage and the output current to a neighborhood of the value of the desired references. It has been noticed that the used inverter delivers a high-quality signal with a very low THD (0.13%). As the topology is able to deliver power with less

Figure 11. Floating voltages evolution

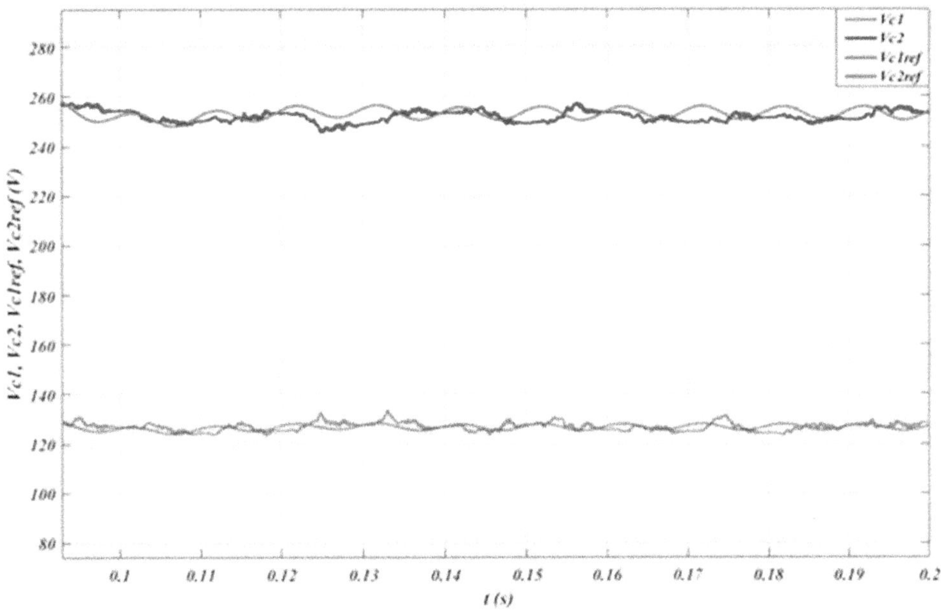

Figure 12. Load current evolution

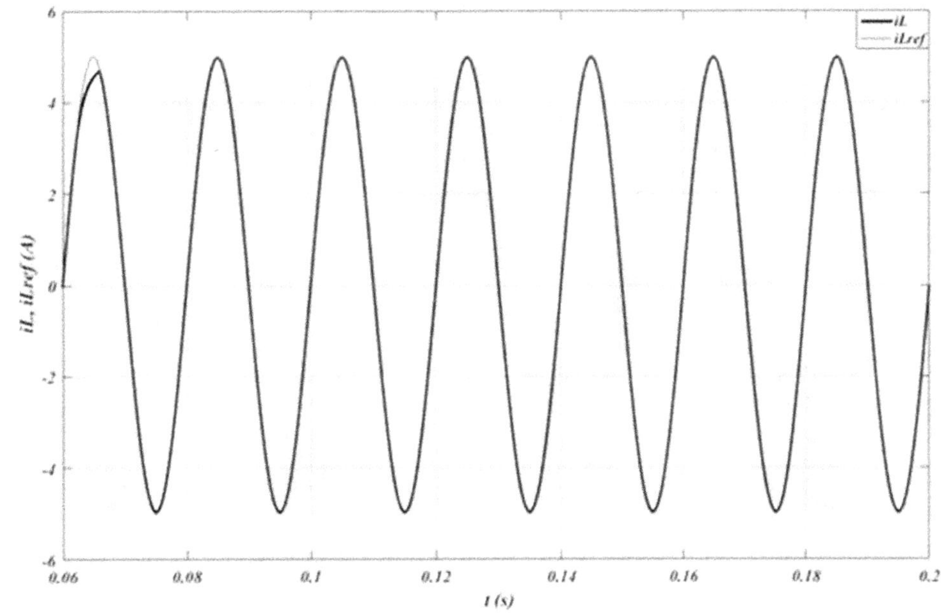

Figure 13. FFT harmonic spectrum of the load current

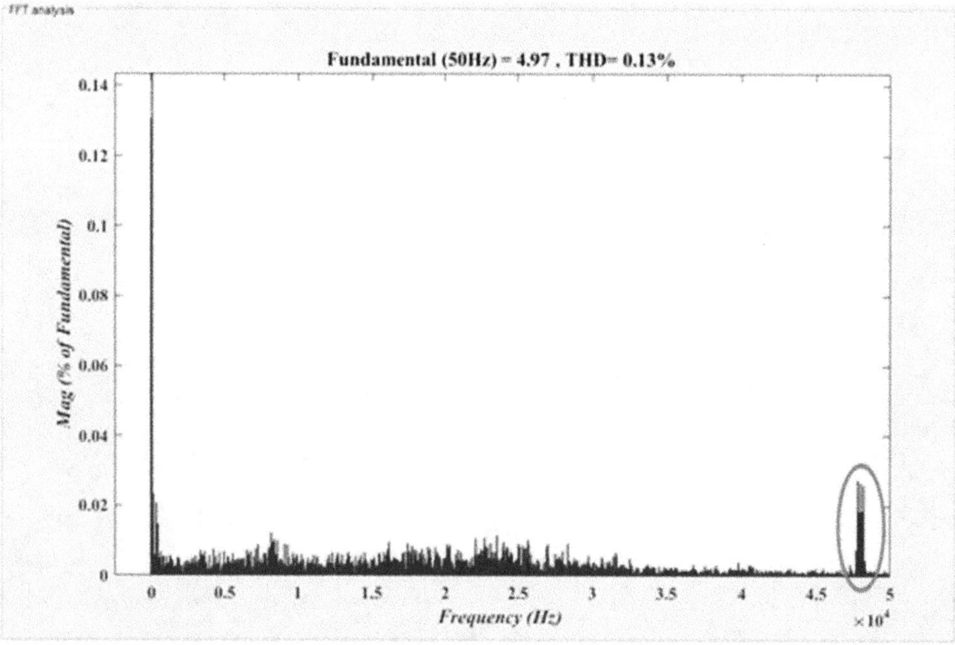

Figure 14. Load voltage evolution

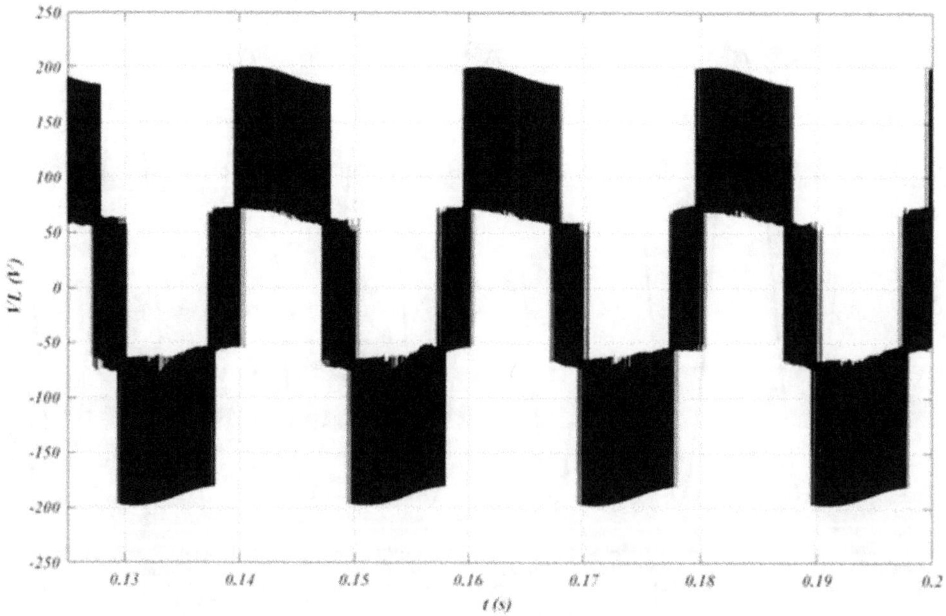

harmonic content, this is permissible according to IEEE standards, this can efficiently be applied to the grid connected systems and is well-suited to the renewable energy sources. Further, the objective is to connect the PV system to the grid through a three-phase flying capacitor inverter.

REFERENCES

Aboubakeur, H., Mohamed, B., & Nabil, A. (2018). A Fast MPPT Control Technique Using PID Controller in a Photovoltaic System. *2018 International Conference on Applied Smart Systems (ICASS)*, 1-5.

Amara, K., Fekik, A., Hocine, D., Bakir, M. L., Bourennane, E.-B., Malek, T. A., & Malek, A. (2018). Improved performance of a PV solar panel with adaptive neuro fuzzy inference system ANFIS based MPPT. *2018 7th International Conference on Renewable Energy Research and Applications (ICRERA)*, 1098-1101.

Amghar, B., El Abbassi, I., Mladjo, M. A. M., & Darcherif, A. (2014). A new hybrid control method for multi-cell choppers. Application to power renewable energy systems. *2014 International Renewable and Sustainable Energy Conference (IRSEC)*, 447-452. 10.1109/IRSEC.2014.7059839

Bahgat, A. B. G., Helwa, N. H., Ahmad, G. E., & El Shenawy, E. T. (2005). Maximum power point traking controller for PV systems using neural networks. *Renewable Energy*, *30*(8), 1257–1268. doi:10.1016/j.renene.2004.09.011

Bakas, P., Okazaki, Y., Shukla, A., Patro, S. K., Ilves, K., Dijkhuizen, F., & Nami, A. (2020). Review of hybrid multilevel converter topologies utilizing thyristors for HVDC applications. *IEEE Transactions on Power Electronics*, *36*(1), 174–190. doi:10.1109/TPEL.2020.2997961

Ciobotaru, M., Teodorescu, R., & Blaabjerg, F. (2006). Control of single-stage single-phase PV inverter. *EPE Journal*, *16*(3), 20–26. doi:10.1080/09398368.2006.11463624

di Benedetto, M., Lidozzi, A., Solero, L., Crescimbini, F., & Grbović, P. J. (2021). High-Performance 3-Phase 5-Level E-Type Multilevel–Multicell Converters for Microgrids. *Energies*, *14*(4), 843. doi:10.3390/en14040843

Escalante, M. F., & Erdmann, A. P. (2021). Estimation of the capacitor voltages in flying capacitor multi-level converters. *IET Power Electronics*, *14*(3), 651–665. doi:10.1049/pel2.12053

Fekik, A., Azar, A. T., Kamal, N. A., Serrano, F. E., Hamida, M. L., Denoun, H., & Yassa, N. (2020). Maximum Power Extraction from a Photovoltaic Panel Connected to a Multi-cell Converter. *International Conference on Advanced Intelligent Systems and Informatics*, 873-882.

Fekik, A., Denoun, H., Azar, A. T., Kamal, N. A., Zaouia, M., Benyahia, N., Hamida, M. L., Benamrouche, N., & Vaidyanathan, S. (2021). Direct power control of three-phase PWM-rectifier with backstepping control. In *Backstepping Control of Nonlinear Dynamical Systems* (pp. 215–234). Elsevier. doi:10.1016/B978-0-12-817582-8.00017-9

Fekik, A., Denoun, H., Azar, A. T., Zaouia, M., Benyahia, N., Hamida, M. L., Benamrouche, N., & Vaidyanathan, S. (2018). Artificial neural network for PWM rectifier direct power control and DC voltage control. In *Advances in System Dynamics and Control* (pp. 286–316). IGI Global. doi:10.4018/978-1-5225-4077-9.ch010

Fekik, A., Denoun, H., Benamrouche, N., Benyahia, N., Badji, A., & Zaouia, M. (2016). Comparative analysis of direct power control and direct power control with space vector modulation of PWM rectifier. *2016 4th International Conference on Control Engineering & Information Technology (CEIT)*, 1-6.

Fekik, A., Denoun, H., Benamrouche, N., Benyahia, N., & Zaouia, M. (2015). A fuzzy–logic based controller for three phase PWM rectifier with voltage oriented control strategy. *International Journal Of Circuits, Systems And. Signal Processing*, *9*, 412–419.

Fekik, A., Denoun, H., Zaouia, M., Benyahia, N., Benamrouche, N., Badji, A., & Vaidyanathan, S. (2017). Improvement of the performances of the direct power control using space vector modulation of three phases PWM-rectifier. *International Journal of Control Theory and Applications*, *10*, 133–145.

Fekik, A., Denoun, H., Zaouia, M., Hamida, M. L., & Vaidyanathan, S. (2019). A new switching table based neural network for direct power control of three-phase PWM-rectifier. *International Journal of Modelling. Identification and Control*, *33*(4), 358–368. doi:10.1504/IJMIC.2019.107484

Fujita, H., Yamasaki, T., & Akagi, H. (1998). A hybrid active filter for damping of harmonic resonance in industrial power systems. *IEEJ Transactions on Industry Applications*, *118*(10), 1193–1200. doi:10.1541/ieejias.118.1193

Green, T. C., & Marks, J. H. (2005). Control techniques for active power filters. *IEE Proceedings. Electric Power Applications*, *152*(2), 369–381. doi:10.1049/ip-epa:20040759

Guo, P., Xu, Q., Yue, Y., Ma, F., He, Z., Luo, A., & Guerrero, J. M. (2019). Analysis and control of modular multilevel converter with split energy storage for railway traction power conditioner. *IEEE Transactions on Power Electronics*, *35*(2), 1239–1255. doi:10.1109/TPEL.2019.2917065

Hamida, M. L., Denoun, H., Fekik, A., Benyahia, N., & Benamrouche, N. (2018). Cyclic reports modulation control strategy for a five cells inverter. *2018 International Conference on Electrical Sciences and Technologies in Maghreb (CISTEM)*, 1-5.

Hamida, M. L., Denoun, H., Fekik, A., Kais, D., Benamrouche, N., & Amara, K. (2019). Fuzzy Logic-Feedback Linearization Control for a Three-Cell Flying Capacitor Converter. *2019 International Conference on Advanced Electrical Engineering (ICAEE)*, 1-5. 10.1109/ICAEE47123.2019.9015121

Hamida, M. L., Denoun, H., Fekik, A., & Vaidyanathan, S. (2019). Control of separately excited DC motor with series multi-cells chopper using PI-Petri nets controller. *Nonlinear Engineering*, *8*(1), 32–38. doi:10.1515/nleng-2017-0174

Hanafi, S., Fellah, M. K., Guebli, A., & Chiali, E. (2014). Control with cyclic reports modulation of 3x2 stacked multicellular DC/AC converter. *Cell*, *11*, 21.

Hassan, A. S., Cipcigan, L., & Jenkins, N. (2017). Optimal battery storage operation for PV systems with tariff incentives. *Applied Energy*, *203*, 422–441. doi:10.1016/j.apenergy.2017.06.043

Kesarwani, K., & Stauth, J. T. (2015). Resonant and multi-mode operation of flying capacitor multi-level dc-dc converters. *2015 IEEE 16th Workshop on Control and Modeling for Power Electronics (COMPEL)*, 1-8.

Knez, M., & Jereb, B. (2013). Solar power plants–alternative sustainable approach to greener environment : A case of Slovenia. *Sustainable Cities and Society*, *6*, 27–32. doi:10.1016/j.scs.2012.07.002

Manna, S., & Akella, A. K. (2021). Comparative analysis of various P & O MPPT algorithm for PV system under varying radiation condition. *2021 1st International Conference on Power Electronics and Energy (ICPEE)*, 1-6.

Marzouki, A., Hamouda, M., & Fnaiech, F. (2015). A hybrid controller for PWM active rectifiers based LCL filters. *COMPEL: The International Journal for Computation and Mathematics in Electrical and Electronic Engineering*.

Merabet Boulouiha, H., Khodja, M., Rahiel, D., Allali, A., Kaddour, F., & Denai, M. (2019). Power quality enhancement in electricity grids with wind energy using multicell converters and energy storage. *Journal of Renewable and Sustainable Energy*, *11*(1), 013302. doi:10.1063/1.5043292

Meynard, T. A., Foch, H., Forest, F., Turpin, C., Richardeau, F., Delmas, L., Gateau, G., & Lefeuvre, E. (2002). Multicell converters : Derived topologies. *IEEE Transactions on Industrial Electronics*, *49*(5), 978–987. doi:10.1109/TIE.2002.803189

Meynard, T. A., Foch, H., Thomas, P., Courault, J., Jakob, R., & Nahrstaedt, M. (2002). Multicell converters : Basic concepts and industry applications. *IEEE Transactions on Industrial Electronics*, *49*(5), 955–964. doi:10.1109/TIE.2002.803174

Mosa, M., Shadmand, M. B., Balog, R. S., & Rub, H. A. (2017). Efficient maximum power point tracking using model predictive control for photovoltaic systems under dynamic weather condition. *IET Renewable Power Generation*, *11*(11), 1401–1409. doi:10.1049/iet-rpg.2017.0018

Mousa, H. H., Youssef, A.-R., & Mohamed, E. E. (2021). State of the art perturb and observe MPPT algorithms based wind energy conversion systems : A technology review. *International Journal of Electrical Power & Energy Systems*, *126*, 106598. doi:10.1016/j.ijepes.2020.106598

Othman, S., Allali, M. A.-E., Mrad, I., Chariag, D. E., Ghanes, M., Sbita, L., & Barbot, J.-P. (2020). Robust hybrid control based on Petri Nets for a multicellular inverter. *6th IEEE International Energy Conference (ENERGYCon)*, 143-148.

Rajagopal, R., Palanisamy, K., & Paramasivam, S. (2019). Shunt Active Filter Based on 7-Level Cascaded Multilevel Inverter for Harmonic and Reactive Power Compensation. *Journal of Electrical Engineering & Technology*, *14*(4), 1543–1551. doi:10.100742835-019-00201-1

Rao, B. N., Suresh, Y., Panda, A. K., Naik, B. S., & Jammala, V. (2020). Development of cascaded multilevel inverter based active power filter with reduced transformers. *CPSS Transactions on Power Electronics and Applications*, *5*(2), 147–157. doi:10.24295/CPSSTPEA.2020.00013

Rex, S. R., & Praba, D. M. S. R. (2020). Design of PWM with four transistor comparator for DC–DC boost converters. *Microprocessors and Microsystems*, *72*, 102844. doi:10.1016/j.micpro.2019.07.003

Rosa-Clot, M., & Tina, G. M. (2017). *Submerged and Floating Photovoltaic Systems : Modelling, Design and Case Studies*. Academic Press.

Rosas-Caro, J. C., Ramirez, J. M., Peng, F. Z., & Valderrabano, A. (2010). A DC–DC multilevel boost converter. *IET Power Electronics*, *3*(1), 129–137. doi:10.1049/iet-pel.2008.0253

Sahoo, B., Routray, S. K., & Rout, P. K. (2019). Artificial neural network-based PI-controlled reduced switch cascaded multilevel inverter operation in wind energy conversion system with solid-state transformer. *Iranian Journal of Science and Technology. Transaction of Electrical Engineering*, *43*(4), 1053–1073. doi:10.100740998-019-00215-5

Tan, R. H., & Hoo, L. Y. (2015). DC-DC converter modeling and simulation using state space approach. *2015 IEEE Conference on Energy Conversion (CENCON)*, 42-47. 10.1109/CENCON.2015.7409511

Vangari, A., Haribabu, D., & Sakamuri, J. N. (2015). Modeling and control of DC/DC boost converter using K-factor control for MPPT of solar PV system. *2015 International Conference on Energy Economics and Environment (ICEEE)*, 1-6. 10.1109/EnergyEconomics.2015.7235087

Yadav, I., Maurya, S. K., & Gupta, G. K. (2020). A literature review on industrially accepted MPPT techniques for solar PV system. *International Journal of Electrical & Computer Engineering, 10*(2).

Yofo, E., & Djondiné, P. (2019). Parallel Active Filtering based on a three–phase multicell converter. *SCIREA Journal of Electrical Engineering*, *4*, 48–62.

Compilation of References

Abdalla, O., Rezk, H., & Ahmed, E. M. (2019). Wind driven optimization algorithm based global MPPT for PV system under non-uniform solar irradiance. *Solar Energy*, *180*, 429–444.

AbdelAty, A. M., Azar, A. T., Vaidyanathan, S., Ouannas, A., & Radwan, A. G. (2018). Applications of continuous-time fractional order chaotic systems. In *Mathematical Techniques of Fractional Order Systems', Advances in Nonlinear Dynamics and Chaos (ANDC)* (pp. 409–449). Elsevier. doi:10.1016/B978-0-12-813592-1.00014-3

Abdelmalek, S., Azar, A. T., & Dib, D. (2018). A novel actuator fault-tolerant control strategy of DFIG-based wind turbines using Takagi-Sugeno Multiple models. *International Journal of Control, Automation, and Systems*, *16*(3), 1415–1424. doi:10.100712555-017-0320-y

Abdelsalam, I., Adam, G. P., Holliday, D., & Williams, B. W. (2015, July). Three-phase ac–dc buck–boost converter with a reduced number of switches. *IET Renewable Power Generation*, *9*(5), 494–502.

Abdel-Salam, M., El-Mohandes, M. T., & El-Ghazaly, M. (2020). An Efficient Tracking of MPP in PV Systems Using a Newly-Formulated P&O-MPPT Method Under Varying Irradiation Levels. *Journal of Electrical Engineering & Technology*, *15*(1), 501–513. doi:10.100742835-019-00283-x

Abdul-Adheem, W. R., Azar, A. T., Ibraheem, I. K., & Humaidi, A. J. (2020b). Novel Active Disturbance Rejection Control Based on Nested Linear Extended State Observers. *Applied Sciences (Basel, Switzerland)*, *10*(12), 4069. doi:10.3390/app10124069

Abdul-Adheem, W. R., Ibraheem, I. K., Azar, A. T., & Humaidi, A. J. (2020a). Improved Active Disturbance Rejection-Based Decentralized control for MIMO Nonlinear Systems: Comparison with The Decoupled Control Scheme. *Applied Sciences (Basel, Switzerland)*, *10*(7), 2515. doi:10.3390/app10072515

Abe, C. F., Dias, J. B., Notton, G., & Poggi, P. (2020). Computing Solar Irradiance and Average Temperature of Photovoltaic Modules from the Maximum Power Point Coordinates. *IEEE Journal of Photovoltaics*, *10*(2), 655–663. doi:10.1109/JPHOTOV.2020.2966362

Abedi, A., Rezaie, B., Khosravi, A., & Shahabi, M. (2020). DC-bus Voltage Control based on Direct Lyapunov Method for a Converter-based Stand-alone DC Micro-grid. *Electric Power Systems Research*, *187*.

Aboubakeur, H., Mohamed, B., & Nabil, A. (2018). A Fast MPPT Control Technique Using PID Controller in a Photovoltaic System. *2018 International Conference on Applied Smart Systems (ICASS)*, 1-5.

Afshari, E. (2017, July 31). Control Strategy for Three-Phase Grid-Connected PV Inverters Enabling Current Limitation Under Unbalanced Faults. *IEEE Transactions on Industrial Electronics*, *64*(11), 8908–8918.

Ahmad, T., & Zhang, D. (2020). A critical review of comparative global historical energy consumption and future demand: The story told so far. *Energy Reports*, *6*, 1973–1991. doi:10.1016/j.egyr.2020.07.020

Ahmad, Z., & Singh, S. N. (2017). An improved single phase transformerless inverter topology for grid connected PV system with reduce leakage current and reactive power capability. *Solar Energy*, *157*, 133–146. doi:10.1016/j.solener.2017.08.007

Ajeil, F. H., Ibraheem, I. K., Azar, A. T., & Humaidi, A. J. (2020). Grid-Based Mobile Robot Path Planning Using Aging-Based Ant Colony Optimization Algorithm in Static and Dynamic Environments. *Sensors (Basel)*, *20*(7), 1880. doi:10.339020071880 PMID:32231091

Ajeil, F., Ibraheem, I. K., Azar, A. T., & Humaidi, A. J. (2020). Autonomous Navigation and Obstacle Avoidance of an Omnidirectional Mobile Robot Using Swarm Optimization and Sensors Deployment. *International Journal of Advanced Robotic Systems*, *17*(3), 1–15. doi:10.1177/1729881420929498

Alain, K. S. T., Romanic, K., Azar, A. T., Vaidyanathan, S., Bertrand, F. H., & Adele, N. M. (2018a), Dynamics analysis and synchronization in relay coupled fractional order colpitts oscillators. In Advances in Systems Analysis, Software Engineering, and High Performance Computing (ASASEHPC). IGI Global. doi:10.4018/978-1-5225-4077-9.ch011

Alain, K. S. T., Azar, A. T., Bertrand, F. H., & Romanic, K. (2019). Robust observer-based synchronisation of chaotic oscillators with structural perturbations and input nonlinearity. *International Journal of Automation and Control*, *13*(4), 387–412. doi:10.1504/IJAAC.2019.100467

Alain, K. S. T., Azar, A. T., Kengne, R., & Bertrand, F. H. (2020). Stability analysis and robust synchronisation of fractional-order modified colpitts oscillators. *International Journal of Automation and Control*, *14*(1), 52–79. doi:10.1504/IJAAC.2020.103806

Alain, K. S. T., Romanic, K., Azar, A. T., Vaidyanathan, S., Bertrand, F. H., & Adele, N. M. (2018b). *Dynamics Analysis and Synchronization in Relay Coupled Fractional Order Colpitts Oscillators. Advances in System Dynamics and Control.* IGI-Global. doi:10.4018/978-1-5225-4077-9

Al-Barazanchi, S. A. M., & Vural, A. M. (2015, December). Modeling and intelligent control of a stand-alone PV-Wind-Diesel-Battery hybrid system. In *2015 International Conference on Control, Instrumentation, Communication and Computational Technologies (ICCICCT)* (pp. 423-430). IEEE.

Alexander, V. (2015). Application of queuing theory: Analysis of services of commercial banks (A case study of branches of some commercial banks in Abuja, Nigeria). *Journal of Science Teacher Education*, 48.

Ali, Park, Yu, Murata, & Tamura. (2009). Improvement of wind-generator stability by fuzzy-logic-controlled SMES. *Industry Applications, IEEE Transactions on, 45*, 1045–1051.

Ali, A., Hasan, A. N., & Marwala, T. (2014, October). Perturb and observe based on fuzzy logic controller maximum power point tracking (MPPT). In *2014 International Conference on Renewable Energy Research and Application (ICRERA)* (pp. 406-411). IEEE.

Ali, M. M., Youssef, A. R., Ali, A. S., & Abdel-Jaber, G. T. (2020). Variable step size PO MPPT algorithm using model reference adaptive control for optimal power extraction. *International Transactions on Electrical Energy Systems*, *30*(1), e12151. doi:10.1002/2050-7038.12151

Ali, Z. (2018, July). Three-phase phase-locked loop synchronization algorithms for grid-connected renewable energy systems: A review. *Renewable & Sustainable Energy Reviews*, *90*, 434–452.

Aljohani, T. M., Ebrahim, A. F., & Mohammed, O. (2019). Single and Multiobjective Optimal Reactive Power Dispatch Based on Hybrid Artificial Physics–Particle Swarm Optimization. *Energies*, *12*(12), 2333. doi:10.3390/en12122333

Allouche, A. (2018). A PLL based mechanical faults detection in PMSM at variable speed. *IFAC-PapersOnLine*, *51*(24), 1445–1451.

Al-Shetwi, A. Q., Sujod, M. Z., & Blaabjerg, F. (2018). Low voltage ride-through capability control for single-stage inverter-based grid-connected photovoltaic power plant. *Solar Energy*, *159*, 665–681. doi:10.1016/j.solener.2017.11.027

Ama, N. 2013. PLL performance under frequency fluctuation-compliance with standards for distributed generation connected to the grid. *2013 IEEE PES Conference on Innovative Smart Grid Technologies (ISGT Latin America.*

Amara, K., Fekik, A., Hocine, D., Bakir, M. L., Bourennane, E. B., Malek, T. A., & Malek, A. (2018, October). Improved performance of a PV solar panel with adaptive neuro fuzzy inference system ANFIS based MPPT. In *2018 7th International Conference on Renewable Energy Research and Applications (ICRERA)* (pp. 1098-1101). IEEE. 10.1109/ICRERA.2018.8566818

Amara, K., Fekik, A., Hocine, D., Bakir, M. L., Bourennane, E.-B., Malek, T. A., & Malek, A. (2018). Improved performance of a PV solar panel with adaptive neuro fuzzy inference system ANFIS based MPPT. *2018 7th International Conference on Renewable Energy Research and Applications (ICRERA)*, 1098-1101.

Amara, K., Bakir, T., Malek, A., Hocine, D., Bourennane, E. B., Fekik, A., & Zaouia, M. (2019a). An Optimized Steepest Gradient Based Maximum Power Point Tracking for PV Control Systems. *International Journal on Electrical Engineering & Informatics*, *11*(4), 662–683. doi:10.15676/ijeei.2019.11.4.3

Amara, K., Malek, A., Bakir, T., Fekik, A., Azar, A. T., Almustafa, K. M., Bourennane, E. B., & Hocine, D. (2019b). Adaptive neuro-fuzzy inference system based maximum power point tracking for stand-alone photovoltaic system. *International Journal of Modelling. Identification and Control*, *33*(4), 311–321. doi:10.1504/IJMIC.2019.107480

Amara, Y., Boukenoui, R., Bradai, R., & Salhi, H. (2018, December). Design and Control of Two-Stage Standalone Photovoltaic Generation System. In *2018 International Conference on Communications and Electrical Engineering (ICCEE)* (pp. 1-5). IEEE.

Amendola, C., & Gonzaga, D. (2007). Fuzzy-logic control system of a variable-speed variable-pitch wind-turbine and a double-fed induction generator. *Intelligent Systems Design and Applications, 2007. ISDA 2007. Seventh International Conference on*, 252–257.

Amghar, B., El Abbassi, I., Mladjo, M. A. M., & Darcherif, A. (2014). A new hybrid control method for multi-cell choppers. Application to power renewable energy systems. *2014 International Renewable and Sustainable Energy Conference (IRSEC)*, 447-452. 10.1109/IRSEC.2014.7059839

Ammar, H. H., & Azar, A. T. (2020). Robust Path Tracking of Mobile Robot Using Fractional Order PID Controller. *The International Conference on Advanced Machine Learning Technologies and Applications (AMLTA2019)*, 370-381. 10.1007/978-3-030-14118-9_37

Ammar, H. H., Azar, A. T., Tembi, T. D., Tony, K., & Sosa, A. (2018). Design and Implementation of Fuzzy PID Controller into Multi Agent Smart Library System Prototype. *The International Conference on Advanced Machine Learning Technologies and Applications (AMLTA2018)*, 127-137. 10.1007/978-3-319-74690-6_13

Ammar, H., Ibrahim, M., Azar, A. T., & Shalaby, R. (2020b). Gray Wolf Optimization of Fractional Order Control of 3-Omni Wheels Mobile Robot: Experimental Study. *2020 16th International Computer Engineering Conference (ICENCO)*, 147-152. 10.1109/ICENCO49778.2020.9357384

Ammar, H. H., Azar, A. T., Shalaby, R., & Mahmoud, M. I. (2019a). Metaheuristic optimization of fractional order incremental conductance (fo-inc) maximum power point tracking (mppt). *Complexity*, *2019*(7687891), 1–13. doi:10.1155/2019/7687891

Anay, K. L., & Pollitt, M. G. (2020, July 15). Reactive power procurement: A review of current trends. *Applied Energy, 270.*

An, H. (2012). Modeling and voltage-control of variable-speed SCAG-based wind farm. *Renewable Energy, 42*, 28–35.

Antoniewicz, P., & Kazmierkowski, P. (2008). Virtual-flux-based predictive direct power control of AC/DC converters with online inductance estimation. *IEEE Transactions on Industrial Electronics, 55*(12), 4381–4390. doi:10.1109/TIE.2008.2007519

Apata, O. & Oyedokun, D. T. O. (2020). An overview of control techniques for wind turbine systems. *Scientific African, 10.*

Arricibita, D., Sanchis, P., González, R., & Marroyo, L. (2017, December). Impedance Emulation for Voltage Harmonic Compensation in PWM Stand-Alone Inverters. *IEEE Transactions on Energy Conversion, 32*(4), 1335–1344.

Atia, Y., & Salem, M. (2015). Novel deadbeat power control strategy for grid connected systems. *Journal of Electrical Systems and Information Technology, 2*(2), 242–256. doi:10.1016/j.jesit.2015.02.001

Attia, H. (2019). High performance PV system based on artificial neural network MPPT with PI controller for direct current water pump applications. *International Journal of Power Electronics and Drive Systems, 10*(3), 1329. doi:10.11591/ijpeds.v10.i3.pp1329-1338

Azar, A. T., & Kamal, N. A. (2021). Handbook of Research on Modeling, Analysis, and Control of Complex Systems. IGI Global.

Azar, A. T., & Serrano, F. E. (2015a). Stabilization and Control of Mechanical Systems with Backlash. In Advanced Intelligent Control Engineering and Automation. IGI Global. doi:10.4018/978-1-4666-7248-2.ch001

Azar, A. T., & Serrano, F. E. (2015b). Design and Modeling of Anti Wind Up PID Controllers. In Complex system modelling and control through intelligent soft computations. Springer-Verlag. . 1. doi:10.1007/978-3-319-12883-2_1

Azar, A. T., & Serrano, F. E. (2015c). Adaptive Sliding mode control of the Furuta pendulum. In Advances and Applications in Sliding Mode Control systems. Springer-Verlag GmbH. doi:10.1007/978-3-319-11173-5_1

Azar, A. T., & Serrano, F. E. (2015d). Deadbeat Control for Multivariable Systems with Time Varying Delays. In Chaos Modeling and Control Systems Design. Springer-Verlag GmbH. doi:10.1007/978-3-319-13132-0_6

Azar, A. T., & Serrano, F. E. (2018). Fractional Order Sliding Mode PID Controller/Observer for Continuous Nonlinear Switched Systems with PSO Parameter Tuning. In *The International Conference on Advanced Machine Learning Technologies and Applications (AMLTA2018).* Springer.

Azar, A. T., & Serrano, F. E. (2018). Fractional Order Sliding Mode PID Controller/Observer for Continuous Nonlinear Switched Systems with PSO Parameter Tuning. *The International Conference on Advanced Machine Learning Technologies and Applications (AMLTA2018),* 13-22.

Azar, A. T., & Serrano, F. E. (2019). Fractional Order Two Degree of Freedom PID Controller for a Robotic Manipulator with a Fuzzy Type-2 Compensator. In *Proceedings of the International Conference on Advanced Intelligent Systems and Informatics 2018.* Springer. 10.1007/978-3-319-99010-1_7

Azar, A. T., & Vaidyanathan, S. (2015a). Handbook of Research on Advanced Intelligent Control Engineering and Automation. IGI Global. doi:10.4018/978-1-4666-7248-2

Azar, A. T., Ammar, H. H., de Brito Silva, G., & Razali, M. S. A. B. (2020b). Self-balancing Robot Modeling and Control Using Two Degree of Freedom PID Controller. *The International Conference on Advanced Machine Learning Technologies and Applications (AMLTA2019),* 64-76.

Azar, A. T., Ammar, H. H., Ibrahim, Z. F., Ibrahim, H. A., Mohamed, N. A., & Taha, M. A. (2020f). Implementation of PID Controller with PSO Tuning for Autonomous Vehicle. *The International Conference on Advanced Intelligent Systems and Informatics AISI 2019,* 288-299. 10.1007/978-3-030-31129-2_27

Azar, A. T., Ammar, H. H., Mayra Beb, M. Y., Garces, S. R., & Boubakarig, A. (2020h). Optimal Design of PID Controller for 2-DOF Drawing Robot using Bat-Inspired Algorithm. *The International Conference on Advanced Intelligent Systems and Informatics AISI 2019,* 175-186.

Azar, A. T., Kumar, J., Kumar, V., & Rana, K. P. S. (2017d). Control of a Two Link Planar Electrically-Driven Rigid Robotic Manipulator Using Fractional Order SOFC. *Proceedings of the International Conference on Advanced Intelligent Systems and Informatics 2017*, 47-56.

Azar, A. T., Kumar, J., Kumar, V., & Rana, K. P. S. (2017d). Control of a Two Link Planar Electrically-Driven Rigid Robotic Manipulator Using Fractional Order SOFC. *Proceedings of the International Conference on Advanced Intelligent Systems and Informatics 2017.* Springer-Verlag.

Azar, A. T., Kumar, J., Kumar, V., & Rana, K. P. S. (2018c). Control of a Two Link Planar Electrically-Driven Rigid Robotic Manipulator Using Fractional Order SOFC. *Proceedings of the International Conference on Advanced Intelligent Systems and Informatics 2017,* 57-68. 10.1007/978-3-319-64861-3_6

Azar, A. T., Madian, A., Ibrahim, H., Taha, M. A., Mohamed, N. A., & Fathy, F. (2020l). Medical nanorobots: Design, applications and future challenges. In Control Systems Design of Bio-Robotics and Bio-mechatronics with Advanced Applications. Elsevier.

Azar, A. T., Ouannas, A., & Singh, S. (2017c). Control of New Type of Fractional Chaos Synchronization. *Proceedings of the International Conference on Advanced Intelligent Systems and Informatics 2017*, 47-56.

Azar, A. T., Ouannas, A., & Singh, S. (2017c). Control of New Type of Fractional Chaos Synchronization. *Proceedings of the International Conference on Advanced Intelligent Systems and Informatics 2017.* Springer-Verlag.

Azar, A. T., Ouannas, A., & Singh, S. (2018b). Control of New Type of Fractional Chaos Synchronization. *Proceedings of the International Conference on Advanced Intelligent Systems and Informatics 2017*, 47-56. 10.1007/978-3-319-64861-3_5

Azar, A. T., Sayed, A. S., Shahin, A. S., Elkholy, H. S., & Ammar, H. H. (2020d). PID Controller for 2-DOFs Twin Rotor MIMO System Tuned with Particle Swarm Optimization. *The International Conference on Advanced Intelligent Systems and Informatics AISI 2019*, 229-242.

Azar, A. T., Serrano, F. E., & Koubaa, A. (2020a). Adaptive Fuzzy Type-2 Fractional Order Proportional Integral Derivative Sliding Mode Controller for Trajectory Tracking of Robotic Manipulators. *2020 IEEE International Conference on Autonomous Robot Systems and Competitions (ICARSC)*, 183-187. 10.1109/ICARSC49921.2020.9096163

Azar, A. T., Serrano, F. E., & Koubaa, A. (2020j). Adaptive Integral Sliding Mode Force Control of Robotic Manipulators with Parametric Uncertainties and Time-Varying Loads. *2020 IEEE International Conference on Autonomous Robot Systems and Competitions (ICARSC),* 188-193. DOI: 10.1109/ICARSC49921.2020.9096110

Azar, A. T., Serrano, F. E., & Vaidyanathan, S. (2018a). Sliding Mode Stabilization and Synchronization of Fractional Order Complex Chaotic and Hyperchaotic Systems. Mathematical Techniques of Fractional Order Systems, 283–317. doi:10.1016/B978-0-12-813592-1.00010-6

Azar, A. T., Serrano, F. E., Hameed, I. A., Kamal, N. A., & Vaidyanathan, S. (2020g). Robust H-Infinity Decentralized Control for Industrial Cooperative Robots. *The International Conference on Advanced Intelligent Systems and Informatics AISI 2019*, 254-265.

Azar, A. T., Serrano, F. E., Kamal, N. A., & Koubaa, A. (2021a). Robust Kinematic Control of Unmanned Aerial Vehicles with Non-holonomic Constraints. In *Proceedings of the International Conference on Advanced Intelligent Systems and Informatics 2020*. Springer. 10.1007/978-3-030-58669-0_74

Azar, A. T., Serrano, F. E., Kamal, N. A., & Koubaa, A. (2021b). Leader-Follower Control of Unmanned Aerial Vehicles with State Dependent Switching. In *Proceedings of the International Conference on Advanced Intelligent Systems and Informatics 2020*. Springer. 10.1007/978-3-030-58669-0_76

Azar, A. T., Serrano, F. E., Koubaa, A., & Kamal, N. A. (2020k). Backstepping H-Infinity Control of Unmanned Aerial Vehicles with Time Varying Disturbances. *The First International Conference of Smart Systems and Emerging Technologies (SMART TECH 2020)*, 243 - 248. 10.1109/SMART-TECH49988.2020.00061

Azar, A. T., Serrano, F. E., Koubaa, A., Kamal, N. A., Vaidyanathan, S., & Fekik, A. (2020e). Adaptive Terminal-Integral Sliding Mode Force Control of Elastic Joint Robot Manipulators in the Presence of Hysteresis. *The International Conference on Advanced Intelligent Systems and Informatics AISI 2019*, 266-276. 10.1007/978-3-030-31129-2_25

Azar, A. T., Serrano, F. E., Vaidyanathan, S., & Albalawi, H. (2020c). Adaptive Higher Order Sliding Mode Control for Robotic Manipulators with Matched and Mismatched Uncertainties. *The International Conference on Advanced Machine Learning Technologies and Applications (AMLTA2019)*, 360-369. 10.1007/978-3-030-14118-9_36

Azar, A. T., Volos, C., Gerodimos, N. A., Tombras, G. S., Pham, V. T., Radwan, A. G., Vaidyanathan, S., Ouannas, A., & Munoz-Pacheco, J. M. (2017b). A novel chaotic system without equilibrium: Dynamics, synchronization and circuit realization. Complexity. doi:10.1155/2017/7871467

Azar, A.T. (2013). Fast neural network learning algorithms for medical applications. *Neural Computing and Applications, 23*(3-4), 1019-1034. . doi:10.1007/s00521-012-1026-y

Azar, A.T., & El-Said, S.A. (2013). Probabilistic neural network for breast cancer classification. *Neural Computing and Applications, 23*(6), 1737-1751. . doi:10.1007/s00521-012-1134-8

Azar, A.T., & Serrano, F.E. (2014). Robust IMC-PID tuning for cascade control systems with gain and phase margin specifications. *Neural Computing and Applications, 25*(5), 983-995. . doi:10.1007/s00521-014-1560-x

Azar, A. T. (2010a). *Fuzzy Systems*. IN-TECH.

Azar, A. T. (2010b). Adaptive Neuro-Fuzzy Systems. In A. T. Azar (Ed.), *Fuzzy Systems*. IN-TECH. doi:10.5772/7220

Azar, A. T. (2012). Overview of Type-2 Fuzzy logic systems. *International Journal of Fuzzy System Applications*, *2*(4), 1–28. doi:10.4018/ijfsa.2012100101

Azar, A. T., & Serrano, F. E. (2016a). Robust control for asynchronous switched nonlinear systems with time varying delays. *Proceedings of the International Conference on Advanced Intelligent Systems and Informatics 2016*, 533, 891-899. 10.1007/978-3-319-48308-5_85

Azar, A. T., & Serrano, F. E. (2016b). Stabilization of Mechanical Systems with Backlash by PI Loop Shaping. *International Journal of System Dynamics Applications*, *5*(3), 20–47. doi:10.4018/IJSDA.2016070102

Azar, A. T., & Serrano, F. E. (2020). Stabilization of Port Hamiltonian Chaotic Systems With Hidden Attractors By Adaptive Terminal Sliding Mode Control. *Entropy (Basel, Switzerland)*, *22*(1), 122. doi:10.3390/e22010122 PMID:33285897

Azar, A. T., & Vaidyanathan, S. (2015b). *Computational Intelligence applications in Modeling and Control.* Springer-Verlag.

Azar, A. T., & Vaidyanathan, S. (2015c). *Chaos Modeling and Control Systems Design*. Springer-Verlag.

Azar, A. T., & Vaidyanathan, S. (2016). *Advances in Chaos Theory and Intelligent Control*. Springer-Verlag. doi:10.1007/978-3-319-30340-6

Azar, A. T., Vaidyanathan, S., & Ouannas, A. (2017a). *Fractional Order Control and Synchronization of Chaotic Systems*. Springer-Verlag. doi:10.1007/978-3-319-50249-6

Azar, A. T., & Zhu, Q. (2015). *Advances and Applications in Sliding Mode Control systems*. Springer-Verlag.

Azizkandi, M. E., Sedaghati, F., Shayeghi, H., & Blaabjerg, F. (2019). A high voltage gain DC–DC converter based on three winding coupled inductor and voltage multiplier cell. *IEEE Transactions on Power Electronics*, *35*(5), 4558–4567. doi:10.1109/TPEL.2019.2944518

Babaie, M., Sharifzadeh, M., & Al-Haddad, K. (2020, August). Three-Phase Grid-Connected NPC Inverter Based on a Robust Artificial Neural Network Controller. In *2020 IEEE Power & Energy Society General Meeting (PESGM)* (pp. 1-5). IEEE.

Babazadeh, M., & Karimi, H. (2013). A robust two-degree-of-freedom control strategy for an islanded microgrid. *IEEE Transactions on Power Delivery*, *28*(3), 1339–1347.

Badihi, H., Zhang, Y., & Hong, H. (2015). Active power control design for supporting grid frequency regulation in wind farms. *Annual Reviews in Control*, *40*, 70–81.

Bagalini, V., Zhao, B. Y., Wang, R. Z., & Desideri, U. (2019). Solar PV-Battery-Electric Grid-Based Energy System for Residential Applications: System Configuration and Viability. *Research*, *2019*, 3838603.

Baghli, L. (1999). *Contribution à la Commande de la Machine Asynchrone, Utilisation de la Logique Floue, des Réseaux de Neurones et des Algorithmes Génétiques*. Thèse de Doctorat, Département de Génie Electrique Université Henri Poincaré, Nancy-I.

Bahgat, A. B. G., Helwa, N. H., Ahmad, G. E., & El Shenawy, E. T. (2005). Maximum power point traking controller for PV systems using neural networks. *Renewable Energy*, *30*(8), 1257–1268. doi:10.1016/j.renene.2004.09.011

Bahrani, B., Saeedifard, M., Karimi, A., & Rufer, A. (2013). A multivariable design methodology for voltage control of a single-DG-unit microgrid. *IEEE Transactions on Industrial Informatics*, *9*(2), 589–599.

Bakas, P., Okazaki, Y., Shukla, A., Patro, S. K., Ilves, K., Dijkhuizen, F., & Nami, A. (2020). Review of hybrid multilevel converter topologies utilizing thyristors for HVDC applications. *IEEE Transactions on Power Electronics*, *36*(1), 174–190. doi:10.1109/TPEL.2020.2997961

Ban, M., Yu, J., Li, Z., Guo, D., & Ge, J. (2019). Battery Swapping: An aggressive approach to transportation electrification. *IEEE Electrification Magazine*, *7*(3), 44–54. doi:10.1109/MELE.2019.2925762

Bao, L., Huang, Z., & Xu, W. (2003). On-line voltage stability monitoring using Var reserves. *IEEE Transactions on Power Systems*, *18*(4), 1461–1469.

Barakat, M. H., Azar, A. T., & Ammar, H. H. (2020) Agricultural Service Mobile Robot Modeling and Control Using Artificial Fuzzy Logic and Machine Vision. *The International Conference on Advanced Machine Learning Technologies and Applications (AMLTA2019)*, 453-465. 10.1007/978-3-030-14118-9_46

Bayat, M., & Karegar, H. (2009). Predictive control of wind energy conversion system. *Developments in Renewable Energy Technology (ICDRET), 2009 1st International Conference on the*, 1–5.

Bayoumi, E. H. E., Soliman, H., & Soliman, M. (2020). Deadbeat PQ Control for Islanded Smart Grids. *EJECE. European Journal of Electrical Engineering and Computer Science*, *4*(5), 1–9. doi:10.24018/ejece.2020.4.5.242

Beatrice, Simone, Marco, Lucio &, & Ivan. (2018). *A queuing theory decision support model and discrete event simulations for the smart charging of electric vehicles.* 23rd Summer School F. Turco of Industrial Systems Engineering.

Behjati, H., Davoudi, A., & Lewis, F. (2014). Modular DC–DC converters on graphs: Cooperative control. *IEEE Transactions on Power Electronics, 29*(12), 6725–6741. doi:10.1109/TPEL.2014.2303192

Beltran, Ahmed-Ali, & Benbouzid. (2008). Sliding mode power control of variable-speed wind energy conversion systems. *Energy Conversion, IEEE Transactions on, 23*, 551–558.

Ben Smida, M., Sakly, A., Vaidyanathan, S., & Azar, A. T. (2018). *Control-Based Maximum Power Point Tracking for a Grid-Connected Hybrid Renewable Energy System Optimized by Particle Swarm Optimization.* IGI-Global. doi:10.4018/978-1-5225-4077-9.ch003

Berkouk, E. M. (1995). *Contribution à la conduite des machines asynchrones monophasée et triphasée alimentées par des convertisseurs directs et indirects. Application aux gradateurs et onduleurs multiniveaux* (Doctoral dissertation). CNAM.

Besheer, Emara, & Abdel Aziz. (2006). Fuzzy based output feedback control for wind energy conversion system: An LMI approach. *Power Systems Conference and Exposition, 2006. PSCE '06. 2006 IEEE PES*, 2030–2037.

Bhattacharjee, A. (2012). Design and comparative study of three photovoltaic battery charge control algorithms in MATLAB/SIMULINK environment. *International Journal of Advanced Computer Research, 2*(3), 129.

Bianchi, F. D., Battista, H. D., & Mantz, R. J. (2007). *Wind Turbine Control System: Principles, Modelling, and Gain Scheduling Design.* SpringerVerlag.

Bidram, A., Nasirian, V., Davoudi, A., & Lewis, F. L. (2017). *Cooperative Synchronization in Distributed Microgrid Control.* Springer.

Bimal, & Bos, K. (1994) Expert system, fuzzy logic, and neural network applications in power electronics and motion control. *Proceedings of the IEEE, 82*(8), 1303 – 1323.

Blaabjerg, F., Teodorescu, R., Liserre, M., & Timbus, A. V. (2006). Overview of control and grid synchronization for distributed power generation systems. *IEEE Transactions on Industrial Electronics, 53*(5), 1398–1409. doi:10.1109/TIE.2006.881997

Bo Jiang, T. X., & Wang, G. (2014). PLL low pass filter design considering unified specification constraints. *Analog Integrated Circuits and Signal Processing, 80*, 113–120.

Bocci, E., Di Carlo, A., McPhail, S. J., Gallucci, K., Foscolo, P. U., Moneti, M., Villarini, M., & Carlini, M. (2014). Biomass to fuel cells state of the art: A review of the most innovative technology solutions. *International Journal of Hydrogen Energy, 39*(36), 21876–21895. doi:10.1016/j.ijhydene.2014.09.022

Boonpirom, N., & Paithoonwattanakij, K. (2006). Wind farm generator control using self-tuning fuzzy controller. *Electrical and Computer Engineering, 2006. ICECE '06. International Conference on*, 221–224.

Bose, B. (2001) Artificial neural network applications in power electronics. *The 27th Annual Conférence of the IEEE Industrial Electronics Society.* 10.1109/IECON.2001.975533

Bouafia, A., Gaubert, J. P., & Krim, F. (2010). Predictive direct power control of three-phase pulsewidth modulation (PWM) rectifier using space-vector modulation (SVM). *IEEE Transactions on Power Electronics, 25*(1), 228–236. doi:10.1109/TPEL.2009.2028731

Bouafia, A., & Krim, F. (2008). A fuzzy-Logic-Based Controller for Three-Phase PWM Rectifier with Unity Power Factor Operation. *J. Electrical Systems, 4*(1), 36–50.

Bouafia, A., Krim, F., & Gaubert, J. P. (2009). Fuzzy-Logic-Based Switching State Selection for Direct Power Control of Three-Phase PWM Rectifier. *IEEE Transactions on Industrial Electronics, 56*(6), 1984–1992. doi:10.1109/TIE.2009.2014746

Bouakrif, F., Azar, A.T., Volos, C.K., Muñoz-Pacheco, J.M., & Pham, V.T. (2019). Iterative Learning and Fractional Order Control for Complex Systems. *Complexity*. . doi:10.1155/2019/7958625

Bouchakour, S. (2005). *Commande Directe de Puissance d'un Convertisseur AC/DC Triphasé Sans Capteurs de Tension.* Memory of Magister Ecole Militaire Polytechnique Algeria.

Bouchama, Z., Khatir, A., Benaggoune, S., & Harmas, M. N. (2020, October). Design and experimental validation of an intelligent controller for DC–DC buck converters. *Journal of the Franklin Institute, 357*(15), 10353–10366.

Boulkroune, A., Bouzeriba, A., Bouden, T., & Azar, A. T. (2016a). Fuzzy Adaptive Synchronization of Uncertain Fractional-order Chaotic Systems. In A. T. Azar & S. Vaidyanathan (Eds.), *Advances in Chaos Theory and Intelligent Control. Studies in Fuzziness and Soft Computing* (Vol. 337). Springer-Verlag. doi:10.1007/978-3-319-30340-6_28

Boulkroune, A., Hamel, S., & Azar, A. T. (2016b). *Fuzzy control-based function synchronization of unknown chaotic systems with dead-zone input.* Springer-Verlag.

Boulouiha, H. M. (2015). Direct torque control of multilevel SVPWM inverter in variable speed SCIG-based wind energy conversion system. *Renewable Energy, 80*, 140–152.

Bourlis, D., & Bleijs, J. (2009). Gain scheduled controller with wind speed estimation via Kalman filtering for a stall regulated variable speed wind turbine. *Universities Power Engineering Conference (UPEC), 2009 Proceedings of the 44th International*, 1–5.

Brahmi, Krichen, & Ouali. (2009). Sensorless control of PMSG in WECS using artificial neural network. *Systems, Signals and Devices, 2009. SSD '09. 6th International Multi-Conference on*, 1–8.

Bratcu, A., Munteanu, I., & Ceanga, E. (2008). Optimal control of wind energy conversion systems: From energy optimization to multi-purpose criteria - a short survey. *Control and Automation, 2008 16th Mediterranean Conference on*, 759 –766.

Bratcu, A., Munteanu, I., Ceanga, E., & Epure, S. (2007). Energetic optimization of variable speed wind energy conversion systems by extremum seeking control. *EUROCON, 2007. The International Conference on 34;Computer as a Tool, 34*, 2536 –2541.

Buehring, I., & Freris, L. (1981). Control policies for wind-energy conversion systems. *Generation, Transmission and Distribution, IEE Proceedings Conference*, 128, 253 –261.

Calderaro, V., Galdi, V., Piccolo, A., & Siano, P. (2007). Design and implementation of a fuzzy controller for wind generators performance optimization. *Power Electronics and Applications, 2007 European Conference on*, 1–10.

Chahwan, J., Abbey, C., Chamberland, M., & Joos, G. (2007, August). Battery storage system modelling, design and operation for wind energy integration in power systems. In *CIGRE Canada Conference on Power Systems* (pp. 1-8). Academic Press.

Chaibi, Y., Allouhi, A., Malvoni, M., Salhi, M., & Saadani, R. (2019). Solar irradiance and temperature influence on the photovoltaic cell equivalent-circuit models. *Solar Energy, 188*, 1102–1110. doi:10.1016/j.solener.2019.07.005

Chapuis, Y. A. (1996). *Contrôle directe du couple d'une machine asynchrone par l'orientation de son flux statorique.* Doctoral thesis of INP-Grenoble, France.

Chen & Yang. (2009). Fuzzy PID controller used in yaw system of wind turbine. *Power Electronics Systems and Applications, 2009. PESA 2009. 3rd International Conference on*, 1–4.

Chen, J.-J., & Ji, Z.-C. (2010). The gain scheduling control for wind energy conversion system based on LPV model. *Networking, Sensing and Control (ICNSC), 2010 International Conference on*, 653–657.

Chen, B., & Joos, G. (2008). Direct power control of active filters with averaged switching frequency regulation. *IEEE Transactions on Power Electronics*, *23*(6), 2729–2737. doi:10.1109/TPEL.2008.2004958

Cheng, X., Lei, Z., & Junqiu, Y. (2009). Fuzzy PID controller for wind turbines. *Intelligent Networks and Intelligent Systems, 2009. ICINIS '09. Second International Conference on*, 74–77.

Chen, S., & Joos, G. (2001) Direct Power Control of Three Active filer with Minimum Energy Storage Components. *APEC 2001. Sixteenth Annual IEEE Applied Power Electronics Conference and Exposition*. 10.1109/APEC.2001.911703

Chen, X. (2018, February). Injected Grid Current Quality Improvement for a Voltage-Controlled Grid-Connected Inverter. *IEEE Transactions on Power Electronics*, *33*(2), 1247–1258.

Chen, Y. W. X., & Gong, Y. W. C. (2019, December). Analysis of frequency characteristics of phase-locked loops and effects on stability of three-phase grid-connected inverter. *International Journal of Electrical Power & Energy Systems*, *113*, 652–663.

Chen, Y. Y., & Man, Y. K. (2009, June). Constant current-based maximum-power-point tracking for photovoltaic power systems. In *2009 Chinese Control and Decision Conference* (pp. 3422-3425). IEEE.

Chinmaya, K. A., & Singh, G. K. (2019, January). Modeling and experimental analysis of grid-connected six-phase induction generator for variable speed wind energy conversion system. *Electric Power Systems Research*, *166*, 151–162.

Cichowlas, M., Malinowski, M., Kazmierkowski, P., Sobczuk, D., & Pou, J. (2005). Active filtering function of three-phase PWM boost rectifier under different line voltage conditions. *IEEE Transactions on Industrial Electronics*, *52*(2), 410–419. doi:10.1109/TIE.2005.843915

Ciobotaru, M., Teodorescu, R., & Blaabjerg, F. (2006). Control of single-stage single-phase PV inverter. *EPE Journal*, *16*(3), 20–26. doi:10.1080/09398368.2006.11463624

Ciocia, A., Di Leo, P., Fichera, S., Giordano, F., Malgaroli, G., & Spertino, F. (2020, June). A Novel Procedure to Adjust the Equivalent Circuit Parameters of Photovoltaic Modules under Shading. In *2020 International Symposium on Power Electronics, Electrical Drives, Automation and Motion (SPEEDAM)* (pp. 711-715). IEEE. 10.1109/SPEEDAM48782.2020.9161878

Cipriani, G., Di Dio, V., Genduso, F., La Cascia, D., Liga, R., Miceli, R., & Galluzzo, G. R. (2014). Perspective on hydrogen energy carrier and its automotive applications. *International Journal of Hydrogen Energy*, *39*(16), 8482–8494. doi:10.1016/j.ijhydene.2014.03.174

Coelho, R. F., Concer, F., & Martins, D. C. (2009, October). A study of the basic DC-DC converters applied in maximum power point tracking. In *2009 Brazilian power electronics conference* (pp. 673-678). IEEE.

Constant, L. (2000). *Modélisation de Dispositifs Electriques par Réseaux de Neurones en Vue de L'émulation Temps Réel.* Thesis Doctorate, Toulouse, France.

Dahbi, A., Nait-Said, N., & Nait-Said, M.-S. (2016). A novel combined MPPT-pitch angle control for wide range variable speed wind turbine based on neural network. *International Journal of Hydrogen Energy*, *41*(2), 9427–9442.

Dalapati, S., & Chakraborty, C. (2009). Dynamic performance of a dead-band controlled capacitor charging type inverter. *Simulation Modelling Practice and Theory*, *17*(5), 911–934. doi:10.1016/j.simpat.2009.02.010

Dang, D., Wang, Y., & Cai, W. (2008). Nonlinear model predictive control of fixed pitch variable speed wind turbine. *Sustainable Energy Technologies, 2008. ICSET 2008. IEEE International Conference on*, 29–33.

Dang, D., Wang, Y., & Cai, W. (2009). A multi-objective optimal nonlinear control of variable speed wind turbine. *Control and Automation, 2009. ICCA 2009. IEEE International Conference on*, 17–22.

De Battista, H., & Mantz, R. (1998). Sliding mode control of torque ripple in wind energy conversion systems with slip power recovery. *Industrial Electronics Society, 1998. IECON '98. Proceedings of the 24th Annual Conference of the IEEE*, 2, 651 –656.

De Battista, Puleston, Mantz, & Christiansen. (2000). Sliding mode control of wind energy systems with DOIG-power efficiency and torsional dynamics optimization. *Power Systems, IEEE Transactions on, 15*, 728–734.

Deb, K. (2001). *Multi-Objective Optimization using Evolutionary Algorithms*. John Wiley & Sons Publications.

Denoun, H., Hamida, M. L., Fekik, A., Dyhia, K. A. I. S., & Ghanes, M. (2018, October). Petri Nets Modeling for Two-Cell Chopper Control Using DSPACE 1104. In *2018 6th International Conference on Control Engineering & Information Technology (CEIT)* (pp. 1-6). IEEE.

Denoun, H., Nabil, B., Mustapha, Z., Nacereddine, B., Haddad, S., & Mamar, S. A. (2015). Modelling and Realisation of a Three-Level PWM Inverter Using a DSP Controller to Feed an Asynchronous Machine. In *Handbook of Research on Advanced Intelligent Control Engineering and Automation* (pp. 687-718). IGI Global.

Derbeli, M., Farhat, M., Barambones, O., & Sbita, L. (2017). Control of PEM fuel cell power system using sliding mode and super-twisting algorithms. *International Journal of Hydrogen Energy*, *42*(13), 8833–8844. doi:10.1016/j.ijhydene.2016.06.103

Dhaouadi, G. U. I. Z. A., Djamel, O. U. N. N. A. S., Youcef, S. O. U. F. I., & Salah, C. H. E. N. I. K. H. E. (2019, September). Implementation of Incremental Conductance Based MPPT Algorithm for Photovoltaic System. In *2019 4th International Conference on Power Electronics and their Applications (ICPEA)* (pp. 1-5). IEEE.

di Benedetto, M., Lidozzi, A., Solero, L., Crescimbini, F., & Grbović, P. J. (2021). High-Performance 3-Phase 5-Level E-Type Multilevel–Multicell Converters for Microgrids. *Energies*, *14*(4), 843. doi:10.3390/en14040843

Dinghui, W., Lili, X., & Zhicheng, J. (2009). Fuzzy adaptive control for wind energy conversion system based on model reference. *Control and Decision Conference, 2009. CCDC '09. Chinese*, 1783–1787.

Djeddi, A., Dib, D., Azar, A. T., & Abdelmalek, S. (2019). Fractional Order Unknown Inputs Fuzzy Observer for Takagi–Sugeno Systems with Unmeasurable Premise Variables. *Mathematics*, *7*(10), 984. doi:10.3390/math7100984

Djoudi, H. (2020). Simulation and Experimental Investigation into a Photovoltaic and Fuel Cell Hybrid Integration Power System for a Typical Small House Application. *Electric Power Components and Systems*, *48*(14-15), 1598–1613.

Dong, S., Kremers, E., Brucoli, M., Rothman, R., & Brown, S. (2020). Techno-enviro-economic assessment of household and community energy storage in the UK. *Energy Conversion and Management*, *205*, 112330. doi:10.1016/j.enconman.2019.112330

Duman, S., Sönmez, Y., Güvenç, U., & Yörükeren, N. (2012). Optimal reactive power dispatch using a gravitational search algorithm. *IET Generation, Transmission & Distribution*, *6*(6), 563–576. doi:10.1049/iet-gtd.2011.0681

Duong, M. Q., Grimaccia, F., Leva, S., & Mussetta, M. (2014). Pitch angle control using hybrid controller for all operating regions of SCIG wind turbine system. *Renewable Energy*, 1–7.

Edwards, P. P., Kuznetsov, V. L., David, W. I., & Brandon, N. P. (2008). Hydrogen and fuel cells: Towards a sustainable energy future. *Energy Policy*, *36*(12), 4356–4362. doi:10.1016/j.enpol.2008.09.036

El Ela, A. A., Abido, M. A., & Spea, S. R. (2011). Differential evolution algorithm for optimal reactive power dispatch. *Electric Power Systems Research*, *81*(2), 458–464. doi:10.1016/j.epsr.2010.10.005

Elkholy, H. A., Azar, A. T., Magd, A., Marzouk, H., & Ammar, H. H. (2020b) Classifying Upper Limb Activities Using Deep Neural Networks. *Proceedings of the International Conference on Artificial Intelligence and Computer Vision (AICV2020)*, 268-282. 10.1007/978-3-030-44289-7_26

Elkholy, H. A., Azar, A. T., Shahin, A. S., Elsharkawy, O. I., & Ammar, H. H. (2020a) Path Planning of a Self Driving Vehicle Using Artificial Intelligence Techniques and Machine Vision. *Proceedings of the International Conference on Artificial Intelligence and Computer Vision (AICV2020)*, 532-542. 10.1007/978-3-030-44289-7_50

El-Khozondar, H. J., El-Khozondar, R. J., Matter, K., & Suntio, T. (2016). A review study of photovoltaic array maximum power tracking algorithms. *Renewables: Wind, Water, and Solar*, *3*(1), 3. doi:10.118640807-016-0022-8

Elmelegi, A., Aly, M., Ahmed, E. M., & Alharbi, A. G. (2019). A simplified phase-shift PWM-based feedforward distributed MPPT method for grid-connected cascaded PV inverters. *Solar Energy*, *187*, 1–12. doi:10.1016/j.solener.2019.05.021

Elyaalaoui, K., Ouassaid, M. & Cherkaoui, M. (2016). *Supervision System of a Wind Farm Based On Squirrel Cage Asynchronous Generator.* Academic Press.

Elyaalaoui, K., Ouassaid, M. & Cherkaoui, M. (2019). Dispatching and Control of Active and Reactive Power for a Wind Farm Considering Fault Ride-Through with a Proposed PI Reactive Power Control. *Renewable Energy Focus, 28*(00), 56-65.

Elyaalaoui, K., Ouassaid, M., & Cherkaoui, M. (2014). Non-linear control and transient stability of SCAG connected to unbalanced AC network. Academic Press.

Elyaalaoui, K., Labbadi, M., Ouassaid, M., & Cherkaoui, M. (2021). Optimal Fractional Order Based on Fuzzy Control Scheme for Wind Farm Voltage Control with Reactive Power Compensation. *Mathematical Problems in Engineering, 2021.*

Elyaalaoui, K., Ouassaid, M., & Cherkaoui, M. (2015). Low Voltage Ride Through for a Wind Turbine Based SCAG. *Third International Conference on Renewable and Sustainable Energy (IRSEC'15)*, 1-6.

Elyaalaoui, K., Ouassaid, M., & Cherkaoui, M. (2018). Primary frequency control using hierarchal fuzzy logic for a wind farm based on SCIG connected to electrical network. *Sustainable Energy. Grids and Networks*, *16*, 188–195.

Elyaalaoui, K., Ouassaid, M., & Cherkaoui, M. (2020). Improvement of THD performance of a robust controller for grid-side energy conversion system based on LCL filter without RC sensor. *International Journal of Electrical Power & Energy Systems*, *121*.

Errami, Y., Ouassaid, M., & Maaroufi, M. (2015). A performance comparison of a nonlinear and a linear control for grid connected PMSG wind energy conversion system. *International Journal of Electrical Power & Energy Systems*, *68*, 180–194.

Escalante, M. F., & Erdmann, A. P. (2021). Estimation of the capacitor voltages in flying capacitor multi-level converters. *IET Power Electronics*, *14*(3), 651–665. doi:10.1049/pel2.12053

Escobar, G., Stankovic, A., Carrasco, M., Galvan, E., & Ortega, R. (2003). Analysis and design of direct power control (DPC) for a three phase synchronous rectifier via output regulation subspaces. *IEEE Transactions on Power Electronics, 18*(3), 823–830. doi:10.1109/TPEL.2003.810862

Etemadi, A. H., Davison, E. J., & Iravani, R. (2012). A Decentralized Robust Control Strategy for Multi-DER Microgrids—Part I: Fundamental Concept. *IEEE Transactions on Power Delivery, 27*(4), 1843–1853.

Etemadi, A. H., Davison, E. J., & Iravani, R. (2014). A generalized decentralized robust control of islanded microgrids. *IEEE Transactions on Power Systems, 29*(6), 3102–3113.

Etxeberria, A., Vechiu, I., Camblong, H., & Vinassa, J. M. (2010, October). Hybrid energy storage systems for renewable energy sources integration in microgrids: A review. In *2010 Conference Proceedings IPEC* (pp. 532-537). IEEE.

Ezzouitine, K., Boulezhar, A., & El Afou, Y. (2019). A cascaded H-bridge multilevel inverter with photovoltaic MPPT control. *Periodicals of Engineering and Natural Sciences, 6*(2), 415–425.

Fadhel, S., Diallo, D., Delpha, C., Migan, A., Bahri, I., Trabelsi, M., & Mimouni, M. F. (2020). Maximum power point analysis for partial shading detection and identification in photovoltaic systems. *Energy Conversion and Management, 224*, 113374. doi:10.1016/j.enconman.2020.113374

Faria, J., Pombo, J., Calado, M. D. R., & Mariano, S. (2019). Power management control strategy based on artificial neural networks for standalone PV applications with a hybrid energy storage system. *Energies, 12*(5), 902.

Fekik, A. (2021). *Direct power control of three-phase PWM-rectifier with backstepping control.* Advances in Nonlinear Dynamics and Chaos (ANDC).

Fekik, A., Denoun, H., Azar, A. T., Hamida, M. L., Atig, M., Ghanes, M., . . . Barbot, J. P. (2018c). Direct power control of a three-phase PWM-rectifier based on Petri nets for the selection of switching states. In *2018 7th International Conference on Renewable Energy Research and Applications (ICRERA)* (pp. 1121-1125). IEEE.

Fekik, A., Denoun, H., Azar, A. T., Hamida, M. L., Atig, M., Ghanes, M., . . . Barbot, J. P. (2018c, October). Direct power control of a three-phase PWM-rectifier based on Petri nets for the selection of switching states. In *2018 7th International Conference on Renewable Energy Research and Applications (ICRERA)* (pp. 1121-1125). IEEE.

Fekik, A., Denoun, H., Azar, A. T., Hamida, M. L., Benamrouche, N., & Vaidyanathan, S. (2017). Direct Power Control of a PWM-Inverter for Grid Connected Photovoltaic System. *5nd international conference on renewable Energy (CIER2017).*

Fekik, A., Denoun, H., Azar, A.T., Hamida, M.L., Benamrouche, N., & Vaidyanathan, S. (2018). Direct Power Control of a PWM-Inverter for Grid Connected Photovoltaic System. *5ème Conférence Internationale des Energies Renouvelables (CIER– 2017) Proceeding of Engineering and Technology –PET, 27*, 1-6.

Fekik, A., Denoun, H., Benamrouche, N., Benyahia, N., Badji, A., & Zaouia, M. (2016). Comparative analysis of direct power control and direct power control with space vector modulation of PWM rectifier. *2016 4th International Conference on Control Engineering & Information Technology (CEIT)*, 1-6.

Fekik, A., Denoun, H., Benamrouche, N., Benyahia, N., Zaouia, M., & Haddad, S. (2015a) Comparative study of PI and FUZZY DC- voltage control for Voltage Oriented Control-PWM rectifier. *The 14th International Conference on Circuits, Systems, Electronics, Control & Signal Processing 2015.*

Fekik, A., Denoun, H., Benamrouche, N., Benyahia, N., Zaouia, M., & Haddad, S. (2015a) Comparative study of PI and FUZZY DC- voltage control for Voltage Oriented Control-PWM rectifier. *The 14th International Conference on Circuits, Systems, Electronics, Control & Signal Processing 2015.CSECS'15.*

Fekik, A., Denoun, H., Benamrouche, N., Benyahia, N., Zaouia, M., & Haddad, S. (2015a). Comparative study of PI and FUZZY DC- voltage control for Voltage Oriented Control-PWM rectifier. *The 14th International Conference on Circuits, Systems, Electronics, Control & Signal Processing 2015.*

Fekik, A., Denoun, H., Benamrouche, N., Benyahia, N., Zaouia, M., & Haddad, S. (2015a). Comparative study of PI and FUZZY DC- voltage control for Voltage Oriented Control-PWM rectifier. *The 14th International Conference on Circuits, Systems, Electronics, Control & Signal Processing 2015.CSECS'15.*

Fekik, A., Denoun, H., Hamida, M. L., Azar, A. T., Atig, M., & Zhu, Q. M. (2018d). Neural Network Based Switching State Selection for Direct Power Control of Three Phase PWM-Rectifier. In *2018 10th International Conference on Modelling, Identification and Control (ICMIC)* (pp. 1-6). IEEE.

Fekik, A., Denoun, H., Hamida, M. L., Azar, A. T., Atig, M., & Zhu, Q. M. (2018d, July). Neural Network Based Switching State Selection for Direct Power Control of Three Phase PWM-Rectifier. In *2018 10th International Conference on Modelling, Identification and Control (ICMIC)* (pp. 1-6). IEEE.

Fekik, A., Azar, A. T., Denoun, H., Kamal, N. A., Hamida, M. L., Kais, D., & Amara, K. (2018e). A Backstepping Direct Power Control of Three Phase Pulse Width Modulated Rectifier. In *International Workshop Soft Computing Applications* (pp. 445-456). Springer.

Fekik, A., Azar, A. T., Kamal, N. A., Denoun, H., Almustafa, K. M., Hamida, L., & Zaouia, M. (2020a). Fractional-Order Control of a Fuel Cell–Boost Converter System. In *International Conference on Advanced Machine Learning Technologies and Applications* (pp. 713-724). Springer.

Fekik, A., Azar, A. T., Kamal, N. A., Serrano, F. E., Hamida, M. L., Denoun, H., & Yassa, N. (2020b, October). Maximum Power Extraction from a Photovoltaic Panel Connected to a Multi-cell Converter. In *International Conference on Advanced Intelligent Systems and Informatics* (pp. 873-882). Springer.

Fekik, A., Denoun, H., Azar, A. T., Hamida, M. L., Zaouia, M., & Benyahia, N. (2018b). Comparative study of two level and three level PWM-rectifier with voltage oriented control. In *International Conference on Advanced Intelligent Systems and Informatics* (pp. 40-51). Springer.

Fekik, A., Denoun, H., Azar, A. T., Kamal, N. A., Zaouia, M., Benyahia, N., & Vaidyanathan, S. (2020d). Direct power control of three-phase PWM-rectifier with backstepping control. In *Backstepping Control of Nonlinear Dynamical Systems* (pp. 215–234). Academic Press.

Fekik, A., Denoun, H., Azar, A. T., Kamal, N. A., Zaouia, M., Yassa, N., & Hamida, M. L. (2019a). Direct Torque Control of Three Phase Asynchronous Motor with Sensorless Speed Estimator. In *International Conference on Advanced Intelligent Systems and Informatics* (pp. 243-253). Springer.

Fekik, A., Denoun, H., Azar, A. T., Koubaa, A., Kamal, N. A., Zaouia, M., ... Yassa, N. (2020c). Adapted Fuzzy Fractional Order proportional-integral controller for DC Motor. In *2020 First International Conference of Smart Systems and Emerging Technologies (SMARTTECH)* (pp. 1-6). IEEE.

Fekik, A., Denoun, H., Azar, A. T., Zaouia, M., Benyahia, N., Hamida, M. L., & Vaidyanathan, S. (2018a). Artificial neural network for PWM rectifier direct power control and DC voltage control. In *Advances in System Dynamics and Control* (pp. 286–316). IGI Global. doi:10.4018/978-1-5225-4077-9.ch010

Fekik, A., Denoun, H., Benamrouche, N., Benyahia, N., Badji, A., & Zaouia, M. (2016*)* Comparative Analysis of Direct Power Control and Direct power control with space vector modulation of PWM rectifier. *4th IEEE International Conference on Control Engineering & Information Technology (CEIT-2016).* 10.1109/CEIT.2016.7929058

Fekik, A., Denoun, H., Benamrouche, N., Benyahia, N., & Zaouia, M. (2015). A fuzzy–logic based controller for three phase PWM rectifier with voltage oriented control strategy. *International Journal Of Circuits, Systems And. Signal Processing*, *9*, 412–419.

Fekik, A., Denoun, H., Benamrouche, N., Benyahia, N., & Zaouia, M. (2015b). A Fuzzy Logic Based Controller For Three Phase PWM Rectifier With Voltage Oriented Control Strategy. *International Journal of Circuits, Systems and Signal Processing*, *9*, 412–419.

Fekik, A., Denoun, H., Benamrouche, N., Benyahia, N., & Zaouia, M. (2015b). A Fuzzy Logic Based Controller For Three Phase PWM Rectifier With Voltage Oriented Control Strategy. International Journal of Circuits, Systems And. *Signal Processing*, *9*, 412–419.

Fekik, A., Denoun, H., Benamrouche, N., Benyahia, N., Zaouia, M., Badji, A., & Vaidyanathan, S. (2017). Improvement of the Performances of the Direct Power Control Using Space Vector Modulation of Three Phases PWM-Rectifier. *International Journal of Control Theory and Applications*, *10*, 125–137.

Fekik, A., Denoun, H., Zaouia, M., Benyahia, N., Benamrouche, N., Badji, A., & Vaidyanathan, S. (2017). Improvement of the performances of the direct power control using space vector modulation of three phases PWM-rectifier. *International Journal of Control Theory and Applications*, *10*, 133–145.

Fekik, A., Denoun, H., Zaouia, M., Hamida, M. L., & Vaidyanathan, S. (2019b). A new switching table based neural network for direct power control of three-phase PWM-rectifier. International Journal of Modelling. *Identification and Control*, *33*(4), 358–368. doi:10.1504/IJMIC.2019.107484

Fischer, J. R., González, S. A., Carugat, I., Herrán, M. A., Judewicz, M. G., & Carrica, D. O. (2014). Robust predictive control of grid-tied converters based on direct power control. *IEEE Transactions on Power Electronics*, *29*(10), 5634–5643. doi:10.1109/TPEL.2013.2294919

Folly, K. A., & Sheetekela, S. P. N. (2009). Impact of Fixed and Variable Speed Wind Generators on the Transient Stability of a Power System Network. Academic Press.

Foti, S., De Caro, S., Scelba, G., Scimone, T., Testa, A., Cacciato, M., & Scarcella, G. (2018). An optimal current control strategy for asymmetrical hybrid multilevel inverters. *IEEE Transactions on Industry Applications*, *54*(5), 4425–4436. doi:10.1109/TIA.2018.2832024

Fujita, H., Yamasaki, T., & Akagi, H. (1998). A hybrid active filter for damping of harmonic resonance in industrial power systems. *IEEJ Transactions on Industry Applications*, *118*(10), 1193–1200. doi:10.1541/ieejias.118.1193

Ganthia, B. P., Barik, S. K., & Nayak, B. (2021). Wind Turbines in Energy Conversion System: Types & Techniques. In V. K. Singh, A. K. Bhoi, A. Saxena, A. F. Zobaa, & S. Biswal (Eds.), Renewable Energy and Future Power Systems. Energy Systems in Electrical Engineering. Springer. https://doi.org/10.1007/978-981-33-6753-1_9

Ganthia, B. P., Pradhan, R., Sahu, R., & Pati, A. K. (2021). Artificial Ant Colony Optimized Direct Torque Control of Mathematically Modeled Induction Motor Drive Using PI and Sliding Mode Controller. In J. Kumar & P. Jena (Eds.), Recent Advances in Power Electronics and Drives. Lecture Notes in Electrical Engineering (Vol. 707). Springer. https://doi.org/10.1007/978-981-15-8586-9_35.

Ganthia. (2020). Application of Hybrid Facts Devices in DFIG Based Wind Energy System for LVRT Capability Enhancements. *Journal of Mechanics of Continua and Mathematical Sciences. J. Mech. Cont. & Math. Sci.*, *15*(6), 139-157. doi:10.26782/jmcms.2020.06.00012

Ganthia. (2020). Application of Hybrid Facts Devices in DFIG Based Wind Energy System for LVRT Capability Enhancements. *Journal of Mechanics of Continua and Mathematical Sciences. J. Mech. Cont. & Math. Sci., 15*(6), 245-256. doi:10.26782/jmcms.2020.06.00019

Ganthia, B. P., Agarwal, V., Rout, K., & Pardhe, M. K. (2017). Optimal control study in DFIG based wind energy conversion system using PI & GA. *International Conference on Power and Embedded Drive Control (ICPEDC)*, 343-347.

Ganthia, B. P., & Barik, S. K. (2020). Steady-State and Dynamic Comparative Analysis of PI and Fuzzy Logic Controller in Stator Voltage Oriented Controlled DFIG Fed Wind Energy Conversion System. *J. Inst. Eng. India Ser. B, 101*, 273–286. https://doi.org/10.1007/s40031-020-00455-8

Ganthia, B. P., Barik, S. K., & Nayak, B. (n.d.). Shunt Connected FACTS Devices for LVRT Capability Enhancement in WECS. *Engineering, Technology & Applied Scientific Research, 10*(3), 5819–5823.

Ganthia, B. P., Mohanty, S., Rana, P. K., & Sahu, P. K. (2016). Compensation of voltage sag using DVR with PI controller. *2016 International Conference on Electrical, Electronics, and Optimization Techniques (ICEEOT)*, 2138-2142. doi: 10.1109/ICEEOT.2016.7755068

Ganthia, B. P., Rana, P. K., Patra, T., Pradhan, R., & Sahu, R. (2018). Design and Analysis of Gravitational Search Algorithm Based TCSC Controller in Power System. *Materials Today: Proceedings, 5*(1), 841–847.

Gargoom, A., Haruni, A., Haque, M., & Negnevitsky, M. (2010). Voltage and frequency stabilization using PI-like fuzzy controller for the load side converters of the stand alone wind energy systems. Applied Power Electronics Conference and Exposition (APEC), 2010 Twenty-Fifth Annual IEEE, 2132–2137.

Geyer, T., & Quevedo, E. (2015). Performance of multistep finite control set model predictive control for power electronics. *IEEE Transactions on Power Electronics, 30*(3), 1633–1644. doi:10.1109/TPEL.2014.2316173

Ghennam, T., Francois, B., & Berkouk, E. M. (2009). Local supervisory algorithm for reactive power dispatching of a wind farm. *13th European Conference on Power Electronics and Applications EPE'09*, 1-10.

Ghoudelbourk, S., Azar, A. T., & Dib, D. (2021). Three-level (NPC) Shunt Active Power Filter Based on Fuzzy Logic and Fractional-order PI Controller. *Int. J. Automation and Control, 15*(2), 149–169. doi:10.1504/IJAAC.2021.113338

Ghoudelbourk, S., Azar, A. T., Dib, D., & Omeiri, A. (2020). Selective Harmonic Elimination Strategy in the Multilevel Inverters for Grid Connected Photovoltaic System. *Int. J. Advanced Intelligence Paradigms, 15*(3), 317–339. doi:10.1504/IJAIP.2020.105826

Ghoudelbourk, S., Dib, D., Omeiri, A., & Azar, A. T. (2016). MPPT Control in wind energy conversion systems and the application of fractional control (PIα) in pitch wind turbine. *International Journal of Modelling Identification and Control, 26*(2), 140–151. doi:10.1504/IJMIC.2016.078329

Giove, S., Azar, A. T., & Nordio, M. (2013). Fuzzy Logic Control for Dialysis Application. In Biofeedback Systems and Soft Computing Techniques of Dialysis. Springer-Verlag GmbH. doi:10.1007/978-3-642-27558-6_9

Gkizas. (2021, February). Optimal robust control of a Cascaded DC–DC boost converter. *Control Engineering Practice, 107*.

Glavin, M. E., Chan, P. K., Armstrong, S., & Hurley, W. G. (2008, September). A stand-alone photovoltaic supercapacitor battery hybrid energy storage system. In *2008 13th International power electronics and motion control conference* (pp. 1688-1695). IEEE.

Golestan, S., Guerrero, J. M., & Vasquez, J. C. (2017, December). Single-Phase PLLs: A Review of Recent Advances. *IEEE Transactions on Power Electronics, 32*(12), 9013–9030.

Gorripotu, T. S., Azar, A. T., Pilla, R., & Kamal, N. A. (2021). Impact of Ultra Capacitor on Automatic Load Frequency Control of Nonlinear Power System. In *Intelligent Computing in Control and Communication. Lecture Notes in Electrical Engineering* (Vol. 702, pp. 333–341). Springer. doi:10.1007/978-981-15-8439-8_27

Gorripotu, T. S., Samalla, H., Jagan Mohana Rao, C., Azar, A. T., & Pelusi, D. (2019). TLBO Algorithm Optimized Fractional-Order PID Controller for AGC of Interconnected Power System. In J. Nayak, A. Abraham, B. Krishna, G. Chandra Sekhar, & A. Das (Eds.), *Soft Computing in Data Analytics. Advances in Intelligent Systems and Computing* (Vol. 758). Springer. doi:10.1007/978-981-13-0514-6_80

Grassi, G., Ouannas, A., Azar, A. T., Radwan, A. G., Volos, C., Pham, V. T., Ziar, T., Madian, A., Kyprianidis, I. M., & Stouboulos, I. N. (2017). Chaos Synchronisation Of Continuous Systems Via Scalar Signal. *The 6th International Conference on Modern Circuits and Systems Technologies (MOCAST)*. 10.1109/MOCAST.2017.7937629

Greco, N. (2018, September). Scalable lumped models of integrated transformers for galvanically isolated power transfer systems. *Integration (Amsterdam)*, *63*, 323–331.

Green, T. C., & Marks, J. H. (2005). Control techniques for active power filters. *IEE Proceedings. Electric Power Applications*, *152*(2), 369–381. doi:10.1049/ip-epa:20040759

Guasch, D., & Silvestre, S. (2003). Dynamic battery model for photovoltaic applications. *Progress in Photovoltaics: Research and Applications*, *11*(3), 193–206.

Guest, E., Jensen, K. H., & Rasmussen, T. W. (2018, June). Sequence Domain Harmonic Modeling of Type-IV Wind Turbines. *IEEE Transactions on Power Electronics*, *33*(6), 4934–4943. doi:10.1109/TPEL.2017.2735303

Guha, S., Yang, J., & Banerjee, A. (2000). Amorphous silicon alloy photovoltaic research—Present and future. *Progress in Photovoltaics: Research and Applications*, *8*(1), 141–150. doi:10.1002/(SICI)1099-159X(200001/02)8:1<141::AID-PIP305>3.0.CO;2-I

Gui, Y. (2018). Improved Direct Power Control for Grid-Connected Voltage Source Converters. *IEEE Transactions on Industrial Electronics*.

Guo, P., Xu, Q., Yue, Y., Ma, F., He, Z., Luo, A., & Guerrero, J. M. (2019). Analysis and control of modular multilevel converter with split energy storage for railway traction power conditioner. *IEEE Transactions on Power Electronics*, *35*(2), 1239–1255. doi:10.1109/TPEL.2019.2917065

GWEC. (2018). *GWEC Global Wind Report_April 2018*. GWEC.

Hadjidemetriou, L., Kyriakides, E. & Blaabjerg, F. (2013). A New Hybrid PLL for Interconnecting Renewable Energy Systems to the Grid. *IEEE Transactions on Industry Applications, 49*(6), 2709-2719.

Hamida, M. L., Denoun, H., Fekik, A., Benyahia, N., Zaouia, M., & Benamrouche, N. (2017). A Comparison Between PWM and Petri Nets Control Applied for Series Multi-cells Chopper. In *International conference on control and signal processing* (*Vol. 26*, pp. 124-128). Academic Press.

Hamida, M. L., Denoun, H., Fekik, A., Benyahia, N., & Benamrouche, N. (2018). Cyclic reports modulation control strategy for a five cells inverter. *2018 International Conference on Electrical Sciences and Technologies in Maghreb (CISTEM)*, 1-5.

Hamida, M. L., Denoun, H., Fekik, A., Kais, D., Benamrouche, N., & Amara, K. (2019b, November). Fuzzy Logic-Feedback Linearization Control for a Three-Cell Flying Capacitor Converter. In *2019 International Conference on Advanced Electrical Engineering (ICAEE)* (pp. 1-5). IEEE. 10.1109/ICAEE47123.2019.9015121

Hamida, M. L., Denoun, H., Fekik, A., & Vaidyanathan, S. (2019a). Control of separately excited DC motor with series multi-cells chopper using PI-Petri nets controller. *Nonlinear Engineering*, *8*(1), 32–38. doi:10.1515/nleng-2017-0174

Hanafi, S., Fellah, M. K., Guebli, A., & Chiali, E. (2014). Control with cyclic reports modulation of 3x2 stacked multicellular DC/AC converter. *Cell*, *11*, 21.

Harifi, S., Mohammadzadeh, J., & Khalilian, M. (2020). *Giza Pyramids Construction: an ancient-inspired metaheuristic algorithm for optimization*. Evol. Intel.

Hasan, M. M., Abu-Siada, A., Islam, S. M., & Dahidah, M. S. A. (2018). A New Cascaded Multilevel Inverter Topology with Galvanic Isolation. *IEEE Transactions on Industry Applications*, *54*(4), 3463–3472.

Hassaine, L., Olias, E., Quintero, J., & Haddadi, M. (2009). Digital power factor control and reactive power regulation for grid-connected photovoltaic inverter. *Renewable Energy*, *34*(1), 315–321. doi:10.1016/j.renene.2008.03.016

Hassan, A. S., Cipcigan, L., & Jenkins, N. (2017). Optimal battery storage operation for PV systems with tariff incentives. *Applied Energy*, *203*, 422–441. doi:10.1016/j.apenergy.2017.06.043

Hassanien, A.E., Moftah, H.M., Azar, A.T., & Shoman, M. (2014). MRI Breast cancer diagnosis hybrid approach using adaptive Ant-based segmentation and Multilayer Perceptron neural networks classifier. *Applied Soft Computing, 14*(Part A), 62-71.

Hassan, M. S., Mughal, S. N., Jarial, R. K., & Sood, Y. R. (2019). A Comparative Analysis of Different Maximum Power Point Tracking Algorithms of Solar Photovoltaic System. In *Applications of Computing, Automation and Wireless Systems in Electrical Engineering* (pp. 217–229). Springer. doi:10.1007/978-981-13-6772-4_20

Hassan, W., Lu, Y., Farhangi, M., Lu, D. D. C., & Xiao, W. (2020). Design, analysis and experimental verification of a high voltage gain and high-efficiency DC–DC converter for photovoltaic applications. *IET Renewable Power Generation*, *14*(10), 1699–1709. doi:10.1049/iet-rpg.2019.1020

Herrera, R. S., SalmerÓn, P., & Kim, H. (2008, January). Instantaneous Reactive Power Theory Applied to Active Power Filter Compensation: Different Approaches, Assessment, and Experimental Results. *IEEE Transactions on Industrial Electronics*, *44*(1), 184–196.

He, S., Xiong, J., Wang, Z., & Lin, S. (2020, May 20). Robust AD for LCL-type grid-connected inverter with capacitor current quasi-integral feedback. *IET Power Electronics*, *13*(7), 1332–1342.

He, Z., & … . (2017, June). Reactive Power Strategy of Cascaded Delta-Connected STATCOM Under Asymmetrical Voltage Conditions. *IEEE Journal of Emerging and Selected Topics in Power Electronics*, *5*(2), 784–795.

Hong, C.-M., Lin, W.-M., & Cheng, F.-S. (2007). Application of fuzzy neural network sliding mode controller for wind driven induction generator system. *Intelligent Systems Applications to Power Systems, 2007. ISAP 2007. International Conference on*, 1–6.

Honrubia-Escribano, A., Jiménez-Buendía, F., Gómez-Lázaro, E., & Fortmann, J. (2018, March). Field Validation of a Standard Type 3 Wind Turbine Model for Power System Stability, According to the Requirements Imposed by IEC 61400-27-1. *IEEE Transactions on Energy Conversion*, *33*(1), 137–145. doi:10.1109/TEC.2017.2737703

Hossam, H. H., Mousa, A.-R. Y., Mohamed, E. M., & March, E. (2021). State of the art perturb and observe MPPT algorithms based wind energy conversion systems: A technology review. *International Journal of Electrical Power & Energy Systems*, 126.

Hui, J., & Bakhshai, A. (2008). A new adaptive control algorithm for maximum power point tracking for wind energy conversion systems. *Power Electronics Specialists Conference, 2008. PESC 2008,* 4003 –4007.

Hui, J., & Bakhshai, A. (2008). Adaptive algorithm for fast maximum power point tracking in wind energy systems. *Industrial Electronics, 2008. IECON 2008. 34th Annual Conference of IEEE*, 2119–2124.

Hui, J., Bakhshai, A., & Jain, P. K. (2010). A master-slave fuzzy logic control scheme for maximum power point tracking in wind energy systems. *Telecommunications Energy Conference (INTELEC), 32nd International*, 1–6.

Hu, J., Shang, L., & He, Y. (2011). Direct Active and Reactive Power Regulation of Grid-Connected DC/AC Converters Using Sliding Mode Control Approach. *IEEE Transactions on Power Electronics*, *26*(1), 210–222.

Hu, J., Sun, L., Yuan, X., Wang, S., & Chi, Y. (2017, July). Modeling of Type 3 Wind Turbines With df/dt Inertia Control for System Frequency Response Study. *IEEE Transactions on Power Systems*, *32*(4), 2799–2809. doi:10.1109/TPWRS.2016.2615631

Hu, J., Zhu, J., Zhang, Y., Platt, G., Ma, Q., & Dorrell, D. G. (2013). Predictive direct virtual torque and power control of doubly fed induction generators for fast and smooth grid synchronization and flexible power regulation. *IEEE Transactions on Power Electronics*, *28*(7), 3182–3194. doi:10.1109/TPEL.2012.2219321

Hu, J., & Zhu, Z. Q. (2011). Investigation on switching patterns of direct power control strategies for grid-connected DC-AC converters based on power variation rates. *IEEE Transactions on Power Electronics*, *26*(2), 3582–3598. doi:10.1109/TPEL.2011.2164812

Humaidi, A. J., Ibraheem, I. K., Azar, A. T., & Sadiq, M. E. (2020). A New Adaptive Synergetic Control Design for Single Link Robot Arm Actuated by Pneumatic Muscles. *Entropy (Basel, Switzerland)*, *22*(7), 723. doi:10.3390/e22070723 PMID:33286496

Huo, Z., Zheng, Y., Xu, C., & Huang, D. (2009). Robust stability research for wind energy conservation systems based on T-S fuzzy models. *Sustainable Power Generation and Supply, 2009. SUPERGEN '09. International Conference on*, 1–4.

Hussain, A. N., Abdullah, A. A., & Neda, O. M. (2018). Modified Particle Swarm Optimization for Solution of Reactive Power Dispatch. *Research Journal of Applied Sciences, Engineering and Technology*, *15*(8), 316–327. doi:10.19026/rjaset.15.5917

Ibraheem, G. A. R., Azar, A. T., Ibraheem, I. K., & Humaidi, A. J. (2020). A Novel Design of a Neural Network based Fractional PID Controller for Mobile Robots Using Hybridized Fruit Fly and Particle Swarm Optimization. *Complexity*. doi:10.1155/2020/3067024

Ibrahim, H. A., Azar, A. T., Ibrahim, Z. F., & Ammar, H. H. (2020) A Hybrid Deep Learning Based Autonomous Vehicle Navigation and Obstacles Avoidance. *Proceedings of the International Conference on Artificial Intelligence and Computer Vision (AICV2020)*, 296-307. 10.1007/978-3-030-44289-7_28

IEEE Recommended Practice for Monitoring Electric Power Quality. (2009). *IEEE Standard 1159.*

Ilyas, A., Khan, M. R., & Ayyub, M. (2020). FPGA based real-time implementation of fuzzy logic controller for maximum power point tracking of solar photovoltaic system. *Optik (Stuttgart)*, *213.*

Jain, S., & Agarwal, V. (2007). Comparison of the performance of maximum power point tracking schemes applied to single-stage grid-connected photovoltaic systems. *IET Electric Power Applications*, *1*(5), 753–762. doi:10.1049/iet-epa:20060475

Jallad, J., Mekhilef, S., & Mokhlis, H. (2017). Frequency Regulation Strategies in Grid Integrated Offshore Wind Turbines via VSC-HVDC Technology: A Review. *Energies*, *10*(9), 1244. doi:10.3390/en10091244

Jana, S., Kumar, N., Mishra, R., Sen, D., & Saha, T. K. (2020). Development and implementation of modified MPPT algorithm for boost converter-based PV system under input and load deviation. *International Transactions on Electrical Energy Systems*, *30*(2), e12190. doi:10.1002/2050-7038.12190

Jha, S. K., Bilalovic, J., Jha, A., Patel, N., & Zhang, H. (2017). Renewable energy: Present research and future scope of Artificial Intelligence. *Renewable & Sustainable Energy Reviews*, *77*, 297–317. doi:10.1016/j.rser.2017.04.018

Jiang, W. (2010) Sliding-mode Control of Single-phase PWM Rectifier for DC Microgrid Applications. *International Conference on Computer and Electrical Engineering*, 53, 82-90. DOI: 10.7763/IPCSIT.2012.V53.No.1.82

Jian-jun, X., Li-mei, Y., Xiao-ning, Q., Chun-lei, J., & Jian-ren, W. (2010). Study of variable-pitch wind turbine based on fuzzy control. *Future Computer and Communication (ICFCC), 2010 2nd International Conference on,* 1.

Jianlin, L., & Honghua, X. (2008). Research on control system of high power DFIG wind power system. *MultiMedia and Information Technology, 2008. MMIT '08. International Conference on*, 669–672.

Joshi, P., & Arora, S. (2017). Maximum power point tracking methodologies for solar PV systems. *RE:view*, *70*, 1154–1177.

Kairous, D., Wamkeue, R., & Belmadani, B. (2010). Advanced control of variable speed wind energy conversion system with DFIG. *Environment and Electrical Engineering (EEEIC), 2010 9th International Conference on*, 41–44.

Kalantar, M., & Sedighizadeh, M. (2005). Adaptive-neural PID control of wind energy conversion systems using wavenets. *Intelligent Control, 2005. Proceedings of the 2005 IEEE International Symposium on, Mediterrean Conference on Control and Automation*, 219–224.

Kamal, N. A., & Ibrahim, A. M. (2018). Conventional, Intelligent, and Fractional-Order Control Method for Maximum Power Point Tracking of a Photovoltaic System: A Review. Mathematical Techniques of Fractional Order Systems, 603-671.

Kamal, N. A., Azar, A. T., Elbasuony, G. S., Almustafa, K. A., & Almakhles, D. (2020). PSO-based Adaptive Perturb and Observe MPPT Technique for Photovoltaic Systems. In *The International Conference on Advanced Intelligent Systems and Informatics AISI 2019*. Springer.

Kamal, N. A., Azar, A. T., Elbasuony, G. S., Almustafa, K. A., & Almakhles, D. (2020). PSO-based Adaptive Perturb and Observe MPPT Technique for Photovoltaic Systems. *The International Conference on Advanced Intelligent Systems and Informatics AISI 2019,* 125-135.

Kamal, N. A., & Ibrahim, A. M. (2018). Conventional, intelligent, and fractionalorder control method for maximum power point tracking of a photovoltaic system: A review. In *Fractional Order Systems', Advances in Nonlinear Dynamics and Chaos (ANDC)* (pp. 603–671). Academic Press.

Kammogne, A. S. T., Kountchou, M. N., Kengne, R., Azar, A. T., Fotsin, H. B., & Ouagni, S. T. M. (2020). Polynomial Robust Observer Implementation based-passive Synchronization of Nonlinear Fractional-Order Systems with Structural Disturbances. *Frontiers of Information Technology & Electronic Engineering.*, *21*(9), 1369–1386. doi:10.1631/FITEE.1900430

Karamanakos, P., Geyer, T., Oikonomou, N., Kieferndorf, F., & Manias, S. (2014). Direct model predictive control: A review of strategies that achieve long prediction intervals for power electronics. *IEEE Industrial Electronics Magazine*, *8*(1), 32–43. doi:10.1109/MIE.2013.2290474

Kasper, M., Bortis, D., & Kolar, J. W. (2013). Classification and comparative evaluation of PV panel-integrated DC–DC converter concepts. *IEEE Transactions on Power Electronics*, *29*(5), 2511–2526. doi:10.1109/TPEL.2013.2273399

Kazim, M., Azar, A. T., Koubaa, A., & Zaidi, A. (2021). Disturbance Rejection Based Optimized Robust Adaptive Controller for UAVs. *IEEE Systems Journal*, *15*(2), 1–12. doi:10.1109/JSYST.2020.3006059

Kesarwani, K., & Stauth, J. T. (2015). Resonant and multi-mode operation of flying capacitor multi-level dc-dc converters. *2015 IEEE 16th Workshop on Control and Modeling for Power Electronics (COMPEL)*, 1-8.

Keshavarzzadeh, A. H., Ahmadi, P., & Safaei, M. R. (2019). Assessment and optimization of an integrated energy system with electrolysis and fuel cells for electricity, cooling and hydrogen production using various optimization techniques. *International Journal of Hydrogen Energy*, *44*(39), 21379–21396. doi:10.1016/j.ijhydene.2019.06.127

Khadkikar, V., Varma, R. K., & Seethapathy, R. (2009, October). Grid voltage regulation utilizing storage batteries in PV solar—Wind plant based distributed generation system. In 2009 IEEE Electrical Power & Energy Conference (EPEC) (pp. 1-6). IEEE.

Khalil, H. K. (1996). *Nonlinear Systems* (2nd ed.). Prentice-Hall. Inc.

Khalilpourazari, S., & Khalilpourazary, S. (2017). An efficient hybrid algorithm based on Water Cycle and Moth-Flame Optimization algorithms for solving numerical and constrained engineering optimization problems. *Journal of Soft Computing*, *23*(5), 1699–1722. doi:10.100700500-017-2894-y

Khan, A., Singh, S., & Azar, A. T. (2020a). Synchronization between a Novel Integer-Order Hyperchaotic System and a Fractional-Order Hyperchaotic System Using Tracking Control. *The International Conference on Advanced Machine Learning Technologies and Applications (AMLTA2019)*, 382-391. 10.1007/978-3-030-14118-9_38

Khan, A., Singh, S., & Azar, A. T. (2020b). Combination-Combination Anti-Synchronization of Four Fractional Order Identical Hyperchaotic Systems. *The International Conference on Advanced Machine Learning Technologies and Applications (AMLTA2019)*, 406-414. 10.1007/978-3-030-14118-9_41

Khan, M. J., & Mathew, L. (2019). Fuzzy logic controller-based MPPT for hybrid photo-voltaic/wind/fuel cell power system. *Neural Computing & Applications*, *31*(10), 6331–6344. doi:10.100700521-018-3456-7

Khan, S., Zehetbauer, P., & Schwalbe, R. (2021, March). Evaluation of sensitivity based coordinated volt-var control and local reactive power for voltage regulation and power exchange across system boundaries in smart distribution networks. *Electric Power Systems Research, 192.*

Khazraj, H., da Silva, F. F., Bak, C. L., & Golestan, S. (2017, June). Analysis and design of notch filter-based PLLs for grid-connected applications. *Electric Power Systems Research*, *147*, 62–69.

Khennaoui, A. A., Ouannas, A., Grassi, G., & Azar, A. T. (2020b) Dynamic Analysis of a Fractional Map with Hidden Attractor. *Proceedings of the International Conference on Artificial Intelligence and Computer Vision (AICV2020)*, 731-739. 10.1007/978-3-030-44289-7_68

Khennaoui, A. A., Ouannas, A., Boulaaras, S., Pham, V. T., & Azar, A. T. (2020a). A fractional map with hidden attractors: Chaos and control. *The European Physical Journal. Special Topics*, *229*(6-7), 1083–1093. doi:10.1140/epjst/e2020-900177-6

Khettab, K., Bensafia, Y., Bourouba, B., & Azar, A. T. (2018) Enhanced Fractional Order Indirect Fuzzy Adaptive Synchronization of Uncertain Fractional Chaotic Systems Based on the Variable Structure Control: Robust H∞ Design Approach. Mathematical Techniques of Fractional Order Systems, 559–595.

Khettab, K., Bensafia, Y., Bourouba, B., & Azar, A. T. (2018). Enhanced Fractional Order Indirect Fuzzy Adaptive Synchronization of Uncertain Fractional Chaotic Systems Based on the Variable Structure Control: Robust H∞ Design Approach. In Mathematical Techniques of Fractional Order Systems. Elsevier.

Kim, Kim, & Ko. (2010). Modeling and control of PMSG-based variable-speed wind turbine. *Electric Power Systems Research, 80*(1), 46-52. doi:10.1016/j.epsr.2009.08.003

Knez, M., & Jereb, B. (2013). Solar power plants–alternative sustainable approach to greener environment : A case of Slovenia. *Sustainable Cities and Society*, *6*, 27–32. doi:10.1016/j.scs.2012.07.002

Ko, Yoon, Kyung, & Hong. (2008). Modeling and control of DFIG-based variable-speed wind-turbine. *Electric Power Systems Research, 78*(11), 1841-1849. doi:10.1016/j.epsr.2008.02.018

Kocalmis, A., & Sunter, S. (2006, November). Simulation of a space vector PWM controller for a three-level voltage-fed inverter motor drive. In *IECON 2006-32nd Annual Conference On IEEE Industrial Electronics* (pp. 1915-1920). IEEE. 10.1109/IECON.2006.347442

Kollimalla, S. K., & Mishra, M. K. (2014, April). Coordinated control and energy management of hybrid energy storage system in PV system. In *2014 International Conference on Computation of Power, Energy, Information and Communication (ICCPEIC)* (pp. 363-368). IEEE.

Konagai, M. (2011). Present status and future prospects of silicon thin-film solar cells. *Japanese Journal of Applied Physics*, *50*(3R), 030001. doi:10.1143/JJAP.50.030001

Koubaa, A., & Azar, A. T. (2021). *Unmanned Aerial Systems: Theoretical Foundation and Applications.* Elsevier.

Krishna, K. S., & Kumar, K. S. (2015). A review on hybrid renewable energy systems. *Renewable & Sustainable Energy Reviews*, *52*, 907–916. doi:10.1016/j.rser.2015.07.187

Krishnamurthy, K., Padmanaban, S., Blaabjerg, F., Neelakandan, R. B., & Prabhu, K. R. (2019). Power Electronic Converter Configurations Integration with Hybrid Energy Sources–A Comprehensive Review for State-of the-Art in Research. *Electric Power Components and Systems*, *47*(18), 1623–1650. doi:10.1080/15325008.2019.1689457

Kumar, J., Azar, A. T., Kumar, V., & Rana, K. P. S. (2018) Design of Fractional Order Fuzzy Sliding Mode Controller for Nonlinear Complex Systems. Mathematical Techniques of Fractional Order Systems, 249–282. doi:10.1016/B978-0-12-813592-1.00009-X

Kumar, P. (2018, April). Power Management in Microgrid Integrated Wind-PV-Battery Generation System. In *2018 International Conference on Power Energy, Environment and Intelligent Control (PEEIC)* (pp. 746-750). IEEE.

Kumar, P., Jain, G., & Palwalia, D. K. (2015, August). Genetic algorithm based maximum power tracking in solar power generation. In *2015 International Conference on Power and Advanced Control Engineering (ICPACE)* (pp. 1-6). IEEE.

Kumar, S. S., Kumar, V., Kumar, R., Malyan, S. K., & Pugazhendhi, A. (2019). Microbial fuel cells as a sustainable platform technology for bioenergy, biosensing, environmental monitoring, and other low power device applications. *Fuel*, *255*, 115682. doi:10.1016/j.fuel.2019.115682

Kundur, P. (1994). *Power System Stability and Control.* McGraw-Hill.

Kyriakopoulos, G. L., & Arabatzis, G. (2016). Electrical energy storage systems in electricity generation: Energy policies, innovative technologies, and regulatory regimes. *Renewable & Sustainable Energy Reviews*, *56*, 1044–1067.

Laks, J., Pao, L., & Wright, A.(2009). Control of wind turbines: Past, present, and future. *American Control Conference, 2009. ACC '09*, 2096–2103.

Lalouni, S., Rekioua, D., Rekioua, T., & Matagne, E. (2009). Fuzzy logic control of stand-alone photovoltaic system with battery storage. *Journal of Power Sources*, *193*(2), 899–907. doi:10.1016/j.jpowsour.2009.04.016

Lamine, M., Hakim, D., Arezki, F., Dyhia, K., Nacereddine, B., & Youssef, B. (2019c, July). Control of Three-Cell Inverter with a Fuzzy Logic-Feedback Linearization Strategy to Reduce the Harmonic Content of the Output Current. In *2019 International Conference of Computer Science and Renewable Energies (ICCSRE)* (pp. 1-5). IEEE.

Lamine, M., Hakim, D., Arezki, F., Nabil, B., & Nacerddine, B. (2018, October). Cyclic Reports Modulation Control Strategy for a Five Cells Inverter. In *2018 International Conference on Electrical Sciences and Technologies in Maghreb (CISTEM)* (pp. 1-5). IEEE.

Lamterkati, J., Khafallah, M., & Ouboubker, L. (2014a). A New DPC for Three-phase PWM rectifier with unity power factor operation. *International Journal of Advanced Research in Electrical Electronics and Instrumentation Engineering*, *3*(4), 8273–8285.

Lamterkati, J., Khafallah, M., & Ouboubker, L. (2014b). Comparison of PI and Fuzzy logic DC-Link Voltage Controller for DPC PWM-Rectifier. *International Journal of Enhanced Research in Science Technology and Engineering*, *3*(4), 321–332.

Larrinaga, S., Vidal, M., Oyarbide, E., & Apraiz, J. (2007). Predictive control strategy of DC/AC converters based on direct power control. *IEEE Transactions on Industrial Electronics*, *54*(3), 1261–1271. doi:10.1109/TIE.2007.893162

Le, L., Ying, L., Tian, P., & Li, P. (2008). Maximum wind capture based on fuzzy PD control strategy. *Electrical Machines and Systems, 2008. ICEMS 2008. International Conference on*, 2542–2546.

Lee, J., Jeong, S., Han, Y. H., & Park, B. J. (2010). Concept of cold energy storage for superconducting flywheel energy storage system. *IEEE Transactions on Applied Superconductivity*, *21*(3), 2221–2224.

Leonardi, B., & Ajjarapu, V. (2008). *Investigation of various generator reactive power reserve (GRPR) definitions for online voltage stability/security assessment. In Power and Energy Society General Meeting-Conversion and Delivery of Electrical Energy in the 21st Century*. IEEE.

Li & Chen. (2008). Overview of different wind generator systems and their comparisons. *Renewable Power Generation, IET, 2*, 123–138.

Li, C. H., Zhu, X. J., Cao, G. Y., Sui, S., & Hu, M. R. (2009). Dynamic modeling and sizing optimization of stand-alone photovoltaic power systems using hybrid energy storage technology. *Renewable Energy*, *34*(3), 815–826. doi:10.1016/j.renene.2008.04.018

Li, H. (2012). Analysis and estimation of transient stability for a grid-connected wind turbine with induction generator. *Renewable Energy*, *36*, 1469–1476.

Li, H., Shi, K., & McLaren, P. (2005, November). Neural-network-based sensorless maximum wind energy capture with compensated power coefficient. *Industry Applications, IEEE Transactions on*, *41*, 1548–1556.

Li, J., & Liu, J. (2019, April). A Novel Buck–Boost Converter With Low Electric Stress on Components. *IEEE Transactions on Industrial Electronics*, *66*(4), 2703–2713.

Lima, M., Silvino, J., & de Resende, P. (1999). H∞ control for a variablespeed adjustable-pitch wind energy conversion system. *Industrial Electronics, 1999. ISIE '99. Proceedings of the IEEE International Symposium on*, 2, 556–561.

Lin, W.-M., Hong, C.-M., Lee, M.-R., Huang, C.-H., Huang, C.-C., & Wu, B.-L. (2010). Fuzzy sliding mode-based control for PMSG maximum wind energy capture with compensated pitch angle. *Computer Communication Control and Automation (3CA), 2010 International Symposium on, 2*, 397–400.

Linyu, W., Wang, X., Sun, Q., & Lai, L. L. (2015). The service modeling of EV charging/swapping station and its cost benefit analysis. *10th International Conference on Advances in Power System Control, Operation & Management*, 1-7.

Li, P., Yang, C., Sun, L., Xiang, J., Wen, X., Zhong, J., & Deng, T. (2020). Dynamic characteristics and operation strategy of the discharge process in compressed air energy storage systems for applications in power systems. *International Journal of Energy Research*, *44*(8), 6363–6382. doi:10.1002/er.5362

Li, Q., & Wolfs, P. (2008). A review of the single phase photovoltaic module integrated converter topologies with three different DC link configurations. *IEEE Transactions on Power Electronics*, *23*(3), 1320–1333. doi:10.1109/TPEL.2008.920883

Liu, C., Wang, Y., Wang, L., & Chen, Z. (2019). Load-adaptive real-time energy management strategy for battery/ultracapacitor hybrid energy storage system using dynamic programming optimization. *Journal of Power Sources*, *438*, 227024. doi:10.1016/j.jpowsour.2019.227024

Liu, L., Ma, D., Azar, A. T., & Zhu, Q. (2020). Neural computing Enhanced Parameter Estimation for Multi-Input and Multi-Output Total Nonlinear Dynamic Models. *Entropy (Basel, Switzerland)*, *22*(5), 510. doi:10.3390/e22050510 PMID:33286282

Li, Z., Amir, K., & Song, J. (2019). A comprehensive review of the key technologies for pure electric vehicles. *Energy, Elsevier*, *182*(C), 824–839.

Lofthouse, J., Simmons, R. T., & Yonk, R. M. (2015). *Reliability of renewable energy: Solar.* Institute of Political Economy, Utah State University.

Luo, X., Wang, J., Dooner, M., & Clarke, J. (2015). Overview of current development in electrical energy storage technologies and the application potential in power system operation. *Applied Energy*, *137*, 511–536.

Lü, X., Wu, Y., Lian, J., Zhang, Y., Chen, C., Wang, P., & Meng, L. (2020). Energy management of hybrid electric vehicles: A review of energy optimization of fuel cell hybrid power system based on genetic algorithm. *Energy Conversion and Management*, *205*, 112474. doi:10.1016/j.enconman.2020.112474

Mahmoud, M. M. (2006). Transient analysis of a PV power generator charging a capacitor for measurement of the I–V characteristics. *Renewable Energy*, *31*(13), 2198–2206. doi:10.1016/j.renene.2005.09.019

Mahmud, N., Zahedi, A., & Mahmud, A. (2017). A cooperative operation of novel PV inverter control scheme and storage energy management system based on ANFIS for voltage regulation of grid-tied PV system. *IEEE Transactions on Industrial Informatics*, *13*(5), 2657–2668. doi:10.1109/TII.2017.2651111

Malakondareddy, B., Kumar, S. S., Gounden, N. A., & Anand, I. (2019). An adaptive PI control scheme to balance the neutral-point voltage in a solar PV fed grid connected neutral point clamped inverter. *International Journal of Electrical Power & Energy Systems*, *110*, 318–331. doi:10.1016/j.ijepes.2019.03.012

Malinowski, M., & Kazmierkowski, M.P. (2000) Simulation Study of Virtual Flux Based Direct Power Control for Three-Phase PWM Rectifiers. *26th Annual Conference of the IEEE Industrial Electronics Society, 2000. IECON 2000.* doi:10.1109/IECON.2000.972411

Malinowski, M., Jasinski, M., & Kazmierkowski, P. (2004). Simple Direct power control of three phase PWM rectifier using space vector modulation (DPC-SVM). *IEEE Transactions on Industrial Electronics*, *51*(2), 447–454. doi:10.1109/TIE.2004.825278

Malinowski, M., Kaz'mierkowski, M. P., Hansen, S., Blaabjerg, F., & Marques, G. D. (2001). Virtual flux based direct power control of three-phase PWM rectifiers. *IEEE Transactions on Industry Applications*, *37*(4), 1019–1027. doi:10.1109/28.936392

Malinowski, M., Kazmierkowski, M. P., & Trzynadlowski, A. M. (2003). A Comparative Study of Control Techniques for PWM Rectifiers in AC Adjustable Speed Drives. *IEEE Transactions on Power Electronics*, *18*(6), 1390–1396. doi:10.1109/TPEL.2003.818871

Manikanta, S., Neha, R., Deepa, K., & Sarada, J. (2020). Application of Aspiration Level Model in determining QoS for an EV battery charging station. *International Conference on Smart Technologies in Computing, Electrical and Electronics*, 71-77.

Manna, S., & Akella, A. K. (2021). Comparative analysis of various P & O MPPT algorithm for PV system under varying radiation condition. *2021 1st International Conference on Power Electronics and Energy (ICPEE)*, 1-6.

Mariano-Hernández, D., Hernández-Callejo, L., Zorita-Lamadrid, A., Duque-Pérez, O., & García, F. S. (2020). A review of strategies for building energy management system: Model predictive control, demand side management, optimization, and fault detect & diagnosis. *Journal of Building Engineering*.

Martinez, M. I., Tapia, G., Susperregui, A., & Camblong, H. (2012, June). Sliding-Mode Control for DFIG Rotor- and Grid-Side Converters Under Unbalanced and Harmonically Distorted Grid Voltage. *IEEE Transactions on Energy Conversion*, *27*(2), 328–339.

Marzouki, A., Hamouda, M., & Fnaiech, F. (2015). A hybrid controller for PWM active rectifiers based LCL filters. *COMPEL: The International Journal for Computation and Mathematics in Electrical and Electronic Engineering*.

Masoum, M. A., Dehbonei, H., & Fuchs, E. F. (2002). Theoretical and experimental analyses of photovoltaic systems with voltage and current-based maximum power-point tracking. *IEEE Transactions on Energy Conversion*, *17*(4), 514–522.

Mathi, D. K., & Chinthamalla, R. (2020). Enhanced leader adaptive velocity particle swarm optimisation based global maximum power point tracking technique for a PV string under partially shaded conditions. *IET Renewable Power Generation*, *14*(2), 243–253.

Mayosky, M., & Cancelo, I. (1999, July). Direct adaptive control of wind energy conversion systems using Gaussian networks. *Neural Networks, IEEE Transactions on*, *10*, 898–906.

Meghni, B., Dib, D., & Azar, A. T. (2017c). A Second-order sliding mode and fuzzy logic control to Optimal Energy Management in PMSG Wind Turbine with Battery Storage. *Neural Computing & Applications*, *28*(6), 1417–1434. doi:10.100700521-015-2161-z

Meghni, B., Dib, D., Azar, A. T., Ghoudelbourk, S., & Saadoun, A. (2017a). Robust Adaptive Supervisory Fractional order Controller For optimal Energy Management in Wind Turbine with Battery Storage. *Studies in Computational Intelligence*, *688*, 165–202. doi:10.1007/978-3-319-50249-6_6

Meghni, B., Dib, D., Azar, A. T., & Saadoun, A. (2017b). *Effective Supervisory Controller to Extend Optimal Energy Management in Hybrid Wind Turbine under Energy and Reliability Constraints. International Journal of Dynamics and Control*. doi:10.100740435-016-0296-0

Mekki, H., Boukhetala, D., & Azar, A. T. (2015). Sliding Modes for Fault Tolerant Control. In Advances and Applications in Sliding Mode Control systems. Springer-Verlag GmbH. doi:10.1007/978-3-319-11173-5_15

Mellit, A., Kalogirou, S. A., Hontoria, L., & Shaari, S. (2009). Artificial intelligence techniques for sizing photovoltaic systems: A review. *Renewable & Sustainable Energy Reviews*, *13*(2), 406–419. doi:10.1016/j.rser.2008.01.006

Merabet Boulouiha, H., Khodja, M., Rahiel, D., Allali, A., Kaddour, F., & Denai, M. (2019). Power quality enhancement in electricity grids with wind energy using multicell converters and energy storage. *Journal of Renewable and Sustainable Energy*, *11*(1), 013302. doi:10.1063/1.5043292

Meynard, T. A., Foch, H., Forest, F., Turpin, C., Richardeau, F., Delmas, L., Gateau, G., & Lefeuvre, E. (2002). Multicell converters : Derived topologies. *IEEE Transactions on Industrial Electronics*, *49*(5), 978–987. doi:10.1109/TIE.2002.803189

Meynard, T. A., Foch, H., Thomas, P., Courault, J., Jakob, R., & Nahrstaedt, M. (2002). Multicell converters : Basic concepts and industry applications. *IEEE Transactions on Industrial Electronics*, *49*(5), 955–964. doi:10.1109/TIE.2002.803174

Mezache, A., & Betto, K. (1997). *Estimation et Commande d'un Réacteur de Fabrication de Pâte a Papier par les Réseaux de Neurones Flous. Memory of Engineer.* University of Constantine.

Mirjalili, S., Gandomi, A. H., Mirjalili, S. Z., Saremi, S., Faris, H., & Mirjalili, S. M. (2017). Salp Swarm Algorithm: A bio-inspired optimizer for engineering design problems. *Advances in Engineering Software*, *114*, 163–191.

Mishra, S., Mishra, Y., Li, F., & Dong, Z. (2009). TS-fuzzy controlled DFIG based wind energy conversion systems. *Power Energy Society General Meeting, 2009. PES '09. IEEE*, 1–7.

Mittal, S., Azar, A. T., & Kamal, N. A. (2021). Nonlinear Fractional Order System Synchronization via Combination-Combination Multi-switching. In *Proceedings of the International Conference on Advanced Intelligent Systems and Informatics 2020*. Springer.

Mittal, S., Azar, A. T., & Kamal, N. A. (2021). Nonlinear Fractional Order System Synchronization via Combination-Combination Multi-switching. In *Proceedings of the International Conference on Advanced Intelligent Systems and Informatics 2020*. Springer. 10.1007/978-3-030-58669-0_75

Moghaddam, S., Bigdeli, M., Moradlou, M., & Siano, P. (2019). Designing of stand-alone hybrid PV/wind/battery system using improved crow search algorithm considering reliability index. *International Journal of Energy and Environmental Engineering*, *10*(4), 429–449.

Mohamed, N. A., Azar, A. T., Abbas, N. E., Ezzeldin, M. A., & Ammar, H. H. (2020). Experimental Kinematic Modeling of 6-DOF Serial Manipulator Using Hybrid Deep Learning. In *Proceedings of the International Conference on Artificial Intelligence and Computer Vision (AICV2020)*. Springer.

Mohamed, N. A., Azar, A. T., Abbas, N. E., Ezzeldin, M. A., & Ammar, H. H. (2020). Experimental Kinematic Modeling of 6-DOF Serial Manipulator Using Hybrid Deep Learning. *Proceedings of the International Conference on Artificial Intelligence and Computer Vision (AICV2020)*, 283-295.

Mohapatra, A., Nayak, B., & Saiprakash, C. (2019). Adaptive Perturb & Observe MPPT for PV System with Experimental Validation. *IEEE International Conference on Sustainable Energy Technologies and Systems (ICSETS).*

Mohapatra, A., Nayak, B., & Saiprakash, C. (2019). Adaptive Perturb & Observe MPPT for PV System with Experimental Validation. In *2019 IEEE International Conference on Sustainable Energy Technologies (ICSET)* (pp. 257-261). IEEE. 10.1109/ICSETS.2019.8744819

Mohd, Mohammad, Adil, & Asghar. (2019). A Comprehensive Review on Electric Vehicles Charging Infrastructures and their Impacts on Power-Quality of the Utility Grid. *Journal of eTransportation*, *1*, 100006.

Mondal, S. K., Pinto, J. O. P., & Bose, B. K. (2002). A Neural-Network-Based Space-Vector PWM Controller for a three-Level Voltage-Fed Inverter Induction Motor Drive. *Industry Applications. IEEE Transactions on.*, *38*(3), 660–669.

Moradi, R., Karimi, H., & Karimi-Ghartemani, M. (2010) Robust decentralized control for islanded operation of two radially connected DG systems. *Proc. IEEE Int. Symp. Ind. Electron. (ISIE)*, 2272–2277.

Moriarty, P., & Butterfield, S. (2009). Wind turbine modeling overview for control engineers. *American Control Conference*, 2090–2095.

Mosa, M., Shadmand, M. B., Balog, R. S., & Rub, H. A. (2017). Efficient maximum power point tracking using model predictive control for photovoltaic systems under dynamic weather condition. *IET Renewable Power Generation*, *11*(11), 1401–1409. doi:10.1049/iet-rpg.2017.0018

Mousa, H. H., Youssef, A.-R., & Mohamed, E. E. (2021). State of the art perturb and observe MPPT algorithms based wind energy conversion systems : A technology review. *International Journal of Electrical Power & Energy Systems*, *126*, 106598. doi:10.1016/j.ijepes.2020.106598

Muhando, E., Senjyu, T., Uchida, K., Kinjo, H., & Funabashi, T. (2010, January). Stochastic inequality constrained closed-loop model-based predictive control of MW-class wind generating system in the electric power supply. *Renewable Power Generation, IET*, *4*, 23–35.

Munteanu, I., Bacha, S., Bratcu, A., Guiraud, J., & Roye, D. (2008, September). Energy reliability optimization of wind energy conversion systems by sliding mode control. *Energy Conversion, IEEE Transactions on*, *23*, 975–985.

Murty, V. V. S. N., & Kumar, A. (2020). Multi-objective energy management in microgrids with hybrid energy sources and battery energy storage systems. *Protection and Control of Modern Power Systems*, *5*(1), 1–20. doi:10.118641601-019-0147-z

Naidu, N. K. S., & Singh, B. (2014). Sensorless control of single voltage source converter-based doubly fed induction generator for variable speed wind energy conversion system. *IET Power Electronics*, *7*(12), 2996–3006. doi:10.1049/iet-pel.2014.0034

Naresh, M., & Tripathi, R. K. (2019, March). *Intelligent control strategy for power management in hybrid renewable energy system. In 2019 Innovations in Power and Advanced Computing Technologies (i-PACT)* (Vol. 1). IEEE.

Natarajan, Sivakumar, & Sharaf. (1991). Multirate PI control system performance for a wind energy conversion scheme. *American Control Conference*, 2774–2775.

Nazin, S. A., Polyak, B. T., & Topunov, M. V. (2007). Rejection of Bounded Exogenous Disturbances by the Method of Invariant Ellipsoids. *Automation and Remote Control*, *68*(3), 467–486.

Neha, R., Manikanta, S., & Deepa, K. (2020). Integration of Battery Charging and Swapping using Metaheuristics: A review. *2nd International Conference on Machine Learning, Advances in Computing, Renewable Energy and Communication*, 1-10.

Nguyen, D., & Lehman, B. (2008). An adaptive solar photovoltaic array using model-based reconfiguration algorithm. *IEEE Transactions on Industrial Electronics*, *55*(7), 2644–2654. doi:10.1109/TIE.2008.924169

Niazy, I. 2010. Participation in reactive power market considering generator aging. *IPEC, 2010 Conference Proceedings*, 1067-1072.

Noguchi, T., Tomiki, H., Kondo, S., & Takahashi, I. (1998). Direct power control of PWM converter without power-source voltage sensors. *IEEE Transactions on Industry Applications*, *34*(3), 473–479. doi:10.1109/28.673716

Olatomiwa, L., Mekhilef, S., Ismail, M. S., & Moghavvemi, M. (2016). Energy management strategies in hybrid renewable energy systems: A review. *Renewable & Sustainable Energy Reviews*, *62*, 821–835. doi:10.1016/j.rser.2016.05.040

Olivares, D. E., Mehrizi-Sani, A., Etemadi, A. H., Canizares, C. A., Iravani, R., Kazerani, M., Hajimiragha, A. H., Gomis-Bellmunt, O., Saeedifard, M., Palma-Behnke, R., Jimenez-Estevez, G. A., & Hatziargyriou, N. D. (2014). Trends in microgrid control. *IEEE Transactions on Smart Grid*, *5*(4), 1905–1919. doi:10.1109/TSG.2013.2295514

Othman, S., Allali, M. A.-E., Mrad, I., Chariag, D. E., Ghanes, M., Sbita, L., & Barbot, J.-P. (2020). Robust hybrid control based on Petri Nets for a multicellular inverter. *6th IEEE International Energy Conference (ENERGYCon)*, 143-148.

Ouannas, A., Azar, A. T., & Radwan, A. G. (2016b). *On Inverse Problem of Generalized Synchronization Between Different Dimensional Integer-Order and Fractional-Order Chaotic Systems.* The 28th International Conference on Microelectronics, Cairo, Egypt. 10.1109/ICM.2016.7847942

Ouannas, A., Azar, A. T., & Ziar, T. (2017h). On Inverse Full State Hybrid Function Projective Synchronization for Continuous-time Chaotic Dynamical Systems with Arbitrary Dimensions. *Differential Equations and Dynamical Systems*. doi:10.1007/s12591-017-0362-x

Ouannas, A., Grassi, G., & Azar, A. T. (2020b). A New Generalized Synchronization Scheme to Control Fractional Chaotic Systems with Non-identical Dimensions and Different Orders. *The International Conference on Advanced Machine Learning Technologies and Applications (AMLTA2019),* 415-424. 10.1007/978-3-030-14118-9_42

Ouannas, A., Grassi, G., & Azar, A. T. (2020c). Fractional-Order Control Scheme for Q-S Chaos Synchronization. *The International Conference on Advanced Machine Learning Technologies and Applications (AMLTA2019),* 434-441. 10.1007/978-3-030-14118-9_44

Ouannas, A., Grassi, G., Azar, A. T., & Singh, S. (2019a). New Control Schemes for Fractional Chaos Synchronization. In *Proceedings of the International Conference on Advanced Intelligent Systems and Informatics 2018.* Springer. 10.1007/978-3-319-99010-1_5

Ouannas, A., Grassi, G., Azar, A. T., Khennaouia, A. A., & Pham, V. T. (2020d). Chaotic Control in Fractional-Order Discrete-Time Systems. *AISI 2019. The International Conference on Advanced Intelligent Systems and Informatics AISI 2019*, 207-217. 10.1007/978-3-030-31129-2_20

Ouannas, A., Grassi, G., Azar, A. T., Khennaouia, A. A., & Pham, V. T. (2020e). Synchronization of Fractional-Order Discrete-Time Chaotic Systems. *The International Conference on Advanced Intelligent Systems and Informatics AISI 2019*, 218-228. 10.1007/978-3-030-31129-2_21

Ouannas, A., Grassi, G., Azar, A. T., Radwan, A. G., Volos, C., Pham, V. T., Ziar, T., Kyprianidis, I. M., & Stouboulos, I. N. (2017j). *Dead-Beat Synchronization Control in Discrete-Time Chaotic Systems.* The 6th International Conference on Modern Circuits and Systems Technologies (MOCAST), Thessaloniki, Greece. 10.1109/MOCAST.2017.7937628

Ouannas, A., Azar, A. T., & Abu-Saris, R. (2016a). A new type of hybrid synchronization between arbitrary hyperchaotic maps. *International Journal of Machine Learning and Cybernetics.* Advance online publication. doi:10.100713042-016-0566-3

Ouannas, A., Azar, A. T., & Vaidyanathan, S. (2017b). A new fractional hybrid chaos synchronization. Int. J. Modelling. *Identification and Control*, *27*(4), 314–322. doi:10.1504/IJMIC.2017.084719

Ouannas, A., Azar, A. T., & Vaidyanathan, S. (2017f). On A Simple Approach for Q-S Synchronization of Chaotic Dynamical Systems in Continuous-Time. *Int. J. Computing Science and Mathematics*, *8*(1), 20–27.

Ouannas, A., Azar, A. T., & Vaidyanathan, S. (2017g). New Hybrid Synchronization Schemes Based on Coexistence of Various Types of Synchronization Between Master-Slave Hyperchaotic Systems. *International Journal of Computer Applications in Technology*, *55*(2), 112–120. doi:10.1504/IJCAT.2017.082868

Ouannas, A., Azar, A. T., & Vaidyanathan, S. (2017i). A Robust Method for New Fractional Hybrid Chaos Synchronization. *Mathematical Methods in the Applied Sciences*, *40*(5), 1804–1812. doi:10.1002/mma.4099

Ouannas, A., Azar, A. T., & Ziar, T. (2017h). *On Inverse Full State Hybrid Function Projective Synchronization for Continuous-time Chaotic Dynamical Systems with Arbitrary Dimensions.* Differential Equations and Dynamical Systems., doi:10.100712591-017-0362-x

Ouannas, A., Azar, A. T., & Ziar, T. (2020a). Fractional inverse full state hybrid projective synchronisation. *Int. J. Advanced Intelligence Paradigms*, *17*(3-4), 270–298.

Ouannas, A., Azar, A. T., Ziar, T., & Radwan, A. G. (2017d). Study On Coexistence of Different Types of Synchronization Between Different dimensional Fractional Chaotic Systems. *Studies in Computational Intelligence*, *688*, 637–669. doi:10.1007/978-3-319-50249-6_22

Ouannas, A., Azar, A. T., Ziar, T., & Radwan, A. G. (2017e). Generalized Synchronization of Different Dimensional Integer-order and Fractional Order Chaotic Systems. *Studies in Computational Intelligence*, *688*, 671–697. doi:10.1007/978-3-319-50249-6_23

Ouannas, A., Azar, A. T., Ziar, T., & Vaidyanathan, S. (2017a). On New Fractional Inverse Matrix Projective Synchronization Schemes. *Studies in Computational Intelligence*, *688*, 497–524. doi:10.1007/978-3-319-50249-6_17

Ouannas, A., Azar, A. T., Ziar, T., & Vaidyanathan, S. (2017b). Fractional Inverse Generalized Chaos Synchronization Between Different Dimensional Systems. *Studies in Computational Intelligence*, *688*, 525–551. doi:10.1007/978-3-319-50249-6_18

Ouannas, A., Azar, A. T., Ziar, T., & Vaidyanathan, S. (2017c). A New Method To Synchronize Fractional Chaotic Systems With Different Dimensions. *Studies in Computational Intelligence*, *688*, 581–611. doi:10.1007/978-3-319-50249-6_20

Ouannas, A., Grassi, G., Azar, A. T., & Khennaoui, A. A. (2021). Synchronization Control in Fractional Discrete-Time Systems with Chaotic Hidden Attractors. In A. Hassanien, R. Bhatnagar, & A. Darwish (Eds.), *Advanced Machine Learning Technologies and Applications. AMLTA 2020. Advances in Intelligent Systems and Computing* (Vol. 1141, pp. 661–669). Springer. doi:10.1007/978-981-15-3383-9_59

Ouassaid, M., Elyaalaoui, K., & Cherkaoui, M. (2016). Sliding mode control of induction generator wind turbine connected to the grid. In Advances and Applications in Nonlinear Control Systems. Springer.

Pandya, S., & Roy, R. (2015). Particle swarm optimization based optimal reactive power dispatch. *Proceeding of the IEEE International Conference on Electrical, Computer and Communication Technologies*, *1*(1), 1-5. 10.1109/ICECCT.2015.7225981

Pappu, H., Kumar, G. S., & Manala, S. (2017). *An FPGA based New Single Phase Five Level Inverter.* Academic Press.

Parida, A., & Subudhi, B. (2020). Modified leaky LMS-based control strategy for reliable operation of single-stage three-phase grid-tied PV system. *IET Renewable Power Generation*, *14*(9), 1453–1462. doi:10.1049/iet-rpg.2019.0776

Parida, B., Iniyan, S., & Goic, R. (2011). A review of solar photovoltaic technologies. *Renewable & Sustainable Energy Reviews*, *15*(3), 1625–1636. doi:10.1016/j.rser.2010.11.032

Pathak, P. K., & Yadav, A. K. (2019). Design of battery charging circuit through intelligent MPPT using SPV system. *Solar Energy*, *178*, 79–89.

Pena, R., & Sbarbaro, D. (1999). Integral variable structure controllers for small wind energy systems. *Industrial Electronics Society, 1999. IECON '99 Proceedings. The 25th Annual Conference of the IEEE*, *3*, 1067 –1072.

Pham, V. T., Gokul, P. M., Kapitaniak, T., Volos, C., & Azar, A. T. (2018) Dynamics, Synchronization and Fractional Order Form of a Chaotic System With Infinite Equilibria. In Mathematical Techniques of Fractional Order Systems. Elsevier.

Pham, V. T., Gokul, P. M., Kapitaniak, T., Volos, C., & Azar, A. T. (2018). Dynamics, Synchronization and Fractional Order Form of a Chaotic System With Infinite Equilibria. In Mathematical Techniques of Fractional Order Systems. Elsevier.

Pham, V. T., Vaidyanathan, S., Volos, C. K., Azar, A. T., Hoang, T. M., & Yem, V. V. (2017a). A Three-Dimensional No-Equilibrium Chaotic System: Analysis, Synchronization and Its Fractional Order Form. Studies in Computational Intelligence, 688, 449-470.

Pham, V. T., Volos, C. K., Vaidyanathan, S., & Azar, A. T. (2017b). Dynamics, Synchronization and Fractional Order Form of a Chaotic System without Equilibrium. In C. K. Volos (Ed.), *Nonlinear Systems: Design, Applications and Analysis*. Nova Science Publishers.

Pilla, R., Azar, A. T., & Gorripotu, T. S. (2019). Impact of Flexible AC Transmission System Devices on Automatic Generation Control with a Metaheuristic Based Fuzzy PID Controller. *Energies*, *12*(21), 4193. doi:10.3390/en12214193

Pilla, R., Botcha, N., Gorripotu, T. S., & Azar, A. T. (2020). Fuzzy PID Controller for Automatic Generation Control of Interconnected Power System Tuned by Glow-Worm Swarm Optimization. In *Applications of Robotics in Industry Using Advanced Mechanisms. ARIAM 2019. Learning and Analytics in Intelligent Systems* (Vol. 5, pp. 140–149). Springer. doi:10.1007/978-3-030-30271-9_14

Pinto, S. F., Aparicio, L., & Esteves, P. (2007). Direct controlled matrix converters in variable speed wind energy generation systems. *Power Engineering, Energy and Electrical Drives, 2007. POWERENG 2007. International Conference on*, 654–659.

Podlubny, I. (1994). Fractional-order systems and fractional-order controllers. Institute of Experimental Physics, Slovak Academy of Sciences.

Poomrittigul, S., Koomsubsiri, A., Sasithong, P., Deenuch, D., & Wuttisittikulkij, L. (2020). The Simulation of Queuing Model for Bangkok Rapid Transit Train Ticket System Using Python. *34th International Technical Conference on Circuits/Systems, Computers and Communications*, 1-4.

Pragati, A., Ganthia, B. P., & Panigrahi, B. P. (2021). Genetic Algorithm Optimized Direct Torque Control of Mathematically Modeled Induction Motor Drive Using PI and Sliding Mode Controller. In J. Kumar & P. Jena (Eds.), Recent Advances in Power Electronics and Drives. Lecture Notes in Electrical Engineering (Vol. 707). Springer. https://doi.org/10.1007/978-981-15-8586-9_32.

Prathapa, Anandhia, & Sivakumaran. (2018). Xilinx Spartan 3A DSP FPGA based DC Voltage Regulators for PV Systems. *Materials Today Proceedings, 5*(1), 1348-1358.

Priyadarshi, N., Padmanaban, S., Bhaskar, M. S., Blaabjerg, F., & Sharma, A. (2018). Fuzzy SVPWM-based inverter control realisation of grid integrated photovoltaic-wind system with fuzzy particle swarm optimisation maximum power point tracking algorithm for a grid-connected PV/wind power generation system: Hardware implementation. *IET Electric Power Applications*, *12*(7), 962–971. doi:10.1049/iet-epa.2017.0804

Qahouq, J. A. A., & Jiang, Y. (2014). Distributed photovoltaic solar system architecture with single-power inductor single-power converter and single-sensor single maximum power point tracking controller. *IET Power Electronics*, *7*(10), 2600–2609. doi:10.1049/iet-pel.2013.0893

Qi, J., & Liu, Y. (2010). PID control in adjustable-pitch wind turbine system based on fuzzy control. *Industrial Mechatronics and Automation (ICIMA), 2010 2nd International Conference on,* 2, 341–344.

Qin, W., Wang, P., Han, X. & Du, X. (2011). Reactive Power Aspects in Reliability Assessment of Power Systems. *IEEE Transactions on Power Systems, 26*(1).

Radwan, A. G., Azar, A. T., Vaidyanathan, S., Munoz-Pacheco, J. M., & Ouannas, A. (2017). Fractional-Order and Memristive Nonlinear Systems: Advances and Applications. *Complexity*, *2017*, 3760121. Advance online publication. doi:10.1155/2017/3760121

Radwan, A. G., Emira, A. A., Abdelaty, A., & Azar, A. T. (2018). Modeling and Analysis of Fractional Order DC-DC Converter. *ISA Transactions*, *82*, 184–199. doi:10.1016/j.isatra.2017.06.024 PMID:28709651

Rahnamayan, S., Mahdavi, S., Deb, K., & Bidgoli, A. (2020). Ranking Multi-Metric Scientific Achievements Using a Concept of Pareto Optimality. *Mathematics*, *8*(6), 956. doi:10.3390/math8060956

Rajagopal, R., Palanisamy, K., & Paramasivam, S. (2019). Shunt Active Filter Based on 7-Level Cascaded Multilevel Inverter for Harmonic and Reactive Power Compensation. *Journal of Electrical Engineering & Technology*, *14*(4), 1543–1551. doi:10.100742835-019-00201-1

Rajendran, S., Govindarajan, U., Reuben, A. B., & Srinivasan, A. (2013). Shunt reactive VAR compensator for grid-connected induction generator in wind energy conversion systems. *IET Power Electronics*, *6*(9), 1872–1883. doi:10.1049/iet-pel.2012.0524

Rao, B. N., Suresh, Y., Panda, A. K., Naik, B. S., & Jammala, V. (2020). Development of cascaded multilevel inverter based active power filter with reduced transformers. *CPSS Transactions on Power Electronics and Applications*, *5*(2), 147–157. doi:10.24295/CPSSTPEA.2020.00013

Raza, K., Goto, H., Guo, H.-J., & Ichinokura, O. (2008). A novel speedsensorless adaptive hill climbing algorithm for fast and efficient maximum power point tracking of wind energy conversion systems. *Sustainable Energy Technologies, 2008. ICSET 2008. IEEE International Conference on*, 628–633.

Reddy. (2014). Faster evolutionary algorithm based optimal power flow using incremental variables. *Electrical Power and Energy Systems*, *54*(1), 198-210.

Reddy. (2017). Optimal Reactive Power Scheduling Using Cuckoo Search Algorithm. *International Journal of Electrical and Computer Engineering*, *7*(5), 2349-2356.

Renaudineau, H., Donatantonio, F., Fontchastagner, J., Petrone, G., Spagnuolo, G., Martin, J. P., & Pierfederici, S. (2014). A PSO-based global MPPT technique for distributed PV power generation. *IEEE Transactions on Industrial Electronics*, *62*(2), 1047–1058.

Ren, Y., & Bao, G. (2010). Control strategy of maximum wind energy capture of direct-drive wind turbine generator based on neural-network. *Power and Energy Engineering Conference (APPEEC)*, 1–4.

Rex, S. R., & Praba, D. M. S. R. (2020). Design of PWM with four transistor comparator for DC–DC boost converters. *Microprocessors and Microsystems*, *72*, 102844. doi:10.1016/j.micpro.2019.07.003

Richardson, R., & McNerney, G. (1993, March). Wind energy systems. *Proceedings of the IEEE*, *81*(3), 378–389. doi:10.1109/5.241490

Riverso, S., Sarzo, F., & Ferrari-Trecate, G. (2015a). Plug-and-play voltage and frequency control of islanded microgrids with meshed topology. *IEEE Transactions on Smart Grid*, *6*(3), 1176–1184.

Rocha, R., & Filho, L. (2003). A multivariable H∞ control for wind energy conversion system. *Control Applications, 2003. CCA 2003. Proceedings of 2003 IEEE Conference on*, 1, 206 – 211.

Rocha, R., Filho, L., & Bortolus, M. (2005). Optimal multivariable control for wind energy conversion system - a comparison between H 2 and H∞ controllers. *Decision and Control, 2005 and 2005 European Control Conference. CDC-ECC '05. 44th IEEE Conference on*, 7906 – 7911.

Rocha, R., Resende, P., Silvino, J., & Bortolus, M. (2001). Control of stall regulated wind turbine through H∞: loop shaping method. *Control Applications, 2001. (CCA '01). Proceedings of the 2001 IEEE International Conference on*, 925–929.

Rodríguez, P. (2007, March). Decoupled Double Synchronous Reference Frame PLL for Power Converters Control. *IEEE Transactions on Power Electronics*, *22*(2), 584–592.

Rodríguez-Valdez, C. D., & Kerkman, R. J. (2010). Phase Locked Loop for unbalanced utility conditions. *2010 Twenty-Fifth Annual IEEE Applied Power Electronics Conference and Exposition (APEC).*

Rosa-Clot, M., & Tina, G. M. (2017). *Submerged and Floating Photovoltaic Systems : Modelling, Design and Case Studies.* Academic Press.

Rosas-Caro, J., Ramirez, J., Peng, F., & Valderrabano, A. (2010, January). A DC–DC multilevel boost converter. *IET Power Electronics, 3*(1), 129–137.

Ruan, X. (2018). Control Techniques for LCL-Type Grid-Connected Inverters. Springer.

Sadabadi, M. S., Karimi, A., & Karimi, H. (2015). Fixed-order decentralized/distributed control of islanded inverter-interfaced microgrids. *Control Engineering Practice, 45*, 174–193.

Sadabadi, M. S., Shafiee, Q., & Karimi, A. (2017). Plug-and-Play Voltage Stabilization in Inverter-Interfaced Microgrids via a Robust Control Strategy. *IEEE Transactions on Control Systems Technology, 25*(3), 781–791. doi:10.1109/TCST.2016.2583378

Sahoo, B., Routray, S. K., & Rout, P. K. (2019). Artificial neural network-based PI-controlled reduced switch cascaded multilevel inverter operation in wind energy conversion system with solid-state transformer. *Iranian Journal of Science and Technology. Transaction of Electrical Engineering, 43*(4), 1053–1073. doi:10.100740998-019-00215-5

Sahu, N., Londhe, N. D., & Kshirsagar, G. (2017). FPGA applications in inverter and converter circuits: A review on technology, benefits and challenges. *2017 International Conference on Innovations in Information, Embedded and Communication Systems (ICIIECS).*

Salas, V., Olias, E., Barrado, A., & Lazaro, A. (2006). Review of the maximum power point tracking algorithms for stand-alone photovoltaic systems. *Solar Energy Materials and Solar Cells, 90*(11), 1555–1578. doi:10.1016/j.solmat.2005.10.023

Salem, M., & Atia, Y. (2015). Control scheme towards enhancing power quality and operational efficiency of single-phase two-stage grid-connected photovoltaic systems. *Journal of Electrical Systems and Information Technology, 2*(3), 314–327. doi:10.1016/j.jesit.2015.05.002

Sallam, O. K., Azar, A. T., Guaily, A., & Ammar, H. H. (2020) Tuning of PID controller using Particle Swarm Optimization for cross flow heat Exchanger based on CFD System Identification. *The International Conference on Advanced Intelligent Systems and Informatics AISI 2019*, 300-312. 10.1007/978-3-030-31129-2_28

Salmi, H., Badri, A., & Zegrari, M. (2016). Maximum Power Point Tracking (MPPT) Using Artificial Bee Colony Based Algorithm for Photovoltaic System. *Int J Intell Inf Syst, 5*, 1–4.

Samavatian, V., & Radan, A. (2014, December). A novel low-ripple interleaved buck–boost converter with high efficiency and low oscillation for fuel-cell applications. *International Journal of Electrical Power & Energy Systems, 63*, 446–454.

Sanatkar-Chayjani, M., & Monfared, M. (2016, July 27). Design of LCL and LLCL filters for single-phase grid connected converters. *IET Power Electronics, 9*(9), 1971–1978.

Sanchez, S., Bueno, M., Delgado, E., & Giraldo, E. (2009). Optimal PI control of a wind energy conversion system using particles swarm. *Electronics, Robotics and Automotive Mechanics Conference, 2009. CERMA '09*, 332–337.

Sanjuan, S. (2010). Voltage Oriented Control of Three-Phase Boost PWM Converters Design, simulation and implementation of a 3-phase boost battery Charger. Chalmers University of Technologie.

Santos-Martin, D., Arnaltes, S., & Amenedo, J. R. (2008). Reactive power capability of doubly fed asynchronous generators. *Electric Power Systems Research, 78*(111), 1837–1840.

Sasikumar, S., & Deepa, K. (2018). LCL topology based single stage boost rectifier topology for wireless EV charging. *Journal of Green Engineering, 8*, 573-596.

Saxena, A., & Deepa, K. (2020). Power quality analysis for electric vehicle charging and its mitigation strategies. *Test Engineering and Management,* 5409 – 5418.

Sayed, A. S., Azar, A. T., Ibrahim, Z. F., Ibrahim, H. A., Mohamed, N. A., & Ammar, H. H. (2020). Deep Learning Based Kinematic Modeling of 3-RRR Parallel Manipulator. In *Proceedings of the International Conference on Artificial Intelligence and Computer Vision (AICV2020).* Springer.

Sayed, A. S., Azar, A. T., Ibrahim, Z. F., Ibrahim, H. A., Mohamed, N. A., & Ammar, H. H. (2020). Deep Learning Based Kinematic Modeling of 3-RRR Parallel Manipulator. *Proceedings of the International Conference on Artificial Intelligence and Computer Vision (AICV2020),* 308-321.

Schiffer, J., Ortega, R., Astolfi, A., Raisch, J., & Sezi, T. (2014). Conditions for stability of droop-controlled inverter-based microgrids. *Automatica, 50*(10), 2457–2469.

Sedighizadeh, Arzaghi-Harris, & Kalantar. (2004). Adaptive PID control of wind energy conversion systems using RASP1 mother wavelet basis function networks. *TENCON 2004. 2004 IEEE Region 10 Conference, C,* 524 – 527.

Sedighizadeh, M., Rezazadeh, A., & Khatibi, M. (2008). A self-tuning PID control for a wind energy conversion system based on the Lyapunov approach. *Universities Power Engineering Conference, 2008. UPEC 2008. 43rd International,* 1–4.

Sefa, I. (2017). Three-phase three-level inverter with reduced number of switches for stand-alone PV systems. *2017 IEEE 6th International Conference on Renewable Energy Research and Applications (ICRERA).*

Senjyu, T., Sakamoto, R., Urasaki, N., Funabashi, T., & Sekine, H. (2006). Output power leveling of wind farm using pitch angle control with fuzzy neural network. *Power Engineering Society General Meeting.*

Shabarish, P. R., Krishna, R. P., Prasanth, B., Deepa, K., Manitha, P. V., & Sailaja, V. (2020). Fuzzy based Approach to Enhance the Wireless Charging System in Electric Vehicles. *4th International Conference on Electronics, Communication and Aerospace Technology,* 176-181.

Shafiee, G., Guerrero, J. M., & Vasquez, J. C. (2014). Distributed secondary control for islanded microgrids—A novel approach. *IEEE Transactions on Power Electronics, 29*(2), 1018–1031.

Shalaby, R., Ammar, H. H., Azar, A. T., & Mahmoud, M. I. (2021). Optimal Fractional-Order Fuzzy-MPPT for solar water pumping system. *Journal of Intelligent & Fuzzy Systems, 40*(1), 1175–1190. doi:10.3233/JIFS-201538

Shao, S., Guo, S., & Qiu, X. (2017). A Mobile battery swapping Service for Electric Vehicles Based on a battery swapping Van. *Energies.*

Sheina, S. G., Pirozhnikova, A. P., & Priss, E. A. (2019, December). The concept of sustainable development of renewable energy in the modern world. *IOP Conference Series. Materials Science and Engineering, 698*(5), 055010. doi:10.1088/1757-899X/698/5/055010

Shireena, W., Vanapallia, S., & Nene, H. (2007). DSP based inverter control for alternate energy systems. *Journal of Power Sources, 166*(2), 445–449.

Shukla, M. K., Sharma, B. B., & Azar, A. T. (2018). Control and Synchronization of a Fractional Order Hyperchaotic System via Backstepping and Active Backstepping Approach. In Mathematical Techniques of Fractional Order Systems. Elsevier. doi:10.1016/B978-0-12-813592-1.00019-2

Singh, S., & Azar, A. T. (2020). Multi-switching Combination Synchronization of Fractional Order Chaotic Systems. *Proceedings of the International Conference on Artificial Intelligence and Computer Vision (AICV2020),* 655-664. 10.1007/978-3-030-44289-7_61

Singh, S., Azar, A. T., Vaidyanathan, S., Ouannas, A., & Bhat, M. A. (2018). Multiswitching Synchronization of Commensurate Fractional Order Hyperchaotic Systems Via Active Control. In Mathematical Techniques of Fractional Order Systems. Elsevier. doi:10.1016/B978-0-12-813592-1.00011-8

Singh, S., Azar, A. T., Ouannas, A., Zhu, Q., Zhang, W., & Na, J. (2017). Sliding Mode Control Technique for Multi-switching Synchronization of Chaotic Systems. *9th International Conference on Modelling, Identification and Control (ICMIC 2017)*, Kunming, China. 10.1109/ICMIC.2017.8321579

Sinha, A., Jana, K. C., & Das, M. K. (2018). An inclusive review on different multi-level inverter topologies, their modulation and control strategies for a grid connected photo-voltaic system. *Solar Energy*, *170*, 633–657. doi:10.1016/j.solener.2018.06.001

Siraj, K., Siraj, H., & Nasir, M. (2014). Modeling and control of a doubly fed induction generator for grid integrated wind turbine. Power Electronics and Motion Control Conference and Exposition (PEMC). IEEE. doi:10.1109/EPEPEMC.2014.6980613

Sitharthan, R., Karthikeyan, M., Sundar, D. S., & Rajasekaran, S. (2020). Adaptive hybrid intelligent MPPT controller to approximate effectual wind speed and optimal rotor speed of variable speed wind turbine. *ISA Transactions*, *96*, 479–489. PMID:31202532

Soliman, N. S., Said, L. A., Azar, A. T., Madian, A. H., Radwan, A. G., & Ouannas, A. (2017). *Fractional Controllable Multi-Scroll V-Shape Attractor with Parameters Effect.* The 6th International Conference on Modern Circuits and Systems Technologies (MOCAST), Thessaloniki Greece.

Soliman, N. S., Said, L. A., Azar, A. T., Madian, A. H., Radwan, A. G., & Ouannas, A. (2017). *Fractional Controllable Multi-Scroll V-Shape Attractor with Parameters Effect.* The 6th International Conference on Modern Circuits and Systems Technologies (MOCAST), Thessaloniki, Greece.

Soliman, N. S., Said, L. A., Azar, A. T., Madian, A. H., Radwan, A. G., & Ouannas, A. (2017). Fractional Controllable Multi-Scroll V-Shape Attractor with Parameters Effect. *The 6th International Conference on Modern Circuits and Systems Technologies (MOCAST).*

Song, C., Fei, S., Cao, J., & Huang, C. (2019). Robust synchronization of fractional-order uncertain chaotic systems based on output feedback sliding mode control. *Mathematics*, *7*(7), 599. doi:10.3390/math7070599

Song, Z., Tian, Y., Chen, W., Zou, Z., & Chen, Z. (2016). Predictive duty cycle control of three-phase active-front-end rectifier. *IEEE Transactions on Power Electronics*, *31*(1), 698–710. doi:10.1109/TPEL.2015.2398872

Spier, D. W., Oggier, G. G., & Silva, S. A. O. d. (2019, August). Dynamic modeling and analysis of the bidirectional DC-DC boost-buck converter for renewable energy applications. *Sustainable Energy Technologies and Assessments*, *34*, 133–145.

Strzelecki, R. M. (Ed.). (2008). *Power electronics in smart electrical energy networks*. Springer Science & Business Media. doi:10.1007/978-1-84800-318-7

Subbaraj, P., & Rajnarayan, P. N. (2009). Optimal reactive power dispatch using self-adaptive real coded Genetic algorithm. *Electric Power Systems Research*, *79*(2), 374–38. doi:10.1016/j.epsr.2008.07.008

Sun, Lin, Li, & Li. (2009). A review on the integration of wind farms with variable speed wind turbine systems into power systems. *Sustainable Power Generation and Supply, 2009. SUPERGEN '09. International Conference on*, 1–6.

Sun, L., Peng, C., Hu, J., & Hou, Y. (2018, May). Application of Type 3 Wind Turbines for System Restoration. *IEEE Transactions on Power Systems*, *33*(3), 3040–3051. doi:10.1109/TPWRS.2017.2762009

Taha, H. A. (2011). Operations Research (9th ed.). Pearson Publication.

Tahir, S., Wang, J., Baloch, M. H., & Kaloi, G. S. (2018). Digital control techniques based on voltage source inverters in renewable energy applications: A review. *Electronics (Basel)*, *7*(2), 18. doi:10.3390/electronics7020018

Tak, S. Z., & Garg, V. (2018). Simulation of Grid Supported PV/Wind/Battery Hybrid System with Variable Load. *2018 Fifth International Conference on Parallel, Distributed and Grid Computing (PDGC)*, 627-632.

Talha, M., & Rahim, S. N. A. R. (2020, December). PV inverter with decoupled active and reactive power control to mitigate grid faults. *Renewable Energy*, *162*, 877–892.

Tang, R., Luo, B., Deng, X., & Shen, Y. (2020). Research on Reactive Power and Voltage Control for Wind Farm Based on coordinate control of DFIGs and SVG. *Procedia Computer Science*, *175*, 460–467.

Tan, K., & Islam, S. (2004, June). Optimum control strategies in energy conversion of PMSG wind turbine system without mechanical sensors. *Energy Conversion, IEEE Transactions on*, *19*, 392–399.

Tan, L. (2020, December). A novel phase-locked loop applied on frequency fluctuation of EAST power supply. *Energy Reports*, *6*(9), 452–461.

Tan, R. H., & Hoo, L. Y. (2015). DC-DC converter modeling and simulation using state space approach. *2015 IEEE Conference on Energy Conversion (CENCON)*, 42-47. 10.1109/CENCON.2015.7409511

Tareen, W. U., Mekhilef, S., Seyedmahmoudian, M., & Horan, B. (2017). Active power filter (APF) for mitigation of power quality issues in grid integration of wind and photovoltaic energy conversion system. *Renewable & Sustainable Energy Reviews*, *70*, 635–655. doi:10.1016/j.rser.2016.11.091

Thiruvonasundari, D., & Deepa, K. (2020). Electric vehicle battery modelling methods based on state of charge– review. *Journal of Green Engineering*, *10*, 24-61.

Thiruvonasundari, D., & Deepa, K. (2021). Optimized Passive Cell Balancing for Fast Charging in Electric Vehicle. *Journal of the Institution of Electronics and Telecommunication Engineers*, 1–9. Advance online publication. doi:10.1080/03772063.2021.1886604

Thongam, J., Bouchard, P., Ezzaidi, H., & Ouhrouche, M. (2009). Artificial neural network-based maximum power point tracking control for variable speed wind energy conversion systems. Control Applications, (CCA) Intelligent Control, (ISIC), 2009 IEEE, 1667–1671.

Tian, J., Su, C., & Chen, Z. (2013). Reactive power capability of the wind turbine with Doubly Fed Induction Generator. *Industrial Electronics Society, IECON 2013-39th Annual Conference of the IEEE*, 5312-5317.

Tien, H., Scherer, C., & Scherpen, J. (2007). Robust performance of self scheduled LPV control of doubly-fed induction generator in wind energy conversion systems. *Power Electronics and Applications, 2007 European Conference on*, 1–10.

Till, Simon, Niklas, Patrick, Frances, & Anders. (2018). Fast charging infrastructure for electric vehicles: Today's situation and future needs. *Journal of Transportation Research Part D Transport and Environment, 8.*

Tolba, M. F., AbdelAty, A. M., Soliman, N. S., Said, L. A., Madian, A. H., Azar, A. T., & Radwan, A. G. (2017). FPGA implementation of two fractional order chaotic systems. *International Journal of Electronics and Communications*, *28*, 162–172. doi:10.1016/j.aeue.2017.04.028

Tolunay, B. (2012). *Space Vector Pulse Width Modulation for Three-Level Converters: a LabVIEW Implementation.* Academic Press.

Tsang, K. M., & Chan, W. L. (2011). Current sensorless quick charger for lithium-ion batteries. *Energy Conversion and Management*, *52*(3), 1593–1595.

Tsengenes, G., & Adamidis, G. (2011). Investigation of the behavior of a three-phase grid-connected photovoltaic system to control active and reactive power. *Electric Power Systems Research*, *81*(1), 177–184. doi:10.1016/j.epsr.2010.08.008

Tsoumas, I., Safacas, A., Tsimplostefanakis, E., & Tatakis, E. (2003). An optimal control strategy of a variable speed wind energy conversion system. *Electrical Machines and Systems, 2003. ICEMS 2003. Sixth International Conference on*, *1*, 274 – 277.

Vaidyanathan, S., & Azar, A. T. (2015a). Anti-Synchronization of Identical Chaotic Systems using Sliding Mode Control and an Application to Vaidyanathan-Madhavan Chaotic Systems. In Advances and Applications in Sliding Mode Control systems. Springer-Verlag GmbH. doi:10.1007/978-3-319-11173-5_19

Vaidyanathan, S., & Azar, A. T. (2015b). Hybrid Synchronization of Identical Chaotic Systems using Sliding Mode Control and an Application to Vaidyanathan Chaotic Systems. In Advances and Applications in Sliding Mode Control systems. Springer-Verlag GmbH. doi:10.1007/978-3-319-11173-5_20

Vaidyanathan, S., & Azar, A. T. (2015c). Analysis, Control and Synchronization of a Nine-Term 3-D Novel Chaotic System. In Chaos Modeling and Control Systems Design. Springer-Verlag GmbH. doi:10.1007/978-3-319-13132-0_1

Vaidyanathan, S., & Azar, A. T. (2015d). Analysis and Control of a 4-D Novel Hyperchaotic System. In Chaos Modeling and Control Systems Design. Springer-Verlag GmbH. doi:10.1007/978-3-319-13132-0_2

Vaidyanathan, S., Idowu, B. A., & Azar, A. T. (2015c). Backstepping Controller Design for the Global Chaos Synchronization of Sprott's Jerk Systems. In Chaos Modeling and Control Systems Design. Springer-Verlag GmbH. doi:10.1007/978-3-319-13132-0_3

Vaidyanathan, S., & Azar, A. T. (2016a). Takagi-Sugeno Fuzzy Logic Controller for Liu-Chen Four-Scroll Chaotic System. *International Journal of Intelligent Engineering Informatics*, *4*(2), 135–150. doi:10.1504/IJIEI.2016.076699

Vaidyanathan, S., & Azar, A. T. (2016b). *Dynamic Analysis, Adaptive Feedback Control and Synchronization of an Eight-Term 3-D Novel Chaotic System with Three Quadratic Nonlinearities.* Springer-Verlag.

Vaidyanathan, S., & Azar, A. T. (2016b). Dynamic Analysis, Adaptive Feedback Control and Synchronization of an Eight-Term 3-D Novel Chaotic System with Three Quadratic Nonlinearities. *Studies in Fuzziness and Soft Computing*, *337*, 155–178. doi:10.1007/978-3-319-30340-6_7

Vaidyanathan, S., & Azar, A. T. (2016c). *Qualitative Study and Adaptive Control of a Novel 4-D Hyperchaotic System with Three Quadratic Nonlinearities.* Springer-Verlag.

Vaidyanathan, S., & Azar, A. T. (2016c). Qualitative Study and Adaptive Control of a Novel 4-D Hyperchaotic System with Three Quadratic Nonlinearities. *Studies in Fuzziness and Soft Computing*, *337*, 179–202. doi:10.1007/978-3-319-30340-6_8

Vaidyanathan, S., & Azar, A. T. (2016d). A Novel 4-D Four-Wing Chaotic System with Four Quadratic Nonlinearities and its Synchronization via Adaptive Control Method. Advances in Chaos Theory and Intelligent Control. *Studies in Fuzziness and Soft Computing*, *337*, 203–224. doi:10.1007/978-3-319-30340-6_9

Vaidyanathan, S., & Azar, A. T. (2016d). *A Novel 4-D Four-Wing Chaotic System with Four Quadratic Nonlinearities and its Synchronization via Adaptive Control Method.* Springer-Verlag.

Vaidyanathan, S., & Azar, A. T. (2016e). *Adaptive Control and Synchronization of Halvorsen Circulant Chaotic Systems. Advances in Chaos Theory and Intelligent Control.* Springer-Verlag.

Vaidyanathan, S., & Azar, A. T. (2016e). *Adaptive Control and Synchronization of Halvorsen Circulant Chaotic Systems.* Springer-Verlag.

Vaidyanathan, S., & Azar, A. T. (2016e). Adaptive Control and Synchronization of Halvorsen Circulant Chaotic Systems. *Studies in Fuzziness and Soft Computing*, *337*, 225–247. doi:10.1007/978-3-319-30340-6_10

Vaidyanathan, S., & Azar, A. T. (2016f). *Adaptive Backstepping Control and Synchronization of a Novel 3-D Jerk System with an Exponential Nonlinearity.* Springer-Verlag.

Vaidyanathan, S., & Azar, A. T. (2016f). Adaptive Backstepping Control and Synchronization of a Novel 3-D Jerk System with an Exponential Nonlinearity. *Studies in Fuzziness and Soft Computing*, *337*, 249–274. doi:10.1007/978-3-319-30340-6_11

Vaidyanathan, S., & Azar, A. T. (2016g). *Generalized Projective Synchronization of a Novel Hyperchaotic Four-Wing System via Adaptive Control Method.* Springer-Verlag.

Vaidyanathan, S., & Azar, A. T. (2016g). Generalized Projective Synchronization of a Novel Hyperchaotic Four-Wing System via Adaptive Control Method. *Studies in Fuzziness and Soft Computing*, *337*, 275–296. doi:10.1007/978-3-319-30340-6_12

Vaidyanathan, S., & Azar, A. T. (2020). *Backstepping Control of Nonlinear Dynamical Systems. Advances in Nonlinear Dynamics and Chaos (ANDC).* Elsevier.

Vaidyanathan, S., Azar, A. T., Akgul, A., Lien, C. H., Kacar, S., & Cavusoglu, U. (2019). A memristor-based system with hidden hyperchaotic attractors, its circuit design, synchronisation via integral sliding mode control and voice encryption. *Int. J. Automation and Control*, *13*(6), 644–667. doi:10.1504/IJAAC.2019.102665

Vaidyanathan, S., Azar, A. T., & Ouannas, A. (2017a). An Eight-Term 3-D Novel Chaotic System with Three Quadratic Nonlinearities, its Adaptive Feedback Control and Synchronization. *Studies in Computational Intelligence*, *688*, 719–746. doi:10.1007/978-3-319-50249-6_25

Vaidyanathan, S., Azar, A. T., & Ouannas, A. (2017c). Hyperchaos and Adaptive Control of a Novel Hyperchaotic System with Two Quadratic Nonlinearities. *Studies in Computational Intelligence*, *688*, 773–803. doi:10.1007/978-3-319-50249-6_27

Vaidyanathan, S., Azar, A. T., Rajagopal, K., & Alexander, P. (2015b). Design and SPICE implementation of a 12-term novel hyperchaotic system and its synchronization via active control (2015). *International Journal of Modelling Identification and Control*, *23*(3), 267–277. doi:10.1504/IJMIC.2015.069936

Vaidyanathan, S., Sampath, S., & Azar, A. T. (2015a). Global chaos synchronisation of identical chaotic systems via novel sliding mode control method and its application to Zhu system. *International Journal of Modelling Identification and Control*, *23*(1), 92–100. doi:10.1504/IJMIC.2015.067495

Vaidyanathan, S., Zhu, Q., & Azar, A. T. (2017b). Adaptive Control of a Novel Nonlinear Double Convection Chaotic System. *Studies in Computational Intelligence*, *688*, 357–385. doi:10.1007/978-3-319-50249-6_12

Valenciaga, F., Puleston, P., & Battaiotto, P. (2004). Variable structure system control design method based on a differential geometric approach: application to a wind energy conversion subsystem. *Control Theory and Applications, IEE Proceedings, 151*, 6 – 12.

Valenciaga, F., & Puleston, P. (2007, June). Variable structure control of a wind energy conversion system based on a brushless doubly fed reluctance generator. *Energy Conversion, IEEE Transactions on*, *22*, 499–506.

Valenciaga, F., & Puleston, P. (2008, September). High-order sliding control for a wind energy conversion system based on a permanent magnet synchronous generator. *Energy Conversion, IEEE Transactions on*, *23*, 860–867.

Vangari, A., Haribabu, D., & Sakamuri, J. N. (2015). Modeling and control of DC/DC boost converter using K-factor control for MPPT of solar PV system. *2015 International Conference on Energy Economics and Environment (ICEEE)*, 1-6. 10.1109/EnergyEconomics.2015.7235087

Vazquez, J., & Salmeron, P. (2003). Active power filter control using neural network technologies. *IEE Proceedings. Electric Power Applications*, *150*(2), 139–145. doi:10.1049/ip-epa:20030009

Vazquez, S., Leon, J. I., Franquelo, L. G., Rodriguez, J., Young, H., Marquez, A., & Zanchetta, P. (2014). Model predictive control: A review of its applications in power electronics. *IEEE Industrial Electronics Magazine*, *8*(1), 16–31. doi:10.1109/MIE.2013.2290138

Vazquez, S., Sanchez, J., Carrasco, J., Leon, J., & Galvan, E. (2008). A model-based direct power control for three-phase power converters. *IEEE Transactions on Industrial Electronics*, *55*(4), 1647–11657. doi:10.1109/TIE.2008.917113

Veerachary, M. (2018). Third-order boost converter. *IET Power Electronics*, *11*(3), 566–575.

Viola, F. (2018). Experimental Evaluation of the Performance of a Three-Phase Five-Level Cascaded H-Bridge Inverter by Means FPGA-Based Control Board for Grid Connected Applications. *Energies*, *11*(12).

Vukosavic, S. N. (2018). *Grid-Side Converters Control and Design: Interfacing Between the AC Grid and Renewable Power Sources*. Academic Press.

Wagner, H. J., & Mathur, J. (2009). *Introduction to Wind Energy Systems: Basics, Technology and Operation*. Springer-Verlag.

Wang, B., Wang, P., Chen, X., & Yuan, X. (2009). Robust control of variable speed adjustable-pitch wind energy conversion system based on Hamiltonian energy theory. *Sustainable Power Generation and Supply, 2009. SUPERGEN '09. International Conference on*, 1 –6.

Wang, D., Lu, K., Rasmussen, P. O. & Yang, Z. (2017). Comparative study of low-pass filter and phase-locked loop type speed filters for sensorless control of AC drives. *CES Transactions on Electrical Machines and Systems, 1*(2), 207 - 215.

Wang, Liu, & Liu. (2010). Intelligent controlling of wind turbine based on fuzzy immunity PID algorithmic. *Biomedical Engineering and Computer Science (ICBECS), 2010 International Conference on*, 1–4.

Wang, Q., He, F., Zhou, X., Geng, C., & Zhang, W. (2019, November). Heat Dissipation System Modeling of Three-level Converter Considering Thermal Coupling Effects. In *2019 IEEE 4th International Future Energy Electronics Conference (IFEEC)* (pp. 1-8). IEEE. 10.1109/IFEEC47410.2019.9014927

Wang, W., Wu, D., Wang, Y., & Ji, Z. (2010). H∞ gain scheduling control of PMSG-based wind power conversion system. *Industrial Electronics and Applications (ICIEA), 2010 the 5th IEEE Conference on*, 712–717.

Wang, Y., Fu, Y., & Li, D. (2009). Synthesized power and frequency control strategies based on fuzzy neural networks for wind power generation systems. *Energy and Environment Technology, 2009. ICEET '09. International Conference on, 1*, 869–872.

Wang, G., Ciobotaru, M., & Agelidis, V. G. (2014). Power smoothing of large solar PV plant using hybrid energy storage. *IEEE Transactions on Sustainable Energy*, *5*(3), 834–842.

Wang, Z., Volos, C., Kingni, S. T., Azar, A. T., & Pham, V. T. (2017). Four-wing attractors in a novel chaotic system with hyperbolic sine nonlinearity. *Optik (Stuttgart)*, *131*, 1071–1078. doi:10.1016/j.ijleo.2016.12.016

Wu, G. C., Deng, Z. G., Baleanu, D., & Zeng, D. Q. (2019). New variable-order fractional chaotic systems for fast image encryption. *Chaos (Woodbury, N.Y.)*, *29*(8), 083103. doi:10.1063/1.5096645 PMID:31472503

Xilinx. (2021). *XA Spartan-3A DSP FPGA Family*. https://www.xilinx.com/products/silicon-devices/fpga/xa-spartan-3a-dsp.html

Xin-Fang, Z., Da-Ping, X., & Yi-Bing, L. (2004). Adaptive optimal fuzzy control for variable speed fixed pitch wind turbines. *Intelligent Control and Automation, 2004. WCICA 2004. Fifth World Congress on, 3*, 2481 – 2485.

Xing, Z., Li, Q., Su, X., & Guo, H. (2009). Application of BP neural network for wind turbines. *Intelligent Computation Technology and Automation, 2009. ICICTA '09. Second International Conference on, 1*, 42–44.

Xu, L., Zhi, D., & Yao, L. (2007). Direct Power Control of Grid Connected Voltage Source Converters. *IEEE Power Engineering Society General Meeting.*

Xu, Q., Wang, P., & Tiayang, Z. (2017). Optimal operation of battery swapping charging systems considering quality-of-service constraints. *IEEE Power & Energy Society General Meeting*, 1-5.

Xuand, Y., & Yi, Z. (2020b). *Optimal distributed secondary control for an islanded microgrid, in the book:Distributed Control Methods and Cyber Security Issues in Microgrids*. Academic Press.

Xu, S., Zhang, J., Huang, Y., & Jatskevich, J. (2019). Dynamic average-value modeling of three-level T-type grid-connected converter system. *IEEE Journal of Emerging and Selected Topics in Power Electronics*, *7*(4), 2428–2442. doi:10.1109/JESTPE.2019.2894057

Yadav, I., Maurya, S. K., & Gupta, G. K. (2020). A literature review on industrially accepted MPPT techniques for solar PV system. *International Journal of Electrical & Computer Engineering, 10*(2).

Yan, J., Chen, J., Ma, H., Liu, H., & Chen, L. (2009). The survey of electrical control systems of wind turbine generators. *Sustainable Power Generation and Supply, 2009. SUPERGEN 09. International Conference on,* 1 –5.

Yang, X., & Liu, X. (2010). Integral variable structure fuzzy adaptive control for variable speed wind power system. *Logistics Systems and Intelligent Management, 2010 International Conference on, 2*, 1247–1250.

Yang, X., Liu, X., & Wu, Q. (2007). Integral fuzzy sliding mode control for variable speed wind power system. *Automation and Logistics, 2007 IEEE International Conference on*, 1289–1294.

Yao, X., Liu, Y., & Guo, C. (2007). Adaptive fuzzy sliding-mode control in variable speed adjustable pitch wind turbine. *Automation and Logistics, 2007 IEEE International Conference on*, 313–318.

Yao, X., Su, X., & Tian, L. (2009). Pitch angle control of variable pitch wind turbines based on neural network PID. *Industrial Electronics and Applications, 2009. ICIEA 2009. 4th IEEE Conference on*, 3235 – 3239.

Yao, X., Su, X., & Tian, L. (2009). Wind turbine control strategy at lower wind velocity based on neural network PID control. *Intelligent Systems and Applications, 2009. ISA 2009. International Workshop on*, 1–5.

Yao, X., Wen, H., Deng, Y., & Zhang, Z. (2007). Research on rotor excitation neural network PID control of variable speed constant frequency wind turbine. *Electrical Machines and Systems, 2007. ICEMS. International Conference on*, 560–565.

Yasmine, A., Rafik, B., Rachid, B., & Adel, M. (2020a). Grid connected photovoltaic system efficiency and quality improvement using fuzzy-incond MPPT. *International Journal of Power Electronics and Drive Systems, 11*(3).

Yasmine, A., Rafik, B., Rachid, B., & Adel, M. (2020b, May). Management of Hybrid PV/Battery Stand-alone System under Partial Shading Conditions. In *1st International Conference on Communications, Control Systems and Signal Processing (CCSSP)* (pp. 429-434). IEEE.

Yazdanian, M., & Mehrizi-Sani, A. (2014). Distributed control techniques in microgrids. *IEEE Transactions on Smart Grid, 5*(6), 2901–2909. doi:10.1109/TSG.2014.2337838

Ye, K., & Li, P. (2020). A new adaptive PSO-PID control strategy of hybrid energy storage system for electric vehicles. *Advances in Mechanical Engineering, 12*(9), 1687814020958574. doi:10.1177/1687814020958574

Yofo, E., & Djondiné, P. (2019). Parallel Active Filtering based on a three–phase multicell converter. *SCIREA Journal of Electrical Engineering, 4*, 48–62.

Youssef, A., El-Telbany, M., & Zekry, A. (2017). The role of artificial intelligence in photo-voltaic systems design and control: A review. *Renewable & Sustainable Energy Reviews, 78*, 72–79. doi:10.1016/j.rser.2017.04.046

Yüksek, H. I., & Arifoğlu, U. (2020). Modeling of three-phase three-level rectifier with space vector pulse width modulation method in matlab/simulink program. *Sigma: Journal of Engineering & Natural Sciences/Mühendislik ve Fen Bilimleri Dergisi, 38*(1).

Yunqi, X. (2010). A novel optimum power fuzzy control strategy for doublyfed wind turbine. *Control and Automation (ICCA), 2010 8th IEEE International Conference on*, 165–170.

Yusheng, P., Xicheng, Z., Qicheng, X., & Hongjiang, L. (2019). Reducing waiting time at an Airport Security Checkpoint based on the queuing model. *Chinese Control And Decision Conference*, 3860-3865.

Zeng, Q., Chang, L., & Shao, R. (2008). Fuzzy-logic-based maximum power point tracking strategy for PMSG variable-speed wind turbine generation systems. *Electrical and Computer Engineering, 2008. CCECE 2008. Canadian Conference on*, 405–410.

Zeng, M., Pan, Y., Zhang, D., Lu, Z., & Li, Y. (2019). Data-Driven Location Selection for Battery Swapping Stations. *IEEE Access: Practical Innovations, Open Solutions, 7*, 133760–133771. doi:10.1109/ACCESS.2019.2941901

Zerroug, N. (2018). DSP-based implementation of fast terminal synergetic control for a DC–DC Buck converter. *Journal of the Franklin Institute, 355*(5), 2329–2343.

Zhai, J., & Liu, H. (2014). Reactive power control strategy of DFIG wind farms for regulating voltage of power grid. PES General Meeting Conference & Exposition, 2014 IEEE, 1-5.

Zhang, H., Zheng, W., Yu, Y., & Li, H. (2019, August). A Hybrid Discontinuous PWM Neutral Point Balance Method Based on Zero-Sequence Component for Three-Level NPC Converter. In *2019 22nd International Conference on Electrical Machines and Systems (ICEMS)* (pp. 1-5). IEEE. 10.1109/ICEMS.2019.8922291

Zhang, L., & Li, H. (2008). Pitch control of large scale wind turbine based on fuzzy-PD method. *Electric Utility Deregulation and Restructuring and Power Technologies, 2008. DRPT 2008. Third International Conference on*, 2447 –2452.

Zhang, X., Wang, W., & Liu, Y. (2006). Fuzzy control of variable speed wind turbine. *Intelligent Control and Automation, 2006. WCICA 2006. The Sixth World Congress on, 1*, 3872–3876.

Zhang, X., Zhang, X., Zhou, P., Cheng, J., Wang, W., Jiale, S., & Liu, Y. (2009). Fuzzy control used in variable speed wind turbine. *Automation and Logistics, 2009. ICAL '09. IEEE International Conference on*, 1194 –1198.

Zhang, X. (2018, December 18). Direct grid-side current model predictive control for grid-connected inverter with LCL filter. *IET Power Electronics*, *11*(15), 2450–2460.

Zhang, Y., Xie, W., Li, Z., & Zhang, Y. (2013). Model predictive direct power control of a PWM rectifier with duty cycle optimization. *IEEE Transactions on Power Electronics*, *28*(11), 5343–5351. doi:10.1109/TPEL.2013.2243846

Zhang, Z., Zhu, Z., Hu, B., Tang, X., Liu, C., & Wang, Z. (2020). Uniform carrier-based PWM method for three-phase three-level three-wire and four-wire converter system with neutral-point balancing. *IET Power Electronics*, *13*(14), 3081–3089. doi:10.1049/iet-pel.2020.0160

Zheng, X., Wei, W., & Xu, D. (2010). Higher-order sliding mode control of DFIG wind energy system under LVRT. *Power and Energy Engineering Conference (APPEEC)*, 1–4.

Zhi, D., Xu, L., & Williams, B. W. (2009). Improved Direct Power Control of Grid-Connected DC/AC Converters. *IEEE Transactions on Power Electronics*, *24*(5), 1280–1292.

Zhou, H., Bhattacharya, T., Tran, D., Siew, T. S. T., & Khambadkone, A. M. (2010). Composite energy storage system involving battery and ultracapacitor with dynamic energy management in microgrid applications. *IEEE Transactions on Power Electronics*, *26*(3), 923–930.

Zhu, Q., & Azar, A. T. (2015). *Complex system modelling and control through intelligent soft computations*. Springer-Verlag.

Zhu, Y., Zhuo, F., & Shi, H. (2013, June). *Power management strategy research for a photovoltaic hybrid energy storage system. In 2013 IEEE ECCE Asia Downunder*. IEEE.

About the Contributors

Arezki Fekik received his BSc, MSc, PhD degrees in Electrical Engineering from Mouloud Mammeri University, Tizi-Ouzou, Algeria, in 2011, 2013 and 2018, respectively. He is currently working at the University of Akli Mohand Oulhadj-Bouira. He is working in the area of power systems, power converters, power electronics, control systems and renewable energy . He became a Lecturer in 2019 and member in the international group of system control (IGCS) since 2018, member of the Springer conference committee (AISI2018, AMLTA2018, AISI2019, AMLTA2019, AISI2020, AMLTA2020, AICV_2020, AMLTA2021) and member in the IEEE SMARTTECH 2020 conference. His current research interests include power electronics and its applications such as wind turbines, photovoltaic systems, Power Quality, harmonics, microgrids and variable speed drives. He has published 30 journal and conference articles and book chapters in the fields of power electronics and its applications and renewable energy.

* * *

Aissa Bokhtache Aicha was born in Chlef, Algeria, in 1971. She received her engineer degree in electrical Engineering from Hassiba Ben Bouali (UHBC) University, Faculty of Technology, Chlef, Algeria, in 1994. She obtained her magister degree from Hassiba Ben Bouali (UHBC) University, Faculty of Technology, Chlef, Algeria, in 2005. She obtained her Doctorate degree from Polytechnic school (ENP) Algiers (Algeria) In scientific research in 2017, since 2015, she is member and contribute in a team specialized in PV systems at LGEER laboratory in UHBC University.

Yasmine Amara received the Bachelor in automatic and Master in system control in an engineering degree from Mouloud Memmeri University, Tizi-Ouzou, Algeria in 2013 and 2015, respectively. She is currently working towards a Ph.D. degree at the university Saad Dahleb, Blida, Algeria. Her research interests focus on Photovoltaic power conversion systems, power quality, and efficiency improvement in Micro-Grids based on renewable energy and smart grids using intelligent and advanced controllers.

Aghiles Ardjal of Electrical Engineering Department, Mouloud Mammeri University of Tizi-Ouzou, Algeria, received the Bachelor, Master of Science and Ph.D degrees in electrical engineering from the Mouloud Mammeri University of Tizi-Ouzou, Algeria in 2011, 2013 and 2018 respectively. His research interests are control systems, sliding mode controllers, fractional controllers, and wind turbine systems.

Ahmad Taher Azar has received the M.Sc. degree in 2006 and Ph.D degree in 2009 from Faculty of Engineering, Cairo University, Egypt and he got his post- doctoral from USA. He is a research Professor at Prince Sultan University, Riyadh, Kingdom Saudi Arabia. He is also a professor at the Faculty of Computers and Artificial intelligence, Benha University, Egypt. Prof. Azar is the Editor in Chief of International Journal of System Dynamics Applications (IJSDA) and International Journal of Service Science, Management, Engineering, and Technology (IJSSMET) published by IGI Global, USA. Also, he is the Editor in Chief of International Journal of Intelligent Engineering Informatics (IJIEI), Inderscience Publishers, Olney, UK. Prof. Azar has worked as associate editor of IEEE Trans. Neural Networks and Learning Systems from 2013 to 2017 and Associate Editor of ISA Transactios, Elsevier from 2018 to 2020. He is currently associate editor of IEEE systems journal and Human-centric Computing and Information Sciences, Springer.

Subrat Kumar Barik has thirteen years of experience in teaching in undergraduate and postgraduate level. He has also an excellent research career in the area of Solar Photovoltaic Grid Integrated System. He has a number of research publications in SCI, Scopus indexed journals and international conferences .He graduated from the VSSUT, Burla in Electrical Engineering discipline, PhD from SOA University and currently working as an Associate Professor in School of Electrical Engineering, KIIT Deemed to be University.

Ehab H. E. Bayoumi received his BSc, MSc, and PhD degrees in electrical power engineering from Helwan University, Ain Shams University and Cairo University, Egypt in 1988, 1996 and 2001 respectively. He has been with the Electronic Research Institute (ERI), Cairo, Egypt, since 1990. From 2000 to 2001 he joined LUT, Finland, as a Visiting Researcher. He was appointed as an Assistant Professor at Chalmers University of Technology, Sweden from 2003 to 2005. From 2005 to 2010 he was Appointed as Associate Professor in ERI Egypt. In 2010 he was appointed as Full Professor in ERI Egypt. Currently, he is a Full Professor in Egyptian Chinese University. He also served at Yanbu Industrial College, Saudi Arabia and Higher Collages of Technology, United Arab Emirates, University of Eswatini, Eswatini. His research interests include high-performance ac machines, power quality, switching power converters, DSP-based control applications, and nonlinear control applications in power electronics, smart-grid, and electric drive systems. Prof. Bayoumi was appointed as an Editor-in Chief of the International Journal in Industrial Electronics and Drives in 2013.

Khalid Chigane was born in Morocco in 1993. He received the Bachelor's degree in Engineering Sciences from the Faculty of Sciences and Techniques, Moulay Ismaïl University, Errachidia, Morocco, in 2014 and the Master degree in Renewable Energies and Storage from the Faculty of Sciences, Mohammed V University, Rabat, Morocco, in 2016. He is currently working toward the Ph.D. degree in Engineering for Smart and Sustainable Systems Research Center (ESSSRC), Electrical Engineering Department, Mohammadia School of Engineers (EMI), Mohamed V University, Rabat, Morocco, his research interests include control and optimization of grid-connected PV systems.

Hakim Denoun received his BSc degree in Electrical Engineering in 1997 from the University of Tizi Ouzou, Algeria, and the D.E.A (Academic year) degree from Paris 6, France and the Magister degree from Polytechnic School, Algiers, Algeria in 1998 and 2001 respectively. He received his PhD degree in Electrical Engineering from The University of Mouloud Mammeri, Tizi-Ouzou. He is a Professor in laboratory (Laboratoire des Technolo-gies avancées du Génie Electrique LATAGE) and a senior lecturer at the same university. He has published articles, and participates in many international scientific conferences. His research interests include electrical machines and drives, power electronics and control systems, Renewable energy, Electric vehicle.

Farag Ali El-Sheikhi obtained M.Sc. degree in the area of power system simulation from University of Missouri at Columbia, USA and Ph.D. degree in the area of power systems reliability evaluation from University of Saskatchewan in Canada. He is a former professor of electrical energy and power system reliability at the University of Benghazi, Libya. His area of interest is power generation systems, transmission and distribution networks planning, operation and maintenance where he has many publications in these areas. Professor FARAG has a special interest in the development and implementation of Reliability Centered Maintenance (RCM) approach, PAS-55 and ISO 55000/1/2 Asset Management International Standards. He has successfully implemented such technique and standards to electrical power utilities. The implementation of such technique and standards to power utilities, petroleum, petrochemical and heavy industries will result in useful life extension of all equipment and a huge saving in the organization investment and cost effectiveness. Successful implementation of RCM and ISO 55001 Standard will lead to the enhancement of the organization's physical assets reliability and efficiency; making significant impact on profits and shareholders value through productivity, quality, and system reliability. Moreover, it will lead to asset uptime improvement, better system availability, less forced and unnecessary planned outages, and a greater understanding of the level of system risk that the organisation is managing; leading to an upgrading of operating procedures and systems strategies. He is now a teaching staff in Electrical and Electronics Engineering Department at Istanbul Esenyurt University in Istanbul, Turkey.

Bibhu Prasad Ganthia is working as assistant professor in the department of electrical engineering, IGIT, Sarang, Dhenkanal, Odisha, India. He is pursuing his Ph.D degree in the School of Electrical Engineering at Kalinga Institute of Industrial Technology (KIIT), Deemed to be University, Bhubaneswar, Odisha, India. He got the 'Excellence in Research' Award in 'NESIN 2020' for his contribution in the field of Wind Energy Conversion System. He is doing his research under the guidance of Dr. Subrat Kumar Barik and Co-guided by Dr. (Prof.) Byamakesh Nayak in the field of renewable energy system. He has received his Master's degree in the year 2016 from IGIT, Sarang in Power Electronics & Drives specialization and awarded Certificate of Merit for one of the best master degree Research Scholar from BPUT Rourkela. He has both teaching as well as industrial experience in his profession. In the short duration he has published 52 journals, book chapters and conference papers in reputed publishers like IEEE, Elsevier and Springer etc. indexed in Web of Science and Scopus with editor of 8 reputed journals and reviewer of many journal and conferences like IET, WJE, BEEI, Indonesia, MDPI Journal, IASTER, The Institute of Engineers, Springer etc. He has 11 patents in his name published in Indian Patent Rights. He is also the member of national and international scientific associations such as IAENG, SESI, ISRD, IFERP, BEIESP, SCIEI and NSPE and many more. He has software proficiency in MATLAB, ANSYS, LabView, DSpace and VLSI designing. He has got Certificate of Merit for excellence in Mater degree from BPUT, Rourkela, Odisha. He is awarded as "Best Ed Tech Teacher of the Year" by Indian Education Award 2019 sponsored by Techpro Labs and IIT Patna. Moreover he is awarded as "Young Teacher Award" by International School Award 2019, India. He owned Best Researcher Award and Young Researcher Award from DFIRF and Vedant Academics Bangkok Awards in the year 2019 respectively. He has also awarded Innovative Researcher of the Year 2019 in renewable energy for the year 2019 by Research Under Literal Access and World Research Council.

Mohamed Lamine Hamida received his BSc, MSc, PhD degrees in Electrical Engineering from Mouloud Mammeri University, Tizi-Ouzou, Algeria, in 2013, 2015 and 2019, respectively. Member of the Electrical Engineering Advanced Technology Laboratory (LATAGE) since 2015 and he became a Lecturer in 2020 at the same university. His current research interests include power electronics and its applications such as the application of PWM, Fuzzy Logic Control, sliding mode control and Petri Nets modeling to DC/DC and DC/AC multicell converters. He has published many journal and conference articles in the field of power electronics and its applications.

Hamza Houssaine received the Engineer, Magister, and Doctorate degrees in electrical engineering in 2001, 2004, and 2010, respectively, from Mouloud Mammeri University of Tizi-Ouzou, Algeria. He is currently an associate professor in Electrical Engineering Department at University of Bouira, Algeria . His research interests include electrical engineering, surge voltage, electric machines, Electromagnetic, electrical machines diagnosis.

Deepa Kaliyaperumal graduated from Alagappa chettiar college of engineering and Technology, T.N, India in 1998. She obtained M.Tech degree from Anna University, Guindy campus, T.N, India in 2005, and Doctoral degree from Jawaharlal Nehru Technological University, Anantapur, A.P, India in 2016. Her Ph.D. thesis title was "Design, Analysis and Control of DC to DC Quasi Resonant Converters". Currently she is working as Assistant professor in Electrical and Electronics Engineering Department, Amrita School of Engineering, Amrita Vishwa Vidyapeetham University, Bangalore, Karnataka, India. She has 22+ years of teaching experience. She is a life Member of IETE and ISTE, India and a Senior member of IEEE. She has authored 2 textbooks on "Electrical Machines" and "Control Systems". She has published 35+ international journal papers, 2 national journal papers, 35+ papers in international conference and 6 papers in national conference. 19 M.Tech Degrees were awarded under her guidance and she is currently guiding 7 phd scholars. Currently she is guiding her Ph. D., M. Tech. and B. Tech. students in the area of Electric Vehicles, Battery management systems, Smart Grid, Power Electronics Applications to Power Systems & Renewable energy Technologies, Soft Computing, Drives, Intelligent Control Techniques & Control System, Power Quality, FACTS Devices, Multi-level Inverter, Resonant Converters.

Lenin Kanagasabai has received his B.E., Electrical and Electronics Engineering from University of Madras, M.E., Degree in Power Systems from Annamalai University and completed PhD in Electrical Engineering from Jawaharlal Nehru Technological University Hyderabad, India. Published more than 360 international journal research papers and presently working as professor in Prasad V. Potluri Siddhartha Institute of Technology, Kanuru, Vijayawada, Andhra Pradesh -520007.

Zina Larabi received the state engineering degree in electro-technique and magister degree in Electrical Traction from the university of Mouloud Memmeri, Tizi-Ouzou in 2009 and 2013, respectively. She is currently an assistant professor at the University M'Hamed Bougara, Boumerdes, Algeria, and a Ph.D. student at the University of Mouloud Mammeri of Tizi-Ouzou, Algeria. Her research interest includes renewable energy modulation, control, quality improvement and electrical machine design.

Jai Kumar Maherchandani received the B.E. degree in Electrical Engineering from M.B.M. Engineering College, Jodhpur and M. E. degree in Control and Instrumentation from M.N.N.I.T, Allahabad. He received the PhD degree in Electrical Engineering from Maharana Pratap University of Agriculture and Technology, Udaipur. He is presently working as Assistant Professor in Electrical Engineering Department, College of Technology & Engineering, Udaipur -313001, India.

Monalisa Mohanty is a Teaching Associate, Department of Electrical & Electronics Engineering, ITER, SIksha 'O' Anusandhan Deemed to be University, Jagamara, Khandagiri, Bhubaneswar, Odisha 751030.

Byamakesh Nayak has more than 20 years of teaching experience in education, teaching B.Tech in all engineering branches, M.Tech in Electrical engineering and PhD course works in KIIT Univesity, Bhubaneswar .He graduated from the Institution of Engineers,went to complete M.Tech in Machine Drives from IT, Benaras Hidu University,India and earned PhD in Electrical Engineering from KIIT University in 2011.He has a vast knowledge of Electrical Engineering with industry experience. His main research areas are Power Electronics and Electrical Drives, hybrid vehicle, renewable energy and application of PIC Microcontrollers in special drive applications.

Aceng Sambas is currently a Lecturer at the Muhammadiyah University of Tasikmalaya, Indonesia since 2015.His current research focuses on dynamical systems, chaotic signals, electrical engineering, computational science, signal processing, robotics, embedded systems and artificial intelligence. Aceng Sambas is a member of Indonesian Operations Research Association (IORA), and in IAENG is a new member has been received on March 2019.

Hisham M. Soliman was born in Cairo Egypt in 1949. He received the BSc (with honors) and MSc degrees in Electrical Engineering from Cairo University, Egypt, in 1972 and 1975, respectively and the PhD in automatic control from the University of Paul Sababtier, Toulouse, France, in 1980. He is a Professor of Electrical Engineering at Cairo University, Cairo, Egypt. He taught at the universities of: Saskatchewan, Canada; Garyounis, Libya; Emirates, UAE; and Yanbu industrial college, Kingdom of Saudi Arabia; Sultan Qaboos, Oman. His areas of research include Hierarchical/decentralized control, robust control and power system control. He is a member of the Egyptian Engineering Syndicate.

Manikanta Suri is presently pursuing his final year B.Tech in the department of Electrical and Electronics Engineering, Amrita School of Engineering, Amrita Vishwa Vidyapeetham University, Bangalore, Karnataka, India. Currently, he is doing projects in the area of Electric Vehicles.

Zoulikha Tebri received the state engineering degree in Electric machine option and Magister degrees in electrical drives from the University of Mouloud Mammeri, TiziOuzou, Algeria in 2012 and 2017, respectively. She is currently a Ph.D. student at the Electrical Engineering Advanced Technologies Laboratory of Tizi-Ouzou. Her research interests include renewable energy, power electronics, and energy management of hybrid systems.

Sundarapandian Vaidyanathan has received the Doctor of Science (D.Sc.) degree (1996) in the Department of Electrical and Systems Engineering from Washington University, St. Louis, Missouri, U.S.A. He is currently the Professor and Dean, Research and Development Centre at Vel Tech University, Avadi, Chennai, India. Dr. Sundarapandian is the Associate Editor of two journals published by IGI Global, USA entitled International Journal of System Dynamics Applications (IJSDA) and International Journal of Rough Sets and Data Analysis (IJRSDA). He has published over 130 Scopus-indexed research publications and five books. He is an active researcher in the areas of Chaos and Control and constructed many novel chaotic and hyperchaotic systems in the chaos literature. He has given many plenary lectures in International Conferences on Control Engineering, Computer Science and information Technology. Dr. Sundarapandian's research interests include: Control Systems, Dynamical Systems, Stability Theory, Chaos Theory, Computational Science, Mathematical Modelling, Optimal Control, Robotics, Operations Research and Intelligent Control.

Index

A

active and reactive power control (PQ control) 207
aspiration level model 91, 93-94, 97-98, 102-105

C

converter control 143, 147, 151, 165, 208-209, 271, 275-277

D

DC/DC converter 113, 157, 213, 233, 294
decentralized control 1-5, 8, 34, 126, 130, 254
DFIG 145, 169-171, 174, 208, 229, 231, 271-273, 276-277, 281-282, 285
direct power control 82-86, 89-90, 107, 109, 115, 119, 124, 126, 131-134, 137, 147, 160, 198, 200-203, 206, 228, 231, 259-260, 303-304
DSPACE Real-Time implementation of PQ control 207-208, 221, 225

E

electric vehicles 15, 25, 91, 105-106, 263, 269
Enhanced Symbiotic Organisms 175-176, 179, 189-190
Enhanced Tree Squirrel Search 65-67, 71, 74, 78-79

F

fast charging station 91
flying capacitor inverter 287, 289-290, 298, 300, 303
FSIG 271, 280-281
fuel cell 132, 227, 233, 235, 237, 241, 249, 253-254, 258-259, 262-263

H

hybrid controller 143, 151, 227, 305
Hydrological Cycle 175-176, 183, 189-190

I

Inc-Cond 38-39, 51-52, 54-58
invariant ellipsoids 1, 3, 9, 34

M

MATLAB/SIMULINK 36, 46, 55, 61, 91, 112, 154, 209, 240, 270, 298
Maximum Power Point Tracking (MPPT) 38, 61, 63, 107, 128, 213, 256, 287-288, 292
microgrids 1, 3, 34-35, 61, 259, 264, 303
multi objective optimization 91, 94, 104

O

optimal reactive power 65-67, 74, 78, 87-88, 175-176, 190, 198, 200, 204

P

PAC 148, 271
photovoltaic 36-42, 45-48, 51, 55, 60-64, 107-111, 119, 122, 125-127, 132-136, 138-140, 210, 227-229, 233-235, 240-241, 250, 253-255, 258-263, 267, 287-288, 303, 305-306
photovoltaic (PV) system 235, 287
PI controller 36, 39, 54, 57, 60, 128, 133, 148, 169, 215, 218, 225, 233, 237, 239, 253, 260, 285
pitch angle controller 143, 149, 151, 159
PMSG 86, 136, 145, 168, 171-172, 174, 202, 208, 228, 264, 271, 273, 276-277, 279-280, 282-283, 285
power 2-5, 15-17, 21, 23, 25-26, 28, 34-41, 44-48, 51, 55-67, 74, 78-92, 105-110, 112-124, 126-128, 131-137, 139-140, 143-153, 155-162, 164, 167-177, 189-191, 196, 198-204, 206-231, 233-237, 240-241, 245, 253-264, 266-267, 269-283, 285-

289, 291-292, 294-295, 300, 303-306
PV 37-40, 43, 47-48, 51, 55, 57, 60-64, 107-114, 126-128, 132, 135-136, 139, 213-215, 226, 229-230, 233-235, 240, 243-244, 253-255, 261, 264, 266, 287-292, 295, 298-300, 303-306
PV-grid connected system 107
PWM-inverter 15, 107, 119, 123-126, 132, 259

Q

queuing models 91-92, 94, 97, 104

R

robust control 1, 3, 34-35, 83, 115, 129, 147, 149, 173, 199, 228, 237, 256
robust controller 143, 149, 151, 227

S

Salp Swarm 63, 65-66, 74-75, 78-79
shark 65-66, 76-79
SoC 36, 38-40, 47, 54-57, 59-60, 92
Soft computing controller 143
STD 36, 55-57, 60
Swim bladder 65-66, 76, 78-79

T

THD 107, 121, 123-124, 126, 227, 233, 235, 253-254, 287, 300
the experimental test bench 207-208, 221
Three level Converter 233
three-phase grid connected PWM inverter 207
Total Harmonic Distortion (THD) 235, 287
tracker 1, 3, 8-10, 12, 14-15, 32, 108, 213-214
Transmission Loss 65, 175

V

variable weather 36, 58, 60

W

WECS 143-145, 149, 151-152, 157-158, 168-169, 271-273, 285
wind turbine system 145, 147, 149, 171-172, 227, 271-278, 281, 284-285
WTs 143, 280-282

www.ingramcontent.com/pod-product-compliance
Lightning Source LLC
LaVergne TN
LVHW062134300926
849161LV00010B/2529